Modelling Finan㎐
with Matl

Mathematical Models and Benchmark Algorithms

WILLIAM T. SHAW

Quantitative Analysis Group
Nomura International plc, London
and Balliol College, Oxford

CAMBRIDGE
UNIVERSITY PRESS

PUBLISHED BY THE PRESS SYNDICATE OF THE UNIVERSITY OF CAMBRIDGE
The Pitt Building, Trumpington Street, Cambridge CB2 1RP, United Kingdom

CAMBRIDGE UNIVERSITY PRESS
The Edinburgh Building, Cambridge CB2 2RU, UK   http://www.cup.cam.ac.uk
40 West 20th Street, New York, NY 10011-4211, USA   http://www.cup.org
10 Stamford Road, Oakleigh, Melbourne 3166, Australia

First published 1998

Printed in the United Kingdom at the University Press, Cambridge

Typeset in *Mathematica* 3 and TEX

A catalogue record of this book is available from the British Library

ISBN 0 521 59233 X hardback (with CD-ROM)

DISCLAIMER

The information contained herein has been developed by Nomura International plc and is
based on sources which we believe to be reliable. Nomura International plc has
endeavoured to ensure the accuracy of the information, but it does not represent that it is
accurate and complete. Neither Nomura International plc and/or connected persons
nor Cambridge University Press accept any liability whatsoever for any direct, indirect or
consequential loss arising from use of the information or its contents.

To the best of the knowledge of Nomura International plc, the author and Cambridge
University Press, none of the code included in this book in itself contains any date sensitive
elements that will cause Year 2000-related problems. However, this statement implies no warranty
in this matter, on the part of the Nomura International plc, the author or the publisher.

for
Susan Mary Wallace
[1946–1997]
and
Sarah-Jane

Contents

Preface

This text has a number of aims. The first is to show how *Mathematica* (version 3 in particular), can be used as a derivatives modelling tool. Second, it presents a complete if concise development of the mathematical approach to the valuation and hedging of a large class of derivative securities. Third, although the basic mathematical development is oriented towards dynamic hedging and partial differential equations, this book aims to present a balanced approach to algorithm development, in which analytical, finite-difference, tree and Monte Carlo methods are each applied in the appropriate context, without any forced adherence to any particular method. Fourth, it is intended that this text collects together and highlights many of the mathematical pathologies that exist in derivatives modelling problems. This last point is all too frequently ignored, so a discussion here may be appropriate.

Financial analysts use often-complex mathematical models to guide their decisions when trading derivative financial instruments. However, derivative securities are capable of exhibiting some diverse forms of mathematical pathology that confound our intuition and play havoc with standard or even state-of-the-art algorithms. The potential traps fall into two categories. The first category contains problems arising from the complexity of some models, leading to their being seriously error-prone in their implementation, even if not intrinsically flawed. The second category contains algorithms that are intrinsically flawed. Let's take a look at some problems in each category.

An obvious example of a type-one problem relates to the computation of hedge parameters, or "Greeks." These are the partial derivatives of the option value with respect to the underlying price and other variables such as time and interest rates. For all but the simplest vanilla options, the pen-and-paper computation of such entities is very complex and therefore error-prone, leading to the potential of errors in coding. The estimation of such quantities by purely numerical methods (differencing) leads to other types of problems associated with inaccuracies in the estimate of the analytical derivative. Such difficulties can often be eliminated in one swoop with the *Mathematica* system, which is able to compute the symbolic derivatives – and hence the hedge parameters – exactly by analytical differentiation of the option-pricing formula.

A more subtle type-one difficulty relates to the computation of implied volatility, which is a favourite parameter of traders. Implied volatility makes sense only for the simplest vanilla options. In other cases, the implied volatility may be unstable, double-valued, or triple-valued, or may even possess infinitely many values. The implementation must check that the price is a strictly increasing or a strictly decreasing function of volatility; otherwise, nonsense can and will be obtained for the implied volatility. In *Mathematica* the graphical tools can be used to test this very quickly.

Some quite well-known algorithms are intrinsically flawed. Problems which we might identify as a type-two issue can be found in the following models.

(i)   Binomial models
(ii) Implicit finite-difference models
(iii) Monte Carlo simulation models

These are essentially numerical methods, and this book looks in detail at them in comparison with exact solutions for known cases. This is straightforward in a system such as *Mathematica*, where complex, exact solutions can be expressed exactly and worked out to any degree of precision. As numerical methods, they involve an essential discretization of time and other relevant variables such as the underlying asset price. A common theme is what happens when the time-step is taken to be large, which is very tempting in an implementation in order to obtain results quickly.

For example, several of the standard binomial models suffer from the well-known difficulty that as the time-step becomes large, the probabilities associated with the underlying tree model may become negative, which is manifest nonsense. In other types of models, the asset prices can become negative. Both of these effects are well known. What appears not to be understood is that the reason for these difficulties has a common root in the fact that tree models are typically underspecified from a mathematical point of view. A number of constraints can be written down that should apply to a tree. The solution of a full set can be quite hard, so in practice the authors of tree models have worked with a subset and made up one or more missing conditions in order to solve for the tree structure. This leads to the problems with negative probabilities or negative asset prices. When one is armed with *Mathematica*'s symbolic equation-solving capabilities, the solution of a full set of tree constraints is a straightforward matter – and in fact leads to a model where neither the up-and-down tree probabilities nor the asset price can become negative. Other problems with trees, discovered by others in relation to barrier and cap effects, are also discussed.

One of the most surprising and deeply rooted difficulties relates to the use of implicit finite-difference schemes. In principle, these allow a larger numerical time-step to be used than in treelike models and are becoming increasingly popular. When properly used, they combine accuracy with efficiency. There is, however, a major difficulty with them that appears not to have fully migrated in its appreciation from the academic numerical analysis community to the market practitioners. When the initial conditions for the associated partial differential equation (in financial terms, the option payoff) are nice and smooth (in loose terms, continuous with continuous slopes), one can get away with almost any implicit finite-difference scheme. This is emphatically not the case in option-pricing problems, where the payoffs are typically non-smooth and frequently discontinuous. Such "glitches" in the payoff will propagate through the solution, and while they do not necessarily cause a large error in the option value, they can cause significant errors in the Greeks such as Delta, Gamma, and Theta. This will occur with some of the most common schemes in current use for larger time-steps. It can be avoided only with a certain subset of implicit schemes. Which subset works and which does not is in fact well known to the numerical analysis community. In the text this is made crystal clear by comparison with some exact solutions; and the good, but infrequently used, schemes are contrasted with the bad, but widely used, schemes.

Monte Carlo simulation is a popular method for the valuation of options that are European in style but path-dependent. The manner in which simulated solutions converge to the correct answer is investigated for some cases where the exact solution is known. This reveals several difficulties with such numerical simulation methods, and in particular the very slow convergence associated with certain classes of options. We give suggestions for control variates in a number of useful cases but highlight the difference between getting the variance down – but possibly converging to the wrong answer – and getting the right answer.

However, it would be wrong to assume that the purpose of writing this book was merely to discuss what can go wrong! The illumination of pathology is only one of the abilities of *Mathematica*. For example, in addition to being able to do calculus, *Mathematica* has other advantages over traditional modelling environments such as spreadsheets and C/C++. For example, the presence of a vast library of special functions, coupled with the ability to do differentiation and integration, means that novel, exact solutions can be implemented with ease. A beautiful example of this is the exact solution for the Asian option with arithmetic averaging, which requires that one invert the Laplace transform of a hypergeometric function. This requires just a few lines in *Mathematica* and can be directly differentiated to obtain the Greeks. Other areas in which *Mathematica* can be fruitfully applied include novel analytical techniques for double-Barrier options and accurate analytical approximations for American options.

How the Text is Organized

This book is divided into six groups of chapters. The first group establishes the preliminaries in terms of the use of *Mathematica*, the basics of stochastic calculus and the derivation of partial differential equations, and the basic technique for solving the Black-Scholes PDE family. The next group of chapters explores a wide variety of analytical models, from simple vanilla options, through a range of by-now standard "exotics", and also develops more complex analytical models for Asian and American options. Next we take a long hard look at the finite-difference models, including the standard approaches and also novel methods with much better numerical characteristics. This block makes particularly good use of the new features of *Mathematica* 3.0, and it is shown how to use the *Mathematica* compiler to build numerical solutions of the PDEs in an efficient manner.

The fourth group of chapters explores the fundamentals and implementation aspects of binomial and trinomial tree models, using *Mathematica* both to define new tree models, and to implement traditional and novel tree models using the compiler. Group five looks in detail at Monte Carlo simulation and applications in particular to path-dependent and Basket options. Finally we take a brief look at some simpler interest-rate models and related non-log-normal equity models.

Some History

The origins of this text are diverse. Many years ago I began running courses for modelling professionals under the auspices of my consulting firm, Oxford System Solutions. Inspired by Ross Miller's work in *The Mathematica Journal*, I began to look at developing a programme tailored to financial applications, and gave it to several London financial organizations. This course focused largely on the analytical aspects – the limited compilation features of version 2.X of *Mathematica* did not then allow complex numerical models to be developed in an efficient fashion. Later, when employed as a consultant to Nomura Research Institute Europe Ltd., the question of how to carefully test the integrity of the models then being employed by Nomura arose. Although the existing models had been developed and tested with considerable care, I proposed that a systematic sweep through all the existing models be done, using *Mathematica* to independently build all the models, using the basic published mathematical research as a starting point. Furthermore, with one eye on the features of the then forthcoming *Mathematica* 3.0, I realized that one could begin to use *Mathematica* to perform detailed numerical computation, so that the project need not be limited to the simpler models admitting exact solutions. The project scope was then expanded not just to include the existing in-house models, but to explore numerous other models in the literature, with a view to assessing the desirability of their implementation. That extended project led to this text, and continues move to forward now that I am on the staff of Nomura International plc, and involved in the specification, prototyping and testing of a wide range of derivative models.

Technology Aspects

The chapters of this book exist in their entirety as a collection of *Mathematica* 3 Notebooks. All chapter material, including *Mathematica* code, text, graphics and typeset mathematical material, is native to *Mathematica* 3. The front- and end-matter (this preface, contents and index etc.) were prepared in LaTeX using Textures 1.8. The book was produced in final form on Power Macintoshes, in the form of an 8500/120 upgraded with a 266 MHz G3, and a further G3/266, with Notebooks being printed to disk as PostScript files, which were then used by the publisher to produce the final printed version. Timing results are based on the G3/266, running *Mathematica* 3.0.0, which in general is slightly faster on average than a Pentium II at 300 MHz running NT4 (if you are a Windows user make sure that you are using *Mathematica* 3.01 or later, because that version is fully Pentium optimized, and note that NT is significantly more efficient that '95). The timings should therefore be typical of desktop computers in production at the intended publication date of mid-1998. The printed version made use of a few features of the 3.1 or 3.5 system with regard to page layout only. The kernel code is targeted at *Mathematica* 3, though most of the non-compiled material is V2.X friendly. Work in progress in the numerical optimization of future versions of *Mathematica* may modify some of the conclusions regarding numerical efficiency issues.

Accuracy and Errors

In a project of this size and scope it is impossible to guarantee the absolute correctness of all the material and its implementation. I have made significant efforts to check the models contained herein against basic research results and other model implementations, but can make no guarantees regarding these implementations. I have prepared this material both for its educational value, and to provide a set of implementations of valuation models for comparison with other systems. This material should emphatically not be used in isolation for pricing and hedging in real-world applications (see the disclaimer also). Note also that some of the algorithms are highly experimental. Furthermore, it should be noted that all results printed here are those obtained on Apple Power Macintosh systems. A substantial number of the calculations (but not necessarily all) have been re-run on Intel Pentium systems running Microsoft Windows 95 and NT4, and on various UNIX systems from SUN, and have been found to give identical results. However, the author cannot guarantee complete hardware independence. Wolfram Research Inc. make their own best efforts to ensure that the *Mathematica* system operates in a consistent fashion, but there are inevitable minor differences, usually when machine-precision arithmetic is employed.

Stylistic Issues

The coding contained herein is for the most part based on my own efforts, except as explicitly acknowledged within the text. My efforts have focused on accuracy and speed, and I have deemed elegance and compactness to be secondary to transparency of function. In financial applications, for checking purposes, transparency of function is critical, and I hope the code contained here is legible and easy to understand and check. I make no apologies for allegedly ugly code! All that matters to me is getting an accurate answer and getting it efficiently.

Typesetting Issues

Mathematica 3.0 and later versions have a variety of styles for the display of *Mathematica* code and mathematical equations. Except in the early tutorial chapters of this book, where consistency has been the goal in order to avoid confusing the reader, I have been fairly liberal in switching between styles, where it appears to be useful to select a particular style for displaying material. Most *Mathematica* input uses the old version 2.X input form that is pure text, but occasionally, in order, for example, to make it easier to compare input with published research, I have converted input cells to "Standard Form" so that they look more like ordinary mathematics. Similarly, most of the output is in Standard Form, but occasionally it has been converted to "Traditional Form" so that it looks *exactly* like ordinary mathematical notation. Some of the Traditional Form outputs have in addition been typeset as numbered equations. Where there is mathematical material without any related *Mathematica* input or output it is almost all Traditional Form, usually created from Input Form, styled as numbered equations.

One notational point needs to be made here. Mathematica 3 Traditional Form uses a partially double-struck font for symbols such as i and e, and for the d in dS in integrals. I have avoided using this when creating my own equations, e.g. in the stochastic calculus material, but equations that are converted Mathematica output use the default typefaces employed by the software system. Typographical purists may dislike this notation, but I have tried to avoid editing Mathematica-created output wherever possible, in order that "what you see is what *Mathematica* made" or, as we shall remark quickly in the text to remind the reader that something strange and unfamiliar may be about to appear: "WYSIWMAMA".

One decision on presentation was to suppress all the "In" and "Out" numbered statements. This has the benefit of tidyness, but also has the potential for confusion as to what is input and what is output. In the printed form, I have used indentation on most of the outputs to try to indicate their character, but if there is any confusion as to the types or styles of cells, this can be resolved by reference to the electronic form.

Conventions

There are may different issues of convention that plague this subject. For example, how shold Delta be quoted? We could quote the raw partial derivative; the same expressed as a percentage; the same expressed in terms of a one per cent change in the underlying, and so on. The following are the rules, except as explicitly stated in the text:

- All variables are in natural units:
  - the interest-rate and continuous dividend yield are continuously compounded, and expressed in absolute terms, i.e., an interest-rate of 10 per cent continuously compounded corresponds to $r = 0.10$;
  - the time is in years;
  - the volatility is in absolute annual terms, and will normally (but not always) be a number less than unity, so that $\sigma = 0.25$ corresponds to 25 per cent annualized volatility;
- All Greeks are based on the raw partial derivatives with respect to absolute quantities in natural units, so that, e.g.,
  - Delta corresponds to the instantaneous rate of change of option value with respect to the underlying price, with the latter expressed in currency terms – for a vanilla Call Delta lies between zero and one;
  - Rho is rate of change with respect to absolute continuously compounded interest rates;
  - Vega is rate of change with respect to absolute volatility;
  - Theta is rate of change with respect to time in years.

These are most convenient for the mathematical description, as it means there are very few occurences of factors of $100, 365, 1/365$ and so on. In making comparisons with your own on-desk systems, this may require various conversion factors to be applied. Note that if you have numerical differencing algorithms in place, you may have made a choice to calculate actual changes rather than rates of change.

Feedback

Comments are actively sought on this material, especially if material errors are discovered. I also wish to hear about how things could have been done better, particularly with regard to speed and/or accuracy. I am not representing this text as necessarily the best way of implementing models in *Mathematica*, and have not doubt that many others will be able to improve on the material here.

Feedback to: `william.shaw@nomura.co.uk`

All trademarks are acknowledged.

Acknowledgements

I have to begin this list by apologizing to anyone I leave out. Over the past few years, I have had numerous discussions with many colleagues inside and outside Nomura regarding the use of *Mathematica* in both financial and non-financial applications, and I am not going to be able to remember everybody! I will therefore keep this list short. Within the Quantitative Analysis Group in London, my special thanks go to Reza Ghassemieh for his unflagging support throughout the project and to Roger Wilson for helping to solve numerous implementation problems. In the derivatives team, I have to acknowledge the infinite patience of David Kelly, Ben Mohamed and Dominic Pang, for their diverse contributions in the various testing and prototyping phases of the project. Marta Garcia has consistently brought me down to earth with reminders of the complex real world of convertibles and of the limitations of mathematics (and mathematicians). A special thanks goes to James Hutton, for many useful discussions on general points, and for making available early copies of the research on LP methods. Numerous members of the Risk Management teams have provided valuable feedback on model test reports that formed the basis for early drafts of this work. Valuable comments on draft chapters at various stages of development have been received from colleagues inside and outside Nomura, including: Martin Baxter, Ian Buckley, Asif Khan, Jason Tigg, Rachel Pownall, Hideki Shimamoto and my anonymous reviewers. My relatively recent education in finance has benefited from countless discussions with other colleagues at Nomura, and Nick Knight and Allison Southey deserve a special mention, along with numerous past and present members of the equity and strategy teams.

At Wolfram research in the US and the UK, Stephen Wolfram, Conrad Wolfram, Magnus Germandson, Theodore Gray, Rachel Leaver, Claire Miller, Tom Wickham-Jones and many others have provided a mixture of support including enthusiastic noises, organizing presentations, fixing my page layout headaches, fixing my code, and telling bad jokes to warm up my audience before presentations on aspects of this material.

With regard to the book production aspects, David Tranah and the Cambridge University Press team displayed chronic enthusiasm and tolerance.

While this book was being edited for final production, I learnt of the sudden death of my eldest sister Susan. This book is dedicated to her memory and to my niece Sarah-Jane.

Chapter 1.
Advanced Tools for Rocket Science

Mathematica and Mathematical Finance

1.1 Why You Should Use *Mathematica* for Derivatives Modelling

When expressed in mathematical terms, the modelling of a derivative security amounts to understanding the behaviour of a function of several variables in considerable detail. We need not just to know how to work out the function (the value of an option), but also to secure an appreciation of the sensitivity of the value to changes in any of the parameters of the function. Such an appreciation is most reliable when we have both a view of the local sensitivities, expressed through partial derivatives, and a global view, expressed by graphical means. The partial derivatives, expressed in financial terms, are the "Greeks" of the option value, and may be passive sensitivity variables, or be active hedging parameters. Such option valuations may, in simpler cases, be based on analytical closed forms involving special functions, or, failing that, may require intensive numerical computation requiring some extensive programming.

If one were asked afresh what sort of a system would combine together the ability to

(i) deal with a myriad of special functions, and do symbolic calculus with them,
(ii) manage advanced numerical computation,
(iii) allow complex structures to be programmed,
(iv) visualize functions in two, three or more dimensions,

and to do so on a range of computer platforms, one would come up with a very short list of modelling systems. Neither spreadsheets nor C/C++ would feature on the list, due to their fundamental failure to cope with point (i) - symbolic algebra and calculus.

One of the very few complete solutions to this list of requirements is the *Mathematica* system. Originally subtitled as "A System for Doing Mathematics by Computer", it is uniquely able as a derivatives modelling tool. In particular, release 3.0 and later versions include substantial new numerical functionality, and are capable of efficient compiled numerical modelling on large structures. *Mathematica* can manage not just scalar, vector and matrix data, but tensors of arbitrary rank, and its performance scales well to large and realistic problems. This book is, in part, an exploration of these capabilities. It is also an exploration of valuation algorithms and what can go wrong with them.

Exploring the Fundamentals with *Mathematica*

This book is not just an exploration of a few well-known models. We shall also explore the matter of solving the Black-Scholes differential equation in a *Mathematica*-assisted fashion. We can explore

(i) separation of variables,
(ii) similarity methods,
(iii) Green's functions and images,
(iv) finite-difference numerics,
(v) tree numerics,
(vi) Monte Carlo simulation

as solution techniques, all within the *Mathematica* environment. A comparison of these approaches and their outcomes is more than just an interesting academic exercise, as we shall consider next.

1.2 The Concept of Model Risk and the Need for Verification

Informal Definitions of Model and Algorithm Risk

Model risk can arise in a variety of ways. Given a financial instrument, we generally formulate a three-stage valuation process:

(a) conceptual description;
(b) mathematical formulation of the conceptual description;
(c) solution of the mathematical model through an analytical and/or numerical process.

It is possible to make mistakes, or perhaps unreasonable simplifying assumptions, at any one of these three stages. Each of these gives rise to a form of model risk - the risk that the process from reading a contract to responding with a set of answers (fair value, Delta, Gamma, ..., implied volatility) has led to the wrong answer being supplied.

This book uses *Mathematica* to look intelligently at the third point, which we might call *algorithm risk*. One might think that this can be eliminated just by being careful - the real story is that derivative securities are capable of exhibiting some diverse forms of mathematical pathology that confound our intuition and play havoc with standard or even state-of-the-art algorithms. This book will present some familiar and rather less familiar examples of trouble.

I wish to emphasize that this is not just a matter of worrying about a level of accuracy that is below the "noise level" of traders. It is possible to get horribly wrong answers even when one thinks that one is being rather careful. I also tend to think that it is the job of a Quants team to supply as accurate an answer as possible to trading and sales teams. If they want to modify your answer by 50 basis points, for whatever reasons, that is their responsibility!

Verification and Resolving Disagreement amongst Experts

A good rule of thumb is that in general you can give what you think is the same problem to six different groups and obtain at least three materially different answers. In derivatives modelling you can usually get four different answers, especially if you include

(a) one analytics fanatic,
(b) one tree-model fanatic,
(c) one Monte-Carlo fanatic,
(d) one finite-difference fanatic

among the group of people you choose. I use the term "fanatic" quite deliberately. I once had a party where I was told categorically that "you absolutely had to use martingales - these PDE people have got it all wrong", and I know of at least one PDE book that makes almost no reference to simulation, and one excellent martingale book that went so far as to describe the PDE approach as "notorious". In fact, all these methods should give the same answer when applied carefully to a problem to which they can all be applied. You can get small but irritating non-material differences out of models because of differences of view over things like calendar management, units, whether to treat parameters on a continuously com-pounded or annual basis, and so on. That sort of thing can and should generally be resolved by getting the risk management team to bash heads together. The issues arising from genuine differences between models of a different type, and more worryingly still, between models of a similar type, require a much more disciplined approach.

The only way to sort out material differences in answers is to perform some form of *verification study*. This can be done in various ways, of which the two most important ones are:

a) comparison of modelling systems with some form of absolutely accurate benchmark (typically only available for simpler models, but we must check that complex algorithms do work correctly when special-ized to simple cases);
b) intercomparison of models and codes when applied to complex instruments.

The basic idea is to define a comprehensive set of test problems, and run off a benchmark model and operational systems against as many others as are available.

In other fields of study this is actually quite a common process. I hope this text encourages similar efforts within the derivatives modelling community.

Resource Management

The essential requirements of such an approach include *duplication of effort*. My view is that no number should be believed unless it can be justified as the result of at least two different modelling efforts (or at least that the same system has been through such an intercomparison or benchmarking study in the past, and that it has been documented). This is generally not what management involved in resource allocation want to hear! The point is that quality is only achieved through careful and critical testing and appraisal.

1.3 The Power of Symbolic Computer Calculus

In recent years a number of tools have emerged that extend the utility of computers from "number crunch-ing" to the domain of carrying out algebra and calculus. More mature systems combine such symbolic capability with advanced numerical algorithms, 2D, 3D and animated graphics, and a programming environment. Such tools are particularly relevant to the derivatives modelling and testing area due to

(a) the fact that they have built-in routines for most basic common operations;
(b) the ability to do calculus with simple and complex functions, which can eliminate many of the head-aches of computing Greeks;
(c) the ability to compare analytical symbolic models directly with the results from a numerical algorithm;
(d) the ability to visualize the results, and literally see the errors if there are any.

Time to Wean Your Team from Spreadsheets and C/C++?

It is my firm belief that in the long term such hybrid symbolic-numeric-graphical systems will replace the spreadsheet and languages such as C or C++ for derivatives modelling. Their primary advantage lies in the provision of advanced symbolic capabilities. In a numerically based system the only two ways to differentiate, to get a Greek such as Delta, are (a) to work out the derivative on paper and code it up, (b) to work out the value of an option for one or two neighbouring values of the parameter, and take differences. Both of these have drawbacks, and the potential for human error at the pen and paper stage, and numerical errors arising from numerical differentiation problems. Instead we can now ask the computer to do the differentiation for us. Such symbolic environments, rich as they are in special-function libraries, can dramatically reduce development time below that for spreadsheet or C/C++ systems. Furthermore, in the testing and verification context they offer a simple integrated environment where a numerical model can be compared directly with its exact form. And I mean exact - current systems can do arbitrary-precision arithmetic with ease, and can do so with complicated "special functions". The latest versions of these systems, as is the case with *Mathematica* 3.0 or later, can also produce mathematical typeset material, so you can write your code, write down a mathematical model, test them, plot the results, and produce a report for risk management/trading/regulatory bodies all within one environment.

Spreadsheet Woes

There are many problems with spreadsheet environments. An extensive discussion of the issues is given by R. Miller (1990) in Chapter 1 of his text, *Computer-Aided Financial Analysis*, where he outlines several principles to which financial modelling environments should conform, and explains why spreadsheets fail to meet them. My own simplistic interpretation of Miller's views (i.e., the author's own prejudice) is that spreadsheets are the best way yet invented of muddling up input data, models, and output data. More seriously, their fabled capability for doing "what-if" calculations is at once both erroneous and misleading. In the particular context of derivatives modelling, the modelling of Greeks within spreadsheets by revaluing for neighbouring values of the parameters is an abomination. This is not to say that computing differences *in addition* to partial derivatives is not a valuable exercise - this separate information can reveal interesting pathology. The point is that you should use exact calculus wherever possible to extract partial derivatives. As Miller points out, the "what-if" concept is also limited to numerical variations. We want a system where we can also vary structural properties ("what if American rather than European"). This leads rather naturally to the desire for an object-oriented approach.

There is one virtue of the spreadsheet environment, and that is the tabular user interface. There is no doubt, that for instruments requiring a small number of input and output parameters, such a table of data for several such instruments is extremely useful. However, even the interface virtues of spreadsheets are strained by the complexities of instruments such as Convertible Bonds. The information relevant to one such bond can easily require several interlinked sheets.

Flushing POO Away?

Not for nothing, in my view, is "POO" the Spanish acronym for object-oriented programming, rather than OOP. C++ has become entrenched as a modelling system in many finance houses. In the view of the author, the job of a "Quant" is to analyse, test and price financial instruments as quickly and reliably as possible to respond to trading needs. There can be no greater waste of a Quant's time than having to write and debug algorithms in C++, let alone waste further energy embedding them in a spreadsheet. If one needs to do a highly numerically intensive model for which a system such as *Mathematica* may, at least for the present, be inadequate, the solution, at least for now, is to use a system properly designed for the task of numerical computation, such as highly optimized FORTRAN 90. Objections to *Mathematica* based on speed of execution are now essentially irrelevant due to improvements in low-price high-speed CPUs, and the increased efficiency of the internal compiled numerical environment in *Mathematica*. C++ systems are much slower to write, slow to compile, and their execution time advantage is now marginal compared to a model suitably adapted to the *Mathematica* internal compiler. The great dream of system managers is that once the headaches of building a library of core C++ library routines are overcome, modellers can build new algorithms quickly. Sadly, this is only a dream. OOP projects mysteriously seem to require massive increases in headcounts and budgets, yet are nevertheless the beloved of system managers.

This is NOT to say that an object-oriented approach is a bad idea - the problem is getting the dream back on track. My view is that the base-line system for object-oriented derivatives development should be at the level of a system such as *Mathematica*, where there is a symbolic capability, and already a huge collection of basic objects already in place with their mathematics sorted out. An excellent question to ask an OOP-based derivative team is how much time they have wasted coding up interpolation, the cumulative normal distribution, Greeks and all kinds of special functions and capabilities that already exist in *Mathematica*. In this text I have not attempted an object-oriented approach. My focus has been on making sure that the answers are correct. I leave it to others to adopt some of these methods and algorithms into a more structured environment.

The Use of *Mathematica*

The power of *Mathematica* as an investigative tool becomes clear in this context. We can define a series of test problems and use the special-function capabilities of *Mathematica* to characterize the solution exactly, using *Mathematica*'s unlimited-precision arithmetic to get the results with as much accuracy as we desire. Next we use *Mathematica*'s symbolic calculus capabilities to define the Greeks by differentiation. Then, for more complex models, we build tree, Monte Carlo and finite-difference models within *Mathematica*. Then we make a comparison, within *Mathematica*, between these analytical forms and the numerical solutions, and the solutions from other company systems. The results are compared, and the reasons for any differences or other problems are explored and resolved.

A Challenge!

To those who find my comments on traditional modelling systems extreme or ill informed, I issue this challenge. There is now available an exact solution for the continuously and arithmetically averaged Asian option. This is the Geman-Yor model described in Chapter 10. I got this up and running in *Mathematica*, and tested it, in a total of 6 hours. To the first person who can implement this in C++, and get answers accurate to 4 significant figures over a wide range of parameters, I offer a bottle of quality vintage champagne. If they can do it in less than 40 hours of coding time, two bottles. To the first person who can get it working as accurately in a spreadsheet *alone*, I offer a case of good red wine, however long it takes. This offer is conditional on me receiving and satisfactorily testing the implementation, and being allowed to distribute it to others - the mathematics underlying that model is so beautiful it deserves wide appreciation of its power.

1.4 A Case Study in Chaos - Does Implied Volatility Make Any Sense?

As an illustration of something that is easy to explore in a system such as *Mathematica*, we give an example of how one can investigate a topic within a system combining symbolic models and graphics.

The topic chosen is "implied volatility", and the point is to make it clear just how infrequently this concept makes any sense. This is important both from a mathematical point of view, as we need to note when a function cannot be inverted as it is "many to one", and from a trading point of view - implied volatility is often used as a key parameter.

Traders are persistent in asking for implied volatilities. You can give them a number, but it might not make any sense at all. Or rather, you can feed a market price for an instrument to a system which does all sorts of fancy things to find the volatility consistent with the market price given all the other data, and this volatility can make no sense at all. It is a good idea to list the nice situation and some of the horrors that can arise. I emphasize that these are not issues linked to a particular solution method, such as Newton-Raphson or bisection, where particular problems linked to the solution method may arise. These are matters of principle that will cause trouble whatever solution technique is employed. I also wish to emphasize that although the examples presented below may seem somewhat contrived, the type of phenomenon described can easily creep into less obviously pathological derivatives as new features are added.

Things Are OK

When the valuation is a monotonic function of volatility (strictly increasing or decreasing), there is usually no problem. The simple case of a vanilla European Call option, as developed by Black and Scholes (1973), falls into this category. Let's build a symbolic model of this option and plot the value as a function of volatility - we take this opportunity to show how easy it is to write down a model in *Mathematica*:

```
Norm[x_] := 1/2*(1 + Erf[x/Sqrt[2]]);
done[s_, σ_, k_, t_, r_, q_] :=
((r - q)*t + Log[s/k])/(σ*Sqrt[t]) + (σ*Sqrt[t])/2;
dtwo[s_, σ_, k_, t_, r_, q_] :=
done[s, σ, k, t, r, q] - σ*Sqrt[t];

BlackScholesCall[s_, k_, v_, r_, q_, t_] :=
s*Exp[-q*t]*Norm[done[s, v, k, t, r, q]] - k*Exp[-r*t]*Norm[dtwo[s, v,
k, t, r, q]];
```

We plot the option value as a function of the volatility, for an option where the strike is at 10, the underlying at 11, for a vanilla Euro-Call with one year to expiry, 5% risk-free rate (continuously compounded) and zero dividends:

```
Plot[BlackScholesCall[11, 10, vol, 0.05, 0, 1],
  {vol, 0.05, 0.4}, PlotRange -> All];
```

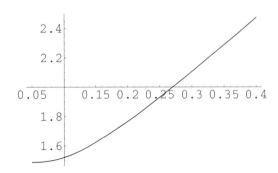

We can also build an implied volatility calculator in a moment (**FindRoot** is a combination Newton-Raphson/secant solver):

```
BlackScholesCallImpVol[s_,k_,r_,q_,t_,price_] :=sd /. FindRoot[BlackSc-
holesCall[s,k, sd, r, q, t]==price,{sd,0.2}];
```

Because the function is strictly increasing, we get a one-one mapping between prices and volatilities:

```
BlackScholesCall[11, 10, 0.25, 0.05, 0, 1]
```

```
    1.93051
```

```
BlackScholesCallImpVol[11, 10,  0.05, 0, 1, 1.93051]
```

```
    0.25
```

The results are also stable - a small change in the market price causes a small shift in the implied volatility:

```
BlackScholesCallImpVol[11, 10,  0.05, 0, 1, 1.92]
```

```
    0.246921
```

Things Are Sometimes OK, Sometimes Highly Unstable

It does not take much to mess up the calculation of implied volatility. Let's add a dilution effect to the Black-Scholes model, providing a simple model for a warrant. This is one of the simpler warrant-pricing models discussed by Lauterbach and Schultz (1990).

```
WarrantEqn[p_, k_, sd_, r_, t_, warprice_, shares_, warrants_,
shperwar_, q_] := (shares * shperwar / (shares + shperwar * warrants)) *
BlackScholesCall[p * Exp[-q * t] + warrants * warprice / shares, k, sd, r, 0, t]

WarrantValue[p_, k_, sd_, r_, t_, shares_, warrants_, shperwar_, q_] :=
warprice /. FindRoot[
warprice == WarrantEqn[p, k, sd, r, t, warprice,
    shares, warrants, shperwar, q], {warprice, 10}]
```

When we plot the result for price against volatility, we see that there is a nice monotonic region for larger volatilities, but that the curve becomes very flat as the volatility dips below 20%. The corresponding inversion for the implied volatility then becomes either unstable or impossible.

```
Plot[WarrantValue[15, 10, vol, 0.05, 1, 1000, 100, 1, 0],
  {vol, 0.05, 0.4}, PlotRange -> All];
```

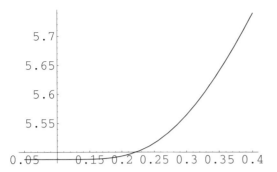

Things Are Definitely Not OK - There Are Two or Even Three Implied Volatilities

You do not have to make the option that much more complicated to really get in a mess. A simple Barrier option will exhibit the phenomenon of there being two volatilities consistent with a given valuation. Suppose we consider an Up-and-Out Call where the underlying is at 45, the strike at 50 and the knock-out barrier at 60. As we first increase the volatility from a very low value the option value increases. Then, as the volatility increases further the probability of knock-out increases, lowering the value of the option. By tweaking in a rebate we can arrange for there to be three solutions for a small range of market option prices. The model of Barriers we use is the analytical model developed by Rubenstein and Reiner (1991) - this has again been built in *Mathematica*, but we suppress the details of the symbolic model here - it is given in Chapter 8. The two graphs following show cases where there are two, and three, implied volatilities, respectively:

```
Plot[UpAndOutCall[0, 45, 50, 60, vol, 0.05, 0.0, 1],
{vol, 0.05, 0.4}, PlotRange -> All];
```

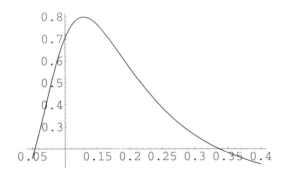

```
Plot[UpAndOutCall[1.7, 45, 50, 60, vol, 0.05, 0.0, 1],
{vol, 0.05, 0.4}, PlotRange -> All];
```

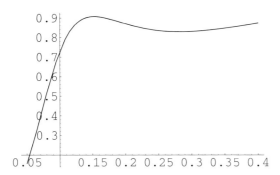

Another way of describing this effect is shown in the next plot, where we see that Vega changes sign twice:

```
Plot[UpAndOutCallVega[1.7, 45, 50, 60, vol, 0.05, 0.0, 1],
{vol, 0.05, 0.4}, PlotRange -> All];
```

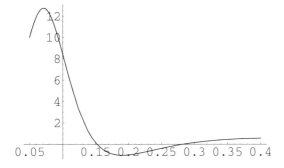

In Chapter 8 you will see how Vega is defined by elementary differentiation using *Mathematica*'s "D" operator.

Things Are Impossible - There Are Infinitely Many Answers

If you want to see just how bad it can get, a Barrier option can do still more interesting things. If we consider an Up-and-Out Put, where the strike coincides with the barrier, and arrange for the risk-free rate and the dividend yield to coincide, we can get a dead zero Vega (and zero Gamma too). So if the market price happens to coincide with the computed value, *you can have any implied volatility you want*. Otherwise there is no implied volatility.

```
Plot3D[UpAndOutPut[0, S, 50, 50, vol, 0.1, 0.1, 1],
{S, 30, 50}, {vol, 0.05, 0.4}, PlotRange -> {-0.01, 0.01},
PlotPoints -> 40];
```

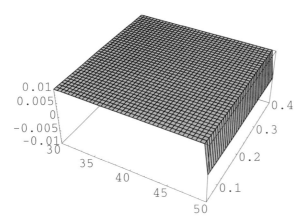

You should be grateful that this is all that goes wrong. Fortunately derivatives modellers tend to use real numbers. In complex function theory, Newton-Raphson and other iterative solution techniques are famous for their generation of fractals and chaotic behaviour. But it is a good idea to avoid quoting implied volatilities unless you have a firm grasp on the characteristics of an instrument. Only the very simplest options behave well, in this respect. More generally, you should not report implied volatility results unless you have first checked that the absolute value of Vega is bounded away from zero. Otherwise the implied volatility is at best unstable and possibly undefined - in such cases it should not be reported.

Remarks

This little study has illustrated several points about the use of *Mathematica*:

(a) you can build symbolic models very quickly;
(b) Greeks can be investigated by symbolic differentiation;
(c) we can visualize our models using two- and three-dimensional graphics;
(d) we can expose peculiar situations with relative ease.

Later in this book we shall see how the exposure of significant problems can be extended to intensively numerical processes.

For now we return to the fundamentals. In the next three chapters we lay the foundations of

(a) using *Mathematica*;
(b) financial mathematics of derivatives pricing;
(c) the mathematics for solving partial differential equations (the Black-Scholes PDE).

A Final Note

We live in a world where applied quantitative science in general, and financial modelling in particular, is increasingly dominated by purely numerical modelling. The view of the author is that the analytical approach is being inappropriately eclipsed by a rather mindless approach involving flinging models at a computer and seeing what numbers come out.

This is not the best way to use computers. We are seeking insight into the behaviour of complex systems, and this is best obtained by first attempting an analytical treatment. This first focuses our instincts, and gives us test cases to check against when we finally come to adopt a purely numerical approach. Symbolic algebra systems such as *Mathematica* allow us to extend the analytical approach to much more complicated systems. The numerical approach should always be the method of absolute last resort, and should then be checked extensively against the analytical results before being used. This is the guiding principle behind the approach taken by this text.

Chapter 1 Bibliography

Black, F. and Scholes, M., 1973, The pricing of options and corporate liabilities, *Journal of Political Economy*, 81, p. 637. (Reprinted and widely accessible as Chapter 1 of *Vasicek and Beyond*, ed. L.P. Hughston, Risk Publications, 1996.)

Lauterbach, B. and Schultz, P., 1990, An empirical study of the Black-Scholes model and its alternatives, *Journal of Finance*, 45, p. 1181.

Miller, R.M., 1990, *Computer-Aided Financial Analysis*, Addison-Wesley.

Rubenstein, M. and Reiner, E., 1991, Breaking down the barriers, *RISK Magazine*, September.

Chapter 2.
An Introduction to *Mathematica*

A System for Doing Mathematics by Computer

2.1 Systems, Kernels and Front Ends

The first thing that you need to do is to learn how to communicate with the computer and the *Mathematica* software system. It is important to realize that the software is in two pieces. The calculations are done by a computational "kernel", but you communicate with the kernel through a front end. The kernel and the front end may be running on the same computer, or on different computers linked by a network. This introduction is targeted at people using a windowing operating system, and the front end is a Notebook interface. This interface creates and uses *Mathematica* Notebook files - you can exchange Notebooks between computer systems such as MacOS, Windows95/NT and UNIX. The Notebooks are just text files, and can be passed between systems using only text-translation utilities.

If you have not used *Mathematica* before, how to proceed depends on whether you have *Mathematica* already installed on your computer. If it is not installed at all, you may need some help if you are working in a multi-computer networked environment running under UNIX, for example - under these circumstances, it is almost certainly best to request help from your System Administrator. If you are running on a solitary system, or otherwise doing an installation local to your computer, it is usually sufficient to follow the instructions in the *Mathematica* installation instructions. These are contained in the booklet "Getting started with *Mathematica* on Macintosh/Windows (etc.) systems", specific to your own computer. If you have never used *Mathematica* before, it will be helpful to run through the examples contained in that booklet. The discussion below is designed to complement that booklet, but is essentially self-contained.

Begin by starting *Mathematica*. You can do this on any system by navigating to the icon for *Mathematica* 3.0, and double-clicking on it. Do not use the "MathKernel" program icon for this. On particular systems there will be other routes to access *Mathematica*. Under Windows 95, the program can be found under the "Start" task. On both Windows 95 and MacOS, it is convenient to make short-cuts/aliases and place them on your desktop - it is really up to you to customize your route to access the system once you have figured out what is most convenient. *Mathematica* will start up, giving the version number, your licence information, and warnings about software piracy.

Input and Output

When a new Notebook is opened you can type straight into the blank window. Type "3 + 3" into the system (not the quotes, just the expression within the quotes). To enter this expression hold down the shift key and hit the "return" key. *Mathematica* labels both your input and its output with numbers, in this case [1], if you are starting a fresh session. You should see the following, with the input prefixed by In[1], the output prefixed by Out[1]:

```
3+3
```

        6

Now type the command "Expand[(2 x + y)^3]", and enter this by SHIFT-RETURN (ENTER, not to be confused with RETURN, and usually located at the bottom right of keyboards, may also work and is a useful short-cut on some computers).

```
Expand[(2 x + y)^3]
```

    $8 x^3 + 12 x^2 y + 6 x y^2 + y^3$

Throughout this chapter, whenever we present examples like this, expressions in bold type are what you should type, followed by SHIFT-RETURN or ENTER. To see some of the capabilities of *Mathematica* try the following sequences of statements. When you enter the commands in bold, they will acquire "In[n] := " labels. Also, *if you make a mistake, just re-type the correct expression*. We will come back to how to edit expressions shortly; the following are all short enough to make re-typing painless.

The first example uses the **N** function to ask for π, denoted in *Mathematica*'s Input Form by **Pi**, to 100-digit precision.

```
N[Pi, 100]
```

    3.1415926535897932384626433832795028841971693993751058209749445923 ∖
    07816406286208998628034825342117068

The next example introduces the plotting capabilities of *Mathematica* by using **Plot**. Given the function and the range over which to draw the graph, the **Plot** function will produce the appropriate graph.

```
Plot[x^2 Sin[x], {x, -3, 3}];
```

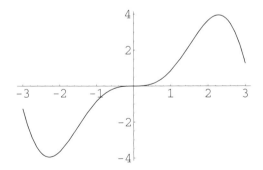

In this simple program we define the function **f** that gives the fourth power of its argument and then apply **f** twice to **y**. Use an ordinary RETURN to get the second line and then SHIFT-RETURN to execute the expression. Note the colon and semicolon which we shall examine later.

```
f[x_] := x^4;
f[f[y]]
```

$$y^{16}$$

The final example shows the incremental capabilities of *Mathematica* by using the **Table** function to produce a list of the cubes of integers up to 10.

```
cubes = Table[ i^3, {i, 0, 10}]
```

$$\{0, 1, 8, 27, 64, 125, 216, 343, 512, 729, 1000\}$$

We can plot given data like the list **cubes** using the function **ListPlot**.

```
ListPlot[cubes];
```

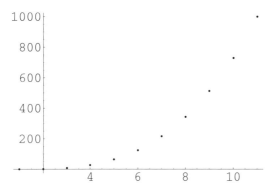

Don't worry about the details of the positioning of the axes for now. But you should note that if asked to plot a simple one-dimensional list, *Mathematica* will use the position in the list as the "x" or horizontal coordinate, so that, e.g., the last element is plotted as the point (11, 1000).

Remarks

In our examples we have seen the four basic areas of *Mathematica*. These are:

Numerical Calculations

For example, we asked for the numerical value of π to 100 digits.

Symbolic or Algebraic Calculations

We asked for the expansion of $(2x + y)^3$.

Graphics

We plotted a function, and a sequence of numbers.

Programming

We defined a function, and asked for it to be applied to a specific argument.

We will explore each of these topics in more detail later. Right now it is necessary to do some groundwork on understanding how *Mathematica* works. From here on, the expressions that we enter may be a little longer, and if we make a mistake we don't want to have to keep re-typing our input. So at this point it will be useful to learn a few more tricks for managing our input.

Using the Cursor

If you move the cursor up and down the Notebook in which you have just done the previous calculations you will see how it changes form dependent on which type of cell it is lying in, or whether it is between cells. A horizontal line under a cell indicates that any text typed will be placed after that cell. To place a horizontal line after any cell you move the cursor so that it indicates that it is lying between cells, i.e. when it is horizontal, and then click. Practice will get you used to being able to place text anywhere in a Notebook.

Opening and Closing Cells

In the Notebook environment all text is held in cells. These can have different styles associated with them, and can be set up to allow different styles of cell to be nested within one another. By double clicking on any cell bracket which is grouping other cells you will be able to hide the text contained in those sub-cells. This can be retrieved by double-clicking on the same cell bracket again. When a cell contains hidden sub-cells you will notice how the lower edge of the bracket changes to a filled triangle. You may find this way of hiding information in the Notebooks useful in organizing a presentation or teaching session. It may take some practice to get used to how these cells are nested.

Editing Text in Cells

Changing the text in a cell is done in a similar way to that used in word-processing applications. The cursor can be placed at any particular point in the text by moving the mouse and clicking on that spot. Selecting text with the mouse allows that text to be removed or typed over. Notice how you can only edit text in certain cells. You will be able to edit your input cells but not in the cells containing the output. This has been set by the way in which the different cells are defined. Here are some handy tips:

(1) use the mouse and scroll bars to get up and down the Notebook;
(2) use an ordinary RETURN to enter the next line of a multi-line expression;
(3) use SHIFT-RETURN (or ENTER) to execute an expression;
(4) use the mouse COPY and PASTE to grab previous formulae.

Quitting and Panicking

Sometimes, say if you make a mistake, *Mathematica* may go into a long internal session and all you want it to do is stop. If you need to stop *Mathematica* in the middle of a calculation, go to the menu and choose "Kernel". You then get some options, of which the ones to pick are "Abort Evaluation" or "Interrupt Evaluation...". The menus also offer keyboard short-cuts specific to your operating system, which you should learn. The kernel can sometimes take time to abort a calculation. It is better to wait for it to finish sorting itself out than to keep thumping the keyboard! If you lose patience waiting, or otherwise wish to exit *Mathematica* completely, you can click on "Quit" having interrupted the session, or type **Quit** at an "In" prompt. There are other options in the "Kernel" menu for quitting and re-starting.

Brackets Galore

Our first examples show that we need to understand how *Mathematica* reads expressions, and especially different types of brackets: [ ], { }, ( ). Sometimes people use square brackets [], round brackets () and curly brackets { } more or less interchangeably. In order to ensure that the computer understands unambiguously what we mean, it is necessary to introduce some conventions.

Arguments of Functions

Arguments of functions are always enclosed in square brackets. This applies just as much to standard mathematical functions like **Sin[x]** as to built-in *Mathematica* graphics functions like **Plot3D[]**, which take another function and some ranges as a minimum set of arguments, and user-defined functions like **f** above.

Grouping

When we wish to group together some expressions for applying a subsequent operation, round brackets are employed. For example **(x + y)^3** means the cube of all of **x + y**. Note that this distinction gives the expressions **f[x+y]** and **f(x+y)** two quite different meanings, and at the same time we have acquired the freedom to use a space between adjacent quantities to denote multiplication. If you wish to you can explicitly type a * to denote multiplication. For example **2 3** is identical in meaning to **2*3**.

Note that you don't always need to use round brackets for grouping, since the usual rules of precedence apply. For example **x^3 + 4** means **(x^3) + 4** and not **x^(3 + 4)**. A good policy is to use grouping brackets in any situation where there is a potential for ambiguity.

List Brackets

Curly brackets **{ }** and double square brackets **[[ ]]** are also used in *Mathematica*. These are used with list structures to denote a list, and an indexed member of a list, respectively. We shall look specifically at these structures later on. For now it will suffice to appreciate that a list consists of a number of elements, separated by commas, with a curly bracket at each end.

Input/Output, Standard and Traditional Forms

Versions of *Mathematica* prior to version 3.0 used styles where inputs and outputs were both based on pure text representations. *Mathematica* 3.0 adds new forms called "Standard Form" and "Traditional Form". These are much more like traditional mathematical notation - indeed, this is exactly the idea of "Traditional Form". "Standard Form" is a half-way form that goes as far as possible towards traditional mathematical notation without introducing any ambiguities. Throughout this chapter the default output format type is "Standard". You can ask for explicit conversion as follows:

```
expression = Exp[-x^2];
StandardForm[expression]
```

$$E^{-x^2}$$

```
TraditionalForm[expression]
```

$$e^{-x^2}$$

```
InputForm[expression]
```

```
E^(-x^2)
```

These conversions can be done using the menus, and defaults can also be set. It is impossible to give a hard and fast rule that has been applied throughout this book, for how expressions are represented, as the choice is rather sensitive to the context, but we try to adhere to the "WYSIWMAMA" (what you see is what *Mathematica* made) principle - note the partially double-struck notation. Generally, chapters involving more mathematical development will make more use of the traditional form for outputs and equations.

2.2 Basic Graphics

Before we launch into investigating each of the main areas of *Mathematica* in detail, it will be useful to look at some of the basic graphic commands used in *Mathematica*. While working in *Mathematica* it is very often helpful to involve graphics in a calculation. This is never more true than when looking at new ideas and will be used extensively through this chapter and the rest of the book, to illustrate points more clearly. For this reason, it will be useful if you feel reasonably comfortable with the most basic graphics commands at the outset. Do not be too concerned about the details of their operation at this point. We shall look more carefully at the range of graphics capabilities available in *Mathematica* later.

Producing a Plot

We have already seen the function `Plot` which will display the plot for a given function. This needs to have the range of values over which the plot is to be made specified. You supply `Plot` with two arguments - the first is the function, and the second is a list in the form `{variable, start, end}`.

```
Plot[x^3 - 8 x^2 + 19 x - 8, {x, 0, 5}];
```

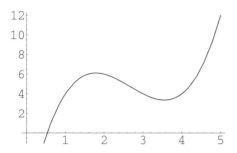

Given a set of values expressed as a list **data**, the function **ListPlot** will plot **data** assuming that **data** is regularly spaced, and will use the position in the list as the horizontal coordinate. The command which changes the **PlotStyle** plots the points large so that we can see them clearly.

```
data = {3,-0.5,2,0,1,0.5};
points = ListPlot[data,PlotStyle -> PointSize[0.03]];
```

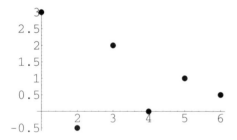

Displaying a Plot Again

All graphics can be displayed again using the function **Show**. We have given our last plot the name **points**. We can now ask to see that particular plot again.

```
Show[points];
```

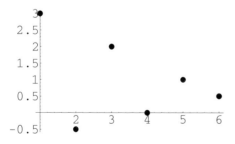

All of the graphics displayed can be altered to suit particular needs. In our plot for **ListPlot** we specified a particular size for the points to be plotted. Similarly, we can ask for those points to be joined by a continuous line.

```
pointsb = ListPlot[data, PlotJoined -> True];
```

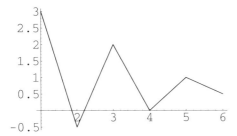

If you have executed the previous few commands on a computer you may have received a spelling warning. Here and elsewhere in this text we have suppressed messages from *Mathematica* warning us about possible spelling mistakes suggested by a new variable name being very much like one previously introduced. The **Show** command can be used to combine plots:

```
Show[points, pointsb];
```

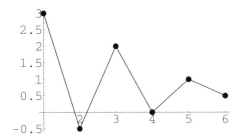

2.3 Numerical Calculations

Exact Arithmetic

Degrees of Precision

With most computer languages you are restricted to certain levels of precision in the values of numbers. This is not the case when you are using *Mathematica*. Try the following series of inputs.

```
30!
```

```
265252859812191058636308480000000
```

```
{N[30!], N[30!, 23]}
```

$$\{2.65253 \times 10^{32}, 2.6525285981219105863631 \times 10^{32}\}$$

The first result is exact, the second is approximate and has used the degree of precision given by default. In the last expression we have specified a precision of 23 digits.

Exact and Approximate Numbers

When we ask for the square root of a number, for example, *Mathematica* will return the most exact version.

```
Sqrt[51]
```

$$\sqrt{51}$$

Note that if you are using *Mathematica* 2.2 or earlier, the output here will differ - you will still see the letters "Sqrt" in the output. Only when we ask explicitly for that to be given in a numerical form will *Mathematica* return a number which approximates this square root.

```
N[Sqrt[51]]
```

```
7.14143
```

In evaluating the last expression *Mathematica* displays the answer to six significant figures. It will have been calculated, in its numerical form, to a higher precision, which will be dependent on the machine you are using. We can find the degree of precision by asking for the value of **Precision** on that particular output line,

```
Precision[%]
```

```
16
```

So in this case, the result will have been calculated to 16 SF, but only displayed to six figures. If that result were to be used in any further calculations, the full 16 figures would be used.

How *Mathematica* Decides on Exact or Approximate

This following example highlights the different routes *Mathematica* will take when dealing with numerical expressions.

```
(3/4)^100
```

$$\frac{515377520732011331036461129765621272702107522001}{1606938044258990275541962092341162602522202993782792835301376}$$

In this expression the answer has been returned in the most exact form possible. *Mathematica* will keep to that form unless explicitly asked otherwise, by the use of **N[ ]**, for example. An alternative way of presenting the above calculation is

```
0.75^100
```

$$3.2072 \times 10^{-13}$$

This second expression has numbers given in their floating point or approximate form, the expression is then kept in this approximate form and the answer given in that form. *Any* number with a decimal point in it will be taken to be approximate. Of course, we can always ask for that approximation to be more precise:

```
N[%,10]
```

$$3.207202185 \times 10^{-13}$$

A Practical Example with Exact and Approximate Numbers

For another illustration of these two paths down which numbers may be sent, we can look at solutions to equations. Giving the equation in exact form:

```
Solve[{x^2 + 3 x + 11/10 == 0}, x]
```

$$\left\{\left\{x \to \frac{1}{10}\left(-15 - \sqrt{115}\right)\right\}, \left\{x \to \frac{1}{10}\left(-15 + \sqrt{115}\right)\right\}\right\}$$

```
N[%]
```

$$\{\{x \to -2.57238\}, \{x \to -0.427619\}\}$$

Alternatively, in an approximate form:

```
Solve[{x^2 + 3 x + 1.1 == 0}, x]
```

$$\{\{x \to -2.57238\}, \{x \to -0.427619\}\}$$

Again, we chose the way in which numbers would be treated by the way we presented the equation - the presence of a decimal point implied the numerical treatment. This example raises some other issues that it is a good idea to explore right away.

A Note on "Equals"

You will have noticed that in the last section we used a double equals sign to denote that an expression was equal to zero. The use of the single equals sign will be examined more carefully shortly, where its role in assignment will be illustrated. For the moment it will be sufficient to highlight that when defining a symbolic equation the double equals sign should be used. This does not involve assigning any particular values to any variables.

```
x == 0
```

$$x == 0$$

If actually setting a quantity equal to some value for future substitution the single equals sign is used. This assigns that value to the variable.

```
x = 0
```

$$0$$

(Having set this value for **x** we must clear it for future calculations)

```
Clear[x]
```

Extracting Results from the Solve Function

It can be frustrating for new users that *Mathematica* returns results from **Solve** as some replacement rules indicated by the arrow:

```
Solve[{x^2 + 3 x + 11/10 == 0}, x]
```

$$\left\{\left\{x \to \frac{1}{10}\left(-15 - \sqrt{115}\right)\right\}, \left\{x \to \frac{1}{10}\left(-15 + \sqrt{115}\right)\right\}\right\}$$

What we really want are a list of values of x given these rules. In *Mathematica* the term "given" is invoked by the use of "/.", as shown here:

```
x /. %
```

$$\left\{\frac{1}{10}\left(-15 - \sqrt{115}\right), \frac{1}{10}\left(-15 + \sqrt{115}\right)\right\}$$

You can do the whole thing at once:

```
x /. Solve[{x^2 + 3 x + 11/10 == 0}, x]
```

$$\{\frac{1}{10}\left(-15 - \sqrt{115}\right), \frac{1}{10}\left(-15 + \sqrt{115}\right)\}$$

Some Useful Numerical Functions

The usual arithmetic functions are as you would expect for $+$, $-$, $*$, $/$ etc. Note that multiplication is denoted by a star or a space. A cross or "times" symbol is not used. The square root of a number is calculated by the **Sqrt** function.

```
Sqrt[78543.]
```

```
    280.255
```

Similar functions exist for exponentials, logarithms and many other numerical functions. You should note that sometimes long words are used, and there may be capitals within the name. For example, The function **FactorInteger** produces a list of the prime factors of a number and the power to which they are raised - useful in cryptography!

```
FactorInteger[15931157760]
```

```
    {{2, 8}, {3, 8}, {5, 1}, {7, 1}, {271, 1}}
```

This is telling us that $15931157760 = 2^8 \times 3^8 \times 5 \times 7 \times 271$.

From Data to Functions

Simple Function Fitting

Suppose that we have a data set and wish to represent it in terms of functions by a least-squares fit. This can be accomplished in a very simple way in *Mathematica*. Let's invent some data:

```
fitstuff = {6, 3, 5, 8, 7}
```

```
    {6, 3, 5, 8, 7}
```

The elements of **fitstuff** are interpreted as being the values of a function at equally spaced value points 1, 2, 3, 4, 5. Now we can plot those points. Calling the plot **dataplot** will allow us to use it again later.

```
dataplot = ListPlot[fitstuff, PlotStyle-> PointSize[0.03]];
```

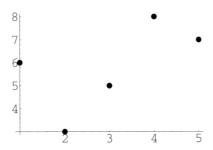

Now type the following, exactly as written. Do not be too concerned about the syntax used to define the function **psi**, we will investigate this later - you are most concerned about the function **Fit**.

```
psi[x_] = Fit[fitstuff, {1, x, x^2, x^3, x^4}, x]
```

$$17. - 15.75\,x + 5.04167\,x^2 - 0.25\,x^3 - 0.0416667\,x^4$$

```
psiplot = Plot[psi[x], {x, 1, 5}];
```

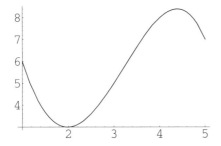

Now we can show the original points of the data set as well as the fitting function, and see that the function passes through them. At the same time we can see how to override *Mathematica*'s conventions on the vertical plot range:

```
Show[psiplot, dataplot, PlotRange -> {0, 10}];
```

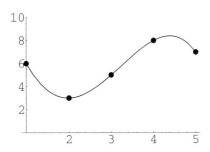

Fitting Functions to an Irregular Array

This approach can be applied to lists of pairs of (*x*, *y*) values also, and any set of functions can be used to construct the function, provided they appear with unknown coefficients in a linear combination. For example for the set of data **tStuff**, we can fit a trigonometric function to the data.

```
tStuff={{0,2.8},{2,11.},{3,5.3},{5,-4.4},{8,11.5}};
```

```
mi[x_] = Fit[tStuff, {1, Sin[x], Cos[x]}, x]
```

$$3.47465 - 0.685387 \, Cos[x] + 7.99654 \, Sin[x]$$

```
miPlot = Plot[mi[x],{x,0,8}];
```

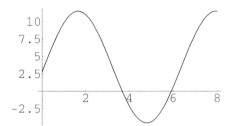

```
tStuffPlot = ListPlot[tStuff,PlotStyle->PointSize[0.03]];
```

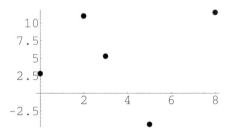

```
Show[miPlot,tStuffPlot];
```

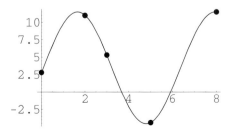

Other Numerical Operations

Numerical Solution of Equations

We have already seen how to solve a quadratic. *Mathematica* can cope with all polynomials up to degree 4 analytically. For degree 5 and above, usually a numerical approach is required. This involves the **NSolve** function.

```
NSolve[5 x^5 - 15 x^3 + 10 x + 1 == 0, x]
```

> {{x → -1.45682}, {x → -0.901736},
>   {x → -0.101566}, {x → 1.11145}, {x → 1.34867}}

NSolve and its symbolic counterpart can be used to solve sets of equations also:

```
NSolve[{y + 3 x == x^3, 4 y^2 - x == 4}, {x, y}]
```

> {{y → -1.17127, x → 1.48747}, {y → -1.04455, x → 0.364297},
>   {y → -0.73426, x → -1.84345}, {y → 0.777157, x → -1.58411},
>   {y → 0.957685, x → -0.331356}, {y → 1.21523, x → 1.90714}}

To obtain the results as a list of (x, y) pairs, you use /. for given, as before:

```
{x, y} /. %
```

> {{1.48747, -1.17127}, {0.364297, -1.04455}, {-1.84345, -0.73426},
>   {-1.58411, 0.777157}, {-0.331356, 0.957685}, {1.90714, 1.21523}}

For functions other than linear and polynomial systems, the **FindRoot** function is better. You give it a starting point and it iterates from there to a root:

```
FindRoot[Cos[x] == 0, {x, 1.5}]
```

> {x → 1.5708}

The result is a numerical approximation to $\pi/2$:

```
N[Pi/2]
```

> 1.5708

However, this function should be used with some care. (The function **FindRoot** operates using Newton's method which, if you have experience with its use, you will know is most effective when supplied with an approximation very near to a root.) For example if we supply a first approximation of 3.1, the root finally given is nowhere near 3.1.

```
FindRoot[Cos[x] == 0, {x, 3.1}]
```

```
{x → -20.4204}
```

The tangent to the curve $y = \cos(x)$, at $x = 3.1$, is nearly horizontal, so that the first iteration takes us to a point far away from the starting value. The process then converges! In financial applications **FindRoot** is particularly useful for solving non-linear equations, such as arise in calculating implied volatility, solving for an internal rate of return, or obtaining a critical price defined by a non-linear equation. We shall see its use many times.

2.4 Lists

There are some recurrent patterns in *Mathematica* of which you will need to be aware, no matter what the type of calculation you are doing. The use of list structures in *Mathematica* occurs in many diverse areas. We have already used them without too much fuss - now is a good time to go into a little more detail.

One-Dimensional Lists

Lists are used when dealing with data, or tables of values. A list is denoted by using curly brackets to enclose the elements of the list, which are separated from each other by commas. We have already used some lists in our examples. An example of a simple list is

```
data = {0,1,4,9,16,25,36,49,64,81,100}
```

```
{0, 1, 4, 9, 16, 25, 36, 49, 64, 81, 100}
```

We can select individual elements of such a list, or index elements of the list using the number which shows their position in the list. This involves the fourth type of bracketing, that of double square brackets. For example we can select the fourth element of the list **data**:

```
data[[4]]
```

```
9
```

There are built-in functions for re-arranging the elements of lists including functions such as **Reverse**, which re-orders the list in the opposite direction, and **RotateRight**, which takes the last element of the list and brings it to the front.

```
Reverse[data]
```

> {100, 81, 64, 49, 36, 25, 16, 9, 4, 1, 0}

```
RotateRight[data]
```

> {100, 0, 1, 4, 9, 16, 25, 36, 49, 64, 81}

The functions **Prepend** and **Append** can be used to add elements to lists.

```
Prepend[data, 1001]
```

> {1001, 0, 1, 4, 9, 16, 25, 36, 49, 64, 81, 100}

```
Append[data, 25000]
```

> {0, 1, 4, 9, 16, 25, 36, 49, 64, 81, 100, 25000}

Note that neither of these changes the meaning of **data**:

```
data
```

> {0, 1, 4, 9, 16, 25, 36, 49, 64, 81, 100}

The functions **PrependTo** and **AppendTo** have the same effect as **Prepend** and **Append**, but also update their argument:

```
PrependTo[data, 1001]
```

> {1001, 0, 1, 4, 9, 16, 25, 36, 49, 64, 81, 100}

```
data
```

> {1001, 0, 1, 4, 9, 16, 25, 36, 49, 64, 81, 100}

We select particular parts using functions, **First**, **Rest** and **Last**.

```
First[data]
```

    1001

```
Rest[data]
```

    {0, 1, 4, 9, 16, 25, 36, 49, 64, 81, 100}

```
Last[data]
```

    100

Our lists may well be generated within *Mathematica* by use of the **Table** command (examined in more detail later), or they may be generated outside *Mathematica*, imported and then manipulated, possibly to obtain a graphic representation.

Arrays, Matrices and Lists of Lists

Next up from a list is the concept of a list of lists. We use this idea in several ways. Here is an example of a list of lists:

```
listb = {{1,2,1,2}, {4,3,4,3}, {7,8,9,6}}
```

    {{1, 2, 1, 2}, {4, 3, 4, 3}, {7, 8, 9, 6}}

We have already seen how the function **FactorInteger** supplies its answer in a list of lists form. Also, lists of lists are how *Mathematica* represents matrices, and lists of (x, y) pairs.

```
MatrixForm[listb]
```

$$\begin{pmatrix} 1 & 2 & 1 & 2 \\ 4 & 3 & 4 & 3 \\ 7 & 8 & 9 & 6 \end{pmatrix}$$

We can ask about the dimensions of this matrix.

```
Dimensions[listb]
```

    {3, 4}

A list of lists is the way we represent any two-dimensional array of numbers. It could be an array for plotting or a matrix, or a rank-2 tensor. Matrix multiplication is specified by a "dot" (.), to distinguish it from element by element multiplication of lists. The simple square matrix **SqMatrix** will illustrate this.

```
SqMatrix={{1,6},{4,2}}
```

          {{1, 6}, {4, 2}}

```
SqMatrix.SqMatrix
```

          {{25, 18}, {12, 28}}

```
SqMatrix SqMatrix
```

          {{1, 36}, {16, 4}}

Selecting out specific rows is done in a similar way to that in which we select individual elements from a list. The second row of **SqMatrix** is therefore denoted by

```
SqMatrix[[2]]
```

          {4, 2}

Similarly we can select an individual element in a matrix by specifying its position in the matrix by row and then column, just as when referring to an entry in the form A_{ij} :

```
SqMatrix[[1,2]]
```

          6

Various other standard matrix operations may be carried out using specific *Mathematica* functions. For example we can transpose a matrix,

```
MatrixForm[Transpose[listb]]
```

$$\begin{pmatrix} 1 & 4 & 7 \\ 2 & 3 & 8 \\ 1 & 4 & 9 \\ 2 & 3 & 6 \end{pmatrix}$$

or find the inverse of a square matrix. To demonstrate this we first add a row to `listb` to make a square matrix:

```
invertible = Prepend[listb, {4, 2, 6, 1}];
inverse = Inverse[invertible];
MatrixForm[inverse]
```

$$\begin{pmatrix} -1 & -\frac{5}{2} & 1 & \frac{1}{2} \\ -1 & -\frac{11}{5} & -\frac{1}{5} & 1 \\ 1 & \frac{19}{10} & -\frac{3}{5} & -\frac{1}{2} \\ 1 & 3 & 0 & -1 \end{pmatrix}$$

```
MatrixForm[inverse . invertible]
```

$$\begin{pmatrix} 1 & 0 & 0 & 0 \\ 0 & 1 & 0 & 0 \\ 0 & 0 & 1 & 0 \\ 0 & 0 & 0 & 1 \end{pmatrix}$$

Flatten and Partition

One of the manipulations which it may be necessary to perform on a list is the conversion from an n-dimensional list to a one-dimensional list and back again. This is easily accomplished using **Flatten** and **Partition**.

```
listc = Flatten[listb]
```

```
{1, 2, 1, 2, 4, 3, 4, 3, 7, 8, 9, 6}
```

```
Partition[listc, 4]
```

```
{{1, 2, 1, 2}, {4, 3, 4, 3}, {7, 8, 9, 6}}
```

Note that the second argument to **Partition** represents the length of the sub-lists generated.

Applying Functions to Lists

There are two different situations under which you may want to apply a function to a list. Either the elements of the list are to be used as the arguments to a function, or a function is to be applied to each element of the list individually, the result being another list. These are two similar processes, but should not be confused. For functions which act on a sequence of arguments we use the **Apply** function to supply the argument set.

```
data
```

```
{1001, 0, 1, 4, 9, 16, 25, 36, 49, 64, 81, 100}
```

```
Apply[Plus,data]
```

```
1386
```

When a function is to be applied to each element of the set we can use the function **Map**, or for certain numerical functions we can use the whole set as the argument.

```
Sqrt[data]
```

$$\left\{\sqrt{1001}, 0, 1, 2, 3, 4, 5, 6, 7, 8, 9, 10\right\}$$

```
Map[Sqrt, data]
```

$$\left\{\sqrt{1001}, 0, 1, 2, 3, 4, 5, 6, 7, 8, 9, 10\right\}$$

Incrementation and Nesting

Incrementation

To perform a simple incremental function there are a few *Mathematica* functions which can be used.

```
Do[Print[{n, n^2, n^3}], {n, 1, 4}]
```

```
{1, 1, 1}
{2, 4, 8}
{3, 9, 27}
{4, 16, 64}
```

In this example the command **Print[{n, n^2, n^3}]** is executed for *n* taking values starting at 1 and being incremented by one each time until *n* equals 4. The function **Table**, used to produce lists, is of a similar form.

```
Table[i^2 +3 i,{i,1,15}]
```

```
{4, 10, 18, 28, 40, 54, 70, 88, 108, 130, 154, 180, 208, 238, 270}
```

In this case the resulting list or table is made up of each of the results obtained when the quantity $i^2 + 3\,i$ is calculated for *i* running from 1 to 15 in steps of 1.

Iterative Application of Functions

Iteration is at the heart of many modern areas of mathematics including dynamical systems, chaos and fractals. *Mathematica* has some simple tools for dealing with iterated systems, based on **Nest** and related functions. **Nest** in particular is a handy way of simulating random walks. We shall give some examples shortly. Before giving some simpler examples, we make sure that we have removed any previous definitions of several handy symbols:

```
Clear[x, y, f, g, h, z]
```

The **Nest** function applies a given function (its first argument) to its second argument a certain number of times, specified by its last argument:

```
Nest[g, x, 3]
```

```
g[g[g[x]]]
```

An interesting related function is **NestList**.

```
NestList[Sin, x, 3]
```

```
{x, Sin[x], Sin[Sin[x]], Sin[Sin[Sin[x]]]}
```

There is also a neat pair of generalizations to functions of two variables:

```
Fold[h, x, {a, b, c}]
```

```
h[h[h[x, a], b], c]
```

```
FoldList[h, x, {a, b, c}]
```

```
{x, h[x, a], h[h[x, a], b], h[h[h[x, a], b], c]}
```

An Example Using Nest

For example, what does `Sin[Sin[Sin ... (100 times)[x]...]]` look like? Ask *Mathematica*:

`Plot[Evaluate[Nest[Sin, x, 100]], {x, -2 Pi, 2 Pi}];`

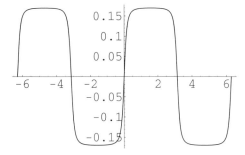

The **Evaluate** function has been used to make the plot easy for *Mathematica* to draw. It prevents *Mathematica* repeating the **Nest** operation at every point at which the function is sampled. With **Evaluate** in place it will only do that once. The example shown, and why it looks like a square wave, are discussed in detail by Gray and Glynn (1990).

A Few Useful Observations

Once or twice we have made use of a couple of features that deserve explicit comment.

Referencing Earlier Results

During a sequence of calculations you will probably want to refer to the results of previous calculations. This is achieved easily using a per cent sign, %, immediately followed by the number of the output corresponding to the result. Recall the definition of data:

`data`

    `{1001, 0, 1, 4, 9, 16, 25, 36, 49, 64, 81, 100}`

A special shorthand for the last computed result is %; %% denotes the next-to-last result.

`%`

    `{1001, 0, 1, 4, 9, 16, 25, 36, 49, 64, 81, 100}`

`%%[[3]]`

    `1`

Suppressing Data Output

Our list **data** is short, so there is no disadvantage in having *Mathematica* output the full result. If you are working with large data sets, however, you will almost certainly not want to see the results at each stage. Consequently, it may sometimes be desirable to suppress large results of purely symbolic operations. In the first of the following operations, the data set **bigdata** is created without being displayed at all. *The semicolon is the key ingredient for suppressing output completely.*

```
bigdata=Table[i^2,{i,0,10000}];
```

The next example shows a way of producing a shortened form of output. Remember this is only shortening what is displayed. The actual result is still a list of 10001 terms.

```
Short[bigdata]
```

```
    {0, 1, 4, 9, 16, «9991», 99920016,
     99940009, 99960004, 99980001, 100000000}
```

You may get different results depending on the exact version of *Mathematica* you are using.

2.5 Algebra

Simple Algebraic Manipulation

We'll begin by doing some simple algebra. Enter the following statements executing each one as you go. You will see that *Mathematica* has its own ideas about how to order the terms in a polynomial – do not let this worry you.

```
y^4 + 4 y^3 + 6 y^2
```

$$6 y^2 + 4 y^3 + y^4$$

Now, add $4\,y$ to this last result:

```
% + 4 y
```

$$4 y + 6 y^2 + 4 y^3 + y^4$$

Add 1 to this last result, and factorize this:

```
Factor[% + 1]
```

$$(1 + y)^4$$

Undo the factorization by expanding this last result:

```
Expand[%]
```

$$1 + 4\,y + 6\,y^2 + 4\,y^3 + y^4$$

Assign *x* to be this last quartic expression in *y*:

```
x = %
```

$$1 + 4\,y + 6\,y^2 + 4\,y^3 + y^4$$

Find the solutions for the quartic expression to be zero, using *x* as a shorter version of the quartic expression:

```
Solve[x == 0, y]
```

$$\{\{y \to -1\}, \{y \to -1\}, \{y \to -1\}, \{y \to -1\}\}$$

Assign the value 2 to *y*, and without displaying that result ask for the resulting value of *x*.

```
y = 2;
x
```

$$81$$

Now cancel any values associated with *y*, and ask for the value of *x*.

```
Clear[y];x
```

$$1 + 4\,y + 6\,y^2 + 4\,y^3 + y^4$$

The three things you should appreciate from this little session are the use of **Expand, Factor** and **Clear**. Also, we used **Solve** as encountered in the Numerical Calculations section, to give us solutions to an algebraic equation. It is a good idea to get into the habit of using **Clear** frequently, to make sure that you do not cause problems simply by trying to define quantities that conflict with earlier definitions. You should clear the definitions of **x** and **y** now, if you are running this session on a computer.

```
Clear[x,y]
```

Simplifying Rational Expressions

It is essential that you should have cleared the values for **x** and **y** from the last sequence of calculations, if you are trying out this chapter on a computer. If you have not done this *Mathematica* will go into an internal loop and will complain about exceeding its recursion depth. If this happens you will want to abort the calculation, which as you will remember from the introductory section is done by selecting "Interrupt Evaluation..." or "Abort Evaluation" from the Kernel menu.

First, we will input this simple rational expression involving x and y.

```
y = 1/(x^3 - 4) - 1/(x^2 - 2)
```

$$-\frac{1}{-2 + x^2} + \frac{1}{-4 + x^3}$$

Now, we can ask for all of the terms to be gathered together over a common denominator, and this expression to be assigned to be **u**.

```
u = Together[y]
```

$$\frac{2 + x^2 - x^3}{(-2 + x^2)(-4 + x^3)}$$

Using **u**, the shorthand for the expression, we can select out the denominator and numerator of the expression.

```
Denominator[u]
```

$$(-2 + x^2)(-4 + x^3)$$

```
Numerator[u]
```

$$2 + x^2 - x^3$$

Expansion can be aimed only at the denominator of the expression.

```
ExpandDenominator[u]
```

$$\frac{2 + x^2 - x^3}{8 - 4x^2 - 2x^3 + x^5}$$

With rational, rather than polynomial expressions, it may help to separate out the top and bottom for manipulation. In most cases you will find **Expand**, **Factor** and **Simplify** sufficient. You should take care in the use of **Simplify**. It can sometimes get into quite a lengthy process, and even have unpredictable results. It is much easier to use more functions but keep control over what changes are taking place.

Applying Simplification Rules Selectively

The following example will give some idea of how to apply operations selectively. We begin by defining the algebraic expression **expr**.

```
expr = (x^3 - 216) (x - 3)^2/((x - 2)(x - 4)(x - 6))
```

$$\frac{(-3 + x)^2 \, (-216 + x^3)}{(-6 + x) \, (-4 + x) \, (-2 + x)}$$

Applying **Expand** to this expression will expand each term in the numerator.

```
Expand[expr]
```

$$-\frac{1944}{(-6 + x) \, (-4 + x) \, (-2 + x)} + \frac{1296\,x}{(-6 + x) \, (-4 + x) \, (-2 + x)} -$$
$$\frac{216\,x^2}{(-6 + x) \, (-4 + x) \, (-2 + x)} + \frac{9\,x^3}{(-6 + x) \, (-4 + x) \, (-2 + x)} -$$
$$\frac{6\,x^4}{(-6 + x) \, (-4 + x) \, (-2 + x)} + \frac{x^5}{(-6 + x) \, (-4 + x) \, (-2 + x)}$$

ExpandAll, however, will take *every* part of the expression.

```
ExpandAll[expr]
```

$$-\frac{1944}{-48 + 44\,x - 12\,x^2 + x^3} + \frac{1296\,x}{-48 + 44\,x - 12\,x^2 + x^3} - \frac{216\,x^2}{-48 + 44\,x - 12\,x^2 + x^3} +$$
$$\frac{9\,x^3}{-48 + 44\,x - 12\,x^2 + x^3} - \frac{6\,x^4}{-48 + 44\,x - 12\,x^2 + x^3} + \frac{x^5}{-48 + 44\,x - 12\,x^2 + x^3}$$

You have already used **Together**, which will gather all terms over a common denominator.

```
Together[%]
```

$$\frac{324 - 162\,x + 9\,x^2 + x^4}{8 - 6\,x + x^2}$$

Splitting this expression into partial fractions uses **Apart**.

```
Apart[%]
```

$$37 + \frac{38}{-4 + x} - \frac{26}{-2 + x} + 6\,x + x^2$$

The **Factor** function will implicitly gather the terms together and factorize the resulting expression:

```
Factor[%]
```

$$\frac{(-3 + x)^2\,(36 + 6\,x + x^2)}{(-4 + x)\,(-2 + x)}$$

Another route to this simplified version of **expr** would have been using the function **Cancel**. This will make sure that any possible cancellations are carried out.

```
Cancel[expr]
```

$$\frac{(-3 + x)^2\,(36 + 6\,x + x^2)}{(-4 + x)\,(-2 + x)}$$

Other Useful Functions

Another useful pair of operations are **Collect** and **Coefficient**. We shall use another polynomial expression to illustrate their use.

```
Clear[x, y];
stuff = (2 x + 3)^2 (x + y)^2
```

$$(3 + 2\,x)^2\,(x + y)^2$$

```
Expand[stuff]
```

$$9\,x^2 + 12\,x^3 + 4\,x^4 + 18\,x\,y + 24\,x^2\,y + 8\,x^3\,y + 9\,y^2 + 12\,x\,y^2 + 4\,x^2\,y^2$$

```
Collect[%, x]
```

$$4\,x^4 + 9\,y^2 + x^3\,(12 + 8\,y) + x^2\,(9 + 24\,y + 4\,y^2) + x\,(18\,y + 12\,y^2)$$

So **Collect** literally collects together terms with the same power of **x**. We can pick out the coefficients by using **Coefficient**. Here we ask for the coefficient of **x^2** (all the terms, including those in **y**) in the expression.

```
Coefficient[Expand[stuff], x^2]
```

$9 + 24\,y + 4\,y^2$

This type of operation can be very useful in doing perturbation theory, and can also be used with the **Series** function which we will come to shortly. An alternative syntax is as follows - this must be used to extract the "constant" term:

```
Coefficient[Expand[stuff], x, 2]
```

$9 + 24\,y + 4\,y^2$

```
Coefficient[Expand[stuff], x, 0]
```

$9\,y^2$

Making Substitutions

Substitutions are a common requirement in manipulation. One way of doing this is to make a simple assignment and then ask for the new value of an expression.

```
x=3 y;
```

```
stuff
```

$16\,y^2\,(3 + 6\,y)^2$

This will mean however that from now on x will always be read as $3\,y$ and stuff will always equal $16\,y^2(3+6\,y)^2$. To prevent this we will have to explicitly remove the assignment of x.

```
Clear[x]
```

The technique to use to make a substitution which will not make a permanent change is to use a replacement rule. These are the same as are given as solutions to **NSolve** and **FindRoot**. To make the same substitution define the rule that x goes to $3\,y$, and apply that using /. to **stuff**.

```
stuff /. x -> 3 y
```

$$16\,y^2\,(3 + 6\,y)^2$$

```
stuff
```

$$(3 + 2\,x)^2\,(x + y)^2$$

You can see that this temporary substitution does not alter the stored definition of **stuff**. This technique can be used for any temporary substitutions whether they are numerical or algebraic.

2.6 Calculus

We now come to a topic that is fundamental to appreciating how useful *Mathematica* can be in modelling financial derivatives. We are interested in finding the values of certain options, together with the derivatives of those values with respect to variables such as the underlying price, its volatility, interest rates, and so on. When there is some form of analytical prescription available, it is very convenient, not to say less error-prone, to perform the computation of such "Greeks" symbolically. Even when one is working with an essentially numerical scheme, we shall still want to test it against known analytic solutions for special cases, or we may wish to differentiate a function based on interpolated numerical values.

Differentiation

Differentiation and integration work in a straightforward way. There are various notations for differentiation, of which the D operator is perhaps the most flexible.

```
Clear[x];
D[x^n, x]
```

$$n\,x^{-1+n}$$

```
Integrate[%, x]
```

$$x^n$$

Partial differentiation can also be carried out by stipulating the variable which you wish to differentiate by.

```
D[x^4 y^2 + x^2, y]
```

$$2\,x^4\,y$$

Mixed and multiple derivatives are also supported. We can differentiate once with respect to *y* and once with respect to *x*.

```
D[x^3 y^2 + x^2, y, x]
```

$$6 \, x^2 \, y$$

Higher derivatives are specified using the list structure to supply the variable and the number of times by which to differentiate with respect to that variable.

```
D[x^3 y^2 + x^2, {x, 3}, {y, 2}]
```

$$12$$

Integration - Symbolic and Numeric

Integrating with respect to y will reverse differentiation up to an arbitrary additive function of x.

```
Integrate[%%, y]
```

$$3 \, x^2 \, y^2$$

Similarly multiple integration can be carried out. As in the examples for differentiation, to integrate with respect to x and y:

```
Integrate[x^2 y,x,y]
```

$$\frac{x^3 \, y^2}{6}$$

Definite integration needs a list supplying the upper and lower limits of the integration variable.

```
Integrate[1/(1-x),{x,2,4}]
```

$$-\text{Log}[3]$$

There is also the facility to calculate integrals numerically. This is available for any functions, including those which *Mathematica* cannot do symbolically. As in the case of definite symbolic integration, we must specify the range of integration. This can frequently be used to compute expected values of complicated payoffs, if they do not work symbolically. This next example shows how symbolic integration is not possible, but numerical integration yields a result.

```
Integrate[Sin[Sin[z]],{z,0,2}]
```

$$\int_0^2 \text{Sin[Sin[z]] d}z$$

```
NIntegrate[Sin[Sin[z]],{z,0,2}]
```

    1.24706

This is another function under which problems can occur, and you should take care. For example:

```
Integrate[Sin[1000 x],{x,0,2}]
```

$$\frac{1}{1000} - \frac{\text{Cos[2000]}}{1000}$$

```
N[%]
```

    0.00136746

```
NIntegrate[Sin[1000 x],{x,0,2}]
```

        NIntegrate::ncvb :
         NIntegrate failed to converge to prescribed accuracy
            after 7 recursive bisections in x near x = 1.4140625`.

    0.00184811

This answer should not be believed because we have been warned about a problem. The ways to get around this problem are covered more fully when we have investigated how built-in *Mathematica* functions can be manipulated in the next section.

Limits

Of course, differentiation is defined by a limiting procedure. *Mathematica* can take limits explicitly:

```
Limit[((x + h)^3 - x^3)/h, h -> 0]
```

    $3 x^2$

Series

Mathematica also knows how to use Taylor's Theorem through the **Series** command. The use of **Series** generates a certain type of expression truncated by an "O" expression. These can be manipulated according to the usual rules of asymptotic series, or converted to a normal form without the "O". Note that the next power in the "O" term is not necessarily computed exactly:

```
Series[Sin[Sin[x]], {x, 0, 5}]
```

$$x - \frac{x^3}{3} + \frac{x^5}{10} + O[x]^6$$

```
%^2
```

$$x^2 - \frac{2\,x^4}{3} + \frac{14\,x^6}{45} + O[x]^7$$

```
Normal[%%]
```

$$x - \frac{x^3}{3} + \frac{x^5}{10}$$

Simple Ordinary Differential Equations

A large part of this book is concerned with various methods for solving partial differential equations, so we shall explore that topic in some detail later. In our introduction we shall take a brief look at ODEs.

Symbolic Solution of Differential Equations

Mathematica is a useful tool for dealing with certain differential equations. The function **DSolve** can deal with a large class of equations. It is used in a similar way to **Solve**, except that you need to specify both the independent and dependent variables. Notice how the equation is specified by using the double = sign.

```
DSolve[y''[x] + y[x] == 0, y[x], x]
```

$$\{\{y[x] \to C[2]\,Cos[x] - C[1]\,Sin[x]\}\}$$

If you give appropriate initial conditions, the arbitrary constants are removed.

```
DSolve[{y'[x] + x y[x] == 0, y[0] == 1}, y[x], x]
```

$$\left\{ \left\{ y[x] \rightarrow E^{-\frac{x^2}{2}} \right\} \right\}$$

To check that this is indeed a solution of the given equation we use the technique of replacement. The solutions to these equations are given as replacement rules presented in a list. We therefore need to extract a particular element of our solution list and apply that rule to get the solution. Setting the solution to be represented by s we select the first element of the list (we only have one in this example), and then apply that rule to $y[x]$ to see the expression which is a solution.

```
s=y[x]/.%[[1]]
```

$$E^{-\frac{x^2}{2}}$$

Now we can check if this is indeed a solution to our original equation.

```
D[s,x] + x s
```

```
0
```

Numerical Solutions to Differential Equations

In a similar way to numerical integration there is a function **NDSolve** which will find numerical solutions to differential equations and extends the scope of the equations to which we can find solutions. This function returns an interpolating function, rather than the usual table of values. This allows you to ask for the values of the solution anywhere in the solution range, for example. As usual, the use of "/." allows for the extraction of numerical values.

```
Clear[x, y, t];
soln = NDSolve[ {x'[t] == y[t],y'[t] == -4.0*x[t] - 0.1*y[t],
x[0] == 1,y[0] == 1},{x, y}, {t, 10}]
```

```
{{x → InterpolatingFunction[{{0., 10.}}, <>],
   y → InterpolatingFunction[{{0., 10.}}, <>]}}
```

Note that the answer is supplied as an **InterpolatingFunction** - this allows a numerical solution computed in steps to be used just like an ordinary function. We can select out the specific solution for any t. At $t = 2$,

```
{x[2], y[2]} /. soln
```

```
{{-0.951524, 0.793321}}
```

We shall use the `InterpolatingFunction` as a means of handling the output of finite-difference solvers for option equations. We shall explore numerical methods for solving PDEs in some detail later in this book, starting in Chapter 13.

Complex Numbers

Built-In Functions for Complex Manipulation

The use of complex numbers and functions is relatively unusual in derivatives modelling, but recent research has introduced Laplace transform methods as a powerful tool for the analytical pricing of options previously thought to be unpriceable except through an essentially numerical scheme (e.g. arithmetically averaged Asian options). So we need to use complex numbers, and to deal with them in *Mathematica*. The symbol I (capital) is the square root of -1, in Standard Form. *Mathematica* operations work on complex numbers in just the same way they work on real numbers. In addition, various special operations for dealing with complex variables are provided.

Let us find the solution set for the following equation.

```
qsoln = Solve[3 x^2 - 4 x + 6 == 0, x]
```

$$\left\{\left\{x \to \frac{1}{3}\left(2 - I\sqrt{14}\right)\right\}, \left\{x \to \frac{1}{3}\left(2 + I\sqrt{14}\right)\right\}\right\}$$

We put that into an approximate numerical form,

```
N[qsoln]
```

$$\{\{x \to 0.666667 - 1.24722\ I\}, \{x \to 0.666667 + 1.24722\ I\}\}$$

and then create the set containing the numerical solutions to the original equation:

```
solutions = x /. %
```

$$\{0.666667 - 1.24722\ I,\ 0.666667 + 1.24722\ I\}$$

As with the exponential function, a special typeface is used for I in Traditional Form notation:

```
TraditionalForm[qsoln]
```

$$\left\{\left\{x \to \frac{1}{3}\left(2 - i\sqrt{14}\right)\right\}, \left\{x \to \frac{1}{3}\left(2 + i\sqrt{14}\right)\right\}\right\}$$

The function **Re** will return the real part of any complex number. We can apply this function to our set **solutions** to get the real parts of our set.

```
Re[solutions]
```

```
{0.666667, 0.666667}
```

Similarly, the function **Im** will give the imaginary part of a complex number. But note that symbols are in general treated as complex - for example:

```
z = x + I y;
```

```
Im[z]
```

```
Im[x] + Re[y]
```

The function **Conjugate** will give the complex conjugate of a complex number:

```
Conjugate[solutions]
```

```
{0.666667 + 1.24722 I, 0.666667 - 1.24722 I}
```

The function **Abs** gives the absolute value of a complex number (i.e. $\sqrt{x^2 + y^2}$):

```
Abs[solutions]
```

```
{1.41421, 1.41421}
```

The function **Arg** will give the argument of the complex number (i.e. ArcTan(y/x))

```
Arg[solutions]
```

```
{-1.07991, 1.07991}
```

There is a useful **ComplexExpand** function, which carries out an expansion assuming that all variables are real:

```
ComplexExpand[Cot[x + I y]]
```

$$-\frac{\text{Sin}[2\,x]}{\text{Cos}[2\,x] - \text{Cosh}[2\,y]} + \frac{\text{I Sinh}[2\,y]}{\text{Cos}[2\,x] - \text{Cosh}[2\,y]}$$

This function is very useful for getting *Mathematica* to output what you expect to see when your mind assumes things are real, but the computer can't read your mind!

Complex Contour Integration

For integration in the complex plane the function **NIntegrate** can take complex numbers to specify the boundary around which the function is to be integrated.

```
NIntegrate[1/z, {z, 2, I, -2, -I, 2}]
```

```
    0. + 6.28319 I
```

We shall use complex contour integration to invert certain types of Laplace transform for option values.

2.7 More about Functions

We have already seen several *Mathematica* functions in the course of our investigations so far. These are only a small proportion of those available. Some of the others are obvious. One example is trigonometric functions. Since you already know we can use **Sin** you would expect **Cos**, **Tan** and all the usual inverse and hyperbolic versions. These can all be used in *Mathematica* in the same way as **Sin**, as can the inverses of these trigonometric functions, as illustrated below.

```
Plot[ArcSin[x], {x, -1, 1}];
```

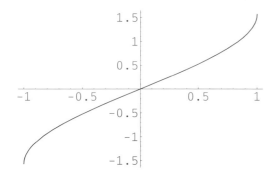

You will have already spotted that all *Mathematica*'s built-in functions begin with a capital letter. Names which can be thought of as composed of two or more pieces may have capitals in the middle. This means that when naming your own functions you can use any name starting with lower-case letters and be sure that you are not duplicating any *Mathematica* built-in function. We'll come back to user-defined functions shortly.

Getting Information about Built-In Functions

As you start to explore the capabilities of *Mathematica* you will need to get information about the functions available. There is the facility to get information about built-in functions from within *Mathematica*. To ask about a function just type a "?" before the function name.

```
?Sqrt
```

```
    Sqrt[z] gives the square root of z.
```

The wild card "*" can also be used to obtain a list of functions, with a similar sequence in common. So we can obtain a list of all built-in functions related to the trigonometric function **Sin**.

```
?*Sin*
```

```
ArcSin                  SingleLetterItalics  Sinh
ArcSinh                 SingularityDepth     SinhIntegral
IncludeSingularTerm     SingularValues       SinIntegral
Sin
```

Many of the built-in functions that you will use have options associated with them. These set default values for various variables associated with the execution of that function. To see how the options are set for a function, **Plot** for example,

```
Options[Plot]
```

$$\left\{ \text{AspectRatio} \rightarrow \frac{1}{\text{GoldenRatio}}, \text{ Axes} \rightarrow \text{Automatic, AxesLabel} \rightarrow \text{None,} \right.$$

```
   AxesOrigin → Automatic, AxesStyle → Automatic, Background → Automatic,
   ColorOutput → Automatic, Compiled → True, DefaultColor → Automatic,
   Epilog → {}, Frame → False, FrameLabel → None, FrameStyle → Automatic,
   FrameTicks → Automatic, GridLines → None, ImageSize → Automatic,
   MaxBend → 10., PlotDivision → 30., PlotLabel → None, PlotPoints → 25,
   PlotRange → Automatic, PlotRegion → Automatic, PlotStyle → Automatic,
   Prolog → {}, RotateLabel → True, Ticks → Automatic,
   DefaultFont :→ $DefaultFont, DisplayFunction :→ $DisplayFunction,
```
$$\left. \text{FormatType} :\rightarrow \text{\$FormatType, TextStyle} :\rightarrow \text{\$TextStyle} \right\}$$

Changing any of these options can be done when executing the function by explicitly stating the desired option value as the final argument to the function. For example, we can change the label points on the *x*-axis for a sine curve plot, and add suitable grid lines.

```
Plot[Sin[x],{x,0,4 Pi},
        Ticks -> {{0, Pi, 2Pi, 3Pi, 4Pi}, Automatic},
        GridLines -> {Table[k*Pi, {k, 0, 4, 1/2}] ,None} ];
```

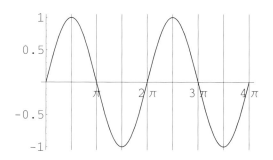

Note that *Mathematica* allows a slight abuse of notation when it comes to pre-multiplying a symbol by a number. In the above, **2Pi** is interpreted as twice π, even in the absence of a star or space. Options may also be used to cure problems with functions such as **NIntegrate**:

```
Options[NIntegrate]
```

```
{AccuracyGoal → ∞, Compiled → True, GaussPoints → Automatic,
 MaxPoints → Automatic, MaxRecursion → 6, Method → Automatic,
 MinRecursion → 0, PrecisionGoal → Automatic,
 SingularityDepth → 4, WorkingPrecision → 16}
```

```
NIntegrate[Sin[1000 x],{x,0,2}, MaxRecursion ->9]
```

```
0.00136746
```

Built-In Special Functions

Examples of Special Functions

Mathematica also knows a phenomenal number of special functions. If you do the kind of work that employs handbooks of functions and their values and properties, you may be surprised by the breadth of *Mathematica*'s built-in capabilities. As well as a host of number-theoretic and combinatorial functions, there are many "classical" functions. Let us take a look at a few examples. First of all, we can use Bessel functions:

```
Plot[{BesselJ[0, x], BesselI[1, x]}, {x, 0, 10},
PlotRange -> {-5, 5}, PlotStyle -> {Thickness[0.015],
Thickness[0.01]}];
```

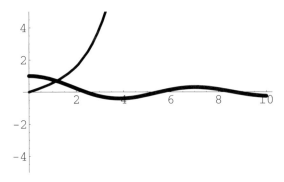

There is a selection of orthogonal polynomials:

```
LegendreP[5, x]
```

$$\frac{15\,x}{8} - \frac{35\,x^3}{4} + \frac{63\,x^5}{8}$$

```
SphericalHarmonicY[1, 1, theta, phi]
```

$$-\frac{1}{2}\,E^{I\,phi}\,\sqrt{\frac{3}{2\,\pi}}\,\text{Sin[theta]}$$

This is probably a good place to remark on how Greek letters are input - previously we have used π without explaining how we wrote it, and our spherical harmonic function looks rather better written as:

```
TraditionalForm[SphericalHarmonicY[1, 1, θ, φ]]
```

$$-\frac{1}{2}\,e^{i\,\phi}\,\sqrt{\frac{3}{2\,\pi}}\,\sin(\theta)$$

The trick is to write "ESC\thetaESC" where ESC stands for the escape key. Occurrences of π are generated with "ESC\piESC".

Naming of Special Functions

This should give you a feel for the way in which these functions are named. Many special functions can be classified under an overall name, like Legendre. Within this there may be more than one type, like the P and Q Legendre functions. And finally, these may be indexed by an integer or some other variable. Hence the Bessel function $K_n(x)$ is denoted by

```
BesselK[n, x]
```

```
BesselK[n, x]
```

Most of these functions are in the core kernel. However, the statistical distributions are accessed from one of the Packages. We shall look at Packages later. Note that *Mathematica* knows some particular values of special functions, and also knows series:

```
Gamma[1/2]
```

$$\sqrt{\pi}$$

```
Series[BesselJ[0, x], {x, 0, 6}]
```

$$1 - \frac{x^2}{4} + \frac{x^4}{64} - \frac{x^6}{2304} + O[x]^8$$

Building Your Own Functions

Defining a Simple Function

When defining a function by using an explicit formula, it is important to specify the variables involved in that function by use of the underscore "_". The delayed assignment sign ":=" then allows you to define the formula of your function.

```
h[x_, y_] := x^2 / y^3
```

This has defined the function h. We can now ask for that function to be evaluated for specific values of x and y.

```
h[1, 2]
```

$$\frac{1}{8}$$

You can now ask about your function:

```
?h
```

```
Global`h

h[x_, y_] := x^2/y^3
```

If you wish to clear that definition, you use **Clear**.

```
Clear[h]
```

Your definition for h no longer exists.

```
?h
```

```
Global`h
```

There are many types of function which can be defined, in the same way as the algebraic one we have just defined. Functions can employ any of the *Mathematica* built-in functions;

```
funcSin[x_,y_] := Sin[x] + Cos[y]^2
```

```
funcSq[x_, y_, z_] := Sqrt[x^2 + y^2 + z^2]
```

Functions can also be defined which will perform a test on the argument(s) returning the result True or False. `testDiv5` will check whether the argument is divisible by 5:

```
testDiv5[t_] := IntegerQ[t/5]
```

```
testDiv5[8]
```

    False

```
testDiv5[15]
```

    True

Defining Functions in Their Pure Form

If you only intend to use the function once, or to use it with operations such as **Map** and **Nest**, there is a more economical way to define it. Instead of using a letter or name for the arguments in the function, the syntax for "pure" functions uses the "#" to indicate the arguments. So the expression which defines the action of a function to add 3 to a number is `#  +  3` instead of `x  +  3`. To actually turn this into a function we use an "&" after the expression of the function. So

```
f[x_] := x + 3
```

can be written as

```
(# + 3) &
```

        #1 + 3 &

without using the definition of `f`. So if we apply that function to 2,

```
(# + 3) & [2]
```

    5

This technique can also be used instead of defining functions of more than one variable. Using the # notation, we now label each of the arguments with a number immediately (i.e. no space) after the #. So instead of

```
g[x_, y_] := x^2 + y
```

```
g[2,3]
```

$$7$$

we use

`(#1^2 + #2) & [2,3]`

$$7$$

Notice how the numbers used for labelling directly correspond to the order in which those arguments will be given. The advantage of this technique for defining functions which will only be used once is that the definitions do not need to be stored, they are just used at the appropriate moment.

Using Pure Functions in Simple Programs

Programs can be shortened in length by using pure functions instead of separately defining a general global function which can be used in those functions which specifically ask for function definitions.

`NestList[#^2 &, x, 5]`

$$\{x, x^2, x^4, x^8, x^{16}, x^{32}\}$$

`Fold[Sqrt[#1 + #2] &, t, {l, m, n}]`

$$\sqrt{n + \sqrt{m + \sqrt{1 + t}}}$$

`Clear[p, q, r, s]`

`NestList[Append[Rest[#],1]&, {p,q,r,s}, 3]`

$$\{\{p, q, r, s\}, \{q, r, s, 1\}, \{r, s, 1, 1\}, \{s, 1, 1, 1\}\}$$

This form of writing functions can also be used for defining functions which will perform a test. As an example we can define a function which will return True if a number is divisible by 3, and use this as the criterion in the function `Select`.

`Select[ {1, 2, 3, 4, 5, 6, 7, 8, 9}, IntegerQ[(#/3)] &]`

$$\{3, 6, 9\}$$

A function which will return True if the second argument is less than the first can be used as the test by which a list can be sorted.

```
Sort[ {8, 34, 9, 2, 7, 4, 75, 4}, (#2 < #1) &]

        {75, 34, 9, 8, 7, 4, 4, 2}
```

This technique can be particularly useful when manipulating lists.

Defining a Function Recursively

There may be functions which you wish to define recursively giving values for the end points. This can be done in the following way.

```
Clear[f]

f[1] = 1;

f[2] = 1;

f[n_] := f[n-1] + f[n-2]
```

We can now evaluate `f[3]`.

```
f[3]

        2
```

There is a useful modification of this function which makes it much more efficient, that forces the function to remember previously computed values.

```
Clear[g]

g[1] = 1;

g[2] = 1;

g[n_] := g[n] =g[n-1] + g[n-2]
```

The following illustration makes it clear that the use of the second function for computing the Fibonacci series is strongly preferred:

```
Timing[f[35]]
```

```
{468. Second, 9227465}
```

```
Timing[g[35]]
```

```
{0. Second, 9227465}
```

Such a device will turn out to be useful in a simple implementation of the binomial tree model, where we recurse back through a tree from the payoff. Note that here and throughout this text, the timings reported are based on the computer used for editing and final testing of this manuscript, which runs a 266 MHz PowerPC 750 processor. You must make allowances for the clock speed and efficiency of your own system when making comparisons or assessments about what computations are feasible.

The Difference between := and =

Constructing functions is dependent on the use of ":=". The use of the colon with equals means that the evaluation of the right side of the equation is delayed, and it will only be evaluated when that function is actually called. As an example let us define some functions which will be dependent on the value of a variable that we call **M** and to which we assign a value of 2 initially:

```
M=2
```

```
2
```

Now if we define **funcOne**, using :=, the right-hand side will not be evaluated yet.

```
funcOne[x_] := M x
```

Defining the function **funcTwo** using = will mean that the right-hand side is evaluated using the present value of **M**.

```
funcTwo[x_] = M x
```

```
2 x
```

With the present definitions both **funcOne[2]** and **funcTwo[2]** will evaluate to 4.

```
{funcOne[2],funcTwo[2]}
```

```
{4, 4}
```

If we were to change the value of **M** however, the definition of **funcOne** would: allow the new value to be used, whereas the definition of **funcTwo** has already been set by the original definition of **M**:

```
M=100
```

```
    100
```

```
{funcOne[2],funcTwo[2]}
```

```
    {200, 4}
```

Assignment and Random Walks

Another way to see the difference between = and := is to define two functions from which we can produce random walks and plot those resultant functions. Firstly let us define two functions which will add a random number between −0.5 and +0.5 to their argument:

```
addrandONE[x_] := x + Random[Real,{-0.5,0.5}]
```

```
addrandTWO[x_] = x + Random[Real,{-0.5,0.5}]
```

```
    -0.365925 + x
```

So if we ask for each of these functions evaluated twice on the same value,

```
{addrandONE[0],addrandONE[0]}
```

```
    {-0.405745, -0.192096}
```

```
{addrandTWO[0],addrandTWO[0]}
```

```
    {-0.365925, -0.365925}
```

The second function does not act as a random function. Now we can form random walks based on these functions using the function NestList,

```
walkONE = NestList[addrandONE,0,1000];
```

```
walkTWO = NestList[addrandTWO,0,1000];
```

and now we can display these lists using ListPlot and asking that the points should be joined.

```
ListPlot[walkONE, PlotJoined->True];
```

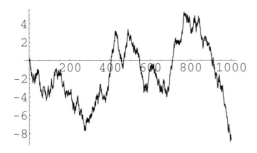

```
ListPlot[walkTWO, PlotJoined->True];
```

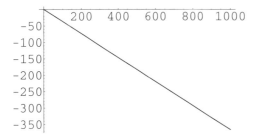

From this it is obvious what has happened in the function **addrandTWO**. The function has added a constant number to the argument of **addrandTWO**, that number being the first random number which was generated when the function was defined. The function **addrandTWO** was displayed with the **Random** function replaced by a number when the function definition was entered and from that point on the function definition remained as such.

Note the ease with which random walks may be constructed - we shall use this facility when carrying out Monte Carlo simulation of paths followed by stock prices.

Packages

Although, as you have seen, there are many functions built-in to *Mathematica*, you may find that to cope with certain situations, you wish that there were functions available which were more specific to your task. Functions which extend the capability of *Mathematica* are available in the Packages. The examples used here are based upon the Packages available with version 3.0 of *Mathematica*. Earlier versions may not have the same capabilities. The principles and techniques will remain the same for all versions of *Mathematica*. If you cannot duplicate these results exactly, investigate the Packages available with your version of *Mathematica*.

A New Function for Fourier Transforms

The most efficient way of loading a Package is to use the **Needs** command. There are other ways but this copes best with accidents, such as trying to load a package twice. Ordinarily, *Mathematica* can do numerical Fourier transforms using the **Fourier** function. When we ask for information about that function we are told that this function operates on a list of numbers.

`?Fourier`

> Fourier[list] finds the discrete
>   Fourier transform of a list of complex numbers.

There are no other built-in functions which will give us an analytical Fourier transform (unless we manage matters manually by using **NIntegrate**).

`?*Fourier*`

> Fourier   InverseFourier

However if we load the Package **FourierTransform** which is in the set of Calculus Packages,

`Needs["Calculus`FourierTransform`"]`

we can now ask for a list of all the functions which that package has introduced.

`?Calculus`FourierTransform`*`

FourierCosSeriesCoefficient	FourierSinTransform	NFourierExpSeriesCoefficient
FourierCosTransform	FourierTransform	NFourierSinSeriesCoefficient
FourierExpSeries	FourierTrigSeries	NFourierTransform
FourierExpSeriesCoefficient	InverseFourierCosTransform	NFourierTrigSeries
FourierFrequencyConstant	InverseFourierSinTransform	NInverseFourierTransform
FourierOverallConstant	InverseFourierTransform	$FourierFrequencyConstant
FourierSample	NFourierCosSeriesCoefficient	$FourierOverallConstant
FourierSinSeriesCoefficient	NFourierExpSeries	

And asking about the function **FourierTransform**, we find that this function will act on an algebraic expression.

`?FourierTransform`

> FourierTransform[expr, t, w] gives a function
>   of w, which is the Fourier transform of expr, a
>   function of t. It is defined by FourierTransform[
>   expr, t, w] = FourierOverallConstant * Integrate[Exp[
>   FourierFrequencyConstant I w t] expr, {t, -Infinity, Infinity}].

We can now perform analytical Fourier transforms - here is a well-known example:

`FourierTransform[Exp[-t^2/s^2], t, w]`

$$E^{-\frac{1}{4} s^2 w^2} \sqrt{\pi}\ s$$

2.8 Further Graphics

In the course of our investigations so far, we have produced a few plots of functions and data. Let's review this in a more systematic fashion. In this brief introduction we will only discuss standard built-in operations that are sufficient for simple graphics.

Colours, Greys, Thickness and Dashing

As a useful preliminary we'll learn about the specification, and indeed the spelling of colours, and so on. This is a tricky section to write as this book is being written and published in Britain but refers to a software system that uses American English. When entering graphical programs, please get used to spelling colour as **Color** and grey as **Gray**! In *Mathematica* colours and greys are parametrized by intensities between 0 and 1. **GrayLevel[0]** denotes black, **GrayLevel[1]** is white, and **GrayLevel[0.5]** is an intermediate grey. Colours may be modelled in a variety of ways. The **RGBColor** function takes three arguments which denote the intensities of the red, green and blue components. So **RGBColor[1, 0, 0]** is a pure bright red. Various other colour models can be used, such as **CMYKColor**, but we will focus on the RGB formulation. Useful graphical distinctions may also be made using dashing and thickness specifications. **Thickness[x]** specifies that lines which follow are to be drawn with a thickness **x**, where **x** is a fraction of the total plot width. **Dashing[{u, v, w, x,...}]** specifies that lines which follow have black and white segments of lengths **u**, **v**, **w**, **x**, . . . etc., repeated cyclically.

Plotting One or More Functions of One Variable

The simplest plot command takes the form **Plot[function, range]**. For example:

```
Plot[Sin[x^2], {x, -4, 4}];
```

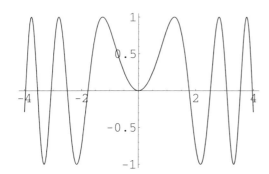

Now let's plot a family of four functions on the same graph, using colour and other attributes and the **Legend** Package to label things properly. For further details of how to manage legends, see Shaw and Tigg (1993), but note that when used with a legend, the thickness and dashing attributes should be used in absolute form. Cloning of the following example is encouraged:

```
Needs["Graphics`Legend`"];
Plot[{Sin[x], Sin[x^2], Cos[x], Cos[x^2]},{x, -4, 4},
PlotStyle -> {{RGBColor[1, 0, 0]}, {GrayLevel[0.2]},
{RGBColor[0,0,1],AbsoluteThickness[2]},
{AbsoluteDashing[{2, 3}]}},
PlotLegend -> {Sin[x], Sin[x^2], Cos[x], Cos[x^2]},
LegendPosition -> {1.2, -0.5},
PlotLabel -> FontForm["Coloured Curves", {"Helvetica", 12}]
];
```

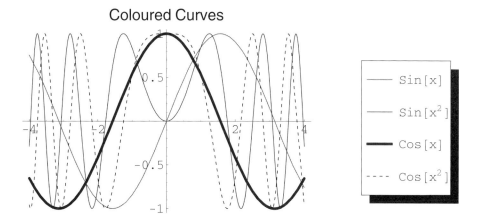

Note that the different specifications can be used singly or together; by combining two specifications we obtained a thick blue line. The use of such specifications give *Mathematica* a capability to represent many data sets on the same graph. There isn't time to discuss legends in detail here, but it is important to note that there is a Package for treating legends. **Plot** also has several options. Let's explore a few of the relevant ones:

```
Plot[{x^2, x^3}, {x, -3, 3}, PlotRange -> {-3, 3},
PlotLabel -> "Squares and Cubes", Frame -> True];
```

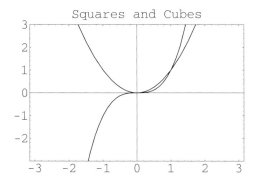

```
Plot[{Sqrt[Abs[x]], 0.5x^2}, {x, -2, 2}, PlotRange -> All,
AspectRatio -> 1, Axes -> None, PlotPoints -> 150];
```

Let's look the idea of a parametric plot, where x and y are given as a function of a third variable, say t:

```
ParametricPlot[{Sin[10 t], Sin[11 t]}, {t, 0, 2 Pi},
AspectRatio -> Automatic];
```

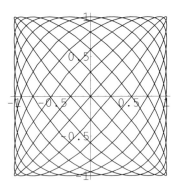

Please do not use this function for doing log plots and similar constructions. The facility for doing these is not part of the core kernel but is in a Package called **Graphics`Graphics`**, along with other useful plots. Here is the command to load the Package and an obvious simple application:

```
Needs["Graphics`Graphics`"];
LogPlot[Exp[-x], {x, 1, 5}];
```

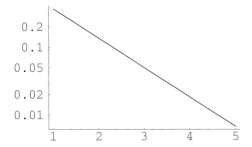

Plotting One or More Data Sets with One Independent Variable

The **ListPlot** function takes data represented as a list and produces scatter and line plots. Let's make up some data (remember the semicolon stops *Mathematica* from producing the lists on the screen):

```
polydata = Table[10*Sqrt[i], {i, 3, 10}];
polydatab = Table[{i, i^(3/2)}, {i, 3, 10}];
redblobs = ListPlot[polydata,
PlotStyle -> {RGBColor[1, 0, 0], PointSize[0.05]},
PlotRange -> {0, 50}];
```

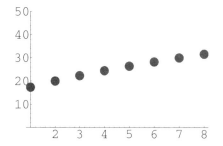

Note the use of **PlotStyle**, and the fact that the first "*x*-value" is 1.

```
ListPlot[polydatab, PlotRange -> {0, 40},
PlotJoined -> True];
```

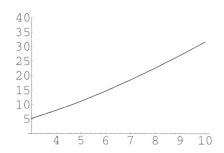

Be careful about whether you are dealing with just values or pairs of values. Similar log plots are available in the graphics Package. Perhaps one of the more irritating features of *Mathematica* is the absence of a core kernel function for plotting several sets of data simultaneously on one graph. There are two ways around this. First, we can always use **Show** to combine previously created plots:

```
polydatac = Table[i, {i, 3, 10}];

blueblobs = ListPlot[polydatac,
PlotStyle -> {RGBColor[0, 0, 1], PointSize[0.02]},
PlotRange -> {0, 40}];
```

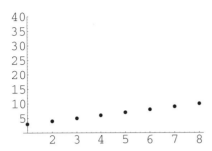

```
Show[redblobs, blueblobs];
```

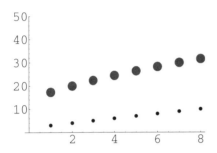

Secondly we can use the function **MultipleListPlot** which is only available in the **MultipleList-Plot** Package.

```
Needs["Graphics`MultipleListPlot`"];
lista = {5, 7, 4, 3}; listb = {9, 8, 7, 6};

MultipleListPlot[lista, listb, listc,
PlotJoined -> True, PlotRange -> {0, 10},
LineStyles -> {{RGBColor[1, 0, 0]}, {RGBColor[0, 1, 0]},
{RGBColor[0, 0, 1]}}];
```

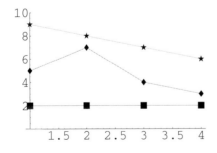

The management of legends in conjunction with this Package is discussed in detail by Shaw and Tigg (1993).

Functions of Two or More Variables

The basic functions to know about are **Plot3D**, **ContourPlot**, **DensityPlot** and their **List** analogues. The forms of the command are very similar to those for **Plot** etc. A few examples will suffice. Let's make a function to use in our examples.

```
func[x_, y_] := Sin[x^2] Cos[y];
Plot3D[func[x, y], {x, -2, 2}, {y, -2, 2}, PlotPoints -> 40];
```

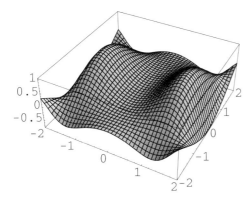

```
ContourPlot[func[x, y], {x, -2, 2}, {y, -2, 2}, PlotPoints -> 40];
```

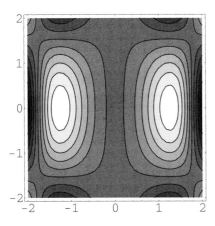

```
DensityPlot[func[x, y], {x, -2, 2},
  {y, -2, 2}, PlotPoints -> 60, Mesh -> False];
```

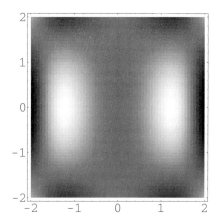

In the same way as for two-dimensional curves, the functions for three-dimensional surfaces also have options associated with them which allow the alteration of graphics to suit your needs.

```
Plot3D[func[x, y], {x, -2, 2}, {y, -2, 2},
ViewPoint -> {0, 1, 0.3},
Axes -> None, Boxed -> False, PlotPoints -> 20];
```

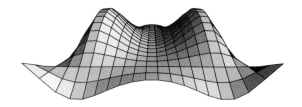

```
ContourPlot[func[x, y], {x, -2, 2}, {y, -2, 2},
Contours -> 40, ContourLines -> False,
ColorFunction -> Hue, Frame -> False];
```

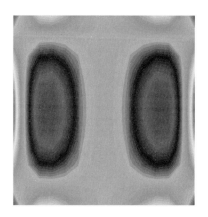

Animation

There is also the facility in *Mathematica* to be able to animate graphics which you have produced. This is done by creating all of the cells which will make up your movie and then asking to see them displayed one after the other. As an example we can see how a three-dimensional surface alters as we step *x* and *y* over slightly larger values. Only the first frame is shown in the book. Such techniques can be used to present plots of option values and their Greeks as functions of other variables, using the animation to represent the evolution through time, or perhaps to vary the angle of view, in order to bring out particular features.

```
Table[Plot3D[Sin[x+t] Cos[y+t], {x, -Pi, Pi}, {y, -Pi, Pi}],
    {t,0,2*Pi-Pi/10,Pi/10}];
```

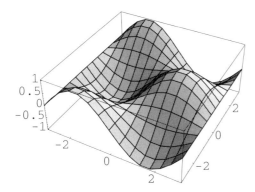

Chapter 2 Bibliography

The following are particularly useful for the topics indicated:

The definitive reference:
Wolfram, S., 1996, *The Mathematica Book*, Wolfram Media-Cambridge University Press.

Data analysis and general scientific work:
Shaw, W.T. and Tigg, J., 1993, *Applied Mathematica*, Addison-Wesley.

Graphics galore:
Wickham-Jones, T., 1994, *Mathematica Graphics, Techniques and Applications*, Springer-TELOS.

Exploring general finance and economics:
Varian, H.R. (editor), 1993, *Economic and Financial Modeling with Mathematica*, Springer-TELOS.

Exploring mathematics:
Gray, T.W. and Glynn, J., 1990, *Exploring Mathematics with Mathematica*, Addison-Wesley.

Chapter 3.
Mathematical Finance Preliminaries

Itô's Lemma and the Principles of Dynamic Hedging

3.1 Introduction

This chapter presents a sketch of the foundations in mathematical finance for what follows. In deciding the content of this chapter, the author has borne in mind several points. First, there exist a number of texts which already do an excellent job of laying the foundations, within various conceptual frameworks (tree, martingale, PDE). Second, and perhaps unfortunately, the existing literature is unfortunately highly factionalized into the writings of those who prefer to regard martingales as fundamental, those who regard the partial differential equations and inequalities as fundamental, and those who regard a decision tree as fundamental. Finally, none of these texts addresses, in a comprehensive fashion, many of the practicalities of implementation. Proponents of the simulation method go rather quiet when asked to price American options accurately, and proponents of the PDE approach are pushed to highly awkward schemes when asked to price path-dependent European options. Hopefully this text, at least when it comes to implementation, will bridge some of the gaps.

That leaves the author with the sensitive task of picking a conceptual framework for setting out basic theory, without annoying anyone, and with something new to say. This is almost certainly impossible, so I have decided instead to pick the approach I understand best, and try to develop it rather more than is commonplace. The approach taken here will be based on dynamic hedging arguments, leading to partial differential equations. I get my apologies in to martingale fanatics right away, and hope that the discussion on practical issues involved in simulation contained later in this text will appease them. I make no apologies to fans of trees for suppressing them as a mainstream approach - the view of the author is that the analytical structure of American-style options has all too often been obscured by their routine presentation through a tree argument, and that there are many cases where anomalous values can be obtained by tree methods that are readily avoided by a more appropriate numerical scheme. In this text, the binomial and trinomial trees are largely relegated to being a very special case of the explicit finite-difference method, when the grid is taken to be triangular. Suitably chosen implicit schemes are both more efficient and more accurate than tree models. I shall nevertheless explore tree models within *Mathematica*, for completeness. It must also be recognized that when one moves beyond the confines of traditional equity derivative analysis, there may be cases where it is difficult to set up a PDE approach, whereas a tree argument may be a valuable place to start. My own perspective is really that tree models have in the past been somewhat over-hyped as an algorithm for problem solution, since carefully posed finite-difference methods are often more appropriate and more accurate. This is not to say that trees are not a useful conceptual framework for developing new difficult models.

For those who prefer a more reasoned *financial* discussion, I can make no better reference than to the classic text by John Hull, now in its third edition (Hull, 1996). For those who prefer a martingale approach, I heartily recommend the excellent text by Baxter and Rennie (1996), who also explain the link between the martingale approach and the PDE approach. An extensive and complementary discussion of the PDE formalism is given by Wilmott, Dewynne and Howison, in their original (1993) and student (1996) texts. The discussion here will attempt to complement that given in other texts by making, in addition to the basic arguments, a clear presentation of the dynamic hedging arguments that apply to options linked simultaneously to assets and foreign exchange rates. This provides a unified theory, from which domestic options, foreign options, FX options, composites, Quantos all emerge naturally with the right parameters as special cases, but are embedded in a general two-factor description. The material described here in terms of similarity solutions of a two-factor PDE was summarized by the author in the article by Ghassemieh *et al* (1997). The two-factor PDE description given by Levy (1996) is entirely consistent with the one given here, provided one is careful with notational conventions.

Having said all this, this discussion is intended to be complete, even if it borrows heavily from elsewhere for results. It is oriented towards equity derivatives, with possible FX involvement. Essentially, we present Itô's Lemma and get on with it!

3.2 Simple Stochastic Processes and Itô's Lemma

For asset prices, indicated by a variable S, we shall assume a traditional type of random walk, where the fractional change in S, dS/S, in a small time interval, dt, is given by a relation of the form

$$\frac{dS}{S} = \mu\,dt + \sigma\,dX \tag{1}$$

where μ is a mean drift rate parameter and σ is a volatility parameter. The term dX describes the randomness of the process, and satisfies two conditions:

(a) dX is a zero-mean normally-distributed random variable;
(b) $E[dX^2] = dt$.

Within these assumptions, it is a simple matter to state Itô's Lemma:

Let $f(S,\ t)$ be a differentiable function of S and t. Then, up to and including terms of order dt,

$$df = \sigma\,S\,dX\,\frac{\partial f}{\partial S} + dt\left(\frac{\partial f}{\partial t} + \frac{1}{2}\,\sigma^2\,S^2\,\frac{\partial^2 f}{\partial S^2} + \mu\,S\,\frac{\partial f}{\partial S}\right) \tag{2}$$

For a rigorous discussion of what lies behind this formula, see Schuss (1980). Informal discussions at various levels of sophistication are given in several of the texts in the bibliography to this chapter. In this text we shall not be addressing the issue of how to use *Mathematica* to use Itô's Lemma directly in various applications, including the derivation of PDEs for option pricing. For one approach see the delightful discussion by Kendall (1993).

3.3 Dynamic Hedging and the Black-Scholes PDE

Under the simplest assumptions for S, we assume that

$$\frac{dS}{S} = \mu\,dt + \sigma\,dX \tag{3}$$

and we write down a portfolio consisting of an instrument V depending on S and t,

$$P = V - \Delta S \tag{4}$$

Application of Ito's Lemma to V gives, for a change in P,

$$dP = dt\left(-D\Delta + S\mu\left(\frac{\partial V}{\partial S} - \Delta\right) + \frac{\partial V}{\partial t} + \frac{1}{2}\sigma^2 S^2 \frac{\partial^2 V}{\partial S^2}\right) + dX S\left(\frac{\partial V}{\partial S} - \Delta\right)\sigma \tag{5}$$

where D is the dividend rate, i.e., the income generated by holding S over a time interval dt. To eliminate the risk associated with the X-process, i.e., to eliminate the random component, we set

$$\Delta = \frac{\partial V}{\partial S} \tag{6}$$

If this is done, we obtain

$$dP = dt\left(-D\frac{\partial V}{\partial S} + \frac{\partial V}{\partial t} + \frac{1}{2}\sigma^2 S^2 \frac{\partial^2 V}{\partial S^2}\right) \tag{7}$$

This is a deterministic change process. The absence of arbitrage opportunities implies that we can equate this to the growth in a portfolio obtained by depositing an equivalent sum of cash in the bank. Under this assumption, the portfolio grows by an amount

$$dP = rP\,dt = r\left(V - S\frac{\partial V}{\partial S}\right)dt \tag{8}$$

in time dt. Equating these two measures for dP yields the following partial differential equation for V.

$$\frac{\partial V}{\partial t} + (rS - D)\frac{\partial V}{\partial S} + \frac{1}{2}\sigma^2 S^2 \frac{\partial^2 V}{\partial S^2} - rV = 0 \tag{9}$$

This is one form of the Black-Scholes partial differential equation. It is sometimes written in terms of the yield variable $q = D/S$, as

$$\frac{\partial V}{\partial t} + S(r - q)\frac{\partial V}{\partial S} + \frac{1}{2}\sigma^2 S^2 \frac{\partial^2 V}{\partial S^2} - rV = 0 \tag{10}$$

The Risk-Neutral World and Simulation

By introducing a dynamic hedge we have eliminated instantaneous stochastic risk from the portfolio. This can all be re-interpreted by considering a new random walk based on the scheme:

$$\frac{ds}{s} = (r - q)\,dt + \sigma\,dX \tag{11}$$

This is referred to as the "risk-neutral" random walk associated with the asset. Let's consider the new quantity, assuming r is constant.

$$v = e^{-rt} V \tag{12}$$

as a function of the variables $s,\ t$. If we apply Itô's Lemma to this case, we obtain

$$d v = e^{-rt} \left(\sigma \, s \, d \, X \, \frac{\partial V}{\partial s} + d \, t \left(\frac{\partial V}{\partial t} - r \, V + \frac{1}{2} \, \sigma^2 \, s^2 \, \frac{\partial^2 V}{\partial s^2} + (r - q) \, s \, \frac{\partial V}{\partial s} \right) \right) \qquad (13)$$

Applying expectations to this result gives, by virtue of the BSPDE,

$$d \, E[v] = 0 \qquad (14)$$

It follows that the Black-Scholes PDE is telling us we can value an option by assuming a risk-neutral random walk on an "asset" s, and value the option by taking expectations of V under many different realizations of the risk-neutral random walk, and discounting back to the present by the discount factor e^{-rt}. This is the theoretical basis for Monte Carlo or simulation-based valuation. So we have two valuation techniques:

(a) solve the BSPDE;
(b) simulate risk-neutralized asset-price paths and take discounted expectations.

The argument we have given shows that these are equivalent, at least for the simple case we have considered. We now turn to computing the forms of the PDEs associated with some more interesting cases. These include, but in real life are not limited to,

(a) coupon payments, as apply to convertible bonds,
(b) risk that is not dynamically hedgeable, such as may apply in cases of default or volatility risk,
(c) foreign-exchange risk, in the case of foreign assets and an option paid in domestic currency.

3.4 Simple Convertible Bonds and a Simple Model of Default

There are a host of complications involved in the valuation of real-world convertible bonds. For our immediate purposes it will be sufficient to consider a financial instrument similar to the ones already considered but with a few extra complications. In addition to incorporating a more appropriate time-dependent model of interest rates, the payment of coupons and the possibility of default have to be modelled, even before one gets into a consideration of scheduled early exercise, and embedded calls, puts and refixes. Here we shall look briefly at coupons and default in a scheme where the dominant stochastic factor is the asset-price random walk - interest rates are here deterministic but time-dependent.

There are several different ways of defining a risk of default associated with a bond or CB. One can write down various schemes such as:

(a) default will occur if the stock price drops to a certain value;
(b) there is a running probability of default, which may be either or both of time-dependent and price-dependent;
(c) the probability of default is a random process correlated with either the stock-price process or the interest-rate process.

The argument presented here is a simple view of approach (b), which is probably the simplest. The idea is to go back to first principles and build a hedge position for the CB, which although it cannot be made risk-neutral, in the presence of default, is assumed to be risk-neutral on average, given a model of the probability of default. The approach is based on an old model of a jump diffusion process for the stock price in valuing an ordinary derivative given by Merton (as described by Jarrow and Rudd (1983, Ch. 12)), except that here we consider jumps in the CB value, not the underlying, modelled as a continuous time Poisson process. With certain auxiliary assumptions, this approach leads in a very clear way to a model with a spread-adjustment to the interest rate used for discounting, but with risk-neutral growth given by the usual formula. It is therefore consistent with what I believe to be common practice, of growing or drifting the underlying at $(r - q)$, but discounting using $(r + \text{spread})$.

So suppose that we have a derivative instrument $V(S, t, Y)$, where S is the underlying, t is the time and Y is the default status, with $Y = 0$ for no default, $Y = 1$ for default. We assume that Y follows a Poisson process of the form

$$
\begin{aligned}
P(Y = 1 \text{ at } (t + d\,t) \mid Y = 0 \text{ at } t) &= \lambda\, d\,t \\
P(Y = 0 \text{ at } (t + d\,t) \mid Y = 0 \text{ at } t) &= 1 - \lambda\, d\,t \\
P(Y = 1 \text{ at } (t + d\,t) \mid Y = 1 \text{ at } t) &= 1
\end{aligned}
\tag{15}
$$

Under the usual assumptions for S, we assume that

$$
\frac{d\,S}{S} = d\,t\,\mu + d\,X\,\sigma
\tag{16}
$$

and we write down a portfolio:

$$
P = V - \Delta\,S
\tag{17}
$$

Application of Ito's Lemma to V gives, for a change in P for a bond:

$$
\begin{aligned}
d\,P = {}& d\,t\left(K - D\,\Delta + S\,\mu\left(\frac{\partial V}{\partial S} - \Delta\right) + \frac{\partial V}{\partial t} + \frac{1}{2}\,\sigma^2\,S^2\,\frac{\partial^2 V}{\partial S^2}\right) + d\,X\,S\left(\frac{\partial V}{\partial S} - \Delta\right)\sigma \\
& + d\,Y\,(V(S, t + d\,t, d\,Y) - V(S, t, 0))
\end{aligned}
\tag{18}
$$

where K is coupon rate, D is dividend rate etc. As usual, to eliminate the risk associated with the X-process, we set

$$
\Delta = \frac{\partial V}{\partial S}
\tag{19}
$$

to obtain

$$
d\,P = d\,t\left(K - D\,\frac{\partial V}{\partial S} + \frac{\partial V}{\partial t} + \frac{1}{2}\,\sigma^2\,S^2\,\frac{\partial^2 V}{\partial S^2}\right) + d\,Y\,(V(S, t + d\,t, Y + d\,Y) - V(S, t, Y))
\tag{20}
$$

The contribution from the Poisson process, as in Merton's jump diffusion model, cannot be eliminated by any further choice of hedge. However, we can proceed by assuming that the process is risk-neutral when averaged over the defaulting process:

$$
E[d\,P] = r\,E[P]\,d\,t
\tag{21}
$$

We further assume that the value of the defaulting CB is some fraction ϕ of the non-defaulting CB, thereby allowing us to approximate the default term. We have

$$
\begin{aligned}
E[d\,Y] &= \lambda\, d\,t \\
V(S, t, 1) &= \phi\,V(S, t, 0)
\end{aligned}
\tag{22}
$$

And we take the expected value of the entire equation, using bars to denote expectations. We represent the default term as

$$
E[d\,Y\,(V(S, t + d\,t, Y + d\,Y) - V(S, t, Y))] \longrightarrow \lambda\, d\,t\,\overline{V}(S, t)\,(\phi - 1)
\tag{23}
$$

and the portfolio grows by

$$r E[P] d t = r \left(\overline{V} - S \frac{\partial \overline{V}}{\partial S} \right) d t \tag{24}$$

Putting this all together gives

$$K + (r S - D) \frac{\partial \overline{V}}{\partial S} + \frac{\partial \overline{V}}{\partial t} + \frac{1}{2} \sigma^2 S^2 \frac{\partial^2 \overline{V}}{\partial S^2} - (r + \lambda (1 - \phi)) \overline{V} = 0 \tag{25}$$

This is a simple version of the default-risk-adjusted Black-Scholes equation. The drift term is

$$(r S - D) \frac{\partial \overline{V}}{\partial S} = S (r - q) \frac{\partial \overline{V}}{\partial S} \tag{26}$$

if the dividend yield is q, and the discounting factor is

$$r + \lambda(1 - \phi) \tag{27}$$

where we interpret sp $= \lambda(1 - \phi)$ as the *spread* term. The spread is zero if there is no risk of default or if the payout is 100%, which is sensible if trivial. Note that the risk-neutral drift or growth rate is still $(r - q)$, but the discounting term is $r +$ sp. For a "straight bond" (non-convertible, independent of the asset price) the equation reduces to

$$K + \frac{\partial \overline{V}}{\partial t} - (r + \text{sp}) \overline{V} = 0 \tag{28}$$

Hypothetically, this allows the spread to be estimated from zero-coupon bonds. Quite what values should be taken for sp is a difficult matter. Another question relates to capturing correlation between underlying price and spread by assuming a certain functional form for the spread as a function of stock price. Such matters take us beyond the scope of this book. The point of our argument is that within the assumptions made we still have the Black-Scholes differential equation to solve, but with different parameters to represent the default risk. In addition the equation has a "source" term representing the coupon payments.

3.5 Asset-FX Coupled Options

The idea is to develop a single conceptual framework within which it is clear how to treat such objects as:

(a) domestic options;
(b) foreign options;
(c) FX Options;
(d) composite options;
(e) Quantos;

as well as other more complex cross-currency instruments.

A useful analysis of several special cases was given by E. Reiner (1992), who gives various cunning arguments for valuing four types of asset-FX linked options, and who comes up with some rules for what to use as effective interest rates, yields and volatilities, based on "domestic market" or "foreign market" analyses, as is most appropriate for each case. The general framework underpinning that analysis was far from clear to me, even after several readings. One may also be interested more generally in the valuation of options involving indices in several currencies, with and without Quanto effects and with various types of payoffs. It is therefore vital to get a grip on the FX effect, and how it applies for various types of payoffs and the different ways of including the FX.

More generally, there is increasing interest in two-factor models for various types of application, where one factor is an asset, and the second is an FX, an interest rate or a volatility. Given that modelling single-factor is a lot quicker than modelling two-factor, it is highly desirable to clarify when a two-factor model can be treated as essentially a single-factor model, with suitable adjustments to parameters. Current research in stochastic volatility, for example, can be based on a two-factor route, or on a single-factor approach with various types of correction to account for some component of the second factor. Given the special character of asset-FX linked derivatives, it is more usual for the reduction for single-factor to be possible, but it is not a foregone conclusion. Another issue is to understand what the correct hedge position should look like, in terms of a suitable mix of the underlying and foreign currency.

Finally, the following allows us to adopt a unified domestic picture of all the derivatives of interest. Reiner's analysis switched between an analysis based on foreign markets and one based on domestic markets according to convenience. When we generalize to multi-asset multi-currency Quantos, for example, it is useful to have a consistent domestic picture, due to the lack of a unique foreign country. Readers with a probabilistic background may wish to explore how these results can be obtained in an alternative manner.

Framework

Suppose that we have two countries, called "Payofferania" and "Assetagonia". These slightly bizarre names are intended to constantly remind the reader that the option payoff is in one currency, while the underlying asset is in another. These terms also avoid all possible confusion that may arise from differing views as to whether the underlying is domestic or foreign, with the option investor being foreign or domestic - in my view these terms are chronically ambiguous, just as is the notion of "FX Rate" (w.r.t. which way up it is!). In order to at least try to avoid confusion, let $X(t)$ denote the spot price in Payofferanian currency of one unit of Assetagonian currency. We have an underlying asset (e.g. equity or index) in Assetagonia with price S' in Assetagonian currency. Here and elsewhere primes denote Assetagonian objects. We can build an equivalent Payofferanian asset with price

$$S = S' X \tag{29}$$

A completely general option, paying off in the Payofferanian currency, can be thought of as a function

$$V = V(S', X, t) \tag{30}$$

and we write it this way to emphasize the fact that in complete generality it is a function separately of the two stochastic factors S' and X, as well as the time t. If things are really this general, there is no alternative to a full two-factor treatment. However, one can conceive of all sorts of special cases where it ought to boil down to just one effective factor. A few types stand out as being important:

Type 1

We have a purely Assetagonian option converted to Payofferanian currency:

$$V = X u(S', t) \tag{31}$$

Type 2

A Payofferanian-denominated option that is a function only of the equivalent Payofferanian asset:

$$V = v(X S', t) \tag{32}$$

Type 3

A Payofferanian-denominated option that is a function only of the Assetagonian asset:

$$V = w(S', t) = w\left(\frac{S}{X}, t\right) \tag{33}$$

Note that "composite" options are an example of Type 2, and "Quanto" options fall into the Type 3 category. We could also build currency options modulated in some way by the asset price, but this is not our primary interest, and would be more properly the domain of an interest-rate derivative model where the two interest rates are stochastic. For completeness we also include Type 0 (zero), a pure domestic option, and Type FX, which is a function only of X.

The Risky Coupled Random Processes

One basic difficulty is that we do not know how to specify in advance what risk-neutral processes are followed, from a domestic point of view, by the foreign asset. What we shall do, therefore, is to specify the real-world risky processes, build a risk-neutral hedge portfolio containing the asset and currency, and introduce an unknown risk-neutral drift for the foreign asset, parametrized by an unknown effective yield. We shall eventually demand that in case (b), the ordinary Black-Scholes single-factor model is recovered. In this way, we can divine the risk-neutral drift in terms of the domestic interest rate, the asset yield and the correlation.

$$\frac{d S'}{S'} = \mu_1 \, dt + \sigma_{S'} \, dZ_1 \tag{34}$$

$$\frac{d X}{X} = \mu_2 \, dt + \sigma_X \, dZ_2 \tag{35}$$

The correlation between the two processes is $\rho_{S'X}$ or just ρ for short. Let's suppose you are the holder of the derivative V, and that we apply Itô's Lemma in two dimensions. Then Itô gives

$$\begin{aligned} dV = & \frac{\partial V}{\partial t} \, dt + \frac{\partial V}{\partial X} \, dX + \frac{\partial V}{\partial S'} \, dS' \\ & + \frac{1}{2} \left(\sigma_X^2 \, X^2 \, \frac{\partial^2 V}{\partial X^2} + 2 \, X S' \, \rho \, \sigma_{S'} \, \sigma_X \, \frac{\partial^2 V}{\partial S' \, \partial X} + \sigma_{S'}^2 \, S'^2 \, \frac{\partial^2 V}{\partial S'^2} \right) dt \end{aligned} \tag{36}$$

Building a Risk-Neutral Portfolio

To eliminate both asset risk and FX risk we build a portfolio of V, the Payofferanian asset and some Assetagonian currency.

$$P = V - \Delta_1 \, (S' \, X) - \Delta_2 \, X \tag{37}$$

We parametrize (what will turn out to be) the risk-neutral drifts in terms of "yields" $q_{S'X}$ and q_X. So the change in the value of the portfolio P is given by

$$dP = dV - \Delta_2\, dX - \Delta_1\, S'\, dX - \Delta_1\, X\, dS' - q_X\, \Delta_2\, X\, dt - q_{S'X}\, \Delta_1\, S'\, X\, dt \tag{38}$$

$$
\begin{aligned}
dP = \left(\frac{\partial V}{\partial X} - \Delta_1\, S' - \Delta_2\right) dX + \left(\frac{\partial V}{\partial S'} - \Delta_1\, X\right) dS' + dt\Bigg(\frac{\partial V}{\partial t} + \frac{1}{2}\, \sigma_X^2\, X^2\, \frac{\partial^2 V}{\partial X^2} \\
+ X\, S'\, \rho\, \sigma_{S'}\, \sigma_X\, \frac{\partial^2 V}{\partial S'\, \partial X} + \frac{1}{2}\, \sigma_{S'}^2\, S'^2\, \frac{\partial^2 V}{\partial S'^2} - q_X\, \Delta_2\, X\, dt - q_{S'X}\, \Delta_1\, S'\, X\Bigg)
\end{aligned}
\tag{39}
$$

The stochastic component of risk can be eliminated by setting

$$\Delta_1 = \frac{1}{X}\frac{\partial V}{\partial S'} \tag{40}$$

$$\Delta_2 = \frac{1}{X}\left(X\frac{\partial V}{\partial X} - S'\frac{\partial V}{\partial S'}\right) \tag{41}$$

Remark on Vanishing FX Hedge Position

The FX component of the hedge position is given by Δ_2. It is a useful check and quite revealing to analyse under what conditions this vanishes automatically. Note that for any function $f = f(S'X)$, then

$$X\frac{\partial f}{\partial X} - S'\frac{\partial f}{\partial S'} = X\, S'\, f' - S'\, X\, f' = 0 \tag{42}$$

Conversely, any function with a zero Δ_2 must be a function only of the combination $S'X$. So there is no currency in the hedge position when the option is a function only of the equivalent Payofferanian underlying. This of course is in accordance with common sense.

Derivation of the Two-Factor PDE

Our portfolio P is now risk-neutral in our Payofferanian environment. We can therefore state that the return on it is no greater than a bank deposit at a rate r, where r, possibly time-dependent, is the Payofferanian risk-free interest rate. For a European-style option, we shall therefore write

$$dP = r\, P\, dt \tag{43}$$

with it being understood that with American options or convertibles, corresponding projection inequalities are to be applied for early exercise/conversion etc. We can now do some algebra to obtain the two-factor Black-Scholes differential equation. There are two versions, depending on what coordinates we use for independent variables.

The S'-X Picture

If we just continue our analysis to its conclusion, we arrive at the following PDE:

$$\frac{\partial V}{\partial t} + \frac{1}{2}\sigma_X^2 X^2 \frac{\partial^2 V}{\partial X^2} + X S' \rho \sigma_{S'} \sigma_X \frac{\partial^2 V}{\partial S' \partial X} + \frac{1}{2}\sigma_{S'}^2 S'^2 \frac{\partial^2 V}{\partial S'^2}$$
$$+ (q_X - q_{S'X}) S' \frac{\partial V}{\partial S'} + (r - q_X) X \frac{\partial V}{\partial X} - rV = 0 \tag{44}$$

The S-X Picture

If, instead, we change variables to use S and X as independent variables, we arrive instead at the following different PDE (this requires some fiddling with the chain rule - basic maths of no financial import, so I omit it):

$$\frac{\partial V}{\partial t} + \frac{1}{2}\sigma_X^2 X^2 \frac{\partial^2 V}{\partial X^2} + X S(\rho \sigma_{S'} \sigma_X + \sigma_X^2) \frac{\partial^2 V}{\partial S \partial X} + \frac{1}{2}\sigma_S^2 S^2 \frac{\partial^2 V}{\partial S^2}$$
$$+ (r - q_{S'X} + \rho \sigma_{S'} \sigma_X) S \frac{\partial V}{\partial S} + (r - q_X) X \frac{\partial V}{\partial X} - rV = 0 \tag{45}$$

The S-volatility used here has been calculated by the chain rule as the expected thing:

$$\sigma_S^2 = \sigma_X^2 + 2\rho \sigma_{S'} \sigma_X + \sigma_{S'}^2 \tag{46}$$

Fixing the Risk-Neutral Drifts in Payofferania

We can now fix the two unknown parameters in this model by appealing to known single-factor problems, using the S-X picture. Suppose first that the option is independent of S. Then we have, by crossing out some terms,

$$\frac{\partial V}{\partial t} + \frac{1}{2}\sigma_X^2 X^2 \frac{\partial^2 V}{\partial X^2} + (r - q_X) X \frac{\partial V}{\partial X} - rV = 0 \tag{47}$$

whence we identify the parametric yield q_X as just the Assetagonian interest rate r_A, by comparison with the FX option equation,

$$\frac{\partial V}{\partial t} + \frac{1}{2}\sigma_X^2 X^2 \frac{\partial^2 V}{\partial X^2} + (r - r_A) X \frac{\partial V}{\partial X} - rV = 0 \tag{48}$$

or by noting that the yield on holding Assetagonian currency is just the Assetagonian interest rate.

So the risk-neutral drift for X is $(r - r_A)$. Next we assume that the option is a function only of S, leading to

$$\frac{\partial V}{\partial t} + K + \frac{1}{2}\sigma_S^2 S^2 \frac{\partial^2 V}{\partial S^2} + (r - q_{S'X} + \rho \sigma_{S'} \sigma_X) S \frac{\partial V}{\partial S} - rV = 0 \tag{49}$$

We write $q = q_{S'X} - \rho \sigma_{S'} \sigma_X$, so that this is equivalent to

$$\frac{\partial V}{\partial t} + K + \frac{1}{2}\sigma_S^2 S^2 \frac{\partial^2 V}{\partial S^2} + (r - q) S \frac{\partial V}{\partial S} - rV = 0 \tag{50}$$

which is the purely domestic Black-Scholes equation, with q the continuously compounded yield. But what exactly is the q that goes in here? It is, in fact, just the local dividend yield of the Assetagonian asset, as this is invariant under changes of market.

The Final Two-Factor PDEs

The S'-X picture (Assetagonian Variable and FX as Independent Variables)

$$
\frac{\partial V}{\partial t} + \frac{1}{2} \sigma_X^2 X^2 \frac{\partial^2 V}{\partial X^2} + X S' \rho \sigma_{S'} \sigma_X \frac{\partial^2 V}{\partial S' \partial X} + \frac{1}{2} \sigma_{S'}^2 S'^2 \frac{\partial^2 V}{\partial S'^2}
$$
$$
+ (r_A - q - \rho \sigma_{S'} \sigma_X) S' \frac{\partial V}{\partial S'} + (r - r_A) X \frac{\partial V}{\partial X} - r V = 0
$$

(51)

The S-X picture (Payofferanian Variable and FX as Independent Variables)

$$
\frac{\partial V}{\partial t} + \frac{1}{2} \sigma_X^2 X^2 \frac{\partial^2 V}{\partial X^2} + X S(\rho \sigma_{S'} \sigma_X + \sigma_X^2) \frac{\partial^2 V}{\partial S \partial X} + \frac{1}{2} \sigma_S^2 S^2 \frac{\partial^2 V}{\partial S^2}
$$
$$
+ (r - q) S \frac{\partial V}{\partial S} + (r - r_A) X \frac{\partial V}{\partial X} - r V = 0
$$

(52)

3.6 Reduction to Single-Factor Models

Type 1 Options - Assetagonian Options Expressed in Payofferanian Currency

To treat the first type of option we write

$$
V = X u(S', t)
$$

(53)

and look at the S'-X equation with this simplification. We just have to note that

$$
X \frac{\partial V}{\partial X} = V
$$

(54)

and simplify the resulting equation. It all boils down to

$$
\frac{\partial u}{\partial t} + \frac{1}{2} \sigma_{S'}^2 S'^2 \frac{\partial^2 u}{\partial S'^2} + (r_A - q) S' \frac{\partial u}{\partial S'} - r_A u = 0
$$

(55)

which is the correct equation for a foreign option. So the value is the current X, the spot rate, times the solution of the Black-Scholes equation in Assetagonia, with the parameters appropriate to Assetagonia.

Effective Parameters for Type 1 Options in Black-Scholes Model

(1) risk-free rate - the Assetagonian interest rate r_A;
(2) volatility - the volatility of the Assetagonian asset;
(3) yield - the local yield of the Assetagonian asset.

Type 2 Options - Composites

In this case the option is a function only of $X S' = S$. We can also look at this type of option in two ways - they are consistent, and we just give the easier view. Consider the S-X two-factor PDE, and cross out the X in the above. Then we obtain

$$\frac{\partial V}{\partial t} + \frac{1}{2} \sigma_S^2 S^2 \frac{\partial^2 V}{\partial S^2} + (r - q) S \frac{\partial V}{\partial S} - r V = 0 \tag{56}$$

This can be valued in the Black-Scholes world of Payofferania.

Effective Parameters for Type 2 Options in Black-Scholes Model

(1) risk-free rate - the Payofferanian interest rate r;
(2) volatility - the volatility of the equivalent Payofferanian asset;
(3) yield - the local yield of the Assetagonian asset.

The volatility to be used is just given by

$$\sigma_S^2 = \sigma_X^2 + 2 \rho \sigma_{S'} \sigma_X + \sigma_{S'}^2 \tag{57}$$

Type 3 Options - Quantos

In this case the option is a function only of $S' = S/X$. We can look at this option in two ways. First consider the $S' - X$ 2-factor picture, and cross out the X in the above. Then we obtain

$$\frac{\partial V}{\partial t} + \frac{1}{2} \sigma_{S'}^2 S'^2 \frac{\partial^2 V}{\partial S'^2} + (r_f - q - \rho \sigma_{S'} \sigma_X) S' \frac{\partial V}{\partial S'} - r V = 0 \tag{58}$$

It is a useful check on our analysis to also work in the S-X picture and write $V = w(S/X, t)$ and to derive the PDE satisfied by $w(z, t)$. The relevant algebra leads to

$$\frac{\partial w}{\partial t} + \frac{1}{2} \sigma_{S'}^2 z^2 \frac{\partial^2 V}{\partial z^2} + (r_f - q - \rho \sigma_{S'} \sigma_X) z \frac{\partial V}{\partial z} - r V = 0 \tag{59}$$

which is precisely the same differential equation as we obtained from the S'-X picture. So we can value Quantos using the Black-Scholes formula, but with some interesting adjustments to the parameters.

$$\frac{\partial V}{\partial t} + \frac{1}{2} \sigma_{S'}^2 S'^2 \frac{\partial^2 V}{\partial S'^2} + (r - Q) S' \frac{\partial V}{\partial S'} - r V = 0 \tag{60}$$

$$Q = q + r - r_A + \rho \sigma_X \sigma_{S'} \tag{61}$$

Effective Parameters for Type 3 Options (Quantos) in Black-Scholes Model

(1) risk-free rate - the Payofferanian interest rate r;
(2) volatility - the local volatility of the Assetagonian asset;
(3) yield - the effective yield is given by

$$Q = q + r - r_A + \rho \sigma_X \sigma_{S'} \tag{62}$$

Type FX Options

In this case the option is a function only of X.

$$\frac{\partial V}{\partial t} + \frac{1}{2}\sigma_X^2 X^2 \frac{\partial^2 V}{\partial X^2} + (r - r_A)X\frac{\partial V}{\partial X} - rV = 0 \tag{63}$$

Effective Parameters for Type FX Options in Black-Scholes Model

(1) risk-free rate - the Payofferanian interest rate r;
(2) volatility - the volatility of the exchange rate;
(3) yield - the effective yield is given by the Assetagonian interest rate r_A.

Remarks

This discussion illustrates the fact that similarity solutions are a powerful way of reducing certain types of two-factor models to single-factor descriptions. It also points out that a wide variety of interesting cases can be treated within the single-factor Black-Scholes framework. One can imagine other types of cross-market option that cannot be reduced to a single-factor view - these would have to be treated using the full three-dimensional partial differential equations. Finally we note that it can often be rather confusing identifying which type of option should be associated with a particular cross-currency contract. Warrants and convertible bonds are rather notorious in this respect - the contracts for both often refer to "fixed exchange rates", but neither is a Quanto option, as the fixed exchange rate is used only to calculate exercise prices or strikes in the appropriate currency. When viewed from the payoff currency, warrants are usually Type 1 options, and convertible bonds are generally Type 2 or composite options.

3.7 Other Higher-Dimensional Partial Differential Equations

There are many other cases where one can obtain three- or higher-dimensional partial differential equations, including, but not limited to, cases

(a) where the asset price and the volatility are stochastic;
(b) where the asset price and interest rates are stochastic;
(c) of combinations of asset price, FX-rate, volatility, interest rates ALL being stochastic;
(d) of many assets and/or currencies;
(e) of path-dependent options.

We will not be attempting to construct models appropriate to all of these cases, and solve high-dimensional PDEs in this text. The case of many assets is very effectively modelled by Monte Carlo simulation, and this will be discussed in Chapter 23. The case of path-dependent options is a special one that deserves comment, as it leads to a useful standard framework for the numerical solution of such problems. Readers interested in a detailed discussion should consult the texts by Wilmott *et al* (1993, 1996), but should also bear in mind that the European forms of such options are more easily treated, at least for continuum models, by analytical methods - most of the important cases are accessible to a "near-closed-form" analysis, albeit using some interesting special functions not usually the domain of option-pricing theory. One of the points of this text is that once one is armed with a system such as *Mathematica*, the use of such objects as are needed - hypergeometric functions in the case of Asian options, for example - is a matter of routine.

An interesting class of path-dependent options may be considered by introducing a further variable that accumulates information about the path taken by the option, in the form

$$J_n = \left(\int_0^t f(S(\tau), \tau)^n \, d\tau \right)^{1/n} \tag{64}$$

for various choices about n and f. The following are interesting cases:

(a) $n = 1$, $f = \log(S)$, J represents the geometric average of the stock price along the path;
(b) $n = 1$, $f = S$, J represents the arithmetic average of the stock price along the path;
(c) $n \to \infty$, $f = S$, J represents the maximum attained along the path.

Suitable boundary conditions involving J must be added, and the payoff needs to be written in terms of it also. Itô's Lemma can be applied as before, leading, in the case $n = 1$, to

$$\frac{\partial V}{\partial t} + f(S) \frac{\partial V}{\partial J} + (rS - D) \frac{\partial V}{\partial S} + \frac{1}{2} \sigma^2 S^2 \frac{\partial^2 V}{\partial S^2} - rV = 0 \tag{65}$$

This is of a rather different character from the asset-FX coupled case - there is no second derivative involving the path variable. However, in common with the asset-FX case, these equations may admit similarity solutions, depending on whether the boundary conditions (including the payoff) are consistent with a similarity approach. Some examples are given by Wilmott *et al* (1993).

3.8 The Greeks and Other Partial Differential Equations

The derivatives of an option value with respect to its various arguments (the "Greeks") play an important role, whether simply as sensitivity parameters, or, additionally, as hedge parameters. This is a convenient place to define these quantities and to explore the relationships between them. There are more links between the Greeks than many workers appear to be aware of - these will be derived here. Similar results can be derived from the probabilistic viewpoint, as has been pointed out by P. Carr (personal communication).

Definitions of Greeks

In what follows we shall give the definition of the Greeks to be used throughout this book. The fundamental sensitivity parameter, Δ, is given as the rate of change of the option price with respect to the underlying asset price S:

$$\Delta = \frac{\partial V}{\partial S} \tag{66}$$

This parameter plays a special role as the hedge position in the underlying to ensure the valuation portfolio is risk-neutral. The second derivative is called Γ:

$$\Gamma = \frac{\partial^2 V}{\partial S^2} \tag{67}$$

The sensitivity with respect to interest rate is called ρ:

$$\rho = \frac{\partial V}{\partial r} \tag{68}$$

The rate of change with respect to *real time* is called θ:

$$\theta = \frac{\partial V}{\partial t} \tag{69}$$

Note that if time is expressed in terms of time to maturity, we need to add a minus sign. The rate of change with respect to volatility is called Vega or Lambda (Λ) (the former is common in speaking about the object, while the latter is quite commonplace in written work):

$$\Lambda = \frac{\partial V}{\partial \sigma} \tag{70}$$

Conventions for Greeks

In the previous sub-section we gave the definition of the Greeks to be used throughout this book. We work exclusively with raw partial derivatives in natural units. If you use this book to check your own risk parameters, you must take the responsibility for mapping these conventions onto your own. Possible variations include:

(a) for Δ, using % terms, and/or expressing Δ with respect to a 1% change in the underlying;
(b) for Γ, using % terms, and /or multiplying by the stock price;
(c) for θ, using days rather than years;
(d) for ρ, using annual rather than continuous compounding, % rather than absolute.

Other variants are possible, and great care should be taken in making detailed comparisons.

Partial Differential Equations for Greeks

The Black-Scholes equation implies a number of differential constraints linking the Greeks. Here is the key one, and two that are not so well known.

The Black-Scholes Equation as a Greek Constraint

First of all, consider the fundamental equation:

$$\frac{\partial V}{\partial t} + (rS - D)\frac{\partial V}{\partial S} + \frac{1}{2}\sigma^2 S^2 \frac{\partial^2 V}{\partial S^2} - rV = 0 \tag{71}$$

If we substitute the definitions of our Greeks, we find that this can be written as

$$\theta + (rS - D)\Delta + \frac{1}{2}\sigma^2 S^2 \Gamma - rV = 0 \tag{72}$$

The Vega-Gamma Constraint

For any function f, we define the Black-Scholes operator acting on f as

$$L_{BS}[f] = \frac{1}{2} S^2 \frac{\partial^2 f}{\partial S^2} \sigma^2 - r f + (r S - D) \frac{\partial f}{\partial S} + \frac{\partial f}{\partial t} \tag{73}$$

Other differential constraints can be obtained by considering whether the Black-Scholes operator commutes with other operators. So define the commutator:

$$[P, Q] f = P[Q[f]] - Q[P[f]] \tag{74}$$

Note that the Black-Scholes operator commutes with the operator

$$S \frac{\partial}{\partial S} \tag{75}$$

so we note first that if V is a solution of the Black-Scholes equation

$$L_{BS}[V] = 0 \tag{76}$$

then

$$L_{BS}[S \Delta] = L_{BS}\left[S \frac{\partial}{\partial S} V\right] = \left[L_{BS}, \ S \frac{\partial}{\partial S}\right] V + S \frac{\partial}{\partial S} L_{BS} V = 0 + 0 = 0 \tag{77}$$

Similarly, we deduce that $S^2 \Gamma$ is also a solution of the Black-Scholes equation. On the other hand,

$$L_{BS}[\Lambda] = L_{BS}\left[\frac{\partial V}{\partial \sigma}\right] = \left[L_{BS}, \frac{\partial}{\partial \sigma}\right] V + \frac{\partial}{\partial \sigma} L_{BS} V = \left[L_{BS}, \frac{\partial}{\partial \sigma}\right] V = -S^2 \sigma \Gamma \tag{78}$$

which follows by differentiation of the Black-Scholes operator. So Vega is not a solution of the Black-Scholes equation, but the following holds instead:

$$L_{BS}[\Lambda - S^2 \sigma\Gamma(T - t)] = -S^2 \sigma \Gamma + L_{BS}[S^2 \sigma\Gamma(t - T)] = 0 \tag{79}$$

So we deduce that the quantity

$$\Lambda - S^2 \sigma\Gamma(T - t) \tag{80}$$

satisfies the Black-Scholes equation. We can argue that under a usefully wide range of circumstances, this means that this quantity is identically zero - that is

$$\Lambda = S^2 \sigma\Gamma(T - t) \tag{81}$$

Clearly this requires that the Black-Scholes equation holds as an equality, which rules out American-style options. At maturity, provided Γ is finite, both sides of this equation are zero, and it also holds for Call- and Put-style payoffs with singularities in Gamma. So we have the Black-Scholes equation with initial data at maturity that are zero - hence the solution is zero for all time. It is also a useful identity for computing Vega for European-style options with many underlyings.

The Rho-Delta Constraint

A similar argument to the one given for Vega shows that the identity

$$\rho = -(T - t)(V - S \Delta) \tag{82}$$

holds for a large class of European-style options. These formulae for ρ and δ may be useful in situations where direct computation is expensive.

Chapter 3 Bibliography

Baxter, M. and Rennie, A., 1996, *Financial Calculus - an Introduction to Derivative Pricing,* Cambridge University Press.

Ghassemieh, R., Shaw, W. and Wilson, R., 1997, Equity-index-linked derivatives, in *The Asian Equity Derivatives Handbook*, Euromoney Publications.

Hull, J.C., 1996, *Options, Futures, and other Derivatives*, 3rd edition, Prentice-Hall.

Jarrow, R.A. and Rudd, A., 1983, *Option Pricing*, Irwin.

Levy, E., 1996, Exotic options I, in *The Handbook of Risk Management and Analysis*, ed. C Alexander, Wiley.

Kendall, W.S., 1993, Itovsn3: doing stochastic calculus with *Mathematica*, in *Economic and Financial Modeling with Mathematica*, ed. H.R. Varian, Springer-TELOS.

Reiner, E., 1992, Quanto mechanics, in *From Black Scholes to Black Holes*, RISK-FINEX Publications.

Schuss, Z., 1980, *Theory and Applications of Stochastic Differential Equations*, Wiley.

Wilmott, P., Dewynne, J. and Howison, S., 1993, *Option Pricing - Mathematical Models and Computation*, Oxford Financial Press.

Wilmott, P., Dewynne, J. and Howison, S., 1996, *The Mathematics of Financial Derivatives*, Cambridge University Press

Chapter 4.
Mathematical Preliminaries

Solving the Partial Differential Equations

4.1 Introduction

In Chapter 3 we saw how to derive partial differential equations for various types of financial instrument. One view of basic option pricing is that one now just needs to solve these equations with various interesting boundary conditions, and we shall motivate our basic solutions via this approach. It has the advantages of simplicity, and calls on rather basic techniques in applied mathematics. A rather gentle approach will be adopted, and we shall gradually build in contributions from *Mathematica* to help solve the relevant equations.

Much of the discussion presented here parallels closely the treatment of the heat equation to be found in many elementary calculus texts. Indeed, we shall see along the way how to convert some of the differential equations directly into the heat equation. The discussion will begin by looking at time-independent and asset-price-independent solutions, and gradually increase the complexity.

4.2 The One-Variable Black-Scholes Equation

We begin by writing down the standard form of our partial differential equation:

$$\frac{\partial V}{\partial t} + S(r-q)\frac{\partial V}{\partial S} + \frac{1}{2}\sigma^2 S^2 \frac{\partial^2 V}{\partial S^2} - rV = 0 \tag{1}$$

One way to simplify matters is to seek solutions that are functions of one variable only. We obtain, for solutions that are functions of time only,

$$\frac{\partial V}{\partial t} - rV = 0 \tag{2}$$

For functions of S only, we get something a little more interesting.

$$\frac{1}{2}\sigma^2 S^2 \frac{\partial^2 V}{\partial S^2} + S(r-q)\frac{\partial V}{\partial S} - rV = 0 \tag{3}$$

We shall suppose, until stated otherwise, that all the parameters in these equations are constants. The first equation has a simple solution of the form

$$V = Z e^{r(t-T)} \tag{4}$$

where we have a boundary condition that $V = Z$ when $t = T$. This represents a trivial form of bond, representing a cash payment of Z at a time T in the future. The value of this instrument is just the discounted, or present, value of this cash payment.

In the case of the time-independent equation, there are in fact two ways of proceeding. The first is to note that each time we differentiate, we also put back a power of S in the equation. So if we were to try a power of S for the solution, there is a good chance that the equation could be solved quite quickly. Matters are much simpler, and just as revealing for future analysis, if we consider the case $q = 0$. So we have

$$\frac{1}{2} \sigma^2 S^2 \frac{\partial^2 V}{\partial S^2} + rS \frac{\partial V}{\partial S} - rV = 0 \tag{5}$$

If we try $V = C S^\alpha$, we obtain the condition

$$\frac{1}{2} \alpha(\alpha - 1) \sigma^2 - r + r\alpha = 0 \tag{6}$$

This has solutions $\alpha = 1$ and $\alpha = -2r/\sigma^2$. The first case gives us CS, which is a multiple of the underlying asset. The second case is a little more interesting, as it corresponds nicely to a situation that is commonplace in the case of the heat equation, i.e., a steady-state solution to which a more general time-dependent solution settles down at very late times. The meaning of this solution will only become clear in Chapter 11 - for now we just give a hint of what is to come. We let

$$B_\infty = \frac{K}{\frac{\sigma^2}{2r} + 1} \tag{7}$$

so that

$$\frac{\sigma^2 B_\infty}{2r} = \frac{K}{\frac{2r}{\sigma^2} + 1} = K - B_\infty \tag{8}$$

and define

$$P(S) = (K - B_\infty) \left(\frac{B_\infty}{S} \right)^{\frac{2r}{\sigma^2}} \tag{9}$$

This is clearly a special case of our second solution. It is easy to check that when $S = B_\infty$, $P(S) = K - B_\infty$. Furthermore, at that same point, we have

$$\frac{\partial P(S)}{\partial S} = -1 \tag{10}$$

and clearly P tends to zero as S becomes large. This is in fact the steady-state solution to the problem of an "American Put" option, whose time-dependent form we shall explore in Chapter 11. The quantity B_t is the critical price separating the regions where early exercise is desirable and where it is not, and for very large times the solution settles down to the one we have given. The solution of the Black-Scholes equation joins smoothly onto the Put option payoff function $K - S$, and this smooth join is characterized by the conditions on our function. We shall explore this more thoroughly in Chapter 11.

Mathematica View of Solution

The *S*-independent problem can be solved in *Mathematica* as follows:

```
V[t] /. DSolve[D[V[t], t] - r V[t] == 0, V[t], t][[1]]
```

$$\mathrm{E}^{r\,t}\,\mathrm{C}[1]$$

Setting $q = 0$, as before, we can get the time-independent solutions just as easily.

```
V[S] /.
  DSolve[1 / 2 S^2 σ^2 D[V[S], {S, 2}] + r S D[V[S], S] - r V[S] == 0, V[S],
    S][[1]]
```

$$\mathrm{S}^{-\frac{2\,r}{\sigma^2}}\,\mathrm{C}[1] + \mathrm{S}\,\mathrm{C}[2]$$

We note that the quadratic equation we derived can also be solved directly:

```
α /. Solve[ 1/2 α (α - 1) σ² - r + r α == 0, α]
```

$$\left\{1,\ -\frac{2\,r}{\sigma^2}\right\}$$

The Constant-Coefficient Picture

The solution of first and second order equations with constant coefficients is familiar to many. Equations of the type

$$\frac{1}{2}\,\sigma^2\,S^2\,\frac{\partial^2 V}{\partial S^2} + (r - q)\,S\,\frac{\partial V}{\partial S} - r\,V = 0 \tag{11}$$

can be reduced to constant coefficient equations by an illuminating change of variable that we shall find exceedingly useful in more general situations. Let K be any suitable base value for S. It could be the strike of an option, for example. Set

$$S = K\,e^x \tag{12}$$

and observe that for any function f

$$S\,\frac{\partial f(S)}{\partial S} = \frac{\partial f(S)}{\partial x} \tag{13}$$

So the time-independent Black-Scholes equation becomes

$$\frac{1}{2}\,\sigma^2\left(\frac{\partial^2 V}{\partial x^2} - \frac{\partial V}{\partial x}\right) + (r-q)\,\frac{\partial V}{\partial x} - rV = 0 \tag{14}$$

or

$$\frac{1}{2}\,\frac{\partial^2 V}{\partial x^2}\,\sigma^2 - rV + \left(r - q - \frac{\sigma^2}{2}\right)\frac{\partial V}{\partial x} = 0 \tag{15}$$

which is a standard constant-coefficient form that can be solved by standard methods, where we look for solutions of the form $A\,e^{\lambda x}$. This leads to exactly the same type of power solutions as before, once the re-substitution

$$x = \log\!\left(\frac{S}{K}\right) \tag{16}$$

has been made. This can be done with "pen and paper", or directly with *Mathematica*.

```
soln = V[x] /.
    DSolve[1 / 2 σ^2 D[V[x], {x, 2}] + (r - q - σ^2 / 2) D[V[x], x] - r V[x] == 0,
      V[x], x][[1]]
```

$$E^{\frac{x\left(2q-2r+\sigma^2 - \sqrt{8r\sigma^2 + (-2q+2r-\sigma^2)^2}\right)}{2\sigma^2}}\ C[1] + E^{\frac{x\left(2q-2r+\sigma^2 + \sqrt{8r\sigma^2 + (-2q+2r-\sigma^2)^2}\right)}{2\sigma^2}}\ C[2]$$

Let's change variables back to S, putting $K = 1$ for illustration:

```
subsoln = Simplify[soln /. x -> Log[S]]
```

$$S^{-\frac{-2q+2r+\sqrt{8r\sigma^2 + (2q-2r+\sigma^2)^2}}{2\sigma^2}}\left(\sqrt{S}\ C[1] + S^{\frac{1}{2} + \frac{\sqrt{8r\sigma^2 + (2q-2r+\sigma^2)^2}}{\sigma^2}}\ C[2]\right)$$

For general q this is rather complicated, so let's re-check the $q = 0$ solution, giving *Mathematica* a kick to do the necessary simplifications:

```
Simplify[ExpandAll[subsoln /. q -> 0]]
```

$$S^{-\frac{2r+\sqrt{(2r+\sigma^2)^2}}{2\sigma^2}}\left(\sqrt{S}\ C[1] + S^{\frac{1}{2} + \frac{\sqrt{(2r+\sigma^2)^2}}{\sigma^2}}\ C[2]\right)$$

```
Simplify[PowerExpand[%]]
```

$$S^{-\frac{2r}{\sigma^2}}\ C[1] + S\ C[2]$$

We recover once again the combination of an asset and a steady-state American Put.

Although in this particular case we have gained nothing new, the fact that we can eliminate the powers of S from the equation by changing variables to x is a useful observation. In the next section we shall proceed to analyse the general case using this observation. First we wish to take a more detailed look at the steady-state solutions for non-zero q.

Steady-State Solutions for Non-zero q

Let's take a closer look at the two steady-state solutions when $q > 0$. This is of importance for FX options, when q is the interest rate associated with the foreign currency.

```
Simplify[ExpandAll[subsoln]]
```

$$S^{-\frac{-2\,q+2\,r+\sqrt{4\,q^2+4\,q\,(-2\,r+\sigma^2)+(2\,r+\sigma^2)^2}}{2\,\sigma^2}} \left(\sqrt{S}\;C[1] + S^{\frac{1}{2}+\frac{\sqrt{4\,q^2+4\,q\,(-2\,r+\sigma^2)+(2\,r+\sigma^2)^2}}{\sigma^2}}\;C[2] \right)$$

```
Expand[%]
```

$$S^{\frac{1}{2}+\frac{q}{\sigma^2}-\frac{r}{\sigma^2}-\frac{\sqrt{4\,q^2+4\,q\,(-2\,r+\sigma^2)+(2\,r+\sigma^2)^2}}{2\,\sigma^2}}\;C[1] + S^{\frac{1}{2}+\frac{q}{\sigma^2}-\frac{r}{\sigma^2}+\frac{\sqrt{4\,q^2+4\,q\,(-2\,r+\sigma^2)+(2\,r+\sigma^2)^2}}{2\,\sigma^2}}\;C[2]$$

We separate them out into two cases:

```
asoln = % /. {C[1] -> 1, C[2] -> 0}
```

$$S^{\frac{1}{2}+\frac{q}{\sigma^2}-\frac{r}{\sigma^2}-\frac{\sqrt{4\,q^2+4\,q\,(-2\,r+\sigma^2)+(2\,r+\sigma^2)^2}}{2\,\sigma^2}}$$

```
bsoln = %% /. {C[2] -> 1, C[1] -> 0}
```

$$S^{\frac{1}{2}+\frac{q}{\sigma^2}-\frac{r}{\sigma^2}+\frac{\sqrt{4\,q^2+4\,q\,(-2\,r+\sigma^2)+(2\,r+\sigma^2)^2}}{2\,\sigma^2}}$$

Let's remind ourselves which is which by looking once more at the zero-yield limit:

```
PowerExpand[Simplify[asoln /. {q -> 0}]]
```

$$S^{-\frac{2\,r}{\sigma^2}}$$

```
PowerExpand[Simplify[bsoln /. {q -> 0}]]
```

        S

So the solution **bsoln** is the one corresponding to the asset, while **asoln** corresponds to the steady state, or "perpetual" Put. When $q > 0$ **bsoln** acquires an interesting new interpretation. To simplify the analysis, let

```
λ = PowerExpand[Log[bsoln /. S -> Exp[1]]]
```

$$\frac{1}{2} + \frac{q}{\sigma^2} - \frac{r}{\sigma^2} + \frac{\sqrt{4 q^2 + 4 q (-2 r + \sigma^2) + (2 r + \sigma^2)^2}}{2 \sigma^2}$$

We define the functions

$$b(K, \lambda) = \frac{K}{1 - \frac{1}{\lambda}} \tag{17}$$

$$\text{PCall}(S, K, \lambda) := (b(K, \lambda) - K)\left(\frac{S}{b(K, \lambda)}\right)^{\lambda} \tag{18}$$

We define them symbolically thus:

```
b[K_, λ_]  := K / (1 - 1 / λ)
```

```
PCall[S_, K_, λ_] := (b[K, λ] - K) * (S / b[K, λ]) ^ λ
```

The function **PCall** is the steady-state solution for an American Call, provided $q > 0$. When $S = b$, we have

```
PCall[b[K, λ], K, λ] == (b[K, λ] - K)
```

        True

```
Simplify[D[PCall[S, K, λ], S] /. S -> b[K, λ]]
```

        1

Thus the function **PCall** joins smoothly onto the surface given by $S - K$ at the point $S = b$. This is the "perpetual Call" solution of the Black-Scholes equation. Note that as $q \to 0$, the value of b becomes infinite. Both the perpetual Call and Put solutions are well known - see the discussion and bibliography of Chapter 11.

4.3 The Black-Scholes Equation and the Diffusion Equation

We return to the standard and more general form of our partial differential equation:

$$\frac{\partial V}{\partial t} + S(r-q)\frac{\partial V}{\partial S} + \frac{1}{2}\sigma^2 S^2 \frac{\partial^2 V}{\partial S^2} - rV = 0 \qquad (19)$$

We have already done the work to establish that by setting, for some base variable K,

$$S = K\,e^x \qquad (20)$$

the time-dependent Black-Scholes equation becomes

$$\frac{\partial V}{\partial t} + \frac{1}{2}\frac{\partial^2 V}{\partial x^2}\sigma^2 - rV + \left(r - q - \frac{\sigma^2}{2}\right)\frac{\partial V}{\partial x} = 0 \qquad (21)$$

Making one further re-arrangement, this becomes, following the notation of Wilmott *et al* (1993), but generalizing to allow for dividends expressed though q,

$$\frac{\partial^2 V}{\partial x^2} - k_1 V + (k_2 - 1)\frac{\partial V}{\partial x} = -\frac{2}{\sigma^2}\frac{\partial V}{\partial t} \qquad (22)$$

$$k_1 = \frac{2r}{\sigma^2} \qquad (23)$$

$$k_2 = \frac{2(r-q)}{\sigma^2} \qquad (24)$$

The next step is to re-scale the time variable. Generally, we are interested in an instrument with an expiry or maturity at a time T in the future. Bearing this in mind, we set, assuming that the volatility is constant,

$$\tau = \frac{1}{2}\sigma^2 (T-t) \qquad (25)$$

and now we have arrived at

$$\frac{\partial^2 V}{\partial x^2} - k_1 V + (k_2 - 1)\frac{\partial V}{\partial x} = \frac{\partial V}{\partial \tau} \qquad (26)$$

How one proceeds next depends on whether we can regard k_i as constant. If we can, matters are rather trivial, for writing

$$V(S, t) = C\,e^{-\frac{1}{2}(k_2 - 1)x - \left(\frac{1}{4}(k_2 - 1)^2 + k_1\right)\tau}\,u(x, \tau) \qquad (27)$$

eliminates several of the remaining terms, so that

$$\frac{\partial^2 u}{\partial x^2} = \frac{\partial u}{\partial \tau} \qquad (28)$$

Various other devices can be used when some or all of the parameters are time-dependent, leading to the same simple diffusion equation. There is a vast amount of standard applied mathematics we can bring to bear on the solution of this equation. In Section 4.4 we shall look at how to solve this equation. First it is convenient to explore some special cases that do not require the power of our general approach.

Special Solutions I - Logs

A few interesting solutions of our PDE can be derived by special methods. Let's first go back to the point where we had this equation:

$$\frac{\partial^2 V}{\partial x^2} - k_1 V + (k_2 - 1) \frac{\partial V}{\partial x} = -\frac{2}{\sigma^2} \frac{\partial V}{\partial t} \tag{29}$$

Matters are simplified if there is no second derivative term:

$$(k_2 - 1) \frac{\partial V}{\partial x} = -\frac{2}{\sigma^2} \frac{\partial V}{\partial t} + k_1 V = -\frac{2}{\sigma^2} \left(\frac{\partial V}{\partial t} - r V \right) \tag{30}$$

We introduce an integrating factor for the right side of this equation, and let

$$V = e^{-r(T-t)} w(x, t) \tag{31}$$

$$(k_2 - 1) \frac{\partial w}{\partial x} = -\frac{2}{\sigma^2} \frac{\partial w}{\partial t} \tag{32}$$

Solutions of this can easily be obtained by assuming that both sides are a constant, and looking for a solution of the form $w = A(x - \alpha t)$. We find that

$$\alpha = r - q - \frac{\sigma^2}{2} \tag{33}$$

Re-assembling it all, we find that, by a suitable re-organization of constants,

$$V = A \left[\left(-\frac{\sigma^2}{2} - q + r \right) (T - t) + \log(S/K) \right] e^{-r(T-t)} \tag{34}$$

which has the property that when $t = T$

$$V = A \log(S/K) \tag{35}$$

This is what is known as the "log contract" - it represents an instrument that pays the log of the asset price at maturity.

Special Solutions II - Separation of Variables and Power Solutions

We have our general solution in the form

$$V = C e^{-\frac{1}{2}(k_2-1)x - \left(\frac{1}{4}(k_2-1)^2 + k_1\right)\tau} u(x, \tau) \tag{36}$$

with the diffusion condition on u

$$\frac{\partial^2 u}{\partial x^2} = \frac{\partial u}{\partial \tau} \tag{37}$$

A standard technique is to suppose that $u = f(x) g(\tau)$. When we substitute into the diffusion equation, we find that, for some constant μ (not necessarily real),

$$\frac{\partial g(\tau)}{\partial \tau} = \mu^2 g \tag{38}$$

$$\frac{\partial^2 f(x)}{\partial x^2} = \mu^2 f \tag{39}$$

Case 1: $\mu^2 > 0$

This yields a family of solutions of the form

$$u = Z\, e^{\mu x + \mu^2 \tau} \tag{40}$$

By writing $\mu = 1/2\,(k_2 - 1) + \lambda$, and doing some reorganization, we obtain the general "power contract"

$$V = Z\left(\frac{S}{K}\right)^{\lambda} e^{\left(\left(-\frac{\sigma^2}{2} - q + r\right)\lambda + \frac{\lambda^2 \sigma^2}{2} - r\right)(T-t)} \tag{41}$$

This contains as special cases some examples we have seen already. When $\lambda = 0$ we obtain the solution

$$V = Z\, e^{-r(T-t)} \tag{42}$$

representing a cash payment at maturity. When $\lambda = 1$ we obtain a slight generalization (yield-adjusted) of the pure asset, in the form

$$V = Z\,\frac{S}{K}\, e^{-q(T-t)} \tag{43}$$

with maturity value $Z S / K$. The steady-state American Put solution is obtained when $q = 0$ and $\lambda = -2\,r/\sigma^2$. More generally, when $t = T$ we have the general power payoff

$$V = Z\left(\frac{S}{K}\right)^{\lambda} \tag{44}$$

Case 2: $\mu^2 = 0$

In this case we get $u = a x + b$. Some re-organization leads to the special log-power solution:

$$V(S, t) = \left(A \log\left(\frac{S}{K}\right) + B\right)\left(\frac{S}{K}\right)^{\frac{1}{2} - \frac{r-q}{\sigma^2}} e^{-(T-t)\left(\frac{1}{2}\left(\frac{\sigma}{2} - \frac{r-q}{\sigma}\right)^2 + r\right)} \tag{45}$$

with payoff

$$V(S, T) = \left(A \log\left(\frac{S}{K}\right) + B\right)\left(\frac{S}{K}\right)^{\frac{1}{2} - \frac{r-q}{\sigma^2}} \tag{46}$$

Case 3: $\mu^2 < 0$

Writing $\mu = i \lambda$ gets us to

$$u = (A \sin(\lambda\,x) + B \cos(\lambda\,x))\, e^{-\lambda^2\,\tau} \tag{47}$$

These Fourier-mode solutions are more important for their role in constructing general solutions of the diffusion equation given an initial (payoff) condition - we shall not consider them in isolation.

Special Solutions III - Similarity Methods

We have already seen how similarity solutions can be used to solve the coupled asset-FX problem. They are exceedingly useful here also. The diffusion equation is invariant under the re-scalings $x \to \alpha\,x$, $\tau \to \alpha^2\,\tau$. To capture this invariance, we change variables to

$$z = \frac{x}{\sqrt{\tau}}, \quad y = \sqrt{\tau} \tag{48}$$

with inverse

$$x = y\,z, \quad \tau = y^2 \tag{49}$$

This change of variables, and the chain rule, lead to the new PDE

$$\frac{\partial^2 u\,(z,\,y)}{\partial z^2} + \frac{z}{2}\,\frac{\partial u\,(z,\,y)}{\partial z} = \frac{1}{2}\,y\,\frac{\partial u\,(z,\,y)}{\partial y} \tag{50}$$

If we consider solutions to this PDE of the form $u = y^n\,h(z)$, we find that h satisfies the ordinary differential equation

$$\frac{\partial^2 h}{\partial z^2} + \frac{1}{2}\,z\,\frac{\partial h}{\partial z} - \frac{1}{2}\,n\,h = 0 \tag{51}$$

For general values of n this is a slightly awkward equation - here is *Mathematica*'s view of the solutions:

```
h[z] /.
  DSolve[D[h[z], {z, 2}] + z / 2 D[h[z], z] - n / 2 h[z] == 0, h[z], z][[1]]
```

$$\sqrt{z^2}\ \text{C[2] Hypergeometric1F1}\left[\frac{1}{2} - \frac{n}{2},\ \frac{3}{2},\ -\frac{z^2}{4}\right] +$$

$$\text{C[1] Hypergeometric1F1}\left[-\frac{n}{2},\ \frac{1}{2},\ -\frac{z^2}{4}\right]$$

The solutions are much more straightforward if $n = 0$,

```
h[z] /. DSolve[D[h[z], {z, 2}] + z / 2 D[h[z], z] == 0, h[z], z][[1]]
```

$$\text{C[2]} + \sqrt{\pi}\ \text{C[1] Erf}\left[\frac{z}{2}\right]$$

or if $n = -1$.

```
h[z] /.
 DSolve[D[h[z], {z, 2}] + z / 2 D[h[z], z] + 1 / 2 h[z] == 0, h[z], z][[1]]
```

$$E^{-\frac{z^2}{4}} C[1] + E^{-\frac{z^2}{4}} C[2] \, \text{Erfi}\Big[\frac{\sqrt{z^2}}{2}\Big]$$

This last case is worth a special look. If $n = -1$ the equation for h can be written as

$$\frac{\partial^2 h}{\partial z^2} + \frac{1}{2} \frac{\partial(z\,h)}{\partial z} = 0 \tag{52}$$

which integrates once to

$$\frac{\partial h(z)}{\partial z} + \frac{1}{2} z\,h = C \tag{53}$$

Introducing a suitable integrating factor, we obtain

$$D\Big[e^{\frac{z^2}{4}} h(z)\Big] = C\,e^{\frac{z^2}{4}} \tag{54}$$

If we now set $C = 0$ (which will in fact remove the **Erfi** solution) we end up with

$$h(z) = A\,e^{-\frac{z^2}{4}} \tag{55}$$

This generates the following solution of the diffusion equation:

$$G = \frac{A\,e^{-\frac{x^2}{4\tau}}}{\sqrt{\tau}} \tag{56}$$

It is convenient to normalize this so that the integral of u is unity. We can get *Mathematica* to do the work to find A:

```
Integrate[A / Sqrt[τ] Exp[-x^2 / (4 τ)], {x, -Infinity, Infinity}]
```

$$\text{If}\Big[\text{Re}[\tau] > 0,\ 2\,A\,\sqrt{\pi},\ \int_{-\infty}^{\infty} \frac{A\,E^{-\frac{x^2}{4\tau}}}{\sqrt{\tau}}\,dx\Big]$$

So the normalized solution is

$$G(x, \tau) = \frac{e^{-\frac{x^2}{4\tau}}}{\sqrt{4\pi\tau}} \tag{57}$$

Let us take a look at this function for various values of τ:

```
G[x_, τ_] := 1 / Sqrt[4 Pi τ] Exp[-x^2 / (4 τ)]
```

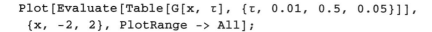

```
Plot[Evaluate[Table[G[x, τ], {τ, 0.01, 0.5, 0.05}]],
 {x, -2, 2}, PlotRange -> All];
```

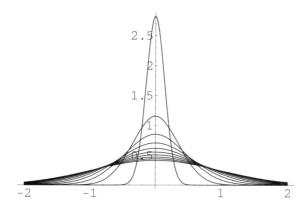

For small τ it is strongly concentrated at the origin, and gradually spreads out. This solution is very special, as we shall see in the next section. It is sometimes called the "Green's function" for the diffusion equation, or the fundamental solution.

We have now generated a wide variety of particular solutions to our PDE by special methods. However, we do not have control over the form of these solutions at $\tau = 0$. What we really want to do is to be able to specify the solution at $\tau = 0$ and evolve the solution from it. The means by which one achieves this makes a subtle combination of the Fourier-mode solutions and the Green's function we have just derived, and will be considered in the next section. Right now we develop a technical result about the Laplace transform of the Green's function. Note that if you are unfamiliar with transform calculus, the more basic results that follow later do not require the transform approach.

Useful Transforms of the Green's Function

Later in our analysis we shall need some technical results about the time-Laplace and spatial-Fourier transforms of the Green's function. This is a good place to get *Mathematica* to help us, as some of the integrals are rather awkward to manage directly. We load a Package to help us:

```
Needs["Calculus`LaplaceTransform`"]
```

```
? LaplaceTransform
```

```
    LaplaceTransform[expr, t, s] gives a function of s,
      which is the Laplace transform of expr, a function of
      t, t >= 0. It is defined by  LaplaceTransform[expr,
      t, s] = Integrate[Exp[-s t] expr, {t, 0, Infinity}].
```

Let's go for the transform of our fundamental solution, with respect to time:

```
LaplaceTransform[Exp[-x^2 / (4 * t)] / Sqrt[4 Pi t], t, s]
```

$$\frac{E^{-\sqrt{s}\ \sqrt{x^2}}}{2\sqrt{s}}$$

```
TraditionalForm[PowerExpand[%]]
```

$$\frac{e^{-\sqrt{s}\ x}}{2\sqrt{s}}$$

What we want to note for future reference is that if we differentiate the fundamental solution with respect to x, we can get rid of the $\sqrt{s}$ in the denominator. As a check on this idea, let's ask directly for the inverse transform of $e^{-x\sqrt{s}}$:

```
InverseLaplaceTransform[Exp[-x Sqrt[s]], s, t]
```

$$\frac{E^{-\frac{x^2}{4t}}\ x}{2\sqrt{\pi}\ t^{3/2}}$$

We shall use these results later. We shall also record a Fourier transform w.r.t. x:

```
Needs["Calculus`FourierTransform`"]
```

```
? FourierTransform
```

```
    FourierTransform[expr, t, w] gives a function
      of w, which is the Fourier transform of expr, a
      function of t.  It is defined by FourierTransform[
      expr, t, w] = FourierOverallConstant * Integrate[Exp[
      FourierFrequencyConstant I w t] expr, {t, -Infinity, Infinity}].
```

Let's go for the transform of our fundamental solution, with respect to x:

```
FourierTransform[Exp[-x^2 / (4 * t)] / Sqrt[4 Pi t], x, w]
```

$$E^{-t\ w^2}$$

Let's go back, just to check:

```
InverseFourierTransform[Exp[-t w^2], w, x]
```

$$\frac{E^{-\frac{x^2}{4t}}}{2\sqrt{\pi}\sqrt{t}}$$

We shall also use these results, and give a mathematical justification of the easier ones, in our subsequent analysis.

4.4 The Diffusion Equation - Solution from Initial Conditions

We now come to an important observation. Suppose the diffusion equation holds for all x,

$$-\infty < x < \infty \qquad (58)$$

and for $t > 0$. Suppose further that we are given that

$$u(x, 0) = f(x) \qquad (59)$$

This is the initial condition for the diffusion equation. In financial terms it represents the option payoff when it is defined for all values of S. Then the solution to the diffusion equation with this initial condition is

$$u(x, \tau) = \int_{-\infty}^{\infty} G(x - y, \tau) f(y) \, dy \qquad (60)$$

In one sense, this result is now almost obvious. Since G is a solution to the diffusion equation, so is any linear combination of Gs - here such a combination is given by integration. Since G integrates to unity, we can write

$$u(x, \tau) - f(x) = \int_{-\infty}^{\infty} G(x - y, \tau) [f(y) - f(x)] \, dy \qquad (61)$$

Now as $\tau \to 0$, the integrand, which has two terms multiplied, vanishes - G concentrates itself around $y = x$, where the other terms vanishes. Hence

$$u(x, \tau) - f(x) \to 0 \qquad (62)$$

as $\tau \to 0$. So the initial conditions are satisfied.

Fourier Transform View

The solution we have written down and justified can also be derived by the use of Fourier transform methods. If you are not familiar with transform methods you may wish to skip this section. We write the spatial Fourier transform as

$$v(\omega, \tau) = \int_{-\infty}^{\infty} e^{i\omega x} u(x, \tau) \, dx \qquad (63)$$

$$u(x, \tau) = \frac{1}{2\pi} \int_{-\infty}^{\infty} e^{-i\omega x} v(\omega, \tau) \, d\omega \tag{64}$$

In *Mathematica,* `FourierOverallConstant` = 1 gives these Fourier conventions. Now as u satisfies the diffusion equation, v satisfies the equation

$$\frac{\partial v(\omega, \tau)}{\partial \tau} + \omega^2 v(\omega, \tau) = 0 \tag{65}$$

This has the simple solution

$$v(\omega, \tau) = v(\omega, 0) e^{-\omega^2 \tau} \tag{66}$$

So using our formula for u at general times:

$$u(x, \tau) = \frac{1}{2\pi} \int_{-\infty}^{\infty} e^{-i\omega x} v(\omega, 0) e^{-\omega^2 \tau} \, d\omega \tag{67}$$

Substituting for the value of v at $\tau = 0$:

$$u(x, \tau) = \frac{1}{2\pi} \int_{-\infty}^{\infty} e^{-i\omega x} \int_{-\infty}^{\infty} e^{i\omega y} u(y, 0) \, dy \, e^{-\omega^2 \tau} \, d\omega \tag{68}$$

Re-organizing gives

$$u(x, \tau) = \int_{-\infty}^{\infty} f(y) H(x - y, \tau) \, dy \tag{69}$$

where

$$H(z, \tau) = \frac{1}{2\pi} \int_{-\infty}^{\infty} e^{-i\omega z - \omega^2 \tau} \, d\omega \tag{70}$$

but we have already established that the quantity $e^{-\omega^2 \tau}$ is the Fourier transform of our fundamental solution - hence the result. (If you wish to prove this transform result yourself, two steps are needed: (a) complete the square in the exponential; (b) use complex contour integration methods to evaluate the remaining integral. We shall not pursue this here.)

4.5 Method of Images: Boundaries and Boundary Conditions

The previous section gave the solution when all values of x are relevant. Sometimes one wishes to consider a situation where the range is limited to domains of one of the following forms:

$$\begin{aligned} &x \geq a \\ &x \leq b \\ &a \leq x \leq b \end{aligned} \tag{71}$$

When this is the case, a condition must be given for the function on the given boundary. An interesting class of options, known as Barrier options, will require an understanding of this type of solution. We consider in detail the simplest case, with $x \geq a$, and $u(a, \tau) = g(\tau)$. We consider the case $g = 0$ first. The solution to this problem can be obtained by a cunning modification to the fundamental solution, based on the "method of images". Consider

$$G_1(x, y, \tau) = G(x - y, \tau) - G(x - (2\,a - y), \tau) \tag{72}$$

which consists of a fundamental solution centred on y, with a second fundamental solution centred on the image point $2\,a - y$, which is the location of a point at y reflected in the boundary. Clearly, G_1 solves the diffusion equation. Furthermore

$$G_1(a, y, \tau) = G(a - y, \tau) - G(a - (2\,a - y), \tau) = G(a - y, \tau) - G(y - a, \tau) = 0 \tag{73}$$

since G is an even function of its spatial argument. So G_1 vanishes on $x = a$ for all values of y. So any linear combination over different y does. Now consider the modified integral

$$u(x, \tau) = \int_a^\infty G_1(x - y, \tau) f(y)\,dy \tag{74}$$

Clearly $u(a, \tau) = 0$. As we approach $\tau = 0$ the contributions to the integrand are focused on the points $y_1 = x$ and $y_2 = 2\,a - x$. But for $x > a$, then $y_2 < a$, which is outside the region of integration, and there is no contribution from $x = a$. Hence, for $x > a$,

$$\lim_{\tau \to 0} u(x, \tau) = f(x) \tag{75}$$

as before. Hence we have solved the problem when the boundary condition is that $u = 0$ on $x = a$. By identical arguments, it follows that, if we have x bounded above by b, where it vanishes,

$$u(x, \tau) = \int_{-\infty}^b G_2(x, y, \tau) f(y)\,dy \tag{76}$$

$$G_2(x, y, \tau) = G(x - y, \tau) - G(x - (2\,b - y), \tau) \tag{77}$$

Now suppose that $a \le x \le b$ and that u vanishes at both ends. The solution method is somewhat analogous to the effect you get when standing between two parallel mirrors - an infinite number of images are generated.

$$\begin{aligned} G_\infty = {}& G(x - y, \tau) - G(x - (2\,a - y), \tau) - G(x - (2\,b - y), \tau) \\ & + G(x - (2\,a - (2\,b - y)), \tau) + {+}G(x - (2\,b - (2\,a - y)), \tau) - \dots \end{aligned} \tag{78}$$

This can be re-organized as

$$G_\infty(x, y, \tau) = \sum_{k=-\infty}^\infty G(x - y + 2\,k\,(b - a),\ \tau) - G(x + y - 2\,a + 2\,k\,(b - a),\ \tau) \tag{79}$$

The solution is then

$$u(x, \tau) = \int_a^b G_\infty(x, y, \tau) f(y)\,dy \tag{80}$$

Addition of a Boundary Term

Sometimes we wish to consider a solution with a given initial condition and a specified function along the boundary. The analysis of this is a little harder - if you do not wish to understand where Barrier option rebate formulae come from, you may skip this section! We can obtain such a solution by adding, to a solution where the boundary is zero, a solution where the initial value is zero, and the given boundary function specified. So, as an example, suppose that

$$x \geq a \tag{81}$$

$$u(a, \tau) = g(\tau)$$
$$u(x, 0) = 0 \text{ for } x \geq a \tag{82}$$

$$\lim_{x \to \infty} u(x, \tau) = 0 \tag{83}$$

This is where a Laplace transform picture becomes useful. We let

$$v(x, p) = \int_0^\infty u(x, \tau) \, e^{-p\tau} \, d\tau \tag{84}$$

Then the Laplace transform of the diffusion equation gives us

$$\frac{\partial^2 v(x, p)}{\partial x^2} = p \, v(x, p) \tag{85}$$

The solution of this vanishing at infinity is

$$v(x, p) = v(a, p) \, e^{-\sqrt{p} \, (x-a)} = \tilde{g}(p) \, e^{-\sqrt{p} \, (x-a)} \tag{86}$$

where the Laplace transform of the boundary condition is

$$\tilde{g}(p) = \int_0^\infty g(\tau) \, e^{-p\tau} \, d\tau \tag{87}$$

It follows that the solution is the Laplace convolution of $g(\tau)$ with the inverse Laplace transform of $e^{-\sqrt{p} \, (x-a)}$, which we have already found with the help of *Mathematica*. The solution is just

$$u(x, \tau) = \frac{1}{2 \sqrt{\pi}} \int_0^\tau \frac{e^{-\frac{(x-a)^2}{4(\tau-t)}} (x-a) \, g(t)}{(\tau-t)^{3/2}} \, dt \tag{88}$$

This may be usefully re-organized by a change of variables:

$$u(x, \tau) = \frac{1}{2 \sqrt{\pi}} \int_{1/\tau}^\infty \frac{e^{-\frac{s(x-a)^2}{4}} (x-a) \, g(\tau - 1/s)}{s^{1/2}} \, ds \tag{89}$$

Let's do a quick check on this rather formidable integral, by setting $g = 1$.

```
Integrate[(Exp[-(1/4)*s(x - a)^2]*(x - a))/Sqrt[s], {s, 1/τ, Infinity}]
```

$$\text{If}\left[\tau > 0 \,\&\&\, \text{Re}[(a-x)^2] > 0 \,\&\&\, \text{Re}\left[\frac{a^2}{4} - \frac{a\,x}{2} + \frac{x^2}{4}\right] > 0,\right.$$

$$\frac{2\sqrt{\pi}\,(a-x)\left(-1 + \text{Erf}\left[\frac{\sqrt{(a-x)^2}}{2\sqrt{\tau}}\right]\right)}{\sqrt{(a-x)^2}}, \quad \left.\int_{\frac{1}{\tau}}^{\infty} \frac{E^{-\frac{1}{4}s\,(-a+x)^2}\,(-a+x)}{\sqrt{s}}\,ds\right]$$

Picking out the relevant piece, and normalizing, we get:

```
%[[2]] / (2 * Sqrt[Pi])
```

$$\frac{(a-x)\left(-1 + \text{Erf}\left[\frac{\sqrt{(a-x)^2}}{2\sqrt{\tau}}\right]\right)}{\sqrt{(a-x)^2}}$$

We are interested in the values of this for $x > a$, so we consider

```
PowerExpand[-%]
```

$$1 - \text{Erf}\left[\frac{a-x}{2\sqrt{\tau}}\right]$$

Note that Erf(0) = 0 and Erf(∞) = 1, so this solution is unity on the boundary $x = a$ and vanishes when $\tau = 0$. This not only is correct, but coincides with our $n = 0$ similarity solution if the arbitrary constants are set suitably. However, the integral for a general g, in the form we have given, can be extremely awkward to manage, and a different representation of the solution is frequently useful. We go back to the Laplace inversion integral in the form

$$q(x, t) = \frac{1}{2\pi i} \int_{c-i\infty}^{c+i\infty} e^{-\sqrt{p}\,(x-a)}\,e^{pt}\,dp \tag{90}$$

and deform the contour to wrap around the negative real axis. Taking care of the branch points gives

$$q(x, t) = \frac{1}{\pi} \int_0^\infty e^{-pt}\,\sin\left(\sqrt{p}\,(x-a)\right)dp \tag{91}$$

The solution with the given boundary condition can then be represented as

$$u = \int_0^t g(\tau)\,q(\tau - t)\,d\tau \tag{92}$$

If you are not familiar with contour integration, *Mathematica* can reassure you that this makes sense!

```
Integrate[Exp[-p t] Sin[Sqrt[p] x], {p, 0, Infinity}]
```

$$
\text{If}\left[\text{Im}[x] == 0 \;\&\&\; \text{Re}[t] > 0,\right.
$$

$$
\frac{E^{-\frac{x^2}{4t}} \sqrt{\pi} \left(\frac{x^2}{t}\right)^{3/2} \text{Sign}[x]}{2 x^2}, \left.\int_0^\infty E^{-p\,t} \text{Sin}\left[\sqrt{p}\; x\right] dp\right]
$$

A minor manipulation reduces this to the result we have used, for $x > 0$.

```
PowerExpand[%[[2]] / Pi]
```

$$
\frac{E^{-\frac{x^2}{4t}} x \, \text{Sign}[x]}{2 \sqrt{\pi} \; t^{3/2}}
$$

Similar formulae may be derived for an upper boundary condition, and a more complicated analysis can be done with boundary conditions supplied on two barriers. This set of basic solutions to the heat equation will serve our main purpose, which is to illustrate how the values of several important contracts may be found.

Neumann and Impedance Boundary Conditions

There are other types of boundary condition that merit consideration, both on general mathematical grounds and on financial grounds. Suppose we consider the region $x > 0$, for simplicity, and consider the general impedance boundary condition

$$
\frac{\partial u}{\partial x} = \alpha u \tag{93}
$$

at $x = 0$. When $\alpha = 0$ this is a "Neumann" boundary condition. This case can easily be solved by the method of images by adding the image Green's function rather than subtracting it. In physics it corresponds to an insulation boundary condition for the heat equation. When α is non-zero matters are a little more complicated. These problems were first studied many years ago as a model of heat conduction when the rate of heat loss at a boundary is proportional to the temperature there. Our purpose here is to show how *Mathematica* can be used to help in the construction of the solution. This section is also quite hard and is in fact not necessary for the more basic option-pricing problems - you may come back to it later if you wish to. We shall give the details of corresponding financial interpretation later - it is related to path-dependent options.

There are various ways of approaching the solution to such a problem. Guided by the cases where simple images work - α infinite is the one we have already treated, $\alpha = 0$ being done by swapping some signs, it is natural to seek a similar solution for general α. We can do so by using Fourier methods. We know, for example, that for all values of ω, a, b, the functions

$$
\left(a\,e^{-i\omega x} + b\,e^{i\omega x}\right)e^{-\omega^2 t} \tag{94}
$$

are solutions of the diffusion equation. As we wish to build a solution that is the fundamental solution plus a correction to satisfy boundary conditions, we set $a = 1$. Then if we apply the boundary condition, it follows that

$$b = \frac{\omega - i\alpha}{\omega + i\alpha} = 1 - \frac{2i\alpha}{\omega + i\alpha} \tag{95}$$

To find out what the new term is, we first confirm the inverse Fourier transform of the second contribution to b - note that x is replaced by $-x$ if we use standard Fourier conventions:

```
FourierTransform[-2 α UnitStep[-x] Exp[α x] ,
  x, ω, FourierFrequencyConstant -> -1]
```

$$-\frac{2\,\alpha}{\alpha - I\,\omega}$$

The following may be more intelligible!

```
Integrate[-2 α Exp[x (α - I ω)], {x, -Infinity, 0}][[2]]
```

$$-\frac{2\,\alpha}{\alpha - I\,\omega}$$

To get the desired result we take the convolution of

$$\theta(-x)\,e^{\alpha x} \tag{96}$$

with the fundamental solution, by ordinary integration with *Mathematica:*

```
Clear[α]
```

```
Integrate[-2 α Exp[α p] 1 / Sqrt[4 Pi τ] Exp[- (p - x)^2 / (4 τ)],
  {p, -Infinity, 0}]
```

$$\text{If}\left[\text{Re}\left[\alpha + \frac{x}{2\,\tau}\right] > 0 \,\&\&\, \text{Re}[\tau] > 0,\right.$$

$$\left. E^{\alpha\,(x+\alpha\,\tau)}\,\alpha\left(-1 + \text{Erf}\left[\frac{x + 2\,\alpha\,\tau}{2\,\sqrt{\tau}}\right]\right), \int_{-\infty}^{0} -\frac{E^{p\,\alpha - \frac{(p-x)^2}{4\,\tau}}\,\alpha}{\sqrt{\pi}\,\sqrt{\tau}}\,dp\right]$$

We want the middle piece of this:

```
GG[x_, τ_, α_] = %[[2]]
```

$$E^{\alpha\,(x+\alpha\,\tau)}\,\alpha\left(-1+\text{Erf}\left[\frac{x+2\,\alpha\,\tau}{2\,\sqrt{\tau}}\right]\right)$$

We now make some checks on this rather odd-looking object. Let's look at it for large α, where we know what the answer should be, by first doing an asymptotic expansion of the Error function:

```
Series[Erf[q], {q, Infinity, 2}] - 1
```

$$E^{-q^2}\left(-\frac{1}{\sqrt{\pi}\,q}+O\left[\frac{1}{q}\right]^3\right)$$

```
Normal[%]
```

$$-\frac{E^{-q^2}}{\sqrt{\pi}\,q}$$

Let's substitute the argument as we found it, and multiply by the factor we have found:

```
% /. q -> (x + 2 α τ) / (2 Sqrt[τ])
```

$$-\frac{2\,E^{-\frac{(x+2\,\alpha\,\tau)^2}{4\,\tau}}\,\sqrt{\tau}}{\sqrt{\pi}\,(x+2\,\alpha\,\tau)}$$

```
% * α * Exp[α (x + α τ)]
```

$$-\frac{2\,E^{\alpha\,(x+\alpha\,\tau)-\frac{(x+2\,\alpha\,\tau)^2}{4\,\tau}}\,\alpha\,\sqrt{\tau}}{\sqrt{\pi}\,(x+2\,\alpha\,\tau)}$$

```
Cancel[ExpandAll[%]]
```

$$-\frac{2\,E^{-\frac{x^2}{4\,\tau}}\,\alpha\,\sqrt{\tau}}{\sqrt{\pi}\,(x+2\,\alpha\,\tau)}$$

```
Series[%, {α, Infinity, 1}]
```

$$-\frac{E^{-\frac{x^2}{4\tau}}}{\sqrt{\pi}\ \sqrt{\tau}} + O\left[\frac{1}{\alpha}\right]^1$$

Thus as α becomes infinite this correction gives us -2 times the fundamental solution, as we require. Now note that

```
ExpandAll[D[GG[x, τ, α], x]  - α GG[x, τ, α]]
```

$$\frac{E^{-\frac{x^2}{4\tau}}\ \alpha}{\sqrt{\pi}\ \sqrt{\tau}}$$

So we write down our full solution, based on a source at y, in the form

```
K[x_, y_, τ_, α_] = Exp[-(x - y)^2 / (4 τ)] / Sqrt[4 Pi τ] +
    Exp[-(x + y)^2 / (4 τ)] / Sqrt[4 Pi τ] + GG[x + y, τ, α]
```

$$\frac{E^{-\frac{(x-y)^2}{4\tau}}}{2\ \sqrt{\pi}\ \sqrt{\tau}} + \frac{E^{-\frac{(x+y)^2}{4\tau}}}{2\ \sqrt{\pi}\ \sqrt{\tau}} + E^{\alpha\ (x+y+\alpha\ \tau)}\ \alpha\left(-1 + Erf\left[\frac{x+y+2\ \alpha\ \tau}{2\ \sqrt{\tau}}\right]\right)$$

This is supposed to be our Green's function. Let's first check that we have satisfied the boundary condition. First we work it out for general position:

```
Simplify[D[K[x, y, τ, α], x]  - α K[x, y, τ, α]]
```

$$-\frac{E^{-\frac{(x+y)^2}{4\tau}}\left(\left(1 + E^{\frac{xy}{\tau}}\right)x - \left(-1 + E^{\frac{xy}{\tau}}\right)(y - 2\ \alpha\ \tau)\right)}{4\ \sqrt{\pi}\ \tau^{3/2}}$$

```
% /. x -> 0
```

```
        0
```

We may as well check that the diffusion equation is satisfied:

```
Simplify[D[K[x, y, τ, α], {x, 2}] - D[K[x, y, τ, α], τ]]
```

```
    0
```

Here is the answer written out carefully in traditional mathematical form (the output of the following has been converted to the style of a numbered equation here):

```
TraditionalForm[K[x, y, τ, α]]
```

$$e^{\alpha\,(x+y+\alpha\,\tau)}\,\alpha\left(\mathrm{erf}\!\left(\frac{x+y+2\,\alpha\,\tau}{2\,\sqrt{\tau}}\right)-1\right)+\frac{e^{-\frac{(x-y)^2}{4\,\tau}}}{2\,\sqrt{\pi}\,\sqrt{\tau}}+\frac{e^{-\frac{(x+y)^2}{4\,\tau}}}{2\,\sqrt{\pi}\,\sqrt{\tau}} \tag{97}$$

This is to be integrated against the initial condition for u, expressed as $f(y)$, over the relevant range, here $y > 0$. If the range of interest is negative, we replace x by $-x$ everywhere, and replace α by $-\alpha$ to preserve the boundary condition. This leads to the solution

```
M[x_, y_, τ_, α_] = Exp[-(x - y)^2 / (4 τ)] / Sqrt[4 Pi τ] +
   Exp[-(x + y)^2 / (4 τ)] / Sqrt[4 Pi τ] + GG[-x - y, τ, -α];
```

```
TraditionalForm[M[x, y, τ, α]]
```

$$-e^{-\alpha\,(-x-y-\alpha\,\tau)}\,\alpha\left(\mathrm{erf}\!\left(\frac{-x-y-2\,\alpha\,\tau}{2\,\sqrt{\tau}}\right)-1\right)+\frac{e^{-\frac{(x-y)^2}{4\,\tau}}}{2\,\sqrt{\pi}\,\sqrt{\tau}}+\frac{e^{-\frac{(x+y)^2}{4\,\tau}}}{2\,\sqrt{\pi}\,\sqrt{\tau}} \tag{98}$$

The reader may check that **M** satisfies the diffusion and boundary conditions.

4.6 Derivation of Basic Option Prices

We have carried out a substantial investigation of the diffusion equation with various types of initial and boundary condition. We shall now map various option-pricing problems of interest into one of the standard forms we have developed. We shall begin by considering options where there are no "barriers".

A General Formula

Suppose that the asset price S may attain any positive value,

$$0 < S < \infty \tag{99}$$

and that the option payoff function

$$V(S_T, T) = P(S_T) \tag{100}$$

We make the change of variables discussed in Section 4.3, including

$$S = K\,e^x,\ S_T = K\,e^y \tag{101}$$

where the subscript T labels the asset value at expiry of the option. Thus we obtain the initial condition for u:

$$f(y) = u(y, 0) = P(K\, e^y)\, e^{\frac{1}{2}(k_2 - 1)\, y} \tag{102}$$

We now write down the solution in the form

$$V(S, t) = e^{-\frac{1}{2}(k_2 - 1)\, x - \left(\frac{1}{4}(k_2 - 1)^2 + k_1\right)\tau}\, u(x, \tau) \tag{103}$$

$$u(x, \tau) = \int_{-\infty}^{\infty} \frac{P(K\, e^y)\, e^{\frac{1}{2}(k_2 - 1)\, y}\, e^{-\frac{(x - y)^2}{4\tau}}}{\sqrt{4\pi\tau}}\, dy \tag{104}$$

Now we note that this pair can be simplified by completing the square in the exponential:

$$V(S, t) = e^{-k_1\,\tau} \int_{-\infty}^{\infty} \frac{P(K\, e^y)\, e^{-\frac{(x - y + \tau(k_2 - 1))^2}{4\tau}}}{\sqrt{4\pi\tau}}\, dy \tag{105}$$

If we change variables in the integration to S_T, we obtain

$$V(S, t) = e^{-t(T - t)} \int_{-\infty}^{\infty} P(S_T)\, \Psi(S_T,\, S,\, r,\, T,\, r,\, q,\, \sigma)\, dy \tag{106}$$

where Ψ is given by

$$\Psi = \frac{e^{-\frac{\left(\log\left(\frac{S_T}{S}\right) - \left(r - q - \frac{\sigma^2}{2}\right)(T - t)\right)^2}{2\sigma^2(T - t)}}}{\left(\sigma\sqrt{2\pi(T - t)}\right) S_T} \tag{107}$$

Now we can view this expression for V as merely an integral, but it is more informative to note that Ψ is the probability density function for a log-normal distribution of S_T, where the log of the expiry price has a normal distribution with mean

$$\log(S) - \left(r - q - \frac{\sigma^2}{2}\right)(T - t) \tag{108}$$

and variance

$$\sigma^2\,(T - t) \tag{109}$$

Our option value can now be seen to be the discounted value of the expected mean payoff, where the expectation is taken over the distribution given by Ψ:

$$V(S, t) = e^{-r(T - t)}\, E\,[P(S_T)] \tag{110}$$

In what follows, the following notation will be useful. The variable y is normalized with respect to a base value K of the underlying asset, usually a strike or some similar variable.

$$y = \log\left(\frac{S}{K}\right) \tag{111}$$

Then y is normally distributed with mean

$$m = \log\!\left(\frac{S}{K}\right) + \left(r - q - \frac{\sigma^2}{2}\right)(T - t) \tag{112}$$

and standard deviation

$$\Sigma = \sigma\sqrt{T - t} \tag{113}$$

So we can write

$$\Psi\,d\,S_T = \frac{e^{-\frac{(y-m)^2}{2\Sigma^2}}\,d\,y}{\sqrt{2\,\pi}\,\Sigma} = n(y, m, \Sigma)\,d\,y \tag{114}$$

where $n(y, m, \Sigma)$ is the probability density function for a standard normal distribution with mean and standard deviation m and Σ.

Cut-Off Power Payoffs

Several cases of interest can be lumped together and based on some standard payoff functions. We consider a power of the stock price times a step function. Two cases are important. The cut-off power Call and the cut-off power Put, with payoffs, respectively,

$$B\,S^\lambda\,\theta(S - K) \tag{115}$$

$$B\,S^\lambda\,\theta(K - S) \tag{116}$$

We can input these to *Mathematica* and plot them:

```
PowerCall[λ_, S_, K_, B_] := B*S^λ*UnitStep[S - K]

PowerPut[λ_, S_, K_, B_] := B*S^λ*UnitStep[K - S]

Plot[PowerCall[0, S, 1, 1], {S, 0, 2}];
```

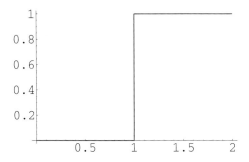

```
Plot[PowerPut[2, S, 1, 1], {S, 0, 2}];
```

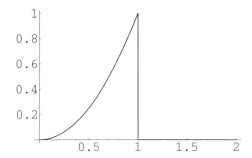

Let's look at the value of a "Call" of this type. We just have to note that

$$S^\lambda = K^\lambda \, e^{\lambda y} \tag{117}$$

and insert the result into our formula. Using the step function to limit the integration range, we get

$$e^{-r(T-t)} \, B \, K^\lambda \int_0^\infty \frac{e^{\lambda y - \frac{(y-m)^2}{2\Sigma^2}}}{\Sigma \sqrt{2\pi}} \, dy \tag{118}$$

We can get *Mathematica* to do the work of evaluating this. If you want to do it yourself, complete the square in the exponential and use the definition of the cumulative normal distribution function to present the answer.

```
Clear[r, y, K, B, m , Σ, λ]

value = Exp[-r (T - t)] B K^λ
    Integrate[1 / (Σ Sqrt[2 Pi]) Exp[λ y - (y - m) ^ 2 / (2 Σ^2)],
      {y, 0, Infinity}][[2]]
```

$$\frac{B \, E^{-r \, (-t+T) + m \, \lambda + \frac{\lambda^2 \, \Sigma^2}{2}} \, K^\lambda \, \sqrt{\Sigma^2} \, \left(1 + \mathrm{Erf}\left[\frac{m + \lambda \, \Sigma^2}{\sqrt{2} \, \sqrt{\Sigma^2}}\right]\right)}{2 \, \Sigma}$$

```
simpvalue = PowerExpand[value]
```

$$\frac{1}{2} \, B \, E^{-r \, (-t+T) + m \, \lambda + \frac{\lambda^2 \, \Sigma^2}{2}} \, K^\lambda \, \left(1 + \mathrm{Erf}\left[\frac{m + \lambda \, \Sigma^2}{\sqrt{2} \, \Sigma}\right]\right)$$

Now we load a Package where the cumulative normal distribution function is defined. See Chapter 7 for a discussion of this Package. For now we just want to relate the Error function **Erf** to the more familiar "N" function used in derivatives modelling. In *Mathematica* "N" is used for numerical evaluation, we shall use the term **Norm** for the cumulative normal distribution. This is defined in *Mathematica* within a Package defined in Chapter 7 - we just look at the definition for now:

```
Needs ["Derivatives`BlackScholes`"];
Norm[z]
```

$$\frac{1}{2}\left(1 + \text{Erf}\left[\frac{z}{\sqrt{2}}\right]\right)$$

The value of the cut-off power Call is then just

$$B K^{\lambda} e^{\frac{\Sigma^2 \lambda^2}{2} + m\lambda - r(T-t)} \text{Norm}\left(\frac{m}{\Sigma} + \lambda \Sigma\right) \tag{119}$$

Similarly the value of the cut-off power Put is given by

$$B K^{\lambda} e^{\frac{\Sigma^2 \lambda^2}{2} + m\lambda - r(T-t)} \text{Norm}\left(-\frac{m}{\Sigma} - \lambda \Sigma\right) \tag{120}$$

We can do a consistency check by adding together the Call and the Put - it is left as an exercise for you to see that the solutions add to the one we have given earlier for an option with payoff S^{λ}.

The Irritating Standard Notation!

It has become conventional to use a rather silly notation for these expressions when $\lambda = 0, 1$. When $\lambda = 0$ we define

$$d_2 = \frac{m}{\Sigma} = \frac{\log(\frac{S}{K}) + \left(r - q - \frac{\sigma^2}{2}\right)(T-t)}{\sigma \sqrt{T-t}} \tag{121}$$

The $\lambda = 1$ definition is

$$d_1 = \frac{m}{\Sigma} + \Sigma = \frac{\log(\frac{S}{K}) + \left(r - q + \frac{\sigma^2}{2}\right)(T-t)}{\sigma \sqrt{T-t}} \tag{122}$$

Binary Options

These are given by the cut-off power payoffs with $\lambda = 0$. The Binary Call therefore has value

$$B e^{-r(T-t)} \text{Norm}(d_2) \tag{123}$$

The Binary Put therefore has value

$$B e^{-r(T-t)} \text{Norm}(-d_2) \tag{124}$$

Put-Call parity for binaries takes the form that the sum of a Binary Call and a Binary Put with the same strike is just the value of a cash payout at expiry:

$$B e^{-r(T-t)} \tag{125}$$

Vanilla Calls and Puts

These are simple linear combinations of the cut-off power options. For a vanilla Call we have the payoff at expiry in the form

$$\text{Max}[S_T - K, 0] = S_T \, \theta(S_T - K) - K \, \theta(S_T - K) \tag{126}$$

whereas for a Put we have

$$\text{Max}[K - S_T, 0] = K \, \theta(K - S_T) - S_T \, \theta(K - S_T) \tag{127}$$

The value of a Call is therefore

$$K \, e^{\frac{\Sigma^2}{2} + m - r(T-t)} \, \text{Norm}\left(\frac{m}{\Sigma} + \Sigma\right) - K \, e^{-r(T-t)} \, \text{Norm}\left(\frac{m}{\Sigma}\right) \tag{128}$$

Simplification of the exponentials reduces this to

$$C = S \, e^{-q(T-t)} \, \text{Norm}(d_1) - K \, e^{-r(T-t)} \, \text{Norm}(d_2) \tag{129}$$

For a Put, we obtain, similarly,

$$P = K \, e^{-r(T-t)} \, \text{Norm}(-d_2) - S \, e^{-q(T-t)} \, \text{Norm}(-d_1) \tag{130}$$

This time Call-Put parity takes the form

$$C - P = S \, e^{-q(T-t)} - K \, e^{-r(T-t)} \tag{131}$$

Note that although we shall only consider Binaries and vanilla Calls/Puts in this particular form, *the methods we have developed allow the valuation of other exotics based on any linear combination of cut-off power options.*

Simple Barrier Options

Now suppose that there is a special value of S on which the option value is set to a given value. If we set this value to zero, we obtain a standard "knock-out" option. If we prescribe a "rebate" function if the asset price reaches this point, we need to impose a boundary condition along the barrier. Both of these cases fit into our model of boundary conditions at a finite point. Let's consider the case of "down-and-out" options, where the rebate at the lower barrier, $S = H$, is zero and there is a given payoff function defined for $S_T > H$. We set

$$H = K \, e^a \tag{132}$$

$$a = \log\left(\frac{H}{K}\right) \tag{133}$$

Then completing the square in the exponential functions gives us

$$V(S, t) = e^{-k_1 \tau} \int_a^\infty \frac{P(K \, e^y) \, e^{-\frac{(x-y+\tau(k_2-1))^2}{4\tau}}}{\sqrt{4 \pi \tau}} \, dy$$

$$- \; e^{-k_1 \tau + (a-x)(k_2-1)} \int_a^\infty \frac{P(K \, e^y) \, e^{-\frac{(2a-x-y+\tau(k_2-1))^2}{4\tau}}}{\sqrt{4 \pi \tau}} \, dy \qquad (134)$$

If the option is such that the barrier is below the strike, so that $a < 0$, then the first of these integrals is the solution in the absence of the barrier. The second term is then given by the first term with the corrections

(a) replace x by $2a - x$ within the solution,
(b) then multiply the solution by

$$e^{(a-x)(k_2-1)} \qquad (135)$$

In terms of standard financial variables, this just says that we make the replacement

$$\log\left(\frac{S}{K}\right) \to 2 \log\left(\frac{H}{K}\right) - \log\left(\frac{S}{K}\right) = \log\left(\frac{H^2}{K S}\right) \qquad (136)$$

equivalent to

$$S \to \frac{H^2}{S} \qquad (137)$$

and multiply by

$$e^{\left(\log\left(\frac{H}{K}\right) - \log\left(\frac{S}{K}\right)\right)(k_2-1)} = \left(\frac{H}{S}\right)^{k_2-1} = \left(\frac{H}{S}\right)^{\frac{2(r-q)}{\sigma^2}-1} \qquad (138)$$

For example, for a down-and-out call, with $H < K$, we have

$$C_{\text{do}} = S \, e^{-q(T-t)} \, \text{Norm}(d_1) - K \, e^{-r(T-t)} \, \text{Norm}(d_2)$$

$$- \left(\frac{H}{S}\right)^{\frac{2(r-q)}{\sigma^2}-1} \left(\frac{H^2 \, e^{-q(T-t)} \, \text{Norm}(y)}{S} - K \, e^{-r(T-t)} \, \text{Norm}\left(y - \sigma \sqrt{T-t}\right)\right) \qquad (139)$$

$$y = \frac{\log\left(\frac{H^2}{K S}\right) + \left(r - q + \frac{\sigma^2}{2}\right)(T - t)}{\sigma \sqrt{T - t}} \qquad (140)$$

A host of variants of this case can be considered:

(a) "in" variants, defined by in+out = Vanilla;
(b) Put variants;
(c) cases where $H > K$;
(d) the payment of a rebate when the barrier is reached.

We shall not go into a detailed analysis of the cases (a), (b), (c), each of which can be obtained by a minor modification of the arguments given. All the resulting formulae are spelt out in Chapter 8, in any case. However, we do need to explore case (d) - the payment of a rebate on knock-out.

Rebate Management

The introduction of a rebate requires some careful integral analysis. Suppose that we specify a boundary condition $u = g(\tau)$ along the line $x = a$. Then in addition to the solution we have given, there is an extra contribution u_R given by

$$u_R[x, \tau] = \int_0^\tau g(s) \, q(\tau - s) \, ds \qquad (141)$$

$$q(t) = \frac{1}{\pi} \int_0^\infty e^{-pt} \sin\left(\sqrt{p} \ (x - a)\right) dp \qquad (142)$$

Suppose that we consider rebate functions of the general form

$$\text{option value} = R(\tau) = R_0 \, e^{-\mu \tau} \qquad (143)$$

along the line $x = a$. Then g has the form

$$g(\tau) = A \, e^{B\tau} \qquad (144)$$

where

$$A = R_0 \, e^{\frac{1}{2}(k_2 - 1)a} = R_0 \left(\frac{H}{K}\right)^{\frac{1}{2}(k_2 - 1)} \qquad (145)$$

$$B = \frac{1}{4} (k_2 - 1)^2 + k_1 - \mu \qquad (146)$$

and we assume that $B > 0$. Then the rebate contribution to u is given by

$$u_R = \frac{A}{\pi} \int_0^\infty \int_0^\tau \sin\left(\sqrt{p} \ (x - a)\right) e^{(B+p)s - p\tau} \, ds \, dp \qquad (147)$$

The s integral can be done, leaving

$$u_R = \frac{A}{\pi} \int_0^\infty \frac{\sin\left(\sqrt{p} \ (x - a)\right)\left(e^{B\tau} - e^{-p\tau}\right)}{B + p} \, dp \qquad (148)$$

There are two terms in this integral. The first is given, for $x > a$, by

$$\frac{A}{\pi} \, e^{B\tau} \int_0^\infty \frac{\sin\left(\sqrt{p} \ (x - a)\right)}{B + p} \, dp = A \, e^{B\tau} \, e^{-\sqrt{B} \ (x-a)} \qquad (149)$$

This can be verified with *Mathematica*:

```
Integrate[Sin[Sqrt[p] (x - a)] / (B + p), {p, 0, Infinity}]
```

$$\text{If}\left[\text{Im}[a-x] == 0 \;\&\&\; \text{Arg}[B] \neq \pi,\right.$$

$$-E^{-\sqrt{B\,(a-x)^2}}\;\pi\,\text{Sign}[a-x],\quad \int_0^\infty \frac{\text{Sin}\left[\sqrt{p}\;(-a+x)\right]}{B+p}\,dp\right]$$

The second term, with a change of variables, can be written as

$$-\frac{2\,A}{\pi}\int_0^\infty \frac{p\,\sin(p\,(x-a))\,e^{-p^2\,\tau}}{p^2+B}\,dp$$

$$= -\frac{A}{\pi}\int_{-\infty}^\infty \frac{p\,\sin(p\,(x-a))\,e^{-p^2\,\tau}}{p^2+B}\,dp = -\frac{A}{\pi}\,\text{Im}[D]$$

(150)

where D is the Fourier integral

$$\int_{-\infty}^\infty \frac{e^{i\,p\,(x-a)-p^2\,\tau}\,p}{p^2+B}\,dp$$

(151)

This can now be evaluated as a convolution integral. The two terms in the convolution product are given by

```
InverseFourierTransform[Exp[-p^2 τ],
  p, y, FourierFrequencyConstant -> -1]
```

$$\frac{E^{-\frac{y^2}{4\,\tau}}}{2\,\sqrt{\pi}\,\sqrt{\tau}}$$

and

```
InverseFourierTransform[p / (p^2 + B),
  p, y, FourierFrequencyConstant -> -1]
```

$$-\frac{1}{2}\,I\,E^{-\sqrt{B}\;\text{Abs}[y]}\,\text{Sign}[y]$$

The form of this last term requires that we divide the convolution integration region into two (intermediate results are not shown):

```
piecea =
  Integrate[Exp[- (x - a - y) ^ 2 / (4 τ) - Sqrt[B] y], {y, 0, Infinity}][[2]];
```

```
pieceb = -Integrate[
    Exp[-(x - a - y)^2 / (4 t) + Sqrt[B] y], {y, -Infinity, 0}][[2]];
```

The imaginary part of the convolution integral is therefore given by

```
ImD = Simplify[2 Pi / (4 Sqrt[Pi t]) (piecea + pieceb)];
```

Adding up all the pieces, the total contribution from the rebate to the solution of the heat equation is therefore

```
u_R = Simplify[-A / Pi ImD + A * Exp[B * t] * Exp[-Sqrt[B] * (x - a)]]
```

$$\frac{1}{2} A E^{-\sqrt{B}\,(a+x-\sqrt{B}\,\tau)}$$

$$\left(E^{2\,a\,\sqrt{B}} + E^{2\,\sqrt{B}\,x} + E^{2\,\sqrt{B}\,x}\,\mathrm{Erf}\left[\frac{a - x - 2\,\sqrt{B}\,\tau}{2\,\sqrt{\tau}}\right] + E^{2\,a\,\sqrt{B}}\,\mathrm{Erf}\left[\frac{a - x + 2\,\sqrt{B}\,\tau}{2\,\sqrt{\tau}}\right] \right)$$

This rebate function looks a little complicated, so we had better check it - first the boundary condition:

```
Simplify[u_R /. x -> a]
```

$$A\,E^{B\,\tau}$$

We also check that the diffusion equation is satisfied:

```
Simplify[D[u_R, {x, 2}] - D[u_R, t]]
```

$$0$$

Finally we need to transform back to financial variables, and straighten out our notation. We do this in two steps. First we convert from x variables to S variables and replace the Error functions by the cumulative normal distribution. The rebate contribution to the option value is then:

$$R_0\,e^{-\mu\tau}\left(\frac{H}{S}\right)^{\frac{1}{2}(k_2-1)}\left(\left(\frac{S}{H}\right)^{\sqrt{B}}\mathrm{Norm}\left(\frac{\log(\frac{H}{S}) - 2\sqrt{B}\,\tau}{\sqrt{2\tau}}\right) + \left(\frac{H}{S}\right)^{\sqrt{B}}\mathrm{Norm}\left(\frac{\log(\frac{H}{S}) + 2\sqrt{B}\,\tau}{\sqrt{2\tau}}\right)\right) \quad (152)$$

Next, we re-parametrize the rebate by setting $\mu = 2\,\alpha/\sigma^2$, so that the rebate is written as

$$R_0\,e^{-\alpha\,(T-t)} \quad (153)$$

Then

$$B = \frac{1}{4}\,(k_2 - 1)^2 + k_1 - 2\,\alpha/\sigma^2 \quad (154)$$

For general α, and hence general $B > 0$, the rebate is given by

$$
R_0 \, e^{-\alpha (T-t)} \left(\frac{H}{S}\right)^{\frac{1}{2}\left(\frac{2(r-q)}{\sigma^2}-1\right)} \left(\left(\frac{S}{H}\right)^{\sqrt{B}} \mathrm{Norm}\left(\frac{\log(\frac{H}{S}) - \sqrt{B}\,\sigma^2\,(T-t)}{\sigma\sqrt{T-t}}\right)\right.
$$
$$
\left. + \left(\frac{H}{S}\right)^{\sqrt{B}} \mathrm{Norm}\left(\frac{\log(\frac{H}{S}) + \sqrt{B}\,\sigma^2\,(T-t)}{\sigma\sqrt{T-t}}\right)\right)
$$

(155)

If the rebate takes the form of a cash payment R_0 at maturity, the boundary condition is the discounted value of this, so that $\alpha = r$. Then this expressions simplifies considerably, since

$$
\sqrt{B} = \frac{1}{2}(k_2 - 1) = \frac{1}{2}\left(\frac{2(r-q)}{\sigma^2} - 1\right)
$$

(156)

The rebate is then

$$
R_0 \, e^{-r(T-t)} \left(\mathrm{Norm}\left(\frac{\log(\frac{H}{S}) - (r - q - \sigma^2/2)(T-t)}{\sigma\sqrt{T-t}}\right)\right.
$$
$$
\left. + \left(\frac{H}{S}\right)^{\left(\frac{2(r-q)}{\sigma^2}-1\right)} \mathrm{Norm}\left(\frac{\log(\frac{H}{S}) + (r - q - \sigma^2/2)(T-t)}{\sigma\sqrt{T-t}}\right)\right)
$$

(157)

Another important case is when the cash payment is paid when the barrier is reached - the "cash-at-hit" case. Then $\alpha = 0$ and B is more complicated, but our model still applies. This is the case considered in Chapter 8, where references to other approaches to the derivation are given.

Remarks on the Double Barrier

We can analyse the double barrier by similar methods, using the infinite sum of Green's functions we have derived. This leads quickly to one of the results presented by Kunitomo and Ikeda (1992), who derived their result within a probabilistic framework. It is an interesting exercise, which we leave to the reader, to spell out the result based on an infinite family of image points. Our detailed analysis of the double barrier, given in Chapter 12, will use, instead, the closed-form solution given by Geman and Yor (1996) based on Laplace transform methods.

A Financial Analogue of an Impedance Boundary Condition?

It should be by now clear that we can go on combining various techniques in partial differential equation theory to construct solutions for various initial ("payoff") and boundary ("barrier") conditions. When we run out of analytical steam, we can go to numerical methods. At this point, we have constructed financial interpretations of most of the classical solutions of the diffusion equation. People more interested in finance might object that this process is the wrong way around - we should really define problems of interest in finance and then develop the mathematics to solve them. However, it is also important to point out that most of the *basic* option-pricing problems were solved mathematically by people studying the heat equation about a hundred years ago. It is then a matter of applying a rather trivial change of variables to deduce many solutions to the Black-Scholes PDE. Some of these solutions have only been "discovered" very recently, using relatively complicated arguments based on probability and the risk-neutral expectation model. Of course, the translation of problems in finance to the state in which they can be solved by such elementary methods is an art in itself.

We have almost used up our family of classical solutions to the diffusion or heat equation. We have one left for which we have not yet constructed a financial interpretation - the solution to the problem with an impedance boundary condition (IBC). Problems similar to the one we have described were discussed as long ago as 1912, by Riemann, as discussed by Jeffreys and Jeffreys (1946), which contains a discussion of many methods of solving the diffusion problem.

The mapping of an important option-pricing problem into the diffusion equation with IBC was, so far as the author is aware, first given by Wilmott *et al* (1993), and is based on an analysis of certain types of Lookback option. As outlined in Chapter 3, a path-dependency variable is added in the form

$$J_n = \left(\int_0^t f(S(\tau), \tau)^n \, d\tau \right)^{1/n} \tag{158}$$

for various choices about n and f. The case of interest here is given by considering $n \to \infty$, with $f = S$, so that J represents the maximum attained along the path. Recall that suitable boundary conditions involving J must be added, and the payoff needs to be written in terms of it also. When the limit is taken the PDE contains no reference to J:

$$\frac{\partial V}{\partial t} + S(r - q)\frac{\partial V}{\partial S} + \frac{1}{2}\sigma^2 S^2 \frac{\partial^2 V}{\partial S^2} - rV = 0 \tag{159}$$

The quantity J is the maximum attained on the path, and appears in the boundary conditions and the payoff function. Let's take a closer look at the case of a Lookback Put. The analysis given here borrows heavily from the presentation by Wilmott *et al* (1993), but also shows explicitly how to use the Green's function for the impedance boundary condition to solve the problem. This is one approach to exercise 1 in Chapter 12 of Wilmott *et al* (1993)! The solution can be arrived at by some cunning guesswork, but we shall proceed by formally mapping the financial problem into the problem for the heat equation that we have already solved, and directly deriving the solution.

The initial condition for a Lookback Put is

$$P(S, J, T) = \text{Max}[J - S, 0] \tag{160}$$

The boundary conditions are, on $S = 0$,

$$P(0, J, t) = J\,e^{-r(T-t)} \tag{161}$$

On $S = J$, we have

$$\frac{\partial P}{\partial J} = 0 \tag{162}$$

Now we construct a similarity solution of the form

$$z = \frac{S}{J} \tag{163}$$

$$P(S, J, t) = J\,W(z, \ t) \tag{164}$$

The Black-Scholes equation becomes

$$\frac{\partial W}{\partial t} + z(r-q)\frac{\partial W}{\partial z} + \frac{1}{2}\sigma^2 z^2 \frac{\partial^2 W}{\partial z^2} - r W = 0 \qquad (165)$$

on the region $0 < z < 1$. The boundary and initial conditions are transformed according to

$$W(z, T) = \text{Max}[1 - z, 0] \qquad (166)$$

$$W(0, t) = e^{-r(T-t)} \qquad (167)$$

Now, on $z = 1$, we have

$$\frac{\partial W}{\partial z} = W \qquad (168)$$

So we see the IBC emerging. We now transform to the standard coordinates for the heat equation.

$$W(z, t) = e^{-\frac{1}{2}(k_2-1)x - \left(\frac{1}{4}(k_2-1)^2 + k_1\right)\tau} u(x, \tau) \qquad (169)$$

where τ is as usual, but

$$x = \log(z) \qquad (170)$$

We only have to consider the range

$$-\infty < x < 0 \qquad (171)$$

since the "option" *will* be exercised, and in that range the initial condition is

$$u = e^{\frac{1}{2}(k_2-1)x}(1 - e^x) \qquad (172)$$

The impedance boundary condition at $x = 0$ is now given by

$$\frac{\partial u}{\partial x} = \frac{1}{2}(k_2 + 1)u \qquad (173)$$

Thus we have mapped the Lookback Put problem into our IBC model with

$$\alpha = \frac{1}{2}(k_2 + 1) \qquad (174)$$

Let's get *Mathematica* to do the work for us. First the payoff in log variables:

```
f[y_] := Exp[1/2*(Subscript[k, 2] - 1)*y]*(1 - Exp[y])
```

We are to integrate this against the sum of these four functions **mone** etc.:

```
Clear[α];
aux[x_, y_, τ_, α_] := Exp[α (x+y+α*τ)];
nasty[x_, y_, τ_, α_] := Erf[(-x - y - 2*α*τ)/(2*Sqrt[τ])];
mone[x_, y_, τ_, α_] := 1/(E^((x - y)^2/(4*τ))*2*Sqrt[Pi]*Sqrt[τ]);
mtwo[x_, y_, τ_, α_] := 1/(E^((x + y)^2/(4*τ))*2*Sqrt[Pi]*Sqrt[τ]);
mthree[x_, y_, τ_, α_] := α*aux[x,y,τ,α];
mfour[x_, y_, τ_, α_] := -aux[x,y,τ,α]*α*nasty[x, y, τ, α];
```

Here is a reminder of the full fundamental solution for the diffusion equation with an IBC:

```
M[x_, y_, τ_, α_] := mone[x, y, τ, α] +
   mtwo[x, y, τ, α] + mthree[x, y, τ, α] + mfour[x, y, τ, α];
```

To isolate any problems, we split the integral into eight pieces:

```
fone[t_] := Exp[1 / 2 * (Subscript[k, 2] - 1) * y];
ftwo[t_] := -Exp[1 / 2 * (Subscript[k, 2] + 1) * y];
```

Six of them evaluate immediately into simple functions of types we have already encountered:

```
α = 1 / 2 (k₂ + 1);
```

```
int[1] =
Integrate[fone[y] * mone[x, y, τ, α], {y, -Infinity, 0}][[2]];
int[2] =
  Integrate[ftwo[y] * mone[x, y, τ, α], {y, -Infinity, 0}][[2]];
int[3] = Integrate[fone[y] * mtwo[x, y, τ, α], {y, -Infinity, 0}][[2]];
int[4] = Integrate[ftwo[y] * mtwo[x, y, τ, α], {y, -Infinity, 0}][[2]];
int[5] =
  Integrate[fone[y] * mthree[x, y, τ, α], {y, -Infinity, 0}][[2]];
int[6] = Integrate[ftwo[y] * mthree[x, y, τ, α], {y, -Infinity, 0}][[2]];
```

The Error function fools the symbolic integrator at first, e.g. (similarly for `int[8]`)

```
int[7] = Integrate[fone[y] * mfour[x, y, τ, α], {y, -Infinity, 0}]
```

$$-\alpha \int_{-\infty}^{0} E^{\alpha\,(x+y+\alpha\,\tau)+\frac{1}{2}\,y\,(-1+k_2)}\;\mathrm{Erf}\Big[\frac{-x-y-2\,\alpha\,\tau}{2\,\sqrt{\tau}}\Big]\,\mathrm{d}y$$

So we integrate by parts semi-manually (you can check that there is no contribution from the lower limit at minus infinity):

```
gone[y_] = Integrate[fone[y] * aux[x, y, τ, α], y];
gtwo[y_] = Integrate[ftwo[y] * aux[x, y, τ, α], y];
dmfour[x_, y_, τ_, α_] = D[-α * nasty[x, y, τ, α], y];
```

```
int[7] = -α * gone[0] * nasty[x, 0, τ, α] -
   Integrate[gone[y] * dmfour[x, y, τ, α], {y, -Infinity, 0}][[2]];
int[8] = -α * gtwo[0] * nasty[x, 0, τ, α] -
   Integrate[gtwo[y] * dmfour[x, y, τ, α], {y, -Infinity, 0}][[2]];
```

To get the result we add them all up and normalize for the valuation. On the way we check that the diffusion equation and IBC remain satisfied:

```
v = Simplify[Sum[int[aa], {aa, 1, 8}]];

Simplify[D[v, {x, 2}] - D[v, τ]]
```

   0

```
Simplify[Simplify[D[v, x] - α v] /. {x -> 0}]
```

   0

Normalizing:

```
w = Simplify[Exp[-(1 / 2) * (k₂ - 1) * x - (1 / 4 * (k₂ - 1) ^ 2 + k₁) * τ] * v];
```

We can simplify the presentation of the result somewhat if we set the yield to zero, so that $k_2 = k_1$. Finally, we multiply back up by J:

```
simpw = Simplify[ExpandAll[PowerExpand[w /. k₂ -> k₁]]]
```

$$\frac{1}{2\,k_1}$$

$$\left(E^{x-(x+\tau)\,k_1} \left(-1 + E^{(x+\tau)\,k_1} - \text{Erf}\left[\frac{x + \tau - \tau\,k_1}{2\,\sqrt{\tau}} \right] + E^{(x+\tau)\,k_1}\,\text{Erf}\left[\frac{x + \tau + \tau\,k_1}{2\,\sqrt{\tau}} \right] \right) + \right.$$

$$\left. \left(-E^x + E^{-\tau\,k_1} - E^{-\tau\,k_1}\,\text{Erf}\left[\frac{x - \tau + \tau\,k_1}{2\,\sqrt{\tau}} \right] + E^x\,\text{Erf}\left[\frac{x + \tau + \tau\,k_1}{2\,\sqrt{\tau}} \right] \right) k_1 \right)$$

Finally, we multiply back up by J, and we package in terms of the Norm function, obtaining

$$-J\,e^x\,\text{Norm}\left(\frac{-x - \tau - \tau\,k_1}{\sqrt{2\,\tau}} \right) + J\,e^{-\tau\,k_1}\,\text{Norm}\left(\frac{-x + \tau - \tau\,k_1}{\sqrt{2\,\tau}} \right)$$

$$+ \frac{1}{k_1}\left(J\,e^x\,\text{Norm}\left(\frac{x + \tau + \tau\,k_1}{\sqrt{2\,\tau}} \right) - J\,e^{x(1-k_1)-\tau\,k_1}\,\text{Norm}\left(\frac{x + \tau - \tau\,k_1}{\sqrt{2\,\tau}} \right) \right) \tag{175}$$

One further clean-up helps to recognize this - we set

$$d' = \frac{x + \tau + \tau\,k_1}{\sqrt{2\,\tau}} = \frac{\log\left(\frac{S}{J}\right) + \left(\frac{\sigma^2}{2} + r\right)(T - t)}{\sigma\,\sqrt{T - t}} \tag{176}$$

and obtain for the value

$$-S \, \text{Norm}(-d') + J \, e^{-r(T-t)} \, \text{Norm}\left(-d' + \sigma \sqrt{T-t}\right)$$

$$+ \frac{\sigma^2}{2r} \left(S \, \text{Norm}(d') - S \, e^{-r(T-t)} \left(\frac{J}{S}\right)^{\frac{2r}{\sigma^2}} \text{Norm}\left(d' - \frac{2r}{\sigma} \sqrt{T-t}\right) \right) \tag{177}$$

This is the formula for a standard Lookback Put, as given by Conze and Viswanathan (1991).

Our Initial Plan of Attack

We now have a collection of example solution methods for representatives of several families of options:

(a) simple log, power and related solutions;
(b) Binary option solutions;
(c) Call and Put solutions;
(d) Barrier solutions;
(e) Lookback solutions.

These will form the topics of the next five chapters respectively. Before proceeding to a detailed analysis of each type, we look briefly at how the diffusion equation representation may be generalized to handle:

(a) discrete dividends;
(b) time-dependent interest rates;
(c) coupon payments.

Inclusion of (a) is important for a wide range of instruments - the inclusion of (b) and (c) is particularly relevant to convertible bond modelling.

4.7 Dividends, Term Structures and Coupons

In a single-factor model of a convertible bond, the convertible bond differential equation is taken to be the following:

$$\frac{1}{2} S^2 \frac{\partial^2 V}{\partial S^2} \sigma^2 - r(t) V + \frac{\partial V}{\partial t} + (r(t) S - \text{Div}(S, t)) \frac{\partial V}{\partial S} + K(S, t) = 0 \tag{178}$$

This is the natural time-dependent version of what we have presented in Chapter 3. If we try the change of variables used in the previous analysis of European options, w soon find that this fails unless r and q are constant. This is not what we want for the case of a convertible bond, or for any option on an asset where the dividends are time-dependent, as they are when they are discrete. The change of variables that does the job, at the expense of some slight complication with trigger levels, if they are present, is one which is part of a family of tricks set out by Harper (1994). The following discussion standardizes the problem in two stages. This analysis is useful both for simple single-factor convertibles, with deterministic time-dependent interest rates, and, with no coupons, for a standard option on an asset where the dividends are discrete.

In particular, we set, if E is a base value of the underlying (to be chosen appropriately), and T is the expiry time, t the current time,

$$x = \log\left(\frac{S}{E}\right)$$

$$\tau = \frac{1}{2}\,\sigma^2\,(T - t)$$

(179)

If $q(\tau)$ is the τ-dependent function representing the continuously compounded dividend yield $DY(t)$ (this may contain discrete pieces),

$$k_1[\tau] = \frac{2\,(R\,\tau)}{\sigma^2} = \frac{2\,(r\,t)}{\sigma^2}$$

$$k_2[\tau] = \frac{2\,(R\,\tau - q\,\tau)}{\sigma^2} = \frac{2\,(r\,t - DY\,t)}{\sigma^2}$$

(180)

Then we obtain

$$\frac{\partial V}{\partial t} + k_1[\tau]\,V = \frac{\partial^2 V}{\partial x^2} + (k_2[\tau] - 1)\,\frac{\partial V}{\partial x} + \frac{2}{\sigma^2}\,K$$

(181)

Thus far the change of variables has proceeded as for the European options discussed previously. The final part of the previous analysis does not work for the present problem unless $k_2[\tau]$ is constant. To treat the time-dependent case, let

$$z = x + F(\tau)$$

(182)

and set

$$V(x,\,\tau) = u(z,\,\tau)\,e^{-B(\tau)}$$

(183)

Making the required change of variable leads to

$$u\left(k_1[\tau] - \frac{\partial B}{\partial\tau}\right) + \frac{\partial u}{\partial t} = \frac{\partial^2 u}{\partial z^2} + \frac{2\,K e^{B(\tau)}}{\sigma^2} + \left(-\frac{\partial F}{\partial\tau} + k_2[\tau] - 1\right)\frac{\partial V}{\partial x}$$

(184)

so we can reduce the problem by making the choices

$$k_1[\tau] = \frac{\partial B}{\partial\tau}$$

(185)

$$k_2[\tau] - 1 = \frac{\partial F}{\partial\tau}$$

(186)

whence we get

$$\frac{\partial u}{\partial t} = \frac{\partial^2 u}{\partial z^2} + Q(\tau)$$

(187)

where the diffusion equation now has a "source term" Q representing the coupon payments in these coordinates:

$$Q(\tau) = \frac{2\,K\,e^{B(\tau)}}{\sigma^2}$$

(188)

The quantities B and F are like integrating factors (B is precisely this for the straight bond). The general plan is therefore:

(1) Integrate to obtain F and B;
(2) Solve the PDE given the coupon payments (and for a CB, call/put/exercise conditions), using any suitable analytical or numerical scheme.

In practice we need some supplementary analysis to understand how to complete these steps, and to handle the boundary condition arising from considering the straight bond.

Note how we have made yet another link with the heat equation in a traditional form. The coupon payments in a bond are like a source of body heat for the heat equation.

Solution of the Auxiliary Conditions

A simplified form of the CB equation applies to the lower boundary condition, where we have a straight bond valuation governed by the equations

$$\frac{\partial u}{\partial t} = Q(\tau)$$

$$k_1[\tau] = \frac{\partial B}{\partial \tau}$$

(189)

Integration of the latter of these two relations gives

$$B(\tau) = \int_0^\tau k_1[p]\,dp = \int_t^T r(s)\,ds$$

(190)

Then integration of the fomer, and conversion back to our original coordinates, give

$$V_0 = e^{-\int_t^T r(s)\,ds}\left(Z + \int_t^T K[s]\,e^{\int_s^T r(p)\,dp}\,ds\right)$$

(191)

Now the coupon payments are typically discrete, so we write, accordingly,

$$K[t] = \sum_{t \le t_i \le T} K_i\,\delta(t - t_i)$$

(192)

representing the coupon payments as a general cash flow of payments K_i at times t_i.

$$V_0 = e^{-\int_t^T r(s)\,ds}\left(Z + \sum_{t \le t_i \le T} K_i\,e^{\int_{t_i}^T r(p)\,dp}\right)$$

(193)

This boils down to

$$V_0 = Ze^{-\int_t^T r(s)\,ds} + \sum_{t \le t_i \le T} K_i\,e^{-\int_t^{t_i} r(p)\,dp}$$

(194)

If the interest rate were constant at a value r_0, this would simplify further to

$$V_0 = Z e^{-r_0(T-t)} + \sum_{t \le t_i \le T} K_i \, e^{-r_0(t_i-t)} \tag{195}$$

but we do not assume this here. In general, we require knowledge of, or an estimate of, the integrals of the form

$$\int_{t_i}^{T} r(p) \, d p \tag{196}$$

and this must be based on yield-curve information. When we come to solve the full CB partial differential equation, the effect of the coupon payments is to define jump conditions for the value. If we integrate the PDE in time through a coupon payment date τ_i with payment K_i, then in a neighbourhood of the coupon payment date, the source to the diffusion equation is given by

$$Q(\tau) = K_i \, e^{B(\tau_i)} \, \delta(\tau - \tau_i) \tag{197}$$

So the valuation jump condition takes the form

$$u(\tau_i{}^+) = u(\tau_i{}^-) + K_i \, e^{B(\tau_i)} \tag{198}$$

The F function and Dividend Jump Conditions

The integration leading to the F function takes place and generates further jump conditions whenever there is a discrete dividend payment. If B has been determined from yield-curve data, and the dividend yield function $q(\tau)$ has been specified, the equation

$$k_2[\tau] - 1 = \frac{\partial F}{\partial \tau} \tag{199}$$

can be solved as

$$F(\tau) = -\tau + B(\tau) - \frac{2 \int_0^{\tau} q(s) \, d s}{\sigma^2} \tag{200}$$

In practice, we would imagine q as being composed of discrete values of the yield isolated at dividend payment dates τ_i.

$$F(\tau) = -\tau + B(\tau) - \sum_{0 \le t_i \le \tau} q_i \tag{201}$$

In some ways, that is all there is to it, but it is useful to re-interpret this result in terms of jump conditions on the valuation. Just before (in τ time, in reality just after) a dividend payment date τ_I, the value of F is

$$F(\tau_I -) = -\tau_I + B(\tau_I) - \sum_{0 \le t_i \le \tau_I -} q_i \tag{202}$$

whereas just after it is

$$F(\tau_I +) = -\tau_I + B(\tau_I) - \sum_{0 \le t_i \le \tau_I +} q_i \tag{203}$$

so the difference is given by

$$F(\tau_I +) = F(\tau_I -) - q_I \tag{204}$$

On the valuation, in (x, τ) coordinates, this leads to

$$V(x, \tau_I +) = V(x - q_I, \tau_I -) \tag{205}$$

and in (S, t) coordinates, bearing in mind the time reversal,

$$V(S, t_I -) = V(S\,e^{-q_i}, t_I +) \tag{206}$$

so we recover the condition that the valuation be continuous across the dividend payment date, with the stock price being adjusted downwards based on the dividend payment. This could have been taken as a starting point assumption on financial grounds, but it is reassuring to see it arise naturally from the differential equation.

Chapter 4 Bibliography

Conze, A. and Viswanathan, 1991, Path dependent options: the case of lookback options, *Journal of Finance*, 46, p. 1893.

Geman, H. and Yor, M., 1996, Pricing and hedging double-barrier options: a probabilistic approach, *Mathematical Finance*, 6, p. 365.

Harper, J., 1994, Reducing parabolic partial differential equations to canonical form, *European Journal of Applied Mathematics*, 5, p. 159.

Hull, J., 1996, *Options, Futures and Other Derivatives*, 3rd edition, Prentice-Hall.

Jeffreys, H. and Jeffreys, B.S., 1946, *Methods of Mathematical Physics* (3rd edition, 1972), Cambridge University Press.

Kunitomo, N. and Ikeda, M., 1992, Pricing options with curved boundaries, *Mathematical Finance*, 2, p. 275.

Rubenstein, M. and Reiner, E., 1991, Breaking down the barriers, *RISK Magazine*, September.

Wilmott, P., Dewynne, J. and Howison, S., 1993, *Option Pricing - Mathematical Models and Computation*, Oxford Financial Press.

Chapter 5.
Log and Power Contracts

Simple Analytical Models Explored

5.1 Introduction

It has become rather traditional to begin a discussion of option pricing by starting with the vanilla European Calls and Puts. Other instruments are generally labelled as "exotic". In fact, as you will have seen in Chapter 4, there are simpler solutions to the Black-Scholes equation than those pertaining to the standard Calls and Puts. We shall begin our discussion by investigating these simpler instruments. Mathematically, it makes sense to investigate the simpler cases first. We can get a feeling for the behaviour of options without needing to introduce any "special functions". Rather more importantly, we shall not allow ourselves to be drawn into any rash generalizations or inferences from the vanilla European case by prematurely focusing on those cases. It is a good idea early on to dispel, for example, any notion of Δ as being the probability of anything.

In this chapter, we shall investigate the simple log and power solutions of the Black-Scholes equation. In Chapter 6 we shall introduce the next most complex case, that of Binary options. That will require an appreciation of how to compute the cumulative normal distribution. Then, in Chapter 7, we shall turn to the vanilla European cases.

The simple contracts we shall consider here are certainly not devoid of financial interest. The reader is referred to the excellent discussion by Neuberger (1990). Such contracts are attracting increasing attention as a trading instrument (see e.g. Boussard, 1997). Notably, as we shall see, the hedge position for a log contract is independent of volatility.

5.2 The Log Contract

This was found in Chapter 4 by the use of elementary separation of variables applied to the time-dependent Black-Scholes equation. The value of the contract maturing at time T is given by

$$V = Z\left(\log(S/K) + \left(r - q - \frac{\sigma^2}{2}\right)(T - t)\right)e^{-r(T-t)} \tag{1}$$

Its value at maturity ($t = T$) is

$$V = Z\log(S/K) \tag{2}$$

This is what is known as the "log contract" - it represents an instrument that pays the log of the asset price at maturity. In what follows we shall set $Z = 1$. This is perhaps the simplest interesting solution to the Black-Scholes equation. Its form can be simplified further by working with forward prices, but we shall keep it in the form given. Let's define the *Mathematica* version of this function, defined in terms of time to maturity, and plot it as a function of price and time:

```
LogContract[S_, K_, r_, q_, σ_, t_] :=
(Log[S/K] + (r - q - σ^2/2)*t) Exp[-r t]

Plot3D[LogContract[S, 10, 0.1, 0, 0.2, t],
  {S, 5, 15}, {t, 0, 2}, AxesLabel -> {"S", "t", ""}];
```

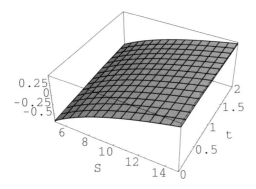

5.3 The Greeks for a Log Contract

Let's compute some symbolic forms for the value and Greeks of a log contract. It is critical that you appreciate what is happening here - *Mathematica* is doing all the differentiation required to evaluate the Greeks. (If you run this notebook in electronic form you should note that the style of the output has been converted to that of a numbered equation.)

```
V = LogContract[S,K,r,q,σ,t];
Δ = LogContractDelta[S_, K_, r_, q_, σ_, t_] =
D[V, S]; TraditionalForm[Δ]
```

$$\frac{e^{-rt}}{S} \qquad\qquad\qquad\qquad (3)$$

```
Γ = LogContractGamma[S_, K_, r_, q_, σ_, t_] =
D[V, {S,2}]; TraditionalForm[Γ]
```

$$-\frac{e^{-rt}}{S^2} \qquad\qquad\qquad\qquad (4)$$

```
θ = LogContractTheta[S_, K_, r_, q_, σ_, t_] =
-Simplify[D[V, t]]; TraditionalForm[θ]
```

$$\frac{1}{2} e^{-rt} \left(2 r \log\left(\frac{S}{K}\right) - (rt-1)(\sigma^2 + 2q - 2r) \right) \tag{5}$$

```
ρ = LogContractRho[S_, K_, r_, q_, σ_, t_] =
Simplify[D[V, r]]; TraditionalForm[ρ]
```

$$-\frac{1}{2} e^{-rt} t \left(-t\sigma^2 - 2qt + 2rt + 2\log\left(\frac{S}{K}\right) - 2 \right) \tag{6}$$

```
Λ = LogContractVega[S_, K_, r_, q_, σ_, t_] =
Simplify[D[V, σ]]; TraditionalForm[Λ]
```

$$-e^{-rt} t \sigma \tag{7}$$

Check of Differential Constraints

To make sure nothing has gone awry, we confirm the differential constraints. First the Black-Scholes equation, expressed in terms of Greeks:

```
Simplify[θ + S (r - q) Δ + 1 / 2 σ^2 S^2 Γ - r V]
```

```
0
```

Next we check the Vega-Gamma and Rho-Delta constraints. We do them simultaneously:

```
Simplify[{Λ - S^2 σ t Γ, ρ + t (V - S Δ)}]
```

```
{0, 0}
```

Observations about Δ for a Log Contract

Let's review the form of the Δ parameter. Remember that this is the amount of underlying needed to construct a risk-neutral portfolio. It is both a sensitivity and a hedge parameter. It is given by

```
TraditionalForm[Δ]
```

$$\frac{e^{-rt}}{S}$$

The first important observation about this quantity is that it is unbounded above. Hypothetically, we can require an arbitrarily large Δ if S is arbitrarily small. There is no condition that

$$|\Delta| < 1 \tag{8}$$

This is one good reason why one should *never* regard Δ as a kind of probability! Second, the hedge parameter depends only on the asset price and the time. Indeed, if we work instead in terms of forward prices:

$$S_F = S \; e^{(r-q)t}$$

(9)

The contract form simplifies to

$$V_F = V \, e^{rt} \; = \; \log(S_F) - \frac{\sigma^2}{2} t$$

(10)

and in these terms the Delta is time-independent.

$$\Delta_F = \frac{\partial V_F}{\partial S_F} = \frac{1}{S_F}$$

(11)

However we view the contract, it is clear that the hedge strategy is stable - there is essentially no material time-dependence in this hedge position.

The next observation about Δ is that it is independent of the volatility, σ. One of the easiest ways to mis-price a contract is to get the wrong volatility. Indeed, the question "What is the right volatility?" is not simply answered. However, a special property of a log contract is that the hedge position is independent of the volatility. So even if we use the "wrong" volatility, we shall not be mis-hedged.

Neither of these properties holds for vanilla European contracts. The hedge parameter Δ is then strongly dependent on both time and volatility.

5.4 Implied Volatility for a Log Contract

The "implied volatility" is that volatility which is consistent with the market price and the model of the price, other variables assumed to be known. It is instructive to ask *Mathematica* to solve this - the results at first seem rather peculiar. We express the answer in percentage terms:

```
ImpliedVolLogContract[MktPrice_, S_, K_, r_, q_, t_] := 100*σ /.Solve[(-
Log[S/K] + (r - q - σ^2/2)*t) Exp[-r t]==MktPrice, σ]
```

Here is a symbolic solution - there are two solutions as the equation involves σ^2:

```
ImpliedVolLogContract[P, S, K, r, q, t]
```

$$\left\{ - \frac{100 \sqrt{2} \sqrt{-\mathrm{E}^{rt} P - q t + r t + \mathrm{Log}\left[\frac{S}{K}\right]}}{\sqrt{t}} \, , \right.$$

$$\left. \frac{100 \sqrt{2} \sqrt{-\mathrm{E}^{rt} P - q t + r t + \mathrm{Log}\left[\frac{S}{K}\right]}}{\sqrt{t}} \right\}$$

Mathematica has, somewhat irritatingly, put the answer in complex form! This is more than a *Mathematica* foible. Here is a sample numerical value, based on $S = K = 1$, $r = 0.1$, $q = 0$ and one year to maturity. *If the price is 2 currency units there is no real solution:*

```
ImpliedVolLogContract[2, 1, 1, 0.1, 0, 1]
```

> {0. - 205.443 I, 0. + 205.443 I}

Let's define a function which selects out solutions that are real and positive, and get some sample results:

```
Select[ImpliedVolLogContract[0.5, 1, 1, 0.1, 0, 1], (Im[#] == 0 && # > 0 &)]
```

> {}

```
Select[ImpliedVolLogContract[0.09048, 1, 1, 0.1, 0, 1],
  (Im[#] == 0 && # > 0 &)]
```

> {0.287588}

```
Select[ImpliedVolLogContract[0.0, 1, 1, 0.1, 0, 1], (Im[#] == 0 && # > 0 &)]
```

> {44.7214}

```
Select[ImpliedVolLogContract[-0.5, 1, 1, 0.1, 0, 1],
  (Im[#] == 0 && # > 0 &)]
```

> {114.244}

We get either no solution or a rapidly varying function of the market price. It is a straightforward matter to work out for what market prices there is a real solution. The maximum allowed market price is given by

```
pcrit[S_, K_, r_, q_, t_] :=
  p /. Solve[q - r + Exp[r t] p / t - Log[S / K] / t == 0, p][[1]]
```

In our example, we have

```
pc = pcrit[1, 1, 0.1, 0, 1]
```

> 0.0904837

Let's plot the implied volatility as a function of the market price of the log contract:

```
Plot[Select[ImpliedVolLogContract[p, 1, 1, 0.1, 0, 1],
(Im[#] == 0 && # > 0 &)][[1]], {p, -3, pc}];
```

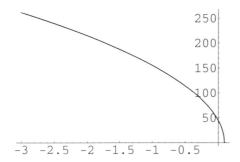

So the implied volatility fails to exist if the market price is greater than **pcrit**. For market prices slightly less than **pcrit**, the implied volatility is a highly unstable function of the market price. Let's make a table of values for implied volatility with the price varying from the critical value down to the critical value minus 0.001:

```
Table[{pc - k / 1000,
  Select[ImpliedVolLogContract[pc - k / 1000, 1, 1, 0.1, 0, 1],
  (Im[#] == 0 && # > 0 &)][[1]]}, {k, 1, 10}]

  {{0.0894837, 4.70143}, {0.0884837, 6.64882},
   {0.0874837, 8.14311}, {0.0864837, 9.40285},
   {0.0854837, 10.5127}, {0.0844837, 11.5161}, {0.0834837, 12.4388},
   {0.0824837, 13.2976}, {0.0814837, 14.1043}, {0.0804837, 14.8672}}
```

So even our simplest possible log contract exhibits some of the implied volatility pathology referred to in the introduction. However, the log contract is special in that it is one of the very few solutions to the Black-Scholes equation where there is a simple explicit formula for the implied volatility.

5.5 The Power Contract

The value of a general power contract is given by

$$V = Z\left(\frac{S}{K}\right)^{\lambda} e^{\left(\left(-\frac{\sigma^2}{2} - q + r\right)\lambda + \frac{\lambda^2 \sigma^2}{2} - r\right)(T-t)} \tag{12}$$

The expiry value of this option, when $t = T$, is given by

$$V = Z\left(\frac{S}{K}\right)^{\lambda} \tag{13}$$

In what follows we set $Z = 1$, as before. First we need the *Mathematica* definition:

```
PowerContract[S_, K_, r_, q_, σ_, t_, λ_] :=
(S/K)^λ*Exp[((-(σ^2/2) - q + r)*λ + (λ^2*σ^2)/2 - r)*t]
```

Recall that there are some notable special cases, given by $\lambda = 0,1$. The first is just a cash payment at maturity.

```
PowerContract[S, K, r, q, σ, t, 0]
```

$$E^{-rt}$$

When $\lambda = 1$ we just get the yield-adjusted asset price:

```
PowerContract[S, K, r, q, σ, t, 1]
```

$$\frac{E^{-qt}\,S}{K}$$

This is time-independent when the yield is zero:

```
Plot3D[PowerContract[S, 10, 0.1, 0, 0.2, t, 1],
  {S, 5, 15}, {t, 0, 2}, AxesLabel -> {"S", "t", ""}];
```

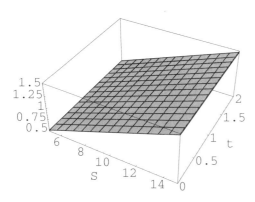

Other special solutions correspond to steady-state American Puts and Calls. More generally, we have profiles as shown, for λ strongly positive or negative. Here is a plot for $\lambda > 0$ - you should explore other values for yourself.

```
Plot3D[PowerContract[S, 10, 0.1, 0, 0.2, t, 4],
  {S, 5, 15}, {t, 0, 2}, AxesLabel -> {"S", "t", ""}];
```

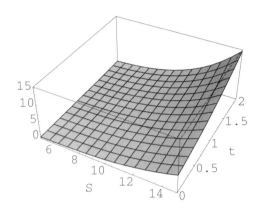

5.6 The Greeks for a Power Contract

Let's compute some symbolic forms for the value and Greeks of a power contract. Note that we are using the same symbols as for the log case. You should check for yourself that the identities relating the Greeks are satisfied.

```
V = PowerContract[S,K,r,q,σ,t,λ];
Δ = PowerContractDelta[S_, K_, r_, q_, σ_, t_, λ_] =
Simplify[D[V, S]]; TraditionalForm[Δ]
```

$$\frac{e^{\frac{1}{2}t(2r(\lambda-1)+\lambda((\lambda-1)\sigma^2-2q))}\left(\frac{S}{K}\right)^{\lambda}\lambda}{S} \tag{14}$$

```
Γ = PowerContractGamma[S_, K_, r_, q_, σ_, t_, λ_] =
Simplify[D[V, {S,2}]]; TraditionalForm[Γ]
```

$$\frac{e^{\frac{1}{2}t(2r(\lambda-1)+\lambda((\lambda-1)\sigma^2-2q))}\left(\frac{S}{K}\right)^{\lambda}(\lambda-1)\lambda}{S^2} \tag{15}$$

```
θ = PowerContractTheta[S_, K_, r_, q_, σ_, t_, λ_] =
-Simplify[D[V, t]]; TraditionalForm[θ]
```

$$-\frac{1}{2}e^{\frac{1}{2}t(2r(\lambda-1)+\lambda((\lambda-1)\sigma^2-2q))}\left(\frac{S}{K}\right)^{\lambda}(2r(\lambda-1)+\lambda((\lambda-1)\sigma^2-2q)) \tag{16}$$

```
ρ = PowerContractRho[S_, K_, r_, q_, σ_, t_, λ_] =
Simplify[D[V, r]]; TraditionalForm[ρ]
```

$$e^{\frac{1}{2}t(2r(\lambda-1)+\lambda((\lambda-1)\sigma^2-2q))}\left(\frac{S}{K}\right)^{\lambda}t(\lambda-1) \tag{17}$$

```
Λ = PowerContractVega[S_, K_, r_, q_, σ_, t_, λ_] =
Simplify[D[V, σ]]; TraditionalForm[Λ]
```

$$e^{\frac{1}{2}t(2r(\lambda-1)+\lambda((\lambda-1)\sigma^2-2q))}\left(\frac{S}{K}\right)^{\lambda}t(\lambda-1)\lambda\sigma \qquad (18)$$

5.7 Power Project

The behaviour of the Greeks, and the hedge parameters in particular, and of the implied volatility, for power contracts exhibits all kinds of weirdness. It is left to the reader to explore the various cases. You should consider separately the cases:

(a) $\lambda < 0$;
(b) $\lambda = 0$;
(c) $0 < \lambda < 1$;
(d) $\lambda = 1$;
(e) $\lambda > 1$.

In particular, find a situation where the implied volatility has infinitely many values!

Chapter 5 Bibliography

Boussard, E., 1997, *Trading and Hedging Volatility*, Presentation at Derivatives '97, Risk Conference, Brussels, February 1997.

Neuberger, A., 1990, The log contract and other power contracts, Chapter 7 in *The Handbook of Exotic Options: Instruments, Analysis and Applications*, ed. I. Nelken, Irwin.

Chapter 6.
Binary Options and the Normal Distribution

When Δ Becomes a Delta Function, Hedging Gets Nasty, and Continued Fractions Make an Elegant Appearance

6.1 Introduction

In Chapter 4, after finding the simpler log and power solutions of the Black-Scholes PDE, we introduced general methods for solving the PDE with a given initial condition. We met some examples of Binary options, based on a simple bet that pays off a fixed amount at maturity if the stock price is above or below a certain value. There are several variations on this theme, some of which are described by Hull (1996), and a substantial catalogue is given by Rubenstein and Reiner (1991). These include cases where the cash is paid when the strike is hit, or if the amount paid at maturity is based on the stock price.

We shall confine the discussion to a careful consideration of the simple Binary Call and Put defined in Section 4.6, where an amount B, or nothing, is paid at maturity based on whether the asset price at maturity is above or below a certain strike level K.

This simple case alone will raise issues of financial and mathematical importance. In particular, we need to address:

(a) the difficulties in hedging binary options;
(b) how to compute the cumulative normal distribution.

Many other cases can be valued using a combination of the rebate solutions and cut-off power calls/puts discussed previously.

6.2 Binary Cash Options Defined and Visualized

We shall need a model of the cumulative normal distribution. This is usually denoted by "N" in standard derivatives texts, but in *Mathematica* the **N** function is reserved for purely numerical modelling. So we shall use the term "Norm", as before, and define it in terms of the Error function, for which *Mathematica* has both a symbolic characterization and a high-precision numerical method for evaluation. At the end of this chapter, in Section 6.5, we shall consider other numerical techniques for estimating the cumulative normal distribution function. So we define

```
Norm[(z_)?NumberQ] := N[0.5*Erf[z/Sqrt[2]] + 0.5];
Norm[x_] := (1 + Erf[x/Sqrt[2]])/2
```

Using the normal distribution, we can define the formulae for Binary Calls and Puts. We do it in two stages - first we introduce the d_2 function defined in Chapter 4:

```
dtwo[S_, σ_, K_, t_, r_, q_] :=
 ((r - q)*t + Log[S/K])/(σ*Sqrt[t]) - (σ*Sqrt[t])/2;
```

The valuation of the simple Binary Call and Put is then

```
BinaryCall[B_, S_, K_, σ_, r_, q_, t_] :=
B*Exp[-r*t]*Norm[dtwo[S, σ, K, t, r, q]];
BinaryPut[B_, S_, K_, σ_, r_, q_, t_] :=
B*Exp[-r*t]*Norm[-dtwo[S, σ, K, t, r, q]];
```

```
Plot3D[BinaryCall[1, S, 100, 0.2, 0.1, 0, t],
 {S, 90, 110}, {t, 0.0001, 1}, PlotPoints -> 40];
```

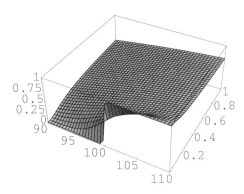

```
Plot3D[BinaryPut[1, S, 100, 0.2, 0.1, 0, t],
 {S, 90, 110}, {t, 0.0001, 1}, PlotPoints -> 40];
```

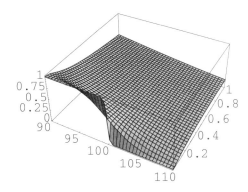

The striking feature of these plots is the fact that the slope becomes vertical at the strike as maturity is approached, and elsewhere becomes zero. This fact plays havoc with the Greeks, as we shall see in Section 6.4.

6.3 Greeks for Binary Cash Options

```
Δ = BinaryCallDelta[B_, S_, K_, σ_, r_, q_, t_] =
D[BinaryCall[B, S, K, σ, r, q, t], S]; TraditionalForm[Δ]
```

$$\frac{B\, e^{-\frac{1}{2}\left(\frac{(r-q)\,t+\log\left(\frac{S}{K}\right)}{\sqrt{t}\,\sigma}-\frac{\sqrt{t}\,\sigma}{2}\right)^2-r\,t}}{\sqrt{2\pi}\,S\sqrt{t}\,\sigma} \tag{1}$$

```
Γ = BinaryCallGamma[B_, S_, K_, σ_, r_, q_, t_] =
D[BinaryCall[B, S, K, σ, r, q, t], {S, 2}];
TraditionalForm[Γ]
```

$$-\frac{e^{-\frac{1}{2}\left(\frac{(r-q)\,t+\log\left(\frac{S}{K}\right)}{\sqrt{t}\,\sigma}-\frac{\sqrt{t}\,\sigma}{2}\right)^2-r\,t}\left(\frac{(r-q)\,t+\log\left(\frac{S}{K}\right)}{\sqrt{t}\,\sigma}-\frac{\sqrt{t}\,\sigma}{2}\right)B}{\sqrt{2\pi}\,S^2\,t\,\sigma^2} - \frac{e^{-\frac{1}{2}\left(\frac{(r-q)\,t+\log\left(\frac{S}{K}\right)}{\sqrt{t}\,\sigma}-\frac{\sqrt{t}\,\sigma}{2}\right)^2-r\,t}\,B}{\sqrt{2\pi}\,S^2\,\sqrt{t}\,\sigma} \tag{2}$$

```
θ = BinaryCallTheta[B_, S_, K_, σ_, r_, q_, t_] =
-D[BinaryCall[B, S, K, σ, r, q, t], t]; TraditionalForm[θ]
```

$$\frac{1}{2}\,B\,e^{-r\,t}\,r\left(\text{erf}\left(\frac{\frac{(r-q)\,t+\log\left(\frac{S}{K}\right)}{\sqrt{t}\,\sigma}-\frac{\sqrt{t}\,\sigma}{2}}{\sqrt{2}}\right)+1\right) - $$
$$\frac{B\,e^{-\frac{1}{2}\left(\frac{(r-q)\,t+\log\left(\frac{S}{K}\right)}{\sqrt{t}\,\sigma}-\frac{\sqrt{t}\,\sigma}{2}\right)^2-r\,t}\left(\frac{r-q}{\sqrt{t}\,\sigma}-\frac{\sigma}{4\sqrt{t}}-\frac{(r-q)\,t+\log\left(\frac{S}{K}\right)}{2\,t^{3/2}\,\sigma}\right)}{\sqrt{2\pi}} \tag{3}$$

```
ρ = BinaryCallRho[B_, S_, K_, σ_, r_, q_, t_] =
D[BinaryCall[B, S, K, σ, r, q, t], r]; TraditionalForm[ρ]
```

$$\frac{B\,e^{-\frac{1}{2}\left(\frac{(r-q)\,t+\log\left(\frac{S}{K}\right)}{\sqrt{t}\,\sigma}-\frac{\sqrt{t}\,\sigma}{2}\right)^2-r\,t}\sqrt{t}}{\sqrt{2\pi}\,\sigma} - \frac{1}{2}\,B\,e^{-r\,t}\,t\left(\text{erf}\left(\frac{\frac{(r-q)\,t+\log\left(\frac{S}{K}\right)}{\sqrt{t}\,\sigma}-\frac{\sqrt{t}\,\sigma}{2}}{\sqrt{2}}\right)+1\right) \tag{4}$$

```
Λ = BinaryCallVega[B_, S_, K_, σ_, r_, q_, t_] =
D[BinaryCall[B, S, K, σ, r, q, t], σ]; TraditionalForm[Λ]
```

$$\frac{B\,e^{-\frac{1}{2}\left(\frac{(r-q)\,t+\log\left(\frac{S}{K}\right)}{\sqrt{t}\,\sigma}-\frac{\sqrt{t}\,\sigma}{2}\right)^2-r\,t}\left(-\frac{(r-q)\,t+\log\left(\frac{S}{K}\right)}{\sqrt{t}\,\sigma^2}-\frac{\sqrt{t}}{2}\right)}{\sqrt{2\pi}} \tag{5}$$

Check of Differential Constraints

We confirm the differential constraints linking the Greeks. First we check that the Black-Scholes equation is indeed satisfied, expressed in terms of Greeks:

```
V = BinaryCall[B, S, K, σ, r, q, t];
Simplify[θ + S (r - q) Δ + 1 / 2 σ^2 S^2 Γ - r V]
```

> 0

We also check the Vega-Gamma and Rho-Delta constraints:

```
Simplify[{Λ - S^2 σ t Γ, ρ + t (V - S Δ)}]
```

> {0, 0}

Note that the fact that these last two identities hold is very significant - it means that the identities linking Vega and Gamma, and Rho and Delta, hold even when these quantities have singular initial data.

Put Variants

The Greeks for the "Put" case can all be defined similarly - we do not repeat all the details and checks:

```
BinaryPutDelta[B_, S_, K_, σ_, r_, q_, t_] =
D[BinaryPut[B, S, K, σ, r, q, t], S];
BinaryPutGamma[B_, S_, K_, σ_, r_, q_, t_] =
D[BinaryPut[B, S, K, σ, r, q, t], {S, 2}];
BinaryPutTheta[B_, S_, K_, σ_, r_, q_, t_] =
-D[BinaryPut[B, S, K, σ, r, q, t], t];
BinaryPutRho[B_, S_, K_, σ_, r_, q_, t_] =
D[BinaryPut[B, S, K, σ, r, q, t], r];
BinaryPutVega[B_, S_, K_, σ_, r_, q_, t_] =
D[BinaryPut[B, S, K, σ, r, q, t], σ];
```

6.4 The Hedge Nightmare

The hedge parameters for the Binary options are particularly interesting. Let's take a look at Δ for a Call, over the range from near expiry to one year:

```
Plot3D[BinaryCallDelta[1, S, 100, 0.2, 0.1, 0, t], {S, 90, 110},
  {t, 0.0001, 1}, PlotPoints -> 41, PlotRange -> All];
```

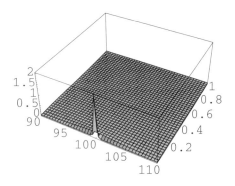

Let's take a closer look at the function:

```
ExpandAll[Δ]
```

$$\frac{B\, E^{-\frac{q\,t}{2}-\frac{r\,t}{2}-\frac{q^2\,t}{2\,\sigma^2}+\frac{q\,r\,t}{\sigma^2}-\frac{r^2\,t}{2\,\sigma^2}-\frac{t\,\sigma^2}{8}-\frac{Log\left[\frac{S}{K}\right]^2}{2\,t\,\sigma^2}}\,\left(\frac{S}{K}\right)^{\frac{1}{2}+\frac{q}{\sigma^2}-\frac{r}{\sigma^2}}}{\sqrt{2\,\pi}\,S\,\sqrt{t}\,\sigma}$$

As we approach maturity, this function becomes increasingly localized at the strike, where it becomes infinite. Elsewhere, it becomes zero. Essentially it has become what mathematicians term a "Delta Function." In fact, the Delta for these options is essentially just the Green's function we wrote down in Chapter 4. This makes the hedging of Binary options that are near the money well-nigh impossible as expiry approaches. If the asset price wanders around the strike, near maturity, huge positions in the underlying can be defined as the hedge position, then bought and sold repeatedly. Coupled with transaction costs, this can wreak havoc with risk management. Gamma is similarly pathological. Just as Delta approaches a Delta Function, Gamma approaches the derivative of a Delta Function at maturity:

```
Plot3D[BinaryCallGamma[1, S, 100, 0.2, 0.1, 0, t], {S, 90, 110},
  {t, 0.0001, 1}, PlotPoints -> 41, PlotRange -> All];
```

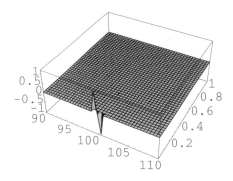

6.5 Computing the Cumulative Normal Distribution

We have been able to define, compute and plot our Binary options within *Mathematica* without any extensive coding. This relied heavily on the existence of accurate built-in routines for the cumulative normal distribution, expressed in terms of the Error function **Erf**:

```
? Erf
```

```
    Erf[z] gives the error function erf(z). Erf[z0, z1]
      gives the generalized error function erf(z1) - erf(z0).
```

```
Erf[{-Infinity, 0, Infinity}]
```

```
    {-1, 0, 1}
```

This was mapped onto the cumulative normal distribution thus:

```
Norm[x_] := (1 + Erf[x/Sqrt[2]])/2
```

Let's check that this is right, by differentiating within *Mathematica:*

```
Norm'[x]
```

$$\frac{E^{-\frac{x^2}{2}}}{\sqrt{2\,\pi}}$$

We get the correct probability density function. This highlights one of the nice features about *Mathematica* - we do not have to worry about coding up "special functions". Such special functions occur in many different cases - for example, hypergeometric functions appear in Asian option pricing; Bessel and chi-square functions appear in bond option pricing, and so on.

If you are using this text to verify numerical models coded up in other systems, you may wonder how to treat such functions. In general, there is quite an art to doing so.

Common Rational Approximations and Issues Raised

There are some standard approximations for the Error function or cumulative normal distribution that have passed into common usage. They are given by Abramowitz and Stegun (1972, Chapter 26) and have been made popular in derivatives modelling as a result of their inclusion in the classic text by John Hull (3rd edition, 1996). Originally due to C. Hastings, and published in 1955, they are presented here as **HFuncOne** and **HFuncTwo**. This pair of functions are familiar to many workers in derivatives.

Approximation Number One

We first introduce the basic function:

```
HFuncOneBase[x_] :=
 Module[
   {γ = 0.33267, a = 0.4361836, b = -0.1201676, c = 0.9372980, k},
   k = 1 / (1 + γ x);
   Exp[-x^2 / 2] / Sqrt[2 Pi] * (a * k + b * k * k + c * k * k * k)]
```

Then the first approximation is given by

```
HFuncOne[x_] :=
 If[x >= 0, 1 - HFuncOneBase[x], HFuncOneBase[-x]]
```

Note first that we have defined the negative case very carefully. Hull gives the formula for $x > 0$ as the function $N(x)$, and then suggests using

$$1 - N(-x) \tag{6}$$

for $x < 0$. This must emphatically NOT be done, for very basic reasons - as x becomes large and positive the N function approaches unity, so that defining the value for negative x this way entails a massive loss of precision as we subtract two nearly identical numbers, both close to unity. The way we have written it, in terms of the "Base" function, nearby numbers are never subtracted. Furthermore, in our discussions below, we are using *Mathematica*'s arbitrary-precision arithmetic to evaluate these approximations. If you use the C programming language, the answers for the errors may be worse than those we present. There is, however, one further potentially nasty issue.

Now we know already what the derivatives of the cumulative normal distribution are, so it may be the case that these are coded up separately. However, if the modelling system is organized in such a way that, for example, central differences of values are used everywhere to compute Delta and Gamma, these approximations may themselves end up being effectively differentiated. Let's do this symbolically:

```
HFuncOneBaseD[x_]  = D[HFuncOneBase[x], x];
HFuncOneBaseD2[x_] = D[HFuncOneBase[x], {x, 2}];

HFuncOneD[x_] :=
 If[x >= 0, - HFuncOneBaseD[x], -HFuncOneBaseD[-x]]

HFuncOneD2[x_] := If[x >= 0, - HFuncOneBaseD2[x], HFuncOneBaseD2[-x]]
```

The function itself looks just fine:

```
Plot[HFuncOne[x], {x, -3, 3}];
```

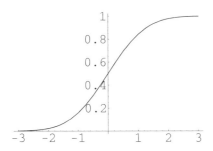

However, the percentage error approaches 4% at large negative arguments (this may be a larger error if single-precision C is used):

```
Plot[100 * (HFuncOne[x] / Norm[x] - 1), {x, -7, 7}, PlotRange -> All];
```

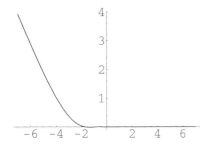

If we look at the difference error, we see that the discrepancy is less than about 0.00001 everywhere:

```
Plot[HFuncOne[x] - Norm[x], {x, -7, 7}, PlotRange -> All];
```

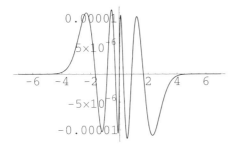

But note the oscillation - it becomes quite severe, and is amplified when we differentiate.

```
Plot[HFuncOneD[x] - Norm'[x], {x, -7, 7}, PlotRange -> All];
```

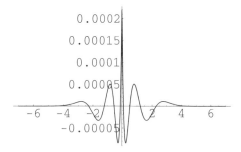

```
Plot[HFuncOneD2[x] - Norm''[x], {x, -7, 7}, PlotRange -> All];
```

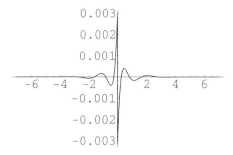

```
{HFuncOneD2[0], Norm''[0]}
```

```
{-0.00327084, 0}
```

So this type of rational approximation cannot be used to get precise answers for large negative arguments, and if differences are taken, the error in Γ may be significant near the strike and at maturity.

Approximation Number Two

We define the corresponding functions by

```
HFuncTwoBase[x_] :=
  Module[{γ = 0.2316419, a = 0.319381530, b = -0.356563782,
    c = 1.781477937, d = -1.821255978, e = 1.330274429, k},
   k = 1 / (1 + γ x);
   u = Exp[-x^2 / 2] / Sqrt[2 Pi] *
     (a * k + b * k^2 + c * k^3 + d k^4 + e k^5)]

HFuncTwo[x_] :=
  If[x >= 0, 1 - HFuncTwoBase[x], HFuncTwoBase[-x]]
```

```
HFuncTwoBaseD[x_] = D[HFuncTwoBase[x], x];
HFuncTwoBaseD2[x_] = D[HFuncTwoBase[x], {x, 2}];

HFuncTwoD[x_] :=
  If[x >= 0, - HFuncTwoBaseD[x], -HFuncTwoBaseD[-x]]

HFuncTwoD2[x_] :=
  If[x >= 0, - HFuncTwoBaseD2[x], HFuncTwoBaseD2[-x]]
```

The function itself looks just fine:

```
Plot[HFuncTwo[x], {x, -3, 3}];
```

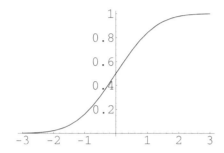

The percentage error is now confined to less than 0.7% at large negative arguments:

```
Plot[100 * (HFuncTwo[x] / Norm[x] - 1), {x, -7, 7}, PlotRange -> All];
```

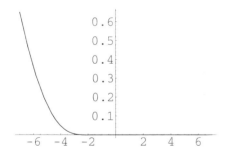

If we look at the difference error, we see that the discrepancy is less than about 0.0000001 everywhere:

```
Plot[HFuncTwo[x] - Norm[x], {x, -7, 7}, PlotRange -> All];
```

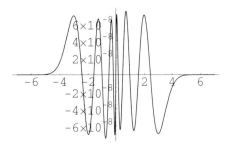

But note the oscillation - it becomes quite severe as we differentiate.

```
Plot[HFuncTwoD[x] - Norm'[x], {x, -7, 7}, PlotRange -> All];
```

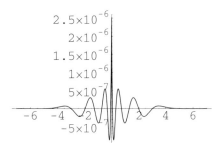

```
Plot[HFuncTwoD2[x] - Norm''[x], {x, -7, 7}, PlotRange -> All];
```

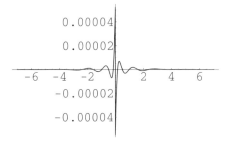

```
{HFuncTwoD2[0], Norm''[0]}
```

```
{-0.0000591723, 0}
```

So the second approximation behaves better than the first, with regard to both large negative arguments and differentiation.

Other Approximations

If you are working in C you may be interested to note that there is an accurate function for the cumulative normal distribution given in the book by Press *et al* (*Numerical Recipes in C*, 2nd edition). The % error plot over the range $-7 < x < 7$ is dead flat for this function when double-precision C is used with care. The code can be embedded in *Mathematica* using *MathLink* should you wish to confirm this directly. The algorithm at the heart of that approximation is a continued fraction expansion. We shall explore this presently. Other approaches can be based on the series method, but it needs to be applied piecewise. First we can develop a power series about the origin. Note that in *Mathematica* the **Normal** function is used to convert a truncated power series to a polynomial.

```
zerotrunc[x_, n_] := Normal[Series[Norm[x], {x, 0, n}]];
zerotrunc[x, 15]
```

$$\frac{1}{2} + \frac{x}{\sqrt{2\pi}} - \frac{x^3}{6\sqrt{2\pi}} + \frac{x^5}{40\sqrt{2\pi}} - \frac{x^7}{336\sqrt{2\pi}} +$$

$$\frac{x^9}{3456\sqrt{2\pi}} - \frac{x^{11}}{42240\sqrt{2\pi}} + \frac{x^{13}}{599040\sqrt{2\pi}} - \frac{x^{15}}{9676800\sqrt{2\pi}}$$

For large values of the argument we can use asymptotic series valid near $+\infty$:

```
inftrunc[x_, n_] :=
  Simplify[1/2 + 1/2 Normal[Series[Erf[y], {y, Infinity, n}]] /.
    y -> x/Sqrt[2]]
```

```
inftrunc[x, 8]
```

$$\frac{1}{2}\left(2 - \frac{E^{-\frac{x^2}{2}}\sqrt{\frac{2}{\pi}}(-15 + 3x^2 - x^4 + x^6)}{x^7}\right)$$

Or near $-\infty$:

```
ninftrunc[x_, n_] :=
  Simplify[1/2 + 1/2 Normal[Series[Erf[y], {y, -Infinity, n}]] /.
    y -> x/Sqrt[2]]
```

```
ninftrunc[x, 8]
```

$$-\frac{E^{-\frac{x^2}{2}}\,(-15 + 3\,x^2 - x^4 + x^6)}{\sqrt{2\,\pi}\,\,x^7}$$

The power series can give excellent results if sufficiently many terms are taken, and works well up to about $x = 2$ with 30 terms:

```
diff[x_] = zerotrunc[x, 30] - Norm[x];
```

```
Plot[diff[x], {x, -2, 2}, PlotRange -> All];
```

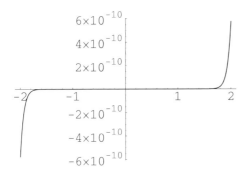

The asymptotic series can be a little hard to handle and control, and so I cannot recommend that combination of power and asymptotic series. An alternative for large positive arguments (large negative dealt with in the usual way) that is both accurate and mind-bogglingly cute is the continued fraction expansion, also from Abramowitz and Stegun. It has the merit that apart from the exponential function, only integer coefficients are used.

```
ContinuedFractionApprox[x_, n_] :=
Module[{u = Range[n], v, w}, v = Join[Reverse[u], {1}];
w = Fold[x + #2 / #1 &, 1, v] - x]
```

```
TraditionalForm[ContinuedFractionApprox[x, 15]]
```

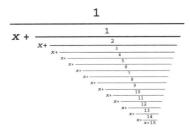

For large positive x we define the following approximation:

```
CFAOne[x_] = 1 - Exp[-x^2 / 2] / Sqrt[2 Pi] ContinuedFractionApprox[x, 35];
```

```
diffCF[x_] = CFAOne[x] - Norm[x];
```

```
Plot[diffCF[x], {x, 1.5, 4}, PlotRange -> All];
```

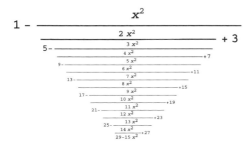

So a careful join of the power series with the continued fraction expansion can give excellent results. In fact we can do rather better still if we replace the power series by another continued fraction. This second series is good in a large neighbourhood of the origin. It is based on the function

```
ContinuedFractionApproxTwo[x_, n_] :=
Module[{u = Range[n], v, w},
v = Reverse[u];
Fold[(2 * #2 - 1) + x^2 (-1)^#2 #2 / #1 &, 1, v]]
```

```
TraditionalForm[ContinuedFractionApproxTwo[x, 15]]
```

$$1 - \cfrac{x^2}{\cfrac{2x^2}{5 - \cfrac{3x^2}{9 - \cfrac{4x^2}{13 - \cfrac{5x^2}{17 - \cfrac{6x^2}{21 - \cfrac{7x^2}{25 - \cfrac{8x^2}{29 - 15x^2 + 27} + 23} + 19} + 15} + 11} + 7} + 3}$$

The approximation to the normal distribution is given by

```
CFATwo[x_] =
  1 / 2 + Exp[-x^2 / 2] / Sqrt[2 Pi] x / ContinuedFractionApproxTwo[x, 20];
```

```
diffCFTwo[x_] = CFATwo[x] - Norm[x];
```

Let's look at the error for $0 < x < 3$:

```
Plot[diffCFTwo[x], {x, 0, 3}, PlotRange -> All];
```

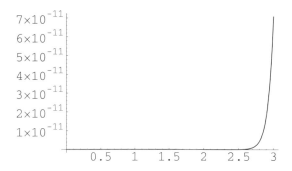

These two continued fraction expansions are what I recommend for accurate computation of the cumulative normal distribution, if you are not using *Mathematica*. They are given by equations (26.2.14) and (26.2.15) of Abramowitz and Stegun (1972). They are easy to code in any language, and give errors significantly smaller than the popular rational approximations.

Chapter 6 Bibliography

Abramowitz, M. and Stegun, I.A., 1972, *Handbook of Mathematical Functions,* Dover 1972 edition.

Hull, J.C., 1996, *Options, Futures, and other Derivatives*, 3rd edition, Prentice-Hall.

Press, W.H., Teukolsky, S.A., Vetterling, W.T. and Flannery, B.P., 1992, *Numerical Recipes in C, the Art of Scientific Computing*, 2nd edition, Cambridge University Press.

Rubenstein, M. and Reiner, E., 1991, Unscrambling the binary code, *RISK Magazine*, October.

Chapter 7.
Vanilla European Calls and Puts

What Can We Learn from a Mathematica Version of Black-Scholes?

7.1 Introduction

Over the last few years, several treatments of the Black-Scholes model (Black and Scholes, 1973) within *Mathematica* have been made available. In 1990 Ross Miller published an article in The *Mathematica* Journal, which he elaborated in a chapter of Varian's book *Economic and Financial Modelling with Mathematica* (Varian, 1993). This by now standard treatment is also available as part of the Finance Pack, from Wolfram Research. We are all acutely aware of the limitations of this model and of the need to produce generalizations. Several are now available in analytical form, but often we need to resort to approximate computation. However, we would be remiss not to review the basic treatment, since there is much to be learnt that is applicable more generally. The points that we would like you to appreciate from this chapter are the following:

(1) the general power of an analytical treatment;

(2) the use of 2D and 3D visualization;

(3) the treatment of sensitivity by differentiation;

(4) the use of **FindRoot** for inverse calculations (here the implied volatility);

(5) the collection of useful functions into a *Mathematica* Package structure.

We shall assume the results of the Black-Scholes analysis, in the form given by Merton (1973), with continuous dividend yield, as was derived in Chapter 3 in a basic application of Itô's Lemma and arbitrage.

7.2 Basic Functions

The following functions are defined directly here. At the end of this chapter we shall see how to bundle them up into a structured Package. First we define some functions that give the intrinsic value of Call and Put options at a given price for the underlying stock:

```
CallPayoff[price_,strike_] = Max[0,price-strike];
PutPayoff[price_,strike_] = Max[0,strike-price];
```

```
CallPayoff[80, 60]
```

    20

```
CallPayoff[60, 80]
```

    0

```
Plot[CallPayoff[x, 60], {x, 0, 100}];
```

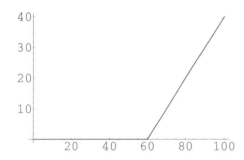

Next we define the fundamental normal distribution functions that we shall need, in both numerical and symbolic form:

```
Norm[(z_)?NumberQ] := N[0.5*Erf[z/Sqrt[2]] + 0.5];
Norm[x_] := (1 + Erf[x/Sqrt[2]])/2
```

Using the normal distribution, we can define the Black-Scholes formula. We do it in two stages:

```
done[s_, σ_, k_, t_, r_, q_] :=
((r - q)*t + Log[s/k])/(σ*Sqrt[t]) + (σ*Sqrt[t])/2;
dtwo[s_, σ_, k_, t_, r_, q_] :=
((r - q)*t + Log[s/k])/(σ*Sqrt[t]) - (σ*Sqrt[t])/2;
```

```
BlackScholesCall[s_, k_, v_, r_, q_, t_] :=
s*Exp[-q*t]*Norm[done[s, v, k, t, r, q]] - k*Exp[-r*t]*Norm[dtwo[s, v,
k, t, r, q]];
BlackScholesPut[s_, k_, v_, r_, q_, t_] :=
k*Exp[-r*t]*Norm[-dtwo[s, v, k, t, r, q]] - s*Exp[-q*t]*Norm[-done[s,
v, k, t, r, q]]
```

We can now use some of *Mathematica*'s tools to explore the properties of this function. In particular, we can use the graphical tools in several ways. First we can look at the payoff at other times:

```
Plot[BlackScholesCall[x, 60., 0.29, 0.04, 0, 0.5], {x, 50, 70}, Plot-
Range -> {{50, 70}, {0, 15}}];
```

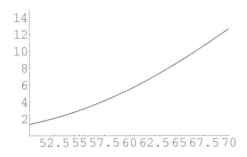

We can loop this plot over *t* to create an animation of the process. Note that only a single frame is shown in the printed version of this text. If you are using the electronic form select the group of cells containing the plot and animate it using the "Animate Selected Graphics" entry in the "Cell" menu.

```
Do[Plot[BlackScholesCall[x, 60., 0.29, 0.04, 0, t], {x, 50, 70}, Plot-
Range -> {{50, 70}, {0, 15}}], {t, 0.001, 1, 0.08}]
```

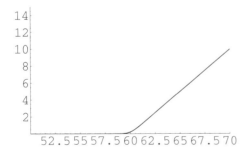

We can also exploit *Mathematica*'s 3D graphics capabilities very effectively. Here we take time and stock price as two coordinates, and construct a surface plot.

```
bssurf = Plot3D[BlackScholesCall[x, 60., 0.29, 0.04,0, t], {x, 50,
70},{t, 0.001, 1}, PlotRange -> {0, 15}, PlotPoints -> 40];
```

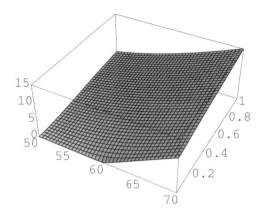

7.3 Sensitivity Analysis

As we have seen with the log, power and Binary options, perhaps the most useful feature of the analytical treatment within *Mathematica* is the ability to work out sensitivity parameters by elementary differentiation. The minimal effort involved is readily contrasted with what would be needed in a numerical approach, and is a good reason for re-considering the use of tools such as spreadsheets, where a set of neighbouring values must be computed and derivatives derived numerically. We can literally just throw the calculation to *Mathematica*. In the following the use of "**Evaluate**" is not strictly necessary; however, it is a good habit to get into using it, as it is most helpful to have it in several other more complicated cases, with option functions defined piecewise.

```
BlackScholesCallDelta[s_,k_, v_, r_, q_, t_]=
Evaluate[D[BlackScholesCall[s,k, v, r, q, t], s]];

BlackScholesPutDelta[s_,k_, v_, r_, q_, t_]=
Evaluate[D[BlackScholesPut[s,k, v, r, q, t], s]];

BlackScholesCallGamma[s_,k_, v_, r_, q_, t_]=
Evaluate[D[BlackScholesCall[s,k, v, r, q, t], {s, 2}]];

BlackScholesPutGamma[s_,k_, v_, r_, q_, t_]=
Evaluate[D[BlackScholesPut[s,k, v, r, q, t], {s, 2}]];

BlackScholesCallTheta[s_,k_, v_, r_, q_, t_]=
-Evaluate[D[BlackScholesCall[s,k, v, r, q, t], t]];

BlackScholesPutTheta[s_,k_, v_, r_, q_, t_]=
-Evaluate[D[BlackScholesPut[s,k, v, r, q, t], t]];

BlackScholesCallRho[s_,k_, v_, r_, q_, t_]=
```

```
Evaluate[D[BlackScholesCall[s,k, v, r, q, t], r]];

BlackScholesPutRho[s_,k_, v_, r_, q_, t_]=
Evaluate[D[BlackScholesPut[s,k, v, r, q, t], r]];

BlackScholesCallVega[s_,k_, v_, r_, q_, t_]=
Evaluate[D[BlackScholesCall[s,k, v, r, q, t], v]];

BlackScholesPutVega[s_,k_, v_, r_, q_, t_]=
Evaluate[D[BlackScholesPut[s,k, v, r, q, t], v]];
```

7.4 Exploring Delta

`BlackScholesCallDelta[p,k,σ,r,q,t]` gives the Black-Scholes Delta of a Call option, its sensitivity to changes in the price of the underlying stock. Let's work it out, and simplify and plot the answer in two ways:

`Simplify[BlackScholesCallDelta[p,k,σ,r,q,t]]`

$$\frac{1}{2} E^{-qt} \left(1 + Erf\left[\frac{t(-2q + 2r + \sigma^2) + 2 Log\left[\frac{p}{k}\right]}{2\sqrt{2}\sqrt{t}\ \sigma}\right]\right)$$

```
Plot[BlackScholesCallDelta[x, 60, 0.29, 0.04, 0, 0.3],
{x, 20., 100.}];
```

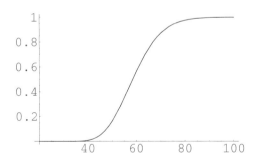

```
Plot3D[BlackScholesCallDelta[p, 60, 0.29, 0.04, 0, t],{p, 50, 70},{t,
0.001, 1}, PlotPoints -> 40];
```

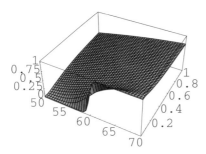

7.5 Exploring Gamma

`BlackScholesCallGamma[p,k,`σ`,r,q,t]` gives the Black-Scholes Gamma of a Call option, the sensitivity of delta to changes in the prices of the underlying stock. Let's plot it.

```
Plot[BlackScholesCallGamma[x, 60, 0.29, 0.04, 0, 0.3], {x, 20., 100.}];
```

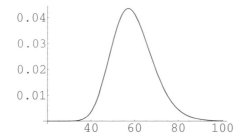

```
Plot3D[BlackScholesCallGamma[p, 60, 0.29, 0.04, 0, t],{p, 50, 70},{t,
0.001, 0.2}, PlotPoints -> 40, PlotRange -> {0, 0.2}];
```

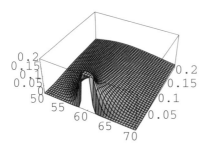

7.6 Exploring Theta

`BlackScholesCallTheta[p,k,sd,r,q,t]` gives the Theta of a Call option, the sensitivity of its value to changes in the time.

```
Plot3D[BlackScholesCallTheta[x, 60, 0.29, 0.04, 0, t], {x,
20.,100.},{t,0.01,0.5}, PlotPoints -> 40];
```

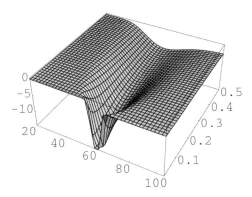

7.7 Exploring Vega

`BlackScholesCallVega[p,k,sd,r,q,t]` gives the Vega of a Call option, the sensitivity of its value to changes in the volatility.

```
Plot3D[BlackScholesCallVega[x, 60, 0.29, 0.04, 0, t], {x,
20.,100.},{t,0.01,0.5}, PlotPoints -> 40];
```

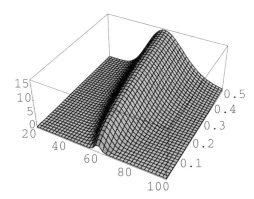

7.8 Exploring Rho

BlackScholesCallRho[p,k,sd,r,q,t] gives the Rho of a Call option, the sensitivity of its value to changes in the (continuously compounded) risk-free interest rate.

```
Plot[BlackScholesCallRho[x, 60, 0.29, 0.04, 0, 0.3], {x, 20.,100.}];
```

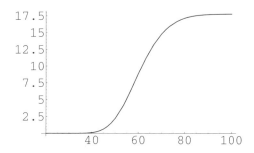

```
Plot3D[BlackScholesCallRho[x, 60, 0.29, 0.04, 0, t], {x, 20.,100.},
{t,0.01,0.5}, PlotPoints -> 40];
```

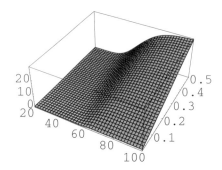

7.9 Inverse Calculations - Implied Volatility

This is a classic application of *Mathematica*'s built-in function **FindRoot**, that implements various algorithms for the solution of non-linear equations. Its default operation is to use Newton-Raphson iteration.

? FindRoot

```
FindRoot[lhs==rhs, {x, x0}] searches for a numerical
   solution to the equation lhs==rhs, starting with x=x0.
```

```
BlackScholesCallImpVol[s_,k_,r_,q_,t_,optionprice_] :=sd /. FindRoot[-
BlackScholesCall[s,k, sd, r, q, t]== optionprice,{sd,0.2}];

BlackScholesPutImpVol[s_,k_,r_,q_,t_,optionprice_] :=sd /. FindRoot[-
BlackScholesPut[s,k, sd, r, q, t]== optionprice,{sd,0.2}];
```

`BlackScholesCallImpVol[s,k,r,q,t,optionprice]` gives the volatility of a vanilla Call option that is implied by its price.

```
BlackScholesCallImpVol[58.5, 60., 0.04, 0, 0.3, 3.34886]
```

> 0.29

When we compute the implied volatility given the price, we are solving a non-linear equation. The function **FindRoot** is very useful generally, especially in finance, for inverting non-linear relations. To reinforce this point, let us look at another familiar relation - that between price and return in a dividend discount model. (It could just as easily be the relationship between value and yield for a bond.)

We are interested in the relationship between price and rate of return based on a dividend stream. We shall make up a price, some forecast dividends, and an associated dividend stream based on future extrapolation and a growth rate that gradually diminishes from 10% to 5%.

```
Price = 100;
Array[div,50];

div[1] = 12;
div[2] = 10;
div[3] = 15;
Do[div[k] = div[k-1]*(1 + 0.1 + (0.05-0.1)*(k-4)/10),{k, 4, 14}];
Do[div[k] = div[14]*1.05*(k - 14), {k, 15, 50}];
```

Now we make up a function that compares a price with a discounted cash flow based on the dividends:

```
f[R_,price_] := price - Sum[div[k]/(1 + R)^k, {k, 1, 50}]
```

Here it is (the text of the output has been reduced in size slightly):

```
f[r,p]
```

$$p - \frac{1254.76}{(1+r)^{50}} - \frac{1219.9}{(1+r)^{49}} - \frac{1185.05}{(1+r)^{48}} - \frac{1150.19}{(1+r)^{47}} - \frac{1115.34}{(1+r)^{46}} - \frac{1080.48}{(1+r)^{45}} - \frac{1045.63}{(1+r)^{44}} -$$

$$\frac{1010.78}{(1+r)^{43}} - \frac{975.921}{(1+r)^{42}} - \frac{941.067}{(1+r)^{41}} - \frac{906.213}{(1+r)^{40}} - \frac{871.358}{(1+r)^{39}} - \frac{836.504}{(1+r)^{38}} - \frac{801.65}{(1+r)^{37}} -$$

$$\frac{766.795}{(1+r)^{36}} - \frac{731.941}{(1+r)^{35}} - \frac{697.087}{(1+r)^{34}} - \frac{662.232}{(1+r)^{33}} - \frac{627.378}{(1+r)^{32}} - \frac{592.524}{(1+r)^{31}} - \frac{557.669}{(1+r)^{30}} -$$

$$\frac{522.815}{(1+r)^{29}} - \frac{487.961}{(1+r)^{28}} - \frac{453.106}{(1+r)^{27}} - \frac{418.252}{(1+r)^{26}} - \frac{383.398}{(1+r)^{25}} - \frac{348.543}{(1+r)^{24}} - \frac{313.689}{(1+r)^{23}} -$$

$$\frac{278.835}{(1+r)^{22}} - \frac{243.98}{(1+r)^{21}} - \frac{209.126}{(1+r)^{20}} - \frac{174.272}{(1+r)^{19}} - \frac{139.417}{(1+r)^{18}} - \frac{104.563}{(1+r)^{17}} - \frac{69.7087}{(1+r)^{16}} -$$

$$\frac{34.8543}{(1+r)^{15}} - \frac{33.1946}{(1+r)^{14}} - \frac{31.6139}{(1+r)^{13}} - \frac{29.9658}{(1+r)^{12}} - \frac{28.2696}{(1+r)^{11}} - \frac{26.5442}{(1+r)^{10}} - \frac{24.8077}{(1+r)^{9}} -$$

$$\frac{23.0769}{(1+r)^{8}} - \frac{21.3675}{(1+r)^{7}} - \frac{19.6936}{(1+r)^{6}} - \frac{18.0675}{(1+r)^{5}} - \frac{16.5}{(1+r)^{4}} - \frac{15}{(1+r)^{3}} - \frac{10}{(1+r)^{2}} - \frac{12}{1+r}$$

So what internal rate of return is implied by the price?

```
100*R /. FindRoot[f[R, Price] == 0, {R,0}]
```

```
    24.3572
```

Just as a check:

```
P - f[0.243572, P]
```

```
    100.
```

What Is FindRoot Doing?

Sometimes the processes going on inside **FindRoot** may be a little obscure, especially when more complicated functions are being used. It is sometimes helpful to elaborate on the operation of the Newton-Raphson algorithm. The following routine allows a picture of the process to be drawn.

```
NewtonRaphson[func_, x_, start_, iter_] :=
Module[{pts, xold = start, xnew, f, df, rangea, rangeb},
pts = {}; f = func[x]; df = D[f, x];
Do[AppendTo[pts, {xold, 0}];
AppendTo[pts, {xold, (f /. x -> xold)}];
xnew = xold - (f /. x -> xold)/(df /. x -> xold);
xold = xnew,{k, 1, iter}]; Print[xnew];
Plot[f, {x, 0.0001, 0.5}, PlotRange -> All,Axes -> True,
PlotStyle -> {{Thickness[0.001], Dashing[{0.02, 0.02}]}},
Epilog -> {Thickness[0.001], Line[pts]}]]
```

```
ShowCallImpliedVolatilityCalc[price_, strike_, rate_, yld_, time_,
optprice_] := NewtonRaphson[(optprice - BlackScholesCall[price,
strike,#,rate,yld,time])& , x, 0.1, 10]
```

```
ShowPutImpliedVolatilityCalc[price_, strike_, rate_, yld_, time_,
optprice_] := NewtonRaphson[(optprice - BlackScholesPut[price,
strike,#,rate,yld,time])& , x, 0.1, 10]
```

```
BlackScholesCallImpVol[56.5, 60., 0.04, 0, 0.3, 3.34886]
```

    0.364928


```
ShowCallImpliedVolatilityCalc[56.5, 60., 0.04, 0, 0.3, 3.34886];
```

    0.364928

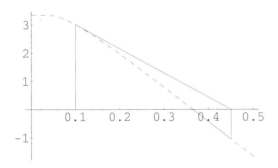

```
ShowPutImpliedVolatilityCalc[61, 60., 0.04, 0, 0.3, 3.34886];
```

    0.316237

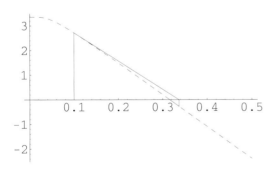

7.10 Packaging It Up

The functions that we have introduced in this chapter can be made into a standard *Mathematica* Package, which can be loaded at will. We shall make extensive use of the basic Black-Scholes functions throughout this text, so this will be very convenient for us. In particular, we shall load it whenever we wish to verify a numerical scheme for solving an option-pricing problem. Any numerical scheme has to be able to get vanilla European Calls and Puts exactly right for it to be credible in more complex applications, so this will be a very useful Package indeed.

It is not our purpose here to explain the general notion of a *Mathematica* Package or contexts. Rather, we wish to exhibit the structure of a Package in a form that is useful for practical applications, and which can be mimicked to provide other Packages containing many other useful functions. Indeed, we shall set up a similar Package for American options in the Chapter 11. For a detailed explanation of Packages and contexts, you are referred to Chapter 13 of Shaw & Tigg (1993), and to the *Mathematica* Packages documentation.

It is assumed that the following Package has been copied as a text file into the appropriate place of your *Mathematica* file structure - see the remarks at the end of this Chapter for how to do this. In order to test out the following functions in a *Mathematica* session, we advise you to quit and re-start the *Mathematica* kernel to clear out the kernel of all definitions we have made. Then load the Package. Note that this may take a few seconds as the computation of the Greeks is carried out.

```
Quit[]
```

```
Needs["Derivatives`BlackScholes`"]
```

First of all we can query what functions have been loaded:

```
? Derivatives`BlackScholes`*
```

BlackScholesCall	BlackScholesCallVega	BlackScholesPutTheta
BlackScholesCallDelta	BlackScholesPut	BlackScholesPutVega
BlackScholesCallGamma	BlackScholesPutDelta	CallPayoff
BlackScholesCallImpVol	BlackScholesPutGamma	Norm
BlackScholesCallRho	BlackScholesPutImpVol	PutPayoff
BlackScholesCallTheta	BlackScholesPutRho	

Next we can query the operation of individual functions:

```
? Norm
```

```
    Norm[x] returns the cumulative
       normal distribution function evaluated at x.
```

```
? BlackScholesCall
```

```
    BlackScholesCall[price, strike, vol, riskfree, divyield, expiry]
       returns the Black-Scholes value of a vanilla European Call.
```

? BlackScholesPutVega

> BlackScholesPutVega[price, strike, vol, riskfree, divyield,
>   expiry] returns the vega (lambda) of a vanilla European Put.

? BlackScholesCallImpVol

> BlackScholesCallImpVol[price, strike,
>   riskfree, divyield, expiry, optionprice] returns
>   the implied volatility of a vanilla European Call.

and so on. It is always a good idea to check that the functions are working!

BlackScholesCall[51, 50, 0.2, 0.03, 0, 3]

> 9.62759

BlackScholesCallImpVol[51, 50, 0.03, 0, 3, 9.62759]

> 0.2

BlackScholesPutGamma[51, 50, 0.2, 0.03, 0, 3]

> 0.0200251

BlackScholesPutImpVol[51, 50, 0.03, 0, 3, 4.32415]

> 0.2

Now you can proceed to use the functions just as if they had been entered in a Notebook. Now we give the complete listing of the package. You can use this as a template for building your own packages. The file is called "BlackScholes.m" and it should be stored in a directory or folder called "Derivatives" in the "Extra-Packages" sub-directory of the "AddOns" directory of the *Mathematica* installation area of your system. If you have any difficulty with the installation of this package, I recommend you copy the text below into a text file, give it the name "BlackScholes.m", remove any line breaks within code, and manually place it in the right place.

Package Listing

```
(* :Title:
   BlackScholes Package *)

(* :Context:
   Derivatives`BlackScholes` *)

(* :Author:
   Dr William T. Shaw *)

(* :Summary:
   *)

(* :Package Version:
   1.1 *)

(* :Mathematica Version:
   3.0, but compatible with 2.X *)

(* :History:
   V.1.0, January 1996, by William T. Shaw *)

BeginPackage["Derivatives`BlackScholes`"]

(* Usage Information *)

CallPayoff::usage =
"CallPayoff[price, strike] returns the payoff value of a vanilla
European Call."

PutPayoff::usage =
"PutPayoff[price, strike] returns the payoff value of a vanilla
European Put."

BlackScholesCall::usage =
"BlackScholesCall[price, strike, vol, riskfree, divyield, expiry]
returns the Black-Scholes value of a vanilla European Call."

BlackScholesPut::usage =
"BlackScholesPut[price, strike, vol, riskfree, divyield, expiry]
returns the Black-Scholes value of a vanilla European Put."

BlackScholesCallDelta::usage =
"BSCallDelta[price, strike, vol, riskfree, divyield, expiry]
returns the delta of a vanilla European Call."

BlackScholesPutDelta::usage =
"BlackScholesPutDelta[price, strike, vol, riskfree, divyield, expiry]
returns the delta of a vanilla European Put."

BlackScholesCallTheta::usage =
"BlackScholesCallTheta[price, strike, vol, riskfree, divyield, expiry]
returns the theta of a vanilla European Call."

BlackScholesPutTheta::usage =
"BlackScholesPutTheta[price, strike, vol, riskfree, divyield, expiry]
returns the theta of a vanilla European Put."

BlackScholesCallRho::usage =
"BlackScholesCallRho[price, strike, vol, riskfree, divyield, expiry]
```

returns the rho of a vanilla European Call."

BlackScholesPutRho::usage =
"BlackScholesPutRho[price, strike, vol, riskfree, divyield, expiry]
returns the rho of a vanilla European Put."

BlackScholesCallGamma::usage =
"BlackScholesCallGamma[price, strike, vol, riskfree, divyield, expiry]
returns the gamma of a vanilla European Call."

BlackScholesPutGamma::usage =
"BlackScholesPutGamma[price, strike, vol, riskfree, divyield, expiry]
returns the gamma of a vanilla European Call."

BlackScholesCallVega::usage =
"BlackScholesCallVega[price, strike, vol, riskfree, divyield, expiry]
returns the vega (lambda) of a vanilla European Call."

BlackScholesPutVega::usage =
"BlackScholesPutVega[price, strike, vol, riskfree, divyield, expiry]
returns the vega (lambda) of a vanilla European Put."

Norm::usage =
"Norm[x] returns the cumulative normal distribution function evaluated at x."

BlackScholesCallImpVol::usage =
"BlackScholesCallImpVol[price, strike, riskfree, divyield, expiry, optionprice]
returns the implied volatility of a vanilla European Call."

BlackScholesPutImpVol::usage =
"BlackScholesPutImpVol[price, strike, riskfree, divyield, expiry, optionprice]
returns the implied volatility of a vanilla European Call."

(* END OF USAGE INFORMATION *)

Begin["`Private`"]

Norm[z_?NumberQ]:= N[0.5 + 0.5 Erf[z/Sqrt[2]]];

Norm[x_] := (1 + Erf[x / Sqrt[2]]) / 2;

dtwo[s_, v_, k_, t_, r_, q_] :=(Log[s / k] + (r-q) t) / (v Sqrt[t]) -1/2 v Sqrt[t];

done[s_, v_, k_, t_, r_, q_] :=(Log[s / k] + (r-q) t) / (v Sqrt[t]) + 1/2 v Sqrt[t];

BlackScholesCall[s_,k_, v_, r_, q_, t_] :=
s Exp[-q*t] Norm[done[s, v, k, t, r, q]] - k*Exp[-r*t] Norm[dtwo[s, v, k, t, r, q]];

BlackScholesPut[s_,k_, v_, r_, q_, t_]:=
k*Exp[-r*t] Norm[-dtwo[s, v, k, t, r, q]]-s Exp[-q*t] Norm[-done[s, v, k, t, r, q]];

BlackScholesCallDelta[s_,k_, v_, r_, q_, t_]=
Evaluate[D[BlackScholesCall[s,k, v, r, q, t], s]];

BlackScholesPutDelta[s_,k_, v_, r_, q_, t_]=
Evaluate[D[BlackScholesPut[s,k, v, r, q, t], s]];

BlackScholesCallGamma[s_,k_, v_, r_, q_, t_]=
Evaluate[D[BlackScholesCall[s,k, v, r, q, t], {s, 2}]];

BlackScholesPutGamma[s_,k_, v_, r_, q_, t_]=

```
Evaluate[D[BlackScholesPut[s,k, v, r, q, t], {s, 2}]];

BlackScholesCallTheta[s_,k_, v_, r_, q_, t_]=
-Evaluate[D[BlackScholesCall[s,k, v, r, q, t], t]];

BlackScholesPutTheta[s_,k_, v_, r_, q_, t_]=
-Evaluate[D[BlackScholesPut[s,k, v, r, q, t], t]];

BlackScholesCallRho[s_,k_, v_, r_, q_, t_]=
Evaluate[D[BlackScholesCall[s,k, v, r, q, t], r]];

BlackScholesPutRho[s_,k_, v_, r_, q_, t_]=
Evaluate[D[BlackScholesPut[s,k, v, r, q, t], r]];

BlackScholesCallVega[s_,k_, v_, r_, q_, t_]=
Evaluate[D[BlackScholesCall[s,k, v, r, q, t], v]];

BlackScholesPutVega[s_,k_, v_, r_, q_, t_]=
Evaluate[D[BlackScholesPut[s,k, v, r, q, t], v]];

CallPayoff[s_, k_] := Max[0, s-k];

PutPayoff[s_, k_] := Max[0, k-s];

BlackScholesCallImpVol[s_,k_,r_,q_,t_,optionprice_] :=
sd /. FindRoot[BlackScholesCall[s,k, sd, r, q, t]== optionprice,{sd,0.2}];

BlackScholesPutImpVol[s_,k_,r_,q_,t_,optionprice_] :=
sd /. FindRoot[BlackScholesPut[s,k, sd, r, q, t]== optionprice,{sd,0.2}];

End[]

EndPackage[]
```

Chapter 7 Bibliography

Black, F. and Scholes, M., 1973, The pricing of assets on corporate liabilities, *Journal of Political Economy,* 81, pp. 637-659.

Mathematica Finance Pack, Wolfram Research, Champaign, Ill., USA.

Mathematica 3.0 Standard Add-on Packages, Wolfram Media-Cambridge University Press, 1996.

Merton, R., 1973, Theory of rational option pricing, *Bell Journal of Economics and Management Science,* 4, p. 141.

Miller, R., 1993, chapter in Varian, 1993.

Shaw, W.T. and Tigg, J., 1993, *Applied Mathematica, Getting Started, Getting it Done*, Addison-Wesley, 1993.

Varian, H., 1993, *Economic and Financial Modelling with Mathematica*, Springer-TELOS.

Chapter 8.
Barrier Options - a Case Study in Rapid Development

Doing it Quickly, and the Curious Properties of Options with Barriers

8.1 Introduction

We can use *Mathematica* in various ways to gain insight into the properties of an option. This chapter illustrates one way of proceeding. Suppose you have done the basic mathematical and financial research to model the option, you have the valuation formula, and what you need is an implementation to get numbers out. What you want is a tool to allow the extraction of results as quickly as possible. When the formulae are in closed form, *Mathematica* is an outstanding tool for this. You literally just have to copy the formulae into a Notebook, and enter the desired calculations. In reality one should be a little more careful - the process consists of:

(a) copying the formulae from the research material into *Mathematica*;
(b) use of symbolic differentiation to get the Greeks;
(c) checking any results given in the basic research against the output from *Mathematica*;
(d) debugging as necessary;
(e) visualization of the results to check that they are intuitive;
(f) investigation of any odd phenomena;
(g) application.

Later on, when one is required to develop a large numerically based system, perhaps in C/C++, to carry out similar calculations (we have to ask, why bother, for this type of calculation - it is all easy in *Mathematica*!), one can refer back to this type of *Mathematica* implementation to check the results, and supply any test documentation for internal or external approval processes.

The Barrier options are sufficiently complex that reliable computation of the Greeks is analytically quite burdensome - we invite you to notice how this is trivialized using the symbolic power of Mathematica.

The material presented in this chapter arose out of a real-world desire to create a benchmark model as quickly as possible, to test other models of Barrier options. It was created literally in the form presented here, by making direct use of known results for European Barrier options.

A good reason for investigating Barrier options is that they represent a very simple situation where it is possible for Gamma and/or Vega to be negative. In simple European vanilla Call or Put options these quantities are strictly non-negative. Roughly speaking, the concave structure of the payoffs is preserved by the time evolution of such vanilla options, so that Gamma is positive, and an increase in volatility increase the chances of a positive payout at maturity. In the case of a Barrier option, the payoff need not be concave, and an increase in volatility can enhance the probability of a knock-out, thereby potentially causing a decrease in value. We shall use *Mathematica* to exhibit these phenomena directly.

The fact that Gamma can have either sign is of considerable significance, especially when one comes to consider the incorporation of transaction costs, for the simple model of costs requiring only an adjustment of volatility fails to work.

The formulae for Barrier options with no rebate are given by Hull in the latest version of his book (Hull, 1996). However, here we shall work with the model with rebates, as given by Rubenstein and Reiner in an article in *RISK Magazine* (1991) and various conferences. A detailed derivation of these results for the case of a down-and-out Call was given in Chapter 4, based on the PDE approach.

8.2 Mathematica Implementation

Loading the Standard Euro-options Package

All the Barrier options involve adding a function to Black-Scholes in order to satisfy the appropriate Barrier condition, and use the cumulative normal distribution. We therefore load the standard European option tools first:

```
Needs["Derivatives`BlackScholes`"]
```

```
?Norm
```

```
    Norm[x] returns the cumulative
      normal distribution function evaluated at x.
```

Modifications for Barriers

The following allow a rebate and are based on Rubenstein and Reiner (1991). We use continuously compounded variables. In all cases, we have:

r = risk-free rate;
q = continuous dividend yield;
sd = volatility;
t = expiry time in years;
p = stock price;
k = strike;
h = barrier.

The Barrier options are all functions of the following basic functions. The *Mathematica* input cells which follow have been converted to Standard Form in order to make them easier to read.

$$\lambda[r\_, q\_, sd\_] := \frac{1}{2} + \frac{r-q}{sd^2};$$

$$x[p\_, k\_, sd\_, r\_, q\_, t\_] := \frac{Log\left[\frac{p}{k}\right]}{sd \sqrt{t}} + \lambda[r, q, sd] \, sd \sqrt{t};$$

$$xone[p\_, h\_, sd\_, r\_, q\_, t\_] := x[p, h, sd, r, q, t];$$

$$y[p\_, k\_, h\_, sd\_, r\_, q\_, t\_] := \frac{Log\left[\frac{h^2}{p k}\right]}{sd \sqrt{t}} + \lambda[r, q, sd] \, sd \sqrt{t};$$

$$yone[p\_, h\_, sd\_, r\_, q\_, t\_] := x[h, p, sd, r, q, t];$$

$$a[r\_, q\_, sd\_] := \frac{r - q}{sd^2} - \frac{1}{2};$$

$$b[r\_, q\_, sd\_] := \frac{\sqrt{\left(r - q - \frac{sd^2}{2}\right)^2 + 2 \, r \, sd^2}}{sd^2};$$

$$z[p\_, h\_, sd\_, r\_, q\_, t\_] := \frac{Log\left[\frac{h}{p}\right]}{sd \sqrt{t}} + b[r, q, sd] \, sd \sqrt{t};$$

```
intone[p_, k_, sd_, r_, q_, t_, φ_] :=
  φ p Exp[-q t] Norm[φ x[p, k, sd, r, q, t]] -
   φ k Exp[-r t] Norm[φ x[p, k, sd, r, q, t] - φ sd √t ]
```

```
inttwo[p_, k_, h_, sd_, r_, q_, t_, φ_] :=
  φ p Exp[-q t] Norm[φ xone[p, h, sd, r, q, t]] -
   φ k Exp[-r t] Norm[φ xone[p, h, sd, r, q, t] - φ sd √t ]
```

$$intthree[p\_, k\_, h\_, sd\_, r\_, q\_, t\_, \phi\_, \eta\_] :=$$
$$\phi \, p \, Exp[-q \, t] \left(\frac{h}{p}\right)^{2 \lambda[r,q,sd]} Norm[\eta \, y[p, k, h, sd, r, q, t]] -$$
$$\phi \, k \, Exp[-r \, t] \left(\frac{h}{p}\right)^{2 \lambda[r,q,sd]-2} Norm\left[\eta \, y[p, k, h, sd, r, q, t] - \eta \, sd \sqrt{t}\right]$$

```
intfour[p_, k_, h_, sd_, r_, q_, t_, φ_, η_] :=
```

$$\phi \, p \, \text{Exp}[-q \, t] \left(\frac{h}{p}\right)^{2\lambda[r,q,sd]} \text{Norm}[\eta \, \text{yone}[p, h, sd, r, q, t]] -$$

$$\phi \, k \, \text{Exp}[-r \, t] \left(\frac{h}{p}\right)^{2\lambda[r,q,sd]-2} \text{Norm}\left[\eta \, \text{yone}[p, h, sd, r, q, t] - \eta \, sd \, \sqrt{t}\,\right]$$

```
intfive[reb_, p_, h_, sd_, r_, q_, t_, η_] :=
```

$$\text{reb} \, \text{Exp}[-r \, t] \left(\text{Norm}\left[\eta \, \text{xone}[p, h, sd, r, q, t] - \eta \, sd \, \sqrt{t}\,\right] - \right.$$

$$\left.\left(\frac{h}{p}\right)^{2\lambda[r,q,sd]-2} \text{Norm}\left[\eta \, \text{yone}[p, h, sd, r, q, t] - \eta \, sd \, \sqrt{t}\,\right]\right)$$

```
intsix[reb_, p_, h_, sd_, r_, q_, t_, η_] :=
```

$$\text{reb} \left(\left(\frac{h}{p}\right)^{a[r,q,sd]+b[r,q,sd]} \text{Norm}[\eta \, z[p, h, sd, r, q, t]] + \right.$$

$$\left.\left(\frac{h}{p}\right)^{a[r,q,sd]-b[r,q,sd]} \text{Norm}\left[\eta \, z[p, h, sd, r, q, t] - 2 \, \eta \, b[r, q, sd] \, sd \, \sqrt{t}\,\right]\right)$$

8.3 Up/Down and In (Knock-In) Calls

Our first task is to define our functions and to do numerical and graphical checks against the original Rubenstein and Reiner paper. First we introduce the value and price sensitivity functions. Readers should note the use of **Evaluate** here. It forces the action of the **D** operator. Note also that the **D** has to be applied within the **If** statement.

Down and In Calls

```
DownAndInCall[reb_, p_, k_, h_, sd_, r_, q_, t_] :=
If[k >= h, intthree[p, k, h, sd, r, q, t, 1, 1] + intfive[reb, p, h, sd, r, q,
t, 1],
intone[p, k, sd, r, q, t, 1] - inttwo[p, k, h, sd, r, q, t, 1] + intfour[p, k,
h, sd, r, q, t, 1, 1] + intfive[reb, p, h, sd, r, q, t, 1]]


DownAndInCallDelta[reb_, p_, k_, h_, sd_, r_, q_, t_] =
If[k >= h, Evaluate[D[intthree[p, k, h, sd, r, q, t, 1, 1] + intfive[reb, p,
h, sd, r, q, t, 1], p]],
Evaluate[D[intone[p, k, sd, r, q, t, 1] - inttwo[p, k, h, sd, r, q, t, 1] +
intfour[p, k, h, sd, r, q, t, 1, 1] + intfive[reb, p, h, sd, r, q, t, 1], p]]];
```

```
DownAndInCallGamma[reb_, p_, k_, h_, sd_, r_, q_, t_] =
If[k >= h, Evaluate[D[intthree[p, k, h, sd, r, q, t, 1, 1] + intfive[reb, p,
h, sd, r, q, t, 1], {p,2}]],
Evaluate[D[intone[p, k, sd, r, q, t, 1] - inttwo[p, k, h, sd, r, q, t, 1] +
intfour[p, k, h, sd, r, q, t, 1, 1] + intfive[reb, p, h, sd, r, q, t, 1],
{p,2}]]];


DownAndInCallTheta[reb_, p_, k_, h_, sd_, r_, q_, t_] =
If[k >= h, Evaluate[-D[intthree[p, k, h, sd, r, q, t, 1, 1] + intfive[reb, p,
h, sd, r, q, t, 1], t]],
Evaluate[-D[intone[p, k, sd, r, q, t, 1] - inttwo[p, k, h, sd, r, q, t, 1] +
intfour[p, k, h, sd, r, q, t, 1, 1] + intfive[reb, p, h, sd, r, q, t, 1], t]]];


DownAndInCallVega[reb_, p_, k_, h_, sd_, r_, q_, t_] =
If[k >= h, Evaluate[D[intthree[p, k, h, sd, r, q, t, 1, 1] + intfive[reb, p,
h, sd, r, q, t, 1], sd]],
Evaluate[D[intone[p, k, sd, r, q, t, 1] - inttwo[p, k, h, sd, r, q, t, 1] +
intfour[p, k, h, sd, r, q, t, 1, 1] + intfive[reb, p, h, sd, r, q, t, 1],
sd]]];


DownAndInCallRho[reb_, p_, k_, h_, sd_, r_, q_, t_] =
If[k >= h, Evaluate[D[intthree[p, k, h, sd, r, q, t, 1, 1] + intfive[reb, p,
h, sd, r, q, t, 1], r]],
Evaluate[D[intone[p, k, sd, r, q, t, 1] - inttwo[p, k, h, sd, r, q, t, 1] +
intfour[p, k, h, sd, r, q, t, 1, 1] + intfive[reb, p, h, sd, r, q, t, 1], r]]];
```

Up and In Calls

```
UpAndInCall[reb_, p_, k_, h_, sd_, r_, q_, t_] :=
If[k >= h, intone[p, k, sd, r, q, t, 1] + intfive[reb, p, h, sd, r, q, t, -1],
inttwo[p, k, h, sd, r, q, t, 1] - intthree[p, k, h, sd, r, q, t, 1,
-1]+intfour[p, k, h, sd, r, q, t, 1, -1] + intfive[reb, p, h, sd, r, q, t,
-1]];


UpAndInCallDelta[reb_, p_, k_, h_, sd_, r_, q_, t_] =
If[k >= h, Evaluate[D[intone[p, k, sd, r, q, t, 1] + intfive[reb, p, h, sd, r,
q, t, -1], p]],
Evaluate[D[inttwo[p, k, h, sd, r, q, t, 1] - intthree[p, k, h, sd, r, q, t, 1,
-1]+ intfour[p, k, h, sd, r, q, t, 1, -1] + intfive[reb, p, h, sd, r, q, t,
-1], p]]];


UpAndInCallGamma[reb_, p_, k_, h_, sd_, r_, q_, t_] =
If[k >= h, Evaluate[D[intone[p, k, sd, r, q, t, 1] + intfive[reb, p, h, sd, r,
q, t, -1], {p,2}]],
Evaluate[D[inttwo[p, k, h, sd, r, q, t, 1] - intthree[p, k, h, sd, r, q, t, 1,
-1]+ intfour[p, k, h, sd, r, q, t, 1, -1] + intfive[reb, p, h, sd, r, q, t,
-1], {p,2}]]];
```

```
UpAndInCallTheta[reb_, p_, k_, h_, sd_, r_, q_, t_] =
If[k >= h, Evaluate[-D[intone[p, k, sd, r, q, t, 1] + intfive[reb, p, h, sd,
r, q, t, -1], t]],
Evaluate[-D[inttwo[p, k, h, sd, r, q, t, 1] - intthree[p, k, h, sd, r, q, t,
1, -1]+ intfour[p, k, h, sd, r, q, t, 1, -1] + intfive[reb, p, h, sd, r, q, t,
-1], t]]];


UpAndInCallVega[reb_, p_, k_, h_, sd_, r_, q_, t_] =
If[k >= h, Evaluate[D[intone[p, k, sd, r, q, t, 1] + intfive[reb, p, h, sd, r,
q, t, -1], sd]],
Evaluate[D[inttwo[p, k, h, sd, r, q, t, 1] - intthree[p, k, h, sd, r, q, t, 1,
-1]+ intfour[p, k, h, sd, r, q, t, 1, -1] + intfive[reb, p, h, sd, r, q, t,
-1], sd]]];


UpAndInCallRho[reb_, p_, k_, h_, sd_, r_, q_, t_] =
If[k >= h, Evaluate[D[intone[p, k, sd, r, q, t, 1] + intfive[reb, p, h, sd, r,
q, t, -1], r]],
Evaluate[D[inttwo[p, k, h, sd, r, q, t, 1] - intthree[p, k, h, sd, r, q, t, 1,
-1]+ intfour[p, k, h, sd, r, q, t, 1, -1] + intfive[reb, p, h, sd, r, q, t,
-1], r]]];
```

The following tables of values can be checked directly against a table given by Rubenstein and Reiner. Note the conversion to annual rates for the interest rates and yield.

```
TableForm[Table[N[DownAndInCall[2,100,k,97,0.20,Log[1.1],Log[1.05],t]],
{k, 90, 110, 10}, {t, 0.5, 1.5, 0.25}]]
```

8.54547	9.96731	11.2001	12.2995	13.296
3.91067	5.30466	6.55243	7.68743	8.73126
1.59872	2.57713	3.57657	4.55321	5.49371

```
TableForm[Table[N[DownAndInCall[2,100,k,100,0.20,Log[1.1],Log[1.05],t]]
, {k, 90, 110, 10}, {t, 0.5, 1.5, 0.25}]]
```

13.0537	14.4511	15.6916	16.8087	17.825
6.63324	8.32391	9.77729	11.0687	12.2376
2.78483	4.27028	5.62437	6.87002	8.02476

```
TableForm[Table[N[UpAndInCall[2,100,k,103,0.20,Log[1.1],Log[1.05],t]],
{k, 90, 110, 10}, {t, 0.5, 1.5, 0.25}]]
```

13.0851	14.5317	15.7835	16.9009	17.9139
6.91517	8.54393	9.9596	11.2249	12.3744
3.07013	4.4921	5.80783	7.02706	8.16217

We can also plot the time-price plots of the value and Delta and Gamma:

```
Plot3D[If[S>95,
DownAndInCall[2,S,100,95,0.2,Log[1.1],Log[1.05],t],
UpAndInCall[2,S,100,95,0.2,Log[1.1],Log[1.05],t]], {S, 80, 120}, {t,
0.0001, 1}, BoxRatios -> {1,1,1}, PlotPoints -> 40];
```

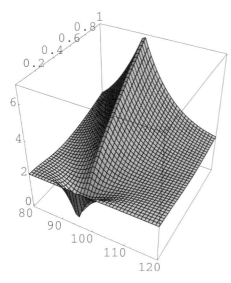

```
Plot3D[If[S>95,
DownAndInCallDelta[2,S,100,95,0.2,Log[1.1],Log[1.05],t],
UpAndInCallDelta[2,S,100,95,0.2,Log[1.1],Log[1.05],t]], {S, 80, 120},
{t, 0.0001, 1}, BoxRatios -> {1, 1, 1}, PlotPoints -> 40];
```

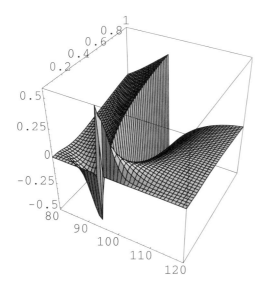

```
Plot3D[If[S>95,
DownAndInCallGamma[2,S,100,95,0.2,Log[1.1],Log[1.05],t],
UpAndInCallGamma[2,S,100,95,0.2,Log[1.1],Log[1.05],t]], {S, 80, 120},
{t, 0.0001, 1},
BoxRatios -> {1, 1, 1}, PlotPoints -> 40];
```

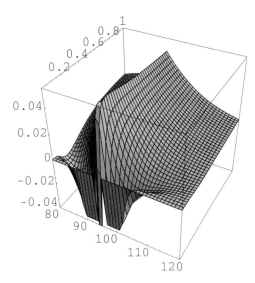

8.4 Up/Down and Out (Knock-Out) Calls

As before, we can do numerical and graphical checks against the original Rubenstein and Reiner paper. First we introduce the value and price sensitivity functions:

Down and Out Calls

```
DownAndOutCall[reb_, p_, k_, h_, sd_, r_, q_, t_] :=
If[k >= h, intone[p, k, sd, r, q, t, 1] - intthree[p, k, h, sd, r, q, t, 1, 1]
+ intsix[reb, p, h, sd, r, q, t, 1],
inttwo[p, k, h, sd, r, q, t, 1] - intfour[p, k, h, sd, r, q, t, 1, 1] +
intsix[reb, p, h, sd, r, q, t, 1]];

DownAndOutCallDelta[reb_, p_, k_, h_, sd_, r_, q_, t_] =
If[k >= h, Evaluate[D[intone[p, k, sd, r, q, t, 1] - intthree[p, k, h, sd, r,
q, t, 1, 1] + intsix[reb, p, h, sd, r, q, t, 1], p]],
Evaluate[D[inttwo[p, k, h, sd, r, q, t, 1] - intfour[p, k, h, sd, r, q, t, 1,
1] + intsix[reb, p, h, sd, r, q, t, 1], p]]];

DownAndOutCallGamma[reb_, p_, k_, h_, sd_, r_, q_, t_] =
If[k >= h, Evaluate[D[intone[p, k, sd, r, q, t, 1] - intthree[p, k, h, sd, r,
q, t, 1, 1] + intsix[reb, p, h, sd, r, q, t, 1], {p,2}]],
Evaluate[D[inttwo[p, k, h, sd, r, q, t, 1] - intfour[p, k, h, sd, r, q, t, 1,
1] + intsix[reb, p, h, sd, r, q, t, 1], {p,2}]]];
```

```
DownAndOutCallTheta[reb_, p_, k_, h_, sd_, r_, q_, t_] =
If[k >= h, Evaluate[-D[intone[p, k, sd, r, q, t, 1] - intthree[p, k, h, sd, r,
q, t, 1, 1] + intsix[reb, p, h, sd, r, q, t, 1], t]]],
Evaluate[-D[inttwo[p, k, h, sd, r, q, t, 1] - intfour[p, k, h, sd, r, q, t, 1,
1] + intsix[reb, p, h, sd, r, q, t, 1], t]]];


DownAndOutCallVega[reb_, p_, k_, h_, sd_, r_, q_, t_] =
If[k >= h, Evaluate[D[intone[p, k, sd, r, q, t, 1] - intthree[p, k, h, sd, r,
q, t, 1, 1] + intsix[reb, p, h, sd, r, q, t, 1], sd]]],
Evaluate[D[inttwo[p, k, h, sd, r, q, t, 1] - intfour[p, k, h, sd, r, q, t, 1,
1] + intsix[reb, p, h, sd, r, q, t, 1], sd]]];


DownAndOutCallRho[reb_, p_, k_, h_, sd_, r_, q_, t_] =
If[k >= h, Evaluate[D[intone[p, k, sd, r, q, t, 1] - intthree[p, k, h, sd, r,
q, t, 1, 1] + intsix[reb, p, h, sd, r, q, t, 1], r]]],
Evaluate[D[inttwo[p, k, h, sd, r, q, t, 1] - intfour[p, k, h, sd, r, q, t, 1,
1] + intsix[reb, p, h, sd, r, q, t, 1], r]]];
```

Up and Out Calls

```
UpAndOutCall[reb_, p_, k_, h_, sd_, r_, q_, t_] :=
If[k >= h, intsix[reb, p, h, sd, r, q, t, -1],
intone[p, k, sd, r, q, t, 1]- inttwo[p, k, h, sd, r, q, t, 1] +
intthree[p, k, h, sd, r, q, t, 1, -1] - intfour[p, k, h, sd, r, q, t, 1, -1] +
intsix[reb, p, h, sd, r, q, t, -1]];


UpAndOutCallDelta[reb_, p_, k_, h_, sd_, r_, q_, t_] =
If[k >= h, Evaluate[D[intsix[reb, p, h, sd, r, q, t, -1], p]],
Evaluate[D[intone[p, k, sd, r, q, t, 1]- inttwo[p, k, h, sd, r, q, t, 1] +
intthree[p, k, h, sd, r, q, t, 1, -1] - intfour[p, k, h, sd, r, q, t, 1, -1] +
intsix[reb, p, h, sd, r, q, t, -1], p]]];


UpAndOutCallGamma[reb_, p_, k_, h_, sd_, r_, q_, t_] =
If[k >= h, Evaluate[D[intsix[reb, p, h, sd, r, q, t, -1], {p,2}]],
Evaluate[D[intone[p, k, sd, r, q, t, 1]- inttwo[p, k, h, sd, r, q, t, 1] +
intthree[p, k, h, sd, r, q, t, 1, -1] - intfour[p, k, h, sd, r, q, t, 1, -1] +
intsix[reb, p, h, sd, r, q, t, -1], {p,2}]]];


UpAndOutCallTheta[reb_, p_, k_, h_, sd_, r_, q_, t_] =
If[k >= h, Evaluate[-D[intsix[reb, p, h, sd, r, q, t, -1], t]],
Evaluate[-D[intone[p, k, sd, r, q, t, 1]- inttwo[p, k, h, sd, r, q, t, 1] +
intthree[p, k, h, sd, r, q, t, 1, -1] - intfour[p, k, h, sd, r, q, t, 1, -1] +
intsix[reb, p, h, sd, r, q, t, -1], t]]];


UpAndOutCallVega[reb_, p_, k_, h_, sd_, r_, q_, t_] =
If[k >= h, Evaluate[D[intsix[reb, p, h, sd, r, q, t, -1], sd]],
Evaluate[D[intone[p, k, sd, r, q, t, 1]- inttwo[p, k, h, sd, r, q, t, 1] +
intthree[p, k, h, sd, r, q, t, 1, -1] - intfour[p, k, h, sd, r, q, t, 1, -1] +
intsix[reb, p, h, sd, r, q, t, -1], sd]]];
```

```
UpAndOutCallRho[reb_, p_, k_, h_, sd_, r_, q_, t_] =
If[k >= h, Evaluate[D[intsix[reb, p, h, sd, r, q, t, -1], r]],
Evaluate[D[intone[p, k, sd, r, q, t, 1]- inttwo[p, k, h, sd, r, q, t, 1] +
intthree[p, k, h, sd, r, q, t, 1, -1] - intfour[p, k, h, sd, r, q, t, 1, -1] +
intsix[reb, p, h, sd, r, q, t, -1], r]]];
```

The following tables of values can be checked directly against a further sample table of Rubenstein and
Reiner. Note again the conversion to annual rates for the interest rates and yield.

```
TableForm[Table[N[DownAndOutCall[2,100,k,97,0.20,Log[1.1],Log[1.05],t]]
, {k, 90, 110, 10}, {t, 0.5, 1.5, 0.25}]]
```

6.47884	6.44667	6.448	6.45987	6.47464
4.69315	4.98216	5.18128	5.332	5.45192
3.15669	3.65604	4.00422	4.26754	4.47666

```
TableForm[Table[N[UpAndOutCall[2,100,k,103,0.20,Log[1.1],Log[1.05],t]],
{k, 90, 110, 10}, {t, 0.5, 1.5, 0.25}]]
```

1.94253	1.88737	1.87128	1.86688	1.86671
1.69197	1.74791	1.78083	1.80287	1.81878
1.6886	1.74611	1.77968	1.80206	1.81819

We can also plot the time-price plots of the value and Delta and Gamma:

```
Plot3D[If[S>95,DownAndOutCall[2,S,100,95,0.2,Log[1.1],Log[1.05],t],
UpAndOutCall[2,S,100,95,0.2,Log[1.1],Log[1.05],t]], {S, 80, 120}, {t,
0.0001, 1}, BoxRatios -> {1, 1, 1}, PlotPoints -> 40];
```

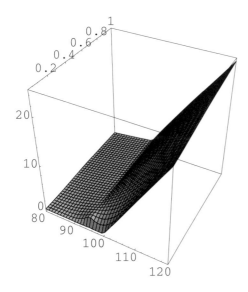

```
Plot3D[If[S>95,
DownAndOutCallDelta[2,S,100,95,0.2,Log[1.1],Log[1.05],t],
UpAndOutCallDelta[2,S,100,95,0.2,Log[1.1],Log[1.05],t]],  {S, 80, 120},
{t, 0.0001, 1}, BoxRatios -> {1, 1, 1}, PlotPoints -> 40];
```

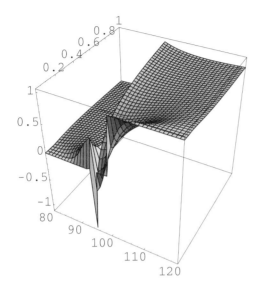

```
Plot3D[If[S>95,
DownAndOutCallGamma[2,S,100,95,0.2,Log[1.1],Log[1.05],t],
UpAndOutCallGamma[2,S,100,95,0.2,Log[1.1],Log[1.05],t]],  {S, 80, 120},
{t, 0.0001, 1},
BoxRatios -> {1, 1, 1}, PlotPoints -> 40];
```

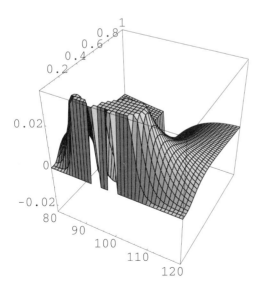

8.5 Up/Down and In (Knock-In) Puts

We can do numerical and graphical checks against the original Rubenstein and Reiner paper. First we introduce the value and price sensitivity functions:

Down and In Puts

```
DownAndInPut[reb_, p_, k_, h_, sd_, r_, q_, t_] :=
If[k >= h, inttwo[p, k, h, sd, r, q, t, -1]-intthree[p, k, h, sd, r, q, t, -1,
1]+ intfour[p, k, h, sd, r, q, t, -1, 1]+ intfive[reb, p, h, sd, r, q, t, 1],
intone[p, k, sd, r, q, t, -1]  + intfive[reb, p, h, sd, r, q, t, 1]];


DownAndInPutDelta[reb_, p_, k_, h_, sd_, r_, q_, t_] =
If[k >= h, Evaluate[D[inttwo[p, k, h, sd, r, q, t, -1]-intthree[p, k, h, sd,
r, q, t, -1, 1]+ intfour[p, k, h, sd, r, q, t, -1, 1]+intfive[reb, p, h, sd,
r, q, t, 1], p]],
Evaluate[D[intone[p, k, sd, r, q, t, -1]  + intfive[reb, p, h, sd, r, q, t,
1], p]]];


DownAndInPutGamma[reb_, p_, k_, h_, sd_, r_, q_, t_] =
If[k >= h, Evaluate[D[inttwo[p, k, h, sd, r, q, t, -1]-intthree[p, k, h, sd,
r, q, t, -1, 1]+ intfour[p, k, h, sd, r, q, t, -1, 1]+intfive[reb, p, h, sd,
r, q, t, 1], {p,2}]],
Evaluate[D[intone[p, k, sd, r, q, t, -1]  + intfive[reb, p, h, sd, r, q, t,
1], {p,2}]]];


DownAndInPutTheta[reb_, p_, k_, h_, sd_, r_, q_, t_] =
If[k >= h, Evaluate[-D[inttwo[p, k, h, sd, r, q, t, -1]-intthree[p, k, h, sd,
r, q, t, -1, 1]+
intfour[p, k, h, sd, r, q, t, -1, 1]+intfive[reb, p, h, sd, r, q, t, 1], t]],
Evaluate[-D[intone[p, k, sd, r, q, t, -1]  + intfive[reb, p, h, sd, r, q, t,
1], t]]];


DownAndInPutVega[reb_, p_, k_, h_, sd_, r_, q_, t_] =
If[k >= h, Evaluate[D[inttwo[p, k, h, sd, r, q, t, -1]-intthree[p, k, h, sd,
r, q, t, -1, 1]+
intfour[p, k, h, sd, r, q, t, -1, 1]+intfive[reb, p, h, sd, r, q, t, 1], sd]],
Evaluate[D[intone[p, k, sd, r, q, t, -1]  + intfive[reb, p, h, sd, r, q, t,
1], sd]]];


DownAndInPutRho[reb_, p_, k_, h_, sd_, r_, q_, t_] =
If[k >= h, Evaluate[D[inttwo[p, k, h, sd, r, q, t, -1]-intthree[p, k, h, sd,
r, q, t, -1, 1]+
intfour[p, k, h, sd, r, q, t, -1, 1]+intfive[reb, p, h, sd, r, q, t, 1], r]],
Evaluate[D[intone[p, k, sd, r, q, t, -1]  + intfive[reb, p, h, sd, r, q, t,
1], r]]];
```

Up and In Puts

```
UpAndInPut[reb_, p_, k_, h_, sd_, r_, q_, t_] :=
If[k >= h, intone[p, k, sd, r, q, t, -1] - inttwo[p, k, h, sd, r, q, t, -1] +
intfour[p, k, h, sd, r, q, t, -1, -1]+intfive[reb, p, h, sd, r, q, t, -1],
intthree[p, k, h, sd, r, q, t, -1, -1]+intfive[reb, p, h, sd, r, q, t, -1]];
```

```
UpAndInPutDelta[reb_, p_, k_, h_, sd_, r_, q_, t_] =
If[k >= h, Evaluate[D[intone[p, k, sd, r, q, t, -1] - inttwo[p, k, h, sd, r,
q, t, -1] + intfour[p, k, h, sd, r, q, t, -1, -1]+intfive[reb, p, h, sd, r, q,
t, -1], p]],
Evaluate[D[intthree[p, k, h, sd, r, q, t, -1, -1]+intfive[reb, p, h, sd, r, q,
t, -1], p]]];
```

```
UpAndInPutGamma[reb_, p_, k_, h_, sd_, r_, q_, t_] =
If[k >= h, Evaluate[D[intone[p, k, sd, r, q, t, -1] - inttwo[p, k, h, sd, r,
q, t, -1] + intfour[p, k, h, sd, r, q, t, -1, -1]+intfive[reb, p, h, sd, r, q,
t, -1], {p,2}]],
Evaluate[D[intthree[p, k, h, sd, r, q, t, -1, -1]+intfive[reb, p, h, sd, r, q,
t, -1], {p,2}]]];
```

```
UpAndInPutTheta[reb_, p_, k_, h_, sd_, r_, q_, t_] =
If[k >= h, Evaluate[-D[intone[p, k, sd, r, q, t, -1] - inttwo[p, k, h, sd, r,
q, t, -1] + intfour[p, k, h, sd, r, q, t, -1, -1]+intfive[reb, p, h, sd, r, q,
t, -1], t]],
Evaluate[-D[intthree[p, k, h, sd, r, q, t, -1, -1]+intfive[reb, p, h, sd, r,
q, t, -1], t]]];
```

```
UpAndInPutVega[reb_, p_, k_, h_, sd_, r_, q_, t_] =
If[k >= h, Evaluate[D[intone[p, k, sd, r, q, t, -1] - inttwo[p, k, h, sd, r,
q, t, -1] + intfour[p, k, h, sd, r, q, t, -1, -1]+intfive[reb, p, h, sd, r, q,
t, -1], sd]],
Evaluate[D[intthree[p, k, h, sd, r, q, t, -1, -1]+intfive[reb, p, h, sd, r, q,
t, -1], sd]]];
```

```
UpAndInPutRho[reb_, p_, k_, h_, sd_, r_, q_, t_] =
If[k >= h, Evaluate[D[intone[p, k, sd, r, q, t, -1] - inttwo[p, k, h, sd, r,
q, t, -1] + intfour[p, k, h, sd, r, q, t, -1, -1]+intfive[reb, p, h, sd, r, q,
t, -1], r]],
Evaluate[D[intthree[p, k, h, sd, r, q, t, -1, -1]+intfive[reb, p, h, sd, r, q,
t, -1], r]]];
```

The following tables of values can be checked directly against a further table of Rubenstein and Reiner.
Note the conversion to annual rates for the interest rates and yield.

```
TableForm[Table[N[DownAndInPut[2,100,k,97,0.20,Log[1.1],Log[1.05],t]],
{k, 90, 110, 10}, {t, 0.5, 1.5, 0.25}]]
```

1.63322	2.12862	2.52532	2.84264	3.09739
4.74369	5.30961	5.70063	5.97865	6.17716
10.177	10.4257	10.5477	10.5925	10.5841

```
TableForm[Table[N[DownAndInPut[2,100,k,100,0.20,Log[1.1],Log[1.05],t]],
{k, 90, 110, 10}, {t, 0.5, 1.5, 0.25}]]
```

1.27535	1.83532	2.27173	2.61689	2.89272
4.38949	5.01828	5.44828	5.75378	5.97313
10.0757	10.2748	10.3863	10.432	10.4282

```
TableForm[Table[N[UpAndInPut[2,100,k,103,0.20,Log[1.1],Log[1.05],t]],
{k, 90, 110, 10}, {t, 0.5, 1.5, 0.25}]]
```

0.879387	1.25265	1.59628	1.89473	2.14931
2.81757	3.46594	3.94601	4.31046	4.59059
7.08064	7.61512	7.96788	8.20427	8.35914

8.6 Up/Down and Out (Knock-Out) Puts

We confine ourselves to numerical checks against the original Rubenstein and Reiner paper. First we introduce the value and price sensitivity functions:

Down and Out Puts

```
DownAndOutPut[reb_, p_, k_, h_, sd_, r_, q_, t_] :=
If[k >= h, intone[p, k, sd, r, q, t, -1]-inttwo[p, k, h, sd, r, q, t,
-1]+intthree[p, k, h, sd, r, q, t, -1, 1]-intfour[p, k, h, sd, r, q, t, -1,
1]+intsix[reb, p, h, sd, r, q, t, 1],
intsix[reb, p, h, sd, r, q, t, 1]];

DownAndOutPutDelta[reb_, p_, k_, h_, sd_, r_, q_, t_] =
If[k >= h, Evaluate[D[intone[p, k, sd, r, q, t, -1]-inttwo[p, k, h, sd, r, q,
t, -1]+
intthree[p, k, h, sd, r, q, t, -1, 1]-intfour[p, k, h, sd, r, q, t, -1, 1]+
intsix[reb, p, h, sd, r, q, t, 1],p]],
Evaluate[D[intsix[reb, p, h, sd, r, q, t, 1], p]]];

DownAndOutPutGamma[reb_, p_, k_, h_, sd_, r_, q_, t_] =
If[k >= h, Evaluate[D[intone[p, k, sd, r, q, t, -1]-inttwo[p, k, h, sd, r, q,
t, -1]+
intthree[p, k, h, sd, r, q, t, -1, 1]-intfour[p, k, h, sd, r, q, t, -1, 1]+
intsix[reb, p, h, sd, r, q, t, 1],{p,2}]],
Evaluate[D[intsix[reb, p, h, sd, r, q, t, 1], {p,2}]]];
```

```
DownAndOutPutTheta[reb_, p_, k_, h_, sd_, r_, q_, t_] =
If[k >= h, Evaluate[-D[intone[p, k, sd, r, q, t, -1]-inttwo[p, k, h, sd, r, q,
t, -1]+intthree[p, k, h, sd, r, q, t, -1, 1]-intfour[p, k, h, sd, r, q, t, -1,
1]+
intsix[reb, p, h, sd, r, q, t, 1],t]],
Evaluate[-D[intsix[reb, p, h, sd, r, q, t, 1], t]]];


DownAndOutPutVega[reb_, p_, k_, h_, sd_, r_, q_, t_] =
If[k >= h, Evaluate[D[intone[p, k, sd, r, q, t, -1]-inttwo[p, k, h, sd, r, q,
t, -1]+intthree[p, k, h, sd, r, q, t, -1, 1]-intfour[p, k, h, sd, r, q, t, -1,
1]+intsix[reb, p, h, sd, r, q, t, 1],sd]],
Evaluate[D[intsix[reb, p, h, sd, r, q, t, 1], sd]]];


DownAndOutPutRho[reb_, p_, k_, h_, sd_, r_, q_, t_] =
If[k >= h, Evaluate[D[intone[p, k, sd, r, q, t, -1]-inttwo[p, k, h, sd, r, q,
t, -1]+ intthree[p, k, h, sd, r, q, t, -1, 1]-intfour[p, k, h, sd, r, q, t,
-1, 1]+intsix[reb, p, h, sd, r, q, t, 1],r]],
Evaluate[D[intsix[reb, p, h, sd, r, q, t, 1], r]]];
```

Up and Out Puts

```
UpAndOutPut[reb_, p_, k_, h_, sd_, r_, q_, t_] :=
If[k >= h, inttwo[p, k, h, sd, r, q, t, -1] -
intfour[p, k, h, sd, r, q, t, -1, -1] + intsix[reb, p, h, sd, r, q, t, -1],
intone[p, k, sd, r, q, t, -1] - intthree[p, k, h, sd, r, q, t, -1, -1] +
intsix[reb, p, h, sd, r, q, t, -1]];


UpAndOutPutDelta[reb_, p_, k_, h_, sd_, r_, q_, t_] =
If[k >= h, Evaluate[D[inttwo[p, k, h, sd, r, q, t, -1] - intfour[p, k, h, sd,
r, q, t, -1, -1] + intsix[reb, p, h, sd, r, q, t, -1], p]],
Evaluate[D[intone[p, k, sd, r, q, t, -1] - intthree[p, k, h, sd, r, q, t, -1,
-1] + intsix[reb, p, h, sd, r, q, t, -1],p]]];


UpAndOutPutTheta[reb_, p_, k_, h_, sd_, r_, q_, t_] =
If[k >= h, Evaluate[-D[inttwo[p, k, h, sd, r, q, t, -1] - intfour[p, k, h, sd,
r, q, t, -1, -1] + intsix[reb, p, h, sd, r, q, t, -1], t]],
Evaluate[-D[intone[p, k, sd, r, q, t, -1] - intthree[p, k, h, sd, r, q, t, -1,
-1] + intsix[reb, p, h, sd, r, q, t, -1],t]]];


UpAndOutPutGamma[reb_, p_, k_, h_, sd_, r_, q_, t_] =
If[k >= h, Evaluate[D[inttwo[p, k, h, sd, r, q, t, -1] - intfour[p, k, h, sd,
r, q, t, -1, -1] + intsix[reb, p, h, sd, r, q, t, -1], {p,2}]],
Evaluate[D[intone[p, k, sd, r, q, t, -1] - intthree[p, k, h, sd, r, q, t, -1,
-1] + intsix[reb, p, h, sd, r, q, t, -1],{p,2}]]];


UpAndOutPutVega[reb_, p_, k_, h_, sd_, r_, q_, t_] =
If[k >= h, Evaluate[D[inttwo[p, k, h, sd, r, q, t, -1] - intfour[p, k, h, sd,
r, q, t, -1, -1] + intsix[reb, p, h, sd, r, q, t, -1], sd]],
Evaluate[D[intone[p, k, sd, r, q, t, -1] - intthree[p, k, h, sd, r, q, t, -1,
-1] + intsix[reb, p, h, sd, r, q, t, -1],sd]]];
```

```
UpAndOutPutRho[reb_, p_, k_, h_, sd_, r_, q_, t_] =
If[k >= h, Evaluate[D[inttwo[p, k, h, sd, r, q, t, -1] - intfour[p, k, h, sd,
r, q, t, -1, -1] + intsix[reb, p, h, sd, r, q, t, -1], r]],
Evaluate[D[intone[p, k, sd, r, q, t, -1] - intthree[p, k, h, sd, r, q, t, -1,
-1] + intsix[reb, p, h, sd, r, q, t, -1],r]]];
```

The following tables of values can also be checked directly against Rubenstein and Reiner. Note the conversion to annual rates for the interest rates and yield.

```
TableForm[
Table[N[DownAndOutPut[2,100,k,97,0.20,Log[1.1],Log[1.05],t]], {k, 90,
110, 10}, {t, 0.5, 1.5, 0.25}]]
```

1.61272	1.6696	1.70282	1.72497	1.74093
1.61638	1.67157	1.70407	1.72585	1.74158
1.86926	1.81193	1.79498	1.79013	1.78968

```
TableForm[
Table[N[DownAndOutPut[2,100,k,100,0.20,Log[1.1],Log[1.05],t]], {k, 90,
110, 10}, {t, 0.5, 1.5, 0.25}]]
```

2.	2.	2.	2.	2.
2.	2.	2.	2.	2.
2.	2.	2.	2.	2.

```
TableForm[
Table[N[UpAndOutPut[2,100,k,103,0.20,Log[1.1],Log[1.05],t]], {k, 90,
110, 10}, {t, 0.5, 1.5, 0.25}]]
```

2.36986	2.55061	2.63858	2.68125	2.69901
3.54582	3.52027	3.46541	3.40241	3.33814
4.96897	4.62758	4.38154	4.18678	4.02461

8.7 An Investigation of Up and Out Puts:

What Happens When k =h (Barrier = Strike) and q > r (Dividend Yield Greater than Risk-Free Rate)?

In the following plots we investigate the structure of the valuation function and associated Delta, Gamma, and Vega. In the first plot we note the almost flat structure of the value, here viewed as a surface $V(S, t)$. The value is zero at the Barrier, here set to 50. The strike is also 50, the volatility is 0.25 (25%), and the continuously compounded interest rate $r = 10\%$. The dividend yield is 15%. The time is measured in years and varies from zero to one year. The stock price varies from 30 to 50 (barrier).

```
Plot3D[UpAndOutPut[0, S, 50, 50, 0.25,0.1, 0.15, t],
{S, 30, 50}, {t, 0.001, 1}, PlotRange -> All];
```

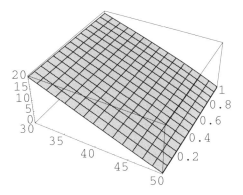

In the following we show Vega, computed symbolically by the computer. *It is nowhere positive.*

```
Plot3D[UpAndOutPutVega[0, S, 50, 50, 0.25,0.1, 0.15, t],
{S, 30, 50}, {t, 0.001, 1}, PlotRange -> All];
```

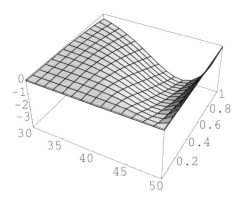

Because of the flatness of the value surface, Delta is close to -1 everywhere, as revealed in the next plot.

```
Plot3D[UpAndOutPutDelta[0, S, 50, 50, 0.25,0.1, 0.15, t],
{S, 30, 50}, {t, 0.001, 1}, PlotRange -> All];
```

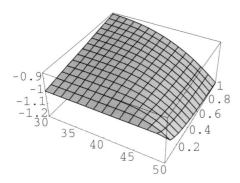

Gamma is very small and negative, as is shown in the following plot (note the vertical axes scale).

```
Plot3D[UpAndOutPutGamma[0, S, 50, 50, 0.25,0.1, 0.15, t],
{S, 30, 50}, {t, 0.001, 1}, PlotRange -> All, PlotPoints -> 30];
```

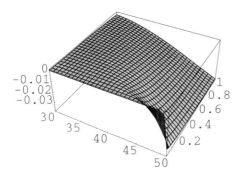

Theta has no definite sign, as revealed by the following plot and subsequent evaluations:

```
Plot3D[UpAndOutPutTheta[0, S, 50, 50, 0.25,0.1, 0.15, t],
{S, 20, 50}, {t, 0.001, 1}, PlotRange -> All];
```

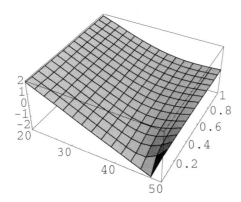

```
UpAndOutPutTheta[0, 30, 50, 50, 0.25,0.1, 0.15, 1]//N
```

```
    0.69521
```

```
UpAndOutPutTheta[0, 40, 50, 50, 0.25,0.1, 0.15, 0.2]//N
```

```
    -0.838731
```

Case Variant: k = h (Barrier = Strike) and q = r (Dividend Yield Equal to Risk-Free Rate.)

The strike is also 50, the volatility is 0.25 (25%), and the continuously compounded interest rate $r = 10\%$. The dividend yield is also 10%. The time is measured in years and varies from zero to 3 years. The stock price varies from 30 to 50 (Barrier).

```
Plot3D[UpAndOutPut[0, S, 50, 50, 0.25,0.1, 0.1, t],
{S, 30, 50}, {t, 0.001, 3}, PlotRange -> All];
```

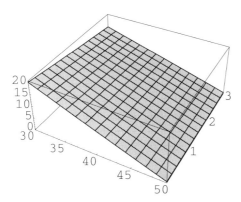

In the following we show Vega, computed symbolically by the computer. It is zero everywhere (to machine precision of the plotted function).

```
Plot3D[UpAndOutPutVega[0, S, 50, 50, 0.25,0.1, 0.10, t],
{S, 30, 50}, {t, 0.001, 3}, PlotRange -> {-0.01, 0.01}];
```

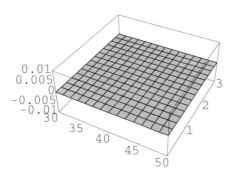

Because of the flatness of the value surface, Delta is close to -1 everywhere, as revealed in the next plot - it is constant in price of the underlying but not constant in time.

```
Plot3D[UpAndOutPutDelta[0, S, 50, 50, 0.25,0.1, 0.1, t],
{S, 30, 50}, {t, 0.001, 3}, PlotRange -> All];
```

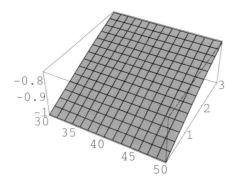

Gamma is zero, as is shown in the next plot, which just reveals the machine-precision rumblings of *Mathematica*!

```
Plot3D[UpAndOutPutGamma[0, S, 50, 50, 0.25,0.1, 0.1, t],
{S, 30, 50}, {t, 0.001, 3}, PlotRange -> All];
```

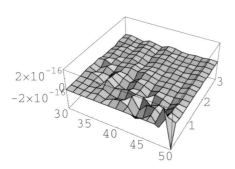

Making Some Symbolic Checks

Is Vega really identically zero? We can get *Mathematica* to help us find out. Note first that our functions are defined by an If statement. We can force inspection of part of the expression as follows. First an example to show us the way:

```
If[a > b, u, v]

    If[a > b, u, v]
```

```
FullForm[%]

    If[Greater[a, b], u, v]
```

```
%[[2]]

    u
```

So we apply this idea to our formula for Vega - let's see it first in its full "glory":

```
UpAndOutPutVega[0, S, K, K, sd, r, r, t][[2]]
```

$$
\frac{E^{-r\,t-\frac{1}{2}\left(-\frac{sd\,\sqrt{t}}{2}-\frac{\log\left[\frac{K}{S}\right]}{sd\,\sqrt{t}}\right)^2}\,K\left(-\frac{\sqrt{t}}{2}+\frac{\log\left[\frac{K}{S}\right]}{sd^2\,\sqrt{t}}\right)}{\sqrt{2\,\pi}} -
$$

$$
\frac{E^{-r\,t-\frac{1}{2}\left(\frac{sd\,\sqrt{t}}{2}-\frac{\log\left[\frac{K}{S}\right]}{sd\,\sqrt{t}}\right)^2}\,S\left(\frac{\sqrt{t}}{2}+\frac{\log\left[\frac{K}{S}\right]}{sd^2\,\sqrt{t}}\right)}{\sqrt{2\,\pi}} -
$$

$$
\frac{E^{-r\,t-\frac{1}{2}\left(-\frac{sd\,\sqrt{t}}{2}-\frac{\log\left[\frac{S}{K}\right]}{sd\,\sqrt{t}}\right)^2}\,S\left(-\frac{\sqrt{t}}{2}+\frac{\log\left[\frac{S}{K}\right]}{sd^2\,\sqrt{t}}\right)}{\sqrt{2\,\pi}} +
$$

$$
\frac{E^{-r\,t-\frac{1}{2}\left(\frac{sd\,\sqrt{t}}{2}-\frac{\log\left[\frac{S}{K}\right]}{sd\,\sqrt{t}}\right)^2}\,K\left(\frac{\sqrt{t}}{2}+\frac{\log\left[\frac{S}{K}\right]}{sd^2\,\sqrt{t}}\right)}{\sqrt{2\,\pi}}
$$

This is a bit of a mess, so let's simplify it:

`Simplify[%]`

$$\frac{E^{-\frac{1}{8}\,(8\,r+sd^2)\,t}\,\left(-E^{-\frac{Log[\frac{K}{S}]^2}{2\,sd^2\,t}}\,\sqrt{\frac{K}{S}}\,S + E^{-\frac{Log[\frac{S}{K}]^2}{2\,sd^2\,t}}\,K\,\sqrt{\frac{S}{K}}\right)\,\sqrt{t}}{\sqrt{2\,\pi}}$$

There are some square roots that should cancel - let's sort this out:

`PowerExpand[%]`

$$\frac{E^{-\frac{1}{8}\,(8\,r+sd^2)\,t}\,\left(-E^{-\frac{(Log[K]-Log[S])^2}{2\,sd^2\,t}}\,\sqrt{K}\,\sqrt{S} + E^{-\frac{(-Log[K]+Log[S])^2}{2\,sd^2\,t}}\,\sqrt{K}\,\sqrt{S}\right)\,\sqrt{t}}{\sqrt{2\,\pi}}$$

We are almost there:

`Simplify[%]`

0

So we can check that Vega is identically zero in this case. What about Gamma? We may as well try the sequence of steps that worked for Vega.

```
Simplify[
  PowerExpand[Simplify[UpAndOutPutGamma[0, S, K, K, sd, r, r, t][[2]]]]]
```

0

So this particular type of option has a Vega and a Gamma which are identically zero!

Chapter 8 Bibliography

Hull, J., 1996, *Options, Futures and Other Derivatives*, 3rd edition, Prentice-Hall.

Rubenstein, M. and Reiner, E., 1991, Breaking down the barriers, *RISK Magazine*, September.

Chapter 9.
Analytical Models of Lookbacks

9.1 Introduction

Here we take our first look at some path-dependent options, where the *payoff depends on the maximum or minimum value attained along the path*. There are other types of path-dependence, based on average values along the path, that will be considered in Chapter 10. Our first task is to establish some unambiguous terminology, as follows, by considering the option payoff as a function of the path taken by the stock price between initiation of the option and expiry.

Standard Lookback Call

Payoff is

$$S[\text{expiry}] - \text{Min}[S[\text{path}]]$$

Standard Lookback Put

Payoff is

$$\text{Max}[S[\text{path}]] - S[\text{expiry}]$$

Lookback Call on Maximum

Payoff is

$$\text{Max}[\text{Max}[S[\text{path}]] - K, \quad 0]$$

Lookback Put on Minimum

Payoff is

$$\text{Max}[0, \quad K - \text{Min}[S[\text{path}]]]$$

For standard Lookback options, we have used

(a) Section 18.1 of the 3rd edition of J. Hull's book and the original research by Goldman, Sosin and Gatto (1979),
(b) A research paper by Conze and Viswanathan "Path dependent options" (1991).

For other lookbacks, notably the Call on maximum and Put on minimum, we have also used Conze and Viswanathan. All of these models will be re-visited later from the point of view of Monte Carlo simulation, as will the Asian options considered in the next chapter. By now the reader may have got the message regarding analytical models - you can literally just pour formulae into *Mathematica* and get on with it. We have made a point of giving a chapter devoted to Lookbacks because the Monte Carlo simulation of such options is highly problematic, and we need an analytical model to do the verification in order to get things straight.

Some Warnings on Issues Regarding Lookbacks.

For lookback options, whether they are of the standard lookback call/put variety or the Call on Max/Put on Min variety, there is at least the potential for confusion in the literature about terminology. The analytical valuation formulae for these options use both the current underlying price and its "historical min/max" or "maximum-to-date". It is important to realize that this is the min/max achieved between the time since the option was initiated and the present valuation time. Historical max or min achieved prior to the initiation is irrelevant. At the time of initiation, we must set Smin = *S* or Smax = *S*; otherwise wrong results will be obtained.

9.2 Analytical Models of Lookbacks 1 - Goldman *et al*/Hull

Here we give the analytical valuation formulae for the standard Lookback Call/Put as given by Goldman *et al*, in the form presented by Hull.

Loading the Standard Euro-options Package

The Lookback options involve some of the auxiliary functions used in Black-Scholes - we load the Package to treat this first.

```
Needs["Derivatives`BlackScholes`"]
```

Additional Auxiliary Functions for Use in Lookbacks

Notation is as used by Hull:

```
aone[p_, pmin_, sd_, r_, q_, t_] :=
(Log[p/pmin]+(r - q + sd^2/2)*t)/(sd*Sqrt[t]);
atwo[p_, pmin_, sd_, r_, q_, t_] :=
aone[p,pmin,sd,r,q,t] - sd*Sqrt[t];
athree[p_, pmin_, sd_, r_, q_, t_] :=
(Log[p/pmin]+(q - r + sd^2/2)*t)/(sd*Sqrt[t]);
yone[p_, pmin_, sd_, r_, q_] :=
-2*(r - q - sd^2/2)*Log[p/pmin]/sd^2;
```

```
bone[p_, pmax_, sd_, r_, q_, t_] :=
(Log[pmax/p]+(-r + q + sd^2/2)*t)/(sd*Sqrt[t]);
btwo[p_, pmax_, sd_, r_, q_, t_] :=
bone[p,pmax,sd,r,q,t] - sd*Sqrt[t];
bthree[p_, pmax_, sd_, r_, q_, t_] :=
(Log[pmax/p]+(r - q - sd^2/2)*t)/(sd*Sqrt[t]);
ytwo[p_, pmax_, sd_, r_, q_] :=
2*(r - q - sd^2/2)*Log[pmax/p]/sd^2;
```

The Lookback Call

First we introduce the value function:

```
LookBackCall[p_, pmin_, sd_, r_, q_, t_] :=
p*Exp[-q*t]*Norm[aone[p, pmin, sd, r, q, t]] -
p*Exp[-q*t]*sd^2*Norm[-aone[p, pmin, sd, r, q, t]]/(2*(r-q)) -
pmin*Exp[-r*t]*(Norm[atwo[p, pmin, sd, r, q, t]]-
sd^2*Exp[yone[p, pmin, sd, r, q]]*Norm[-athree[p,pmin,sd,r,q,t]]/(2*(r-
q)))
```

Now we compute the sensitivities by differentiation:

```
LookBackCallDelta[p_, pmin_, sd_, r_, q_, t_] =
D[LookBackCall[p, pmin, sd, r, q, t], p];
```

```
LookBackCallGamma[p_, pmin_, sd_, r_, q_, t_] =
D[LookBackCall[p, pmin, sd, r, q, t], {p,2}];
```

```
LookBackCallTheta[p_, pmin_, sd_, r_, q_, t_] =
D[-LookBackCall[p, pmin, sd, r, q, t], t];
```

```
LookBackCallVega[p_, pmin_, sd_, r_, q_, t_] =
D[LookBackCall[p, pmin, sd, r, q, t], sd];
```

```
LookBackCallRho[p_, pmin_, sd_, r_, q_, t_] =
D[LookBackCall[p, pmin, sd, r, q, t], r];
```

Lookback Put

First we introduce the value and sensitivity functions:

```
LookBackPut[p_, pmax_, sd_, r_, q_, t_] :=
-p*Exp[-q*t]*Norm[btwo[p, pmax, sd, r, q, t]] +
p*Exp[-q*t]*sd^2*Norm[-btwo[p, pmax, sd, r, q, t]]/(2*(r-q)) +
pmax*Exp[-r*t]*(Norm[bone[p, pmax, sd, r, q, t]]-
sd^2*Exp[ytwo[p, pmax, sd, r, q]]*Norm[-bthree[p,pmax,sd,r,q,t]]/(2*(r-
q)))
```

```
LookBackPutDelta[p_, pmax_, sd_, r_, q_, t_] =
D[LookBackPut[p, pmax, sd, r, q, t], p];

LookBackPutGamma[p_, pmax_, sd_, r_, q_, t_] =
D[LookBackPut[p, pmax, sd, r, q, t], {p,2}];

LookBackPutTheta[p_, pmax_, sd_, r_, q_, t_] =
D[-LookBackPut[p, pmax, sd, r, q, t], t];

LookBackPutVega[p_, pmax_, sd_, r_, q_, t_] =
D[LookBackPut[p, pmax, sd, r, q, t], sd];

LookBackPutRho[p_, pmax_, sd_, r_, q_, t_] =
D[LookBackPut[p, pmax, sd, r, q, t], r];
```

9.3 Analytical Models of Lookbacks 2 - Conze-Viswanathan

In this section we provide analytical functions for all four Lookbacks that we are considering, including the standard Lookbacks considered above, and also the Call on max and Put on min not considered previously. These can be generalized to allow for non-zero yield, and differentiated to get the sensitivity parameters, just as was the case for Barriers.

Details of Auxiliary Functions

This is all based on Conze-Viswanathan and *hence uses zero q* (it is ignored but added as an argument so as to facilitate easy generalizations).

```
dzero[p_, k_, sd_, r_, q_, t_] :=
(Log[p/k] + (r + sd^2/2)*t)/(sd*Sqrt[t]);
done[p_, pmax_, sd_, r_, q_, t_] := dzero[p, pmax, sd, r, q, t];
dtwo[p_, pmin_, sd_, r_, q_, t_] := dzero[p, pmin, sd, r, q, t];
```

The Lookback Call

```
CVLookBackCall[p_, pmin_, sd_, r_, q_, t_] :=
p*Norm[dtwo[p, pmin, sd, r, q, t]] -
Exp[-r*t]*pmin*Norm[dtwo[p, pmin, sd, r, q, t] - sd*Sqrt[t]] +
Exp[-r*t]*(sd^2/(2*r))*p*(
(p/pmin)^(-2*r/sd^2)*Norm[-dtwo[p, pmin, sd, r, q, t] + 2*r*-
Sqrt[t]/sd] -
Exp[r*t]*Norm[-dtwo[p, pmin, sd, r, q, t]])
```

The Lookback Put

```
CVLookBackPut[p_, pmax_, sd_, r_, q_, t_] :=
-p*Norm[-done[p, pmax, sd, r, q, t]] +
Exp[-r*t]*pmax*Norm[-done[p, pmax, sd, r, q, t] + sd*Sqrt[t]] +
Exp[-r*t]*(sd^2/(2*r))*p*(
-(p/pmax)^(-2*r/sd^2)*Norm[done[p, pmax, sd, r, q, t] - 2*r*-
Sqrt[t]/sd] +
Exp[r*t]*Norm[done[p, pmax, sd, r, q, t]])
```

The Call on Maximum

```
CVCallOnMax[p_, pmax_, k_, sd_, r_, q_, t_] :=
If[k >= pmax,
p*Norm[dzero[p, k, sd, r, q, t]] -
Exp[-r*t]*k*Norm[dzero[p, k, sd, r, q, t] - sd*Sqrt[t]] +
Exp[-r*t]*(sd^2/(2*r))*p*(
-(p/k)^(-2*r/sd^2)*Norm[dzero[p, k, sd, r, q, t] - 2*r*Sqrt[t]/sd] +
Exp[r*t]*Norm[dzero[p, k, sd, r, q, t]]),
Exp[-r*t]*(pmax - k) +
p*Norm[done[p, pmax, sd, r, q, t]] -
Exp[-r*t]*pmax*Norm[done[p, pmax, sd, r, q, t] - sd*Sqrt[t]] +
Exp[-r*t]*(sd^2/(2*r))*p*(-(p/pmax)^(-2*r/sd^2)*Norm[done[p, pmax, sd,
r, q, t]
- 2*r*Sqrt[t]/sd] +
Exp[r*t]*Norm[done[p, pmax, sd, r, q, t]])]
```

The Put on Minimum

```
CVPutOnMin[p_, pmin_, k_, sd_, r_, q_, t_] :=
If[k <= pmin,
-p*Norm[-dzero[p, k, sd, r, q, t]] +
Exp[-r*t]*k*Norm[-dzero[p, k, sd, r, q, t] + sd*Sqrt[t]] +
Exp[-r*t]*(sd^2/(2*r))*p*(
(p/k)^(-2r/sd^2)*Norm[-dzero[p, k, sd, r, q, t] + 2*r*Sqrt[t]/sd] -
Exp[r*t]*Norm[-dzero[p, k, sd, r, q, t]]),
Exp[-r*t]*(k - pmin) -
p*Norm[-dtwo[p, pmin, sd, r, q, t]] +
Exp[-r*t]*pmin*Norm[-dtwo[p, pmin, sd, r, q, t] + sd*Sqrt[t]] +
Exp[-r*t]*(sd^2/(2*r))*p*(
(p/pmin)^(-2r/sd^2)*Norm[-dtwo[p, pmin, sd, r, q, t] + 2*r*Sqrt[t]/sd]
-
Exp[r*t]*Norm[-dtwo[p, pmin, sd, r, q, t]])]
```

9.4 A Brief Investigation of Standard Lookback Calls

In this section we evaluate the two analytical functions

`CVLookBackCall[p,pmin,sd,r,q,t]`

  `LookBackCall[p,pmin,sd,r,q,t]`

and show that the results are numerically consistent, provided that pmin is the minimum price between option initiation and valuation, which further implies that we set p = pmin at initiation. You could also investigate this symbolically.

p = pmin = 50;
sd = 20%
t = 1;
r = 5%;
q = 0.

p = pmin

`CVLookBackCall[50., 50., 0.2, 0.05, 0, 1]`

    8.6084

`LookBackCall[50., 50., 0.2, 0.05, 0, 1]`

    8.6084

pmin < p

`CVLookBackCall[50., 40., 0.2, 0.05, 0, 1]`

    12.6869

`LookBackCall[50., 40., 0.2, 0.05, 0, 1]`

    12.6869

9.5 A Brief Investigation of Standard Lookback Puts

In this section we evaluate the two analytical functions

```
CVLookBackPut[p,pmax,sd,r,q,t]
```

```
  LookBackPut[p,pmax,st,r,q,t]
```

and show that the results are numerically consistent, provided that pmax is the maximum price between option initiation and valuation, which further implies that we set p = pmax at initiation.

p = pmax = 50;
sd = 20 %
t = 1;
 r = 5 %;
q = 0.

p = pmax

```
CVLookBackPut[50, 50, 0.2, 0.05, 0, 1]
```

```
    7.14528
```

```
LookBackPut[50, 50, 0.2, 0.05, 0, 1]
```

```
    7.14528
```

p < pmax

```
CVLookBackPut[50, 60., 0.2, 0.05, 0, 1]
```

```
    10.1137
```

```
LookBackPut[50, 60., 0.2, 0.05, 0, 1]
```

```
    10.1137
```

9.6. A Brief Investigation of the Call on Maximum

In this section we explore the CV analytical formula:

```
CVCallOnMax[p,pmax,k,sd,r,q,t]
```

p = pmax = 50;
sd = 20 %
t = 1;

$r = 5\%$;
$q = 0$.

$p = $ pmax

```
CVCallOnMax[50, 50, 50, 0.2, 0.05, 0, 1]
```

> 9.58381

pmax $> p$

```
CVCallOnMax[50., 60., 50., 0.2, 0.05, 0, 1]
```

> 12.5522

$K > $ pmax $> p$

```
CVCallOnMax[50., 60., 70., 0.2, 0.05, 0, 1]
```

> 0.744137

9.7. A Brief Investigation of the Put on Minimum

In this section we explore the analytical functions given by the formulae

```
CVPutOnMin[p,pmin,k,sd,r,q,t]
```

Option at Initiation

$p = $ pmin $= 50$;
sd $= 20\%$
$t = 1$;
$r = 5\%$;
$q = 0$.

$p = $ pmin

```
CVPutOnMin[50, 50, 50, 0.2, 0.05, 0, 1]
```

> 6.16987

$K < $ pmin $< p$

```
CVPutOnMin[50., 45., 40., 0.2, 0.05, 0, 1]
```

```
0.736034
```

$K > \text{pmin} < p$

```
CVPutOnMin[50., 45., 60., 0.2, 0.05, 0, 1]
```

```
16.7804
```

9.8 A Brief Look at Ladders

There is an interesting variation on this general type of path-dependent option that generalizes the Lookback Call on maximum. Rather than use the maximum value of the price over the entire path, we lock in gains associated with the rungs of a ladder, and work with a payoff

$$\text{Max}[S_T - K, L_1 - K, L_2 - K, \ldots, 0] \tag{1}$$

The L_i are the rungs of the ladder, and are predetermined, and ordered: $L_1 < L_2 < \ldots < L_n$. If the stock price reaches L_1, but not L_2, over the life of the option, the payoff based on L_2 and higher ladders is not applied. This was priced by Street by an ingenious argument involving a package consisting of a standard Call, a Put, and a package of Knock-out Puts (Street, 1992). We can build such options very easily by using our existing results. In the following it is assumed that the functions defined in Chapter 8 have been loaded.

Triple-Rung Ladder Option Value Functions

Here is Street's model based on three rungs - it uses a value of **pmax** to describe the maximum value attained since initiation, and also properly knocks out the options:

```
ThreeRungPathDep[p_, pmax_, k_, runga_, rungb_, rungc_, sd_, r_, q_,
t_] :=
Module[{newpmax},
newpmax = Max[p, pmax];
Which[
newpmax <= runga,
(BlackScholesCall[p, k, sd, r, q, t] -
BlackScholesPut[p, k, sd, r, q, t] +
BlackScholesPut[p, rungc, sd, r, q, t] +
UpAndOutPut[0, p, k, runga, sd, r, q, t] -
UpAndOutPut[0, p, runga, runga, sd, r, q, t] +
UpAndOutPut[0, p, runga, rungb, sd, r, q, t] -
UpAndOutPut[0, p, rungb, rungb, sd, r, q, t] +
UpAndOutPut[0, p, rungb, rungc, sd, r, q, t] -
UpAndOutPut[0, p, rungc, rungc, sd, r, q, t]),
newpmax <= rungb,
(BlackScholesCall[p, k, sd, r, q, t] -
```

```
BlackScholesPut[p, k, sd, r, q, t] +
BlackScholesPut[p, rungc, sd, r, q, t] +
UpAndOutPut[0, p, runga, rungb, sd, r, q, t] -
UpAndOutPut[0, p, rungb, rungb, sd, r, q, t] +
UpAndOutPut[0, p, rungb, rungc, sd, r, q, t] -
UpAndOutPut[0, p, rungc, rungc, sd, r, q, t]),
newpmax <= rungc,
(BlackScholesCall[p, k, sd, r, q, t] -
BlackScholesPut[p, k, sd, r, q, t] +
BlackScholesPut[p, rungc, sd, r, q, t] +
UpAndOutPut[0, p, rungb, rungc, sd, r, q, t] -
UpAndOutPut[0, p, rungc, rungc, sd, r, q, t]),
True,
(BlackScholesCall[p, k, sd, r, q, t] -
BlackScholesPut[p, k, sd, r, q, t] +
BlackScholesPut[p, rungc, sd, r, q, t])
]]
```

Examples - Price at Revaluation below Bottom Rung

Max since Existence below Bottom Rung

```
ThreeRungPathDep[50., 55., 50., 60., 70.,
80., 0.4, 0.1, 0, 0.25]
```

```
    5.77933
```

Max since Existence above Bottom Rung and below Next Rung

```
ThreeRungPathDep[50., 65., 50., 60., 70.,
80., 0.4, 0.1, 0, 0.25]
```

```
    11.347
```

Max since Existence above Next to Top Rung and below Top Rung

```
ThreeRungPathDep[50., 75., 50., 60., 70.,
80., 0.4, 0.1, 0, 0.25]
```

```
    19.8489
```

Max since Existence above Top Rung

```
ThreeRungPathDep[50., 85., 50., 60., 70.,
80., 0.4, 0.1, 0, 0.25]

    29.3159
```

Ladder option values can also be calculated by Monte Carlo simulation.

Chapter 9 Bibliography

Conze, A. and Viswanathan, R., 1991, Path dependent options: the case of lookback options, *Journal of Finance*, 46, p. 1893.

Goldman, M.B., Sosin, H.B. and Gatto, M.A., 1979, Path-dependent options: buy at the low, sell at the high, *Journal of Finance*, 34, p. 1111.

Street, A., 1992, Stuck up a ladder, *RISK Magazine*, May.

Chapter 10.
Vanilla Asian Options - Analytical Methods

Extending the Power of Analytical Methods to
Problems Involving the Average

10.1 Introduction

In Chapter 9 we considered path-dependent options involving the maximum or minimum of the underlying price when sampled continuously throughout the life of the option. Another type of path-dependent feature is the *average* of the underlying along the path taken from initiation to expiry. We can consider both the arithmetic average and the geometric average. The latter has the advantage of having a simple closed-form solution. Recent research has provided a means of solving the arithmetic problem almost in closed form. In this chapter we shall explore the mechanics of both types of option, drawing heavily on the existing mathematical knowledge. We shall see also that *Mathematica* can be used to provided some simplifications to the analysis made thus far on the arithmetic case.

The analytical formulation of the pricing problem can be used as is when the option contract is appropriate for such a model. More often, a more complex contract may require Monte Carlo simulation of the stock price path. In these cases these analytical models can both serve as a useful check on the Monte Carlo algorithm (Are there enough paths, and are we sampling the path sufficiently?), and also provide useful control variates for variance reduction in more general cases. A classic example of this is to use the exact solution for a continuously sampled geometric Asian as a control variate for corresponding arithmetic problems. The results presented here will lay the foundations for this approach, which we shall revisit in Chapter 25.

10.2 Definition of Asian Options

Just as in the case of the use of max/min, there are four different ways of feeding an average into an otherwise vanilla object. We can consider Calls and Puts where either stock price at expiry, or the strike, is replaced by the average. Here of course we can consider two further sub-cases according to whether we are working arithmetically or geometrically. In the following definitions, we shall denote the average generically by

$$< S > \tag{1}$$

When it is important to indicate the nature of the averaging, we can use a subscript A or G as follows:

$$< S >_A \qquad < S >_G \tag{2}$$

Average Price Call

Here the payoff is

$$\mathrm{Max}[0, \ < S > - K] \tag{3}$$

Average Price Put

Here the payoff is

$$\mathrm{Max}[0, \ K - < S >] \tag{4}$$

Average Strike Call

Here the payoff is

$$\mathrm{Max}[0, \ S - < S >] \tag{5}$$

Average Strike Put

Here the payoff is

$$\mathrm{Max}[0, \ < S > - S] \tag{6}$$

In what follows we shall focus attention on the average price options, where the strike is fixed and the price used is the average. These are the types most commonly considered.

10.3 Analytical Mathematical Models of Geometric Asian Price Options

For geometric Asian options, we have used the observations made by Kemna and Vorst (1990) who pointed out that the geometric average of a set of log-normally distributed variables is also log-normal. This allows standard formulae to be used, at least for the average price options, with some changes of variables. These can be written down for either discrete or continuous sampling. A large class of such changes of variables are given in Chapter 11 of Wilmott *et al* (1993), to which we refer the reader for the theory, and for a further discussion of average strike options. In the continuous-sampling case, the results are also given by Hull (1996, Chapter 18). It is then just a matter of writing down the effective volatility and yield.

$$\mathrm{VolEff}(\sigma) = \frac{\sigma}{\sqrt{3}} \tag{7}$$

$$\mathrm{QEff}(r, q, \sigma) = \frac{1}{2}\left(\frac{\sigma^2}{6} + q + r\right) \tag{8}$$

Mathematica Implementation of Geometric Asian

There is now very little left to do. We just give definitions for the changes of variables thus:

$$\text{VolEff}[\sigma\_] := \frac{\sigma}{\sqrt{3}}$$

$$\text{QEff}[r\_, q\_, \sigma\_] := \frac{1}{2}\left(r + q + \frac{\sigma^2}{6}\right)$$

Now we load the standard Package for vanilla European options:

```
Needs["Derivatives`BlackScholes`"]
```

Now we define the valuation and Greek functions in the usual way, using symbolic differentiation. First we give the valuations:

```
GeoAsianPriceCall[p_, k_, σ_, r_, q_, t_] := BlackScholesCall[p,k,Vol-
Eff[σ],r,QEff[r,q,σ],t];
GeoAsianPricePut[p_, k_, σ_, r_, q_, t_] := BlackScholesPut[p,k,Vol-
Eff[σ],r,QEff[r,q,σ],t]
```

Next we write down the Greeks for the Call and the Put:

```
GeoAsianPriceCallDelta[p_, k_, σ_, r_, q_, t_] =
  Evaluate[D[GeoAsianPriceCall[p, k, σ, r, q, t], p]];
GeoAsianPriceCallGamma[p_, k_, σ_, r_, q_, t_] =
  Evaluate[D[GeoAsianPriceCall[p, k, σ, r, q, t], {p, 2}]];
GeoAsianPriceCallTheta[p_, k_, σ_, r_, q_, t_] =
  -Evaluate[D[GeoAsianPriceCall[p, k, σ, r, q, t], t]];
GeoAsianPriceCallRho[p_, k_, σ_, r_, q_, t_] =
  Evaluate[D[GeoAsianPriceCall[p, k, σ, r, q, t], r]];
GeoAsianPriceCallVega[p_, k_, σ_, r_, q_, t_] =
  Evaluate[D[GeoAsianPriceCall[p, k, σ, r, q, t], σ]];
GeoAsianPricePutDelta[p_, k_, σ_, r_, q_, t_] =
  Evaluate[D[GeoAsianPricePut[p, k, σ, r, q, t], p]];
GeoAsianPricePutGamma[p_, k_, σ_, r_, q_, t_] =
  Evaluate[D[GeoAsianPricePut[p, k, σ, r, q, t], {p, 2}]];
GeoAsianPricePutTheta[p_, k_, σ_, r_, q_, t_] =
  -Evaluate[D[GeoAsianPricePut[p, k, σ, r, q, t], t]];
GeoAsianPricePutRho[p_, k_, σ_, r_, q_, t_] =
  Evaluate[D[GeoAsianPricePut[p, k, σ, r, q, t], r]];
GeoAsianPricePutVega[p_, k_, σ_, r_, q_, t_] =
  Evaluate[D[GeoAsianPricePut[p, k, σ, r, q, t], σ]];
```

These functions can now be explored in the usual way - for example, all the investigations of Chapter 7 can be carried out just as easily. We shall return to the geometric Asian price options when we consider Monte Carlo pricing methods.

10.4 Approximate Analytical Mathematical Models of Arithmetic Asian Price Options

When we come to consider the *arithmetic* averaging procedure, there is no simple closed-form solution along the lines of the Black-Scholes formula. However, there are both an approximate solution with the simplicity of the geometric solution, and a recently discovered exact analytical solution that uses Laplace transform methods. In this section we shall consider the approximate model - the Laplace approach will be given in Section 10.5.

Here we develop the analytical approximation given by Turnbull and Wakeman (1991). It is based on computing exactly the first two moments of the distribution of the average price, and valuing assuming that the distribution is log-normal with the same two moments. Pros and cons of this are discussed by Levy and Turnbull (1992). That article makes the point that the analytical approximation is very good for volatilities up to about 20%, but has biases for higher volatility. The pattern follows that for geometric Asians, but first we need to write down the appropriate moments. These results are also described by Hull (1996). The first two moments of the distribution are given by $S M_1$ and $S^2 M_2$, where

$$M_1(r, q, t) = \frac{e^{(r-q)t} - 1}{(r-q)t} \tag{9}$$

$$M_2[\sigma, r, q, t] = \frac{2 \left(\frac{1}{\sigma^2 + 2(r-q)} - \frac{e^{(r-q)t}}{\sigma^2 - q + r} \right)}{(r-q)t^2} + \frac{2 e^{(\sigma^2 + 2(r-q))t}}{((\sigma^2 - q + r)(\sigma^2 - 2q + 2r))t^2} \tag{10}$$

The effective yields and volatilities are then given by

$$\text{QEffA}(r, q, t) = r - \frac{\log(M_1)}{t}$$

$$\text{VolEffA}(\sigma, r, q, t) := \sqrt{\frac{\log(M_2)}{t} - 2(r - \text{QEffA})} \tag{11}$$

Mathematica Implementation of Approximate Arithmetic Asian

The pattern for doing this has already been established for the geometric case, and we just parrot it:

```
mone[r_, q_, t_] := Exp[(r - q) t] - 1 ;
                    (r - q) t
```

```
mtwo[σ_, r_, q_, t_] :=
    2 Exp[(2 (r - q) + σ²) t]          2 ( 1         -  Exp[(r-q) t] )
    ─────────────────────────────  +   ( 2 (r-q) +σ²      r-q+σ²    ) ;
    t² ((r - q + σ²) (2 r - 2 q + σ²))      t² (r - q)
```

$$\text{QEffA}[r\_, q\_, t\_] := r - \frac{\text{Log}[\text{mone}[r, q, t]]}{t};$$

$$\text{VolEffA}[\sigma\_, r\_, q\_, t\_] :=$$

$$\sqrt{\frac{\text{Log}[\text{mtwo}[\sigma, r, q, t]]}{t} - 2\,(r - \text{QEffA}[r, q, t])};$$

Now we can give the valuations, using a similar change of variables:

```
AppAriAsianPriceCall[p_, k_, σ_, r_, q_, t_] := BlackScholesCall[p,k,-
VolEffA[σ,r,q,t],r,QEffA[r,q,t],t];
AppAriAsianPricePut[p_, k_, σ_, r_, q_, t_] := BlackScholesPut[p,k,Vol-
EffA[σ,r,q,t],r,QEffA[r,q,t],t]
```

Next we write down the Greeks for the Call:

```
AppAriAsianPriceCallDelta[p_, k_, σ_, r_, q_, t_] =
  Evaluate[D[AppAriAsianPriceCall[p, k, σ, r, q, t], p]];
AppAriAsianPriceCallGamma[p_, k_, σ_, r_, q_, t_] =
  Evaluate[D[AppAriAsianPriceCall[p, k, σ, r, q, t], {p, 2}]];
AppAriAsianPriceCallTheta[p_, k_, σ_, r_, q_, t_] =
  -Evaluate[D[AppAriAsianPriceCall[p, k, σ, r, q, t], t]];
AppAriAsianPriceCallRho[p_, k_, σ_, r_, q_, t_] =
  Evaluate[D[AppAriAsianPriceCall[p, k, σ, r, q, t], r]];
AppAriAsianPriceCallVega[p_, k_, σ_, r_, q_, t_] =
  Evaluate[D[AppAriAsianPriceCall[p, k, σ, r, q, t], σ]];
```

Finally we give the Greeks for the Put:

```
AppAriAsianPricePutDelta[p_, k_, σ_, r_, q_, t_] =
  Evaluate[D[AppAriAsianPricePut[p, k, σ, r, q, t], p]];
AppAriAsianPricePutGamma[p_, k_, σ_, r_, q_, t_] =
  Evaluate[D[AppAriAsianPricePut[p, k, σ, r, q, t], {p, 2}]];
AppAriAsianPricePutTheta[p_, k_, σ_, r_, q_, t_] =
  -Evaluate[D[AppAriAsianPricePut[p, k, σ, r, q, t], t]];
AppAriAsianPricePutRho[p_, k_, σ_, r_, q_, t_] =
  Evaluate[D[AppAriAsianPricePut[p, k, σ, r, q, t], r]];
AppAriAsianPricePutVega[p_, k_, σ_, r_, q_, t_] =
  Evaluate[D[AppAriAsianPricePut[p, k, σ, r, q, t], σ]];
```

Examples

Various computations have appeared in the literature - it is a good idea to make a few simple checks. We make a comparison of results from the approximate formula with the results given by Eydeland and Geman (1995). We fix the risk-free rate at 0.05, the volatility at 0.5, the strike at 2, and consider stock prices of 1.9, 2 and 2.1 - we obtain the following:

```
TableForm[Table[{S, AppAriAsianPriceCall[S, 2, 0.5, 0.05, 0, 1]},
  {S, 1.9, 2.1, 0.1}]]
```

1.9	0.195379
2.	0.249791
2.1	0.310646

Next we fix the price at 2, and consider various values of r, σ, t:

```
TableForm[{{"r", "sigma", "t", "Value"},
  {0.02, 0.1, 1, AppAriAsianPriceCall[2, 2, 0.1, 0.02, 0, 1]},
  {0.18, 0.3, 1, AppAriAsianPriceCall[2, 2, 0.3, 0.18, 0, 1]},
  {0.0125, 0.25, 2, AppAriAsianPriceCall[2, 2, 0.25, 0.0125, 0, 2]},
  {0.05, 0.5, 2, AppAriAsianPriceCall[2, 2, 0.5, 0.05, 0, 2]}}]
```

r	sigma	t	Value
0.02	0.1	1	0.0560537
0.18	0.3	1	0.219829
0.0125	0.25	2	0.17349
0.05	0.5	2	0.359204

These calculations reveal that the method gives values close to the exact answer, but may be biased upwards or downwards depending on the parameters. Finally here is a table of Delta for the first three simulations again:

```
TableForm[
  Table[{S, AppAriAsianPriceCallDelta[S, 2, 0.5, 0.05, 0, 1]},
  {S, 1.9, 2.1, 0.1}]]
```

1.9	0.51008
2.	0.57729
2.1	0.638772

The at-the-money value computed by Eydeland and Geman (1995) is 0.56. So it seems that such methods are actually quite good for giving rough values. If one has a system already geared up to do the Black-Scholes analysis, it is quite tempting to use the simple changes of variables based on the moments. As we shall see in the next section, one can do rather better, while remaining in an essentially analytical framework.

10.5 Exact Analytical Mathematical Models of Arithmetic Asian Price Options

This section gives a brief summary of the exact solution for the Asian Call given by Geman and Yor (1993) (henceforth abbreviated as GY), as also summarized by Eydeland and Geman (1995). Suppose that the current time is t, and that the option matures at a time $T > t$. The averaging is arithmetic, continuous, and began at a time $t_0 \leq t$. Suppose that the known average value of the underlying over the time interval $[t_0, t]$ is ES. GY define the following changes of variables:

$$\tau = \frac{1}{4} \sigma^2 (T - t) \tag{12}$$

$$v = \frac{2(r - q)}{\sigma^2} - 1 \tag{13}$$

$$\alpha = \frac{\sigma^2 (K(T - t_0) - (t - t_0) ES)}{4S} \tag{14}$$

They also define a function of a (transform) variable p as

$$\mu(p) = \sqrt{v^2 + 2p} \tag{15}$$

The value of the average price option is then given by

$$\frac{e^{-r(T-t)} 4S \, C(\tau, v, \alpha)}{(T - t_0) \sigma^2} \tag{16}$$

The remaining function $C[v, \tau, \alpha]$ is not given explicitly, but GY give its Laplace transform,

$$U(p, v, \alpha) = \int_0^\infty C(\tau, v, \alpha) e^{-p\tau} d\tau \tag{17}$$

as an integral:

$$U(p, v, \alpha) = \frac{\int_0^{\frac{1}{2\alpha}} x^{\frac{\mu-v}{2}-2} (1 - 2\alpha x)^{\frac{\mu+v}{2}+1} e^{-x} dx}{p(p - 2v - 2)\Gamma(\frac{\mu-v}{2} - 1)} \tag{18}$$

where μ is as given above as a function of p and v. GY develop a series description of the transform and show how it can be inverted. We shall now explore how this can be managed and simplified in *Mathematica*.

Mathematica Implementation of Exact Arithmetic Asian

The first part of the translation to software is obvious - we first enter the definitions of the various basic functions.

```
τ[T_, t_, σ_] := σ^2 (T - t) / 4;
v[r_, q_, σ_] := 2 (r - q) / σ^2 - 1;
α[S_, ES_, K_, σ_, T_, t_, to_] :=
  σ^2 / (4 * S) (K * (T - to) - (t - to) * ES);
μ[v_, p_] := Sqrt[v^2 + 2 * p]
```

Now we enter the definition of the integral that is part of the transform, and request immediate evaluation:

$$F[p\_, \mu\_, v\_, \alpha\_] = \int_0^{\frac{1}{2\alpha}} x^{\frac{\mu-v}{2}-2} (1 - 2\alpha x)^{\frac{\mu+v}{2}+1} \mathrm{Exp}[-x] \, dx$$

$$\mathrm{If}\left[\alpha > 0 \, \&\& \, \mathrm{Re}\left[\frac{\mu}{2} - \frac{v}{2}\right] > 1 \, \&\& \, \mathrm{Re}[\mu + v] > -4, \ \frac{1}{\mathrm{Gamma}[1+\mu]}\right.$$

$$\left(2^{\frac{1}{2}(2-\mu+v)} \, \alpha^{\frac{1}{2}(2-\mu+v)} \, \mathrm{Gamma}\left[\frac{1}{2}(-2+\mu-v)\right] \mathrm{Gamma}\left[\frac{1}{2}(4+\mu+v)\right]\right.$$

$$\left.\mathrm{Hypergeometric1F1}\left[\frac{1}{2}(-2+\mu-v), 1+\mu, -\frac{1}{2\alpha}\right]\right),$$

$$\left.\int_0^{\frac{1}{2\alpha}} E^{-x} x^{-2+\frac{\mu-v}{2}} (1 - 2x\alpha)^{1+\frac{\mu+v}{2}} \, dx\right]$$

We see that *Mathematica* can actually evaluate the expression in "closed form", albeit in terms of a special function not usually encountered in option pricing. In fact we have received a result dependent on certain conditions - we can extract the answer (which is valid in our case) as follows:

```
G[p_, μ_, v_, α_] = F[p, μ, v, α][[2]]
```

$$\frac{1}{\mathrm{Gamma}[1+\mu]} \left(2^{\frac{1}{2}(2-\mu+v)} \, \alpha^{\frac{1}{2}(2-\mu+v)} \, \mathrm{Gamma}\left[\frac{1}{2}(-2+\mu-v)\right]\right.$$

$$\left.\mathrm{Gamma}\left[\frac{1}{2}(4+\mu+v)\right] \mathrm{Hypergeometric1F1}\left[\frac{1}{2}(-2+\mu-v), 1+\mu, -\frac{1}{2\alpha}\right]\right)$$

There are further cancellations when we insert the other terms that make up the transform:

$$U[p\_, \mu\_, v\_, \alpha\_] = \mathrm{Simplify}\left[\frac{G[p, \mu, v, \alpha]}{p(p - 2v - 2) \, \mathrm{Gamma}\left[\frac{\mu-v}{2} - 1\right]}\right]$$

$$\left(2^{\frac{1}{2}\ (2-\mu+\nu)}\ \alpha^{\frac{1}{2}\ (2-\mu+\nu)}\ \texttt{Gamma}\left[\frac{1}{2}\ (4+\mu+\nu)\right]\right.$$

$$\left.\texttt{Hypergeometric1F1}\left[\frac{1}{2}\ (-2+\mu-\nu),\ 1+\mu,\ -\frac{1}{2\ \alpha}\right]\right)\Big/$$

$$(\texttt{p}\ (-2+\texttt{p}-2\ \nu)\ \texttt{Gamma}[1+\mu])$$

In standard mathematical notation, the transform is just (the output style has been converted here)

`TraditionalForm[U[p, μ, ν, α]]`

$$\frac{2^{\frac{1}{2}(-\mu+\nu+2)}\ \alpha^{\frac{1}{2}(-\mu+\nu+2)}\ \Gamma(\frac{1}{2}(\mu+\nu+4))\ {}_1F_1(\frac{1}{2}(\mu-\nu-2);\mu+1;-\frac{1}{2\alpha})}{p\,(p-2\,\nu-2)\,\Gamma(\mu+1)} \tag{19}$$

Aficionados of obscure integrals might like to know that the evaluation of this integral to hypergeometric functions is a "known fact" - it is quoted, for example, as result 3.383.1 in Gradshteyn and Rhyzik (1980). We now have the ingredients to build the *Mathematica* model of the arithmetic average price Asian Call. In the following the Laplace transform inversion is done by direct numerical integration along the truncated Bromwich contour. The truncation is fixed at a value of 500 - readers may need to experiment and increase the integration range for peculiar parameter values. The author has not obtained satisfactory results with any of the standard numerical approximate methods for Laplace transform inversion.

```
Off[NIntegrate::slwcon]
```

```
AriAsianPriceCall[S_, ES_, K_, r_, q_, σ_, T_, t_, to_] :=
  Module[{ti = τ[T, t, σ], n = ν[r, q, σ],
    a = α[S, ES, K, σ, T, t, to], contour},
  contour = 2 n + 3;
  ac = Re[
    1 / (2 Pi) * NIntegrate[ U[(contour + I p), μ[n, (contour + I p)], n, a] *
      Exp[(contour + I p) * ti], {p, -500, 500}, MaxRecursion -> 11]];
Exp[-r * (T - t)] * 4 * S / ((T - to) * σ^2) * ac]
```

```
AriAsianPriceCall[1.9, 0, 2, 0.05, 0, 0.5, 1, 0, 0]
```

```
    0.193174
```

```
AriAsianPriceCall[2.0, 0, 2, 0.05, 0, 0.5, 1, 0, 0]
```

```
    0.246417
```

```
AriAsianPriceCall[2.1, 0, 2, 0.05, 0, 0.5, 1, 0, 0]
```

```
0.306223
```

For future reference we also record two values when the asset price suggests the option is increasingly well in the money:

```
AriAsianPriceCall[3, 0, 2, 0.05, 0, 0.5, 1, 0, 0]
```

```
1.0405
```

```
AriAsianPriceCall[4, 0, 2, 0.05, 0, 0.5, 1, 0, 0]
```

```
2.00015
```

Poles and the In-the-Money Solution

The transform has some manifest poles in two locations. For the pole at zero, the residue can be identified as

```
PowerExpand[Normal[Series[Sqrt[v^2 + 2 p], {p, 0, 2}]]]
```

$$-\frac{p^2}{2\,v^3} + \frac{p}{v} + v$$

```
Apart[
  Residue[U[p, v, v, α], {p, 0}] /. Gamma[2 + v] -> (1 + v) * Gamma[1 + v]]
```

$$-\alpha - \frac{1}{2\,(1+v)}$$

At the other pole, the residue is just

```
Simplify[Limit[(p - 2 - 2 v) * U[p, v + 2, v, α], p -> 2 + 2 v]]
```

$$\frac{1}{2 + 2\,v}$$

The contribution of the residues is remarkably simple - some algebra just leads to their sum which, when normalized, turns out to be the answer when the Asian is known to be in the money:

```
ITMAsianPriceCall[S_, ES_, K_, r_, q_, σ_, T_, t_, to_] :=
  Exp[-r * (T - t)] * (ES * (t - to) / (T - to) +
    S * ((Exp[(r - q) * (T - t)] - 1) / ((r - q) * (T - to))) - K)
```

We display this in Traditional Form to make the structure clear:

$$e^{-r(T-t)} \left(\frac{(e^{(r-q)(T-t)} - 1) S}{(r-q)(T-\text{to})} + \frac{ES(t-\text{to})}{T-\text{to}} - K \right) \tag{20}$$

Let's take a look at the two values computed previously, when the asset price was rather larger than the strike, using the "in-the-money" formula based on the sum of the residues at the two manifest poles:

```
ITMAsianPriceCall[3, 0, 2, 0.05, 0, 0.5, 1, 0, 0]
```

```
    1.02378
```

```
ITMAsianPriceCall[4, 0, 2, 0.05, 0, 0.5, 1, 0, 0]
```

```
    1.99919
```

The Delta of the Arithmetic Asian

An advantage of the transform method is that the transform is given analytically. In principle, the hedge parameters can be computed by symbolic differentiation. GY point out that this can be done, in principle, for Delta. We can use *Mathematica* to make this all very explicit, using the closed-form expressions for the transform that we have derived. The stock price appears in two places, explicitly, then implicitly through the variable α. The first thing we need to work out is

```
V[p_, μ_, ν_, α_] = Simplify[D[U[p, μ, ν, α], α]];
```

```
TraditionalForm[V[p, μ, ν, α]]
```

$$\left(2^{\frac{1}{2}(-\mu+\nu-2)} \, \alpha^{\frac{1}{2}(-\mu+\nu-2)} \, (\mu-\nu-2) \, \Gamma\!\left(\tfrac{1}{2}(\mu+\nu+4)\right) \right.$$
$$\left. \left(2\alpha(\mu+1) \, {}_1F_1\!\left(\tfrac{1}{2}(\mu-\nu-2); \mu+1; -\tfrac{1}{2\alpha}\right) - {}_1F_1\!\left(\tfrac{\mu-\nu}{2}; \mu+2; -\tfrac{1}{2\alpha}\right) \right) \right) \Big/ \tag{21}$$
$$(p(\mu+1)(-p+2\nu+2)\Gamma(\mu+1))$$

So we need to invert this transform, suitably normalize it, and add the term arising from the explicit appearance of S. Since S appears linearly in the numerator, this term is just the price of the Call divided by S. It is therefore most efficient if we just work out the contribution of the term in V, and add it to the call price divided by S. The inverse Laplace transform of V has to be multiplied by

```
Simplify[
  Exp[-r * (T - t)] * 4 * S / ((T - to) * σ^2) * D[α[S, ES, K, σ, T, t, to], S]]
```

$$\frac{E^{r\,(t-T)}\,(ES\,(t-to)+K\,(-T+to))}{S\,(T-to)}$$

So the new contour integral in Delta is given by (again, experimentation with the truncation may be useful)

```
AAPCDeltaCorr[S_, ES_, K_, r_, q_, σ_, T_, t_, to_] :=
 Module[{ti = τ[T, t, σ], n = ν[r, q, σ],
   a = α[S, ES, K, σ, T, t, to], contour},
  contour = 2 n + 3;
  bc =
   Re[ 1 / (2 Pi) * NIntegrate[V[(contour + I p), μ[n, (contour + I p)], n, a] *
       Exp[(contour + I p) * ti], {p, -500, 500}, MaxRecursion -> 11]];
(E ^ (r * (t - T)) * (ES * (t - to) + K * (-T + to))) /
 (S * (T - to)) * bc]
```

Eydeland and Geman give a sample calculation of Delta in their article in *RISK Magazine* (1995), for the parameters

```
S = 2; ES = 0; K = 2; r = 0.05; q = 0; σ = 0.5; t = 0; to = 0; T = 1;
```

We remind you of the call value:

```
value = AriAsianPriceCall[2.0, 0, 2, 0.05, 0, 0.5, 1, 0, 0]
```

    0.246417

```
AAPCDeltaCorr[S, ES, K, r, q, σ, T, t, to]
```

    0.442851

The value of Delta is this plus the Call value divided by S, i.e.,

```
% + value / S
```

    0.56606

The value quoted in the article is 0.56, so we have good agreement.

Remarks

This type of analytical result can be made the basis of a great deal of further study. It has been used by Fu *et al* (1997) to analyse in detail the efficiency of Monte Carlo simulation. One can also investigate the holomorphic and branching structure of the Laplace transforms, with a view to improving the inversion algorithm, and to investigate the utility of various approximate techniques for transform inversion. One such approximate method was used by Fu *et al* (1997). Others are investigated by this author in the text *Complex Mathematica* (Shaw, 1998).

10.6 Discrete Averaging and Delayed Averaging Start for Geometric Asian Price Options

Matters become more complicated when the sampling of the average is discrete. The author is unaware of any analytical characterization of the problem when the averaging is arithmetic. For geometric Asian options where the sampling is discrete, and sampling begins some time after initiation of the option, a similar procedure can be followed as for the case of continuous sampling. The effective volatility, using arguments given by Levy (1996), is given by the following formula, where $t = 0$ is the initiation time, and averaging is performed at N equal time intervals beginning at $t = $ ts and ending at the maturity date $t = T$. We go straight to the *Mathematica* implementation, as the pattern for this is now clear:

```
Clear[r, q, σ, ts, t, T];
VolEffD[σ_, ts_, T_, N_] := σ*Sqrt[(1-ts/T)*(1/3-1/(6*N)) + ts/T]
```

```
TraditionalForm[VolEffD[σ, ts, T, N]]
```

$$\sigma \sqrt{\frac{\text{ts}}{T} + \left(\frac{1}{3} - \frac{1}{6N}\right)\left(1 - \frac{\text{ts}}{T}\right)} \qquad (22)$$

Some useful checks on this result are given by considering, first, continuous averaging beginning at initiation:

```
VolEffD[σ, 0, T, Infinity]
```

$$\frac{\sigma}{\sqrt{3}}$$

Next, suppose that there is only one averaging point, at maturity:

```
VolEffD[σ, T, T, 1]
```

$$\sigma$$

If we average geometrically over two points, at initiation, where the stock price is known, and at maturity, we are writing an option involving just the square root of the stock price, with effective volatility

```
VolEffD[σ, 0, T, 2]
```

$$\frac{\sigma}{2}$$

This can be seen to be correct by considering the stochastic process followed by $\sqrt{S}$. For the effective yield, some algebra with Levy's analysis leads to the following result:

```
QEffD[r_, q_, σ_, ts_, T_, N_] := 1/2*(σ^2/6*(1-ts/T)*(1+1/N) +
q*(1+ts/T) + r*(1-ts/T))
```

```
TraditionalForm[QEffD[r, q, σ, ts, T, N]]
```

$$\frac{1}{2}\left(\frac{1}{6}\left(1+\frac{1}{N}\right)\left(1-\frac{ts}{T}\right)\sigma^2 + r\left(1-\frac{ts}{T}\right) + q\left(\frac{ts}{T}+1\right)\right) \tag{23}$$

This formula also correctly interpolates between the continuous sampling result and one sampling at maturity:

```
{QEffD[r, q, σ, 0, T, Infinity], QEffD[r, q, σ, T, T, 1]}
```

$$\left\{\frac{1}{2}\left(q + r + \frac{\sigma^2}{6}\right), q\right\}$$

Now we define the valuation and Greek functions in the usual way, using symbolic differentiation. First we give the valuations:

```
DiscreteGeoAsianPriceCall[p_, k_, σ_, r_, q_, ts_, T_, N_] :=
BlackScholesCall[p,k,VolEffD[σ, ts, T, N],r,QEffD[r, q, σ, ts, T,
N],T];
DiscreteGeoAsianPricePut[p_, k_, σ_, r_, q_, ts_, T_, N_] :=
BlackScholesPut[p,k,VolEffD[σ, ts, T, N],r,QEffD[r, q, σ, ts, T, N],t];
```

Next we write down the Greeks for the Call and the Put:

```
DiscreteGeoAsianPriceCallDelta[p_, k_, σ_, r_, q_, ts_, T_, N_] =
 Evaluate[D[DiscreteGeoAsianPriceCall[p, k, σ, r, q, ts, T, N], p]];
DiscreteGeoAsianPriceCallGamma[p_, k_, σ_, r_, q_, ts_, T_, N_] =
 Evaluate[
  D[DiscreteGeoAsianPriceCall[p, k, σ, r, q, ts, T, N], {p, 2}]];
DiscreteGeoAsianPriceCallTheta[p_, k_, σ_, r_, q_, ts_, T_, N_] =
 -Evaluate[D[DiscreteGeoAsianPriceCall[p, k, σ, r, q, ts, T, N], t]];
DiscreteGeoAsianPriceCallRho[p_, k_, σ_, r_, q_, ts_, T_, N_] =
```

```
Evaluate[D[DiscreteGeoAsianPriceCall[p, k, σ, r, q, ts, T, N], r]];
DiscreteGeoAsianPriceCallVega[p_, k_, σ_, r_, q_, ts_, T_, N_] =
  Evaluate[D[DiscreteGeoAsianPriceCall[p, k, σ, r, q, ts, T, N], σ]];
DiscreteGeoAsianPricePutDelta[p_, k_, σ_, r_, q_, ts_, T_, N_] =
  Evaluate[D[DiscreteGeoAsianPricePut[p, k, σ, r, q, ts, T, N], p]];
DiscreteGeoAsianPricePutGamma[p_, k_, σ_, r_, q_, ts_, T_, N_] =
  Evaluate[
  D[DiscreteGeoAsianPricePut[p, k, σ, r, q, ts, T, N], {p, 2}]];
DiscreteGeoAsianPricePutTheta[p_, k_, σ_, r_, q_, ts_, T_, N_] =
  -Evaluate[D[DiscreteGeoAsianPricePut[p, k, σ, r, q, ts, T, N], t]];
DiscreteGeoAsianPricePutRho[p_, k_, σ_, r_, q_, ts_, T_, N_] =
  Evaluate[D[DiscreteGeoAsianPricePut[p, k, σ, r, q, ts, T, N], r]];
DiscreteGeoAsianPricePutVega[p_, k_, σ_, r_, q_, ts_, T_, N_] =
  Evaluate[D[DiscreteGeoAsianPricePut[p, k, σ, r, q, ts, T, N], σ]];
```

Chapter 10 Bibliography

Eydeland, A. and Geman, H., 1995, Asian options revisited: inverting the Laplace transform, *RISK Magazine*, March.

Fu, M.C., Madan, D.B. and Wang, T., 1997, Pricing Asian options: a comparison of analytical and Monte Carlo methods, University of Maryland College of Business and Management preprint, March.

Geman, H. and Yor, M., 1993, Bessel processes, Asian options, and perpetuities, *Mathematical Finance*, 3, p. 349.

Gradshteyn, I.S. and Rhyzik, I.M., 1980, *Tables of Integrals, Series and Products*, Corrected and Enlarged Edition, Translated by A. Jeffrey, Academic Press.

Hull, J.C., 1996, *Options, Futures, and other Derivatives*, 3rd edition, Prentice-Hall.

Kemna, A. and Vorst, A., 1990, A pricing method for options based on average asset values, *Journal of Banking and Finance*, 14, p. 113.

Levy, E., 1996, Exotic options I, in the *Handbook of Risk Management and Analysis*, ed. C. Alexander, J. Wiley and Sons.

Levy, E. and Turnbull, S., 1992, Average intelligence, *RISK Magazine* 5, p. 53.

Shaw, W.T., 1998, *Complex Mathematica*, Addison-Wesley, in preparation.

Turnbull, S.M. and Wakeman, L.M., 1991, A quick algorithm for pricing European average options, *Journal of Financial and Quantitative Analysis*, 26, p. 377.

Wilmott, P., Dewynne, J. and Howison, S., 1993, *Option Pricing - Mathematical Models and Computation*, Oxford Financial Press.

Chapter 11.
Vanilla American Options -
Analytical Methods

Extending the Power of Analytical Methods to Moving Boundary Value Problems

11.1 Introduction

It is common to assume that there is no useful or accurate analytical treatment of options where early or scheduled exercise is permitted. There are two faults with this line of thought. Firstly, it is not correct - more often than not, at least with simpler linear systems, the non-existence of at least a good *approximate* analytical solution has a lot more to do with not trying hard enough to find it, coupled to a premature rush to purely numerical methods. Second, giving up on analytical models forces one to abandon the elegant methods for computing Greeks by ordinary differentiation. In this chapter I will present my current understanding of the state of play regarding the analytical approach to American options. I hope this understanding will be rendered obsolete by further efforts on this topic. I shall focus on the Put case with zero yield.

The general plan of this chapter is as follows. First I shall review one of the standard and well-known analytical approximations - the MacMillan, Barone-Adesi and Whaley (MBAW) models, as given by MacMillan (1986) and Barone-Adesi and Whaley (1987). Then I shall explore a much more interesting and accurate model based on relatively recent work by Carr, Jarrow and Myneni (1992), henceforth abbreviated as just "CJM". Then I shall offer some speculations on how these models might be related, and the numerical treatment of the latter improved. There are other models and approximations, and an excellent survey is given by CJM.

There is a host of general arguments relating to the early exercise of vanilla American options, many of which are covered in various chapters of Hull (1996). Our purpose here is not to regurgitate standard knowledge on the topic, but rather to explore analytical algorithms and their implementation. Our main focus will be on the problem of the American Put, but we shall also investigate the early exercise induced on a Call in the presence of *discrete* dividends. Calls and Puts with continuous dividends can be investigated in a similar manner to that pursued here for the Put. Later in this book we shall explore finite-difference and binomial models, and we shall use the analytical models developed here to check such numerical schemes.

Loading the Vanilla European Package

This chapter will assume that the standard formulae for vanilla European options have been loaded.

```
Needs["Derivatives`BlackScholes`"]
```

11.2 Definition of an American Put via Moving Boundary Condition

The American Put may be defined in various ways. The most sensible definition, at least for proper analysis, is in terms of a moving boundary value problem. This is much more useful than a tree-based definition, and in spite of being around for some time, it has been ignored by most workers. The matter was sorted out by McKean (1965) as follows. We have a function $P(S, t)$ satisfying

$$P(S, T) = \text{Max}[0, K - S] \tag{1}$$

$$\lim_{S \to \infty} P(S, t) = 0 \tag{2}$$

There is a critical stock price B_t such that

$$\lim_{s \to B_t} P(S, t) = K - B_t \tag{3}$$

$$\lim_{S \to B_t} \frac{\partial P(S, t)}{\partial S} = -1 \tag{4}$$

and for values of $S > B_t$, the Black-Scholes differential equation is satisfied:

$$\frac{1}{2} S^2 \frac{\partial^2 P}{\partial S^2} \sigma^2 - r P + (r - q) S \frac{\partial P}{\partial S} + \frac{\partial P}{\partial t} = 0 \tag{5}$$

11.3 Definition of the Early Exercise Premium Function

The definitions above give the full Put value. It has become commonplace, as it is remarkably useful, to write the problem in terms of a premium for early exercise. The subtraction is only explicit for values $S > B_t$, and here one writes

$$P = P_E + \rho \tag{6}$$

where P_E is the vanilla European Put value. In terms of the premium variable ρ, we have the new problem:

$$\rho(S, T) = 0 \tag{7}$$

$$\lim_{S \to \infty} \rho(S, t) = 0 \tag{8}$$

At the critical stock price B_t, we now have

$$\lim_{s \to B_t} \rho(S, t) = K - B_t - P_E(B_t, t) \tag{9}$$

$$\lim_{S \to B_t} \frac{\partial \rho(S, t)}{\partial S} = -1 - \frac{\partial P_E(B_t, t)}{\partial S} \tag{10}$$

and for values of $S > B_t$, the Black-Scholes differential equation is still satisfied:

$$\frac{1}{2} S^2 \frac{\partial^2 \rho}{\partial S^2} \sigma^2 - r\rho + (r-q) S \frac{\partial \rho}{\partial S} + \frac{\partial \rho}{\partial t} = 0 \tag{11}$$

So we have payoff conditions of zero, at the price of more complicated moving boundary conditions.

11.4 The Quasi-stationary Method of Solution and MBAW

There is a standard method for obtaining approximate solutions to differential equations of this type, called the "quasi-stationary method". This, and other, methods of treatment are very well described by Crank (1984) in his classic text. In its raw form, we would first seek a first solution where we can ignore the time-dependence, solving

$$\frac{1}{2} S^2 \frac{\partial^2 \rho}{\partial S^2} \sigma^2 - r\rho + (r-q) S \frac{\partial \rho}{\partial S} = 0 \tag{12}$$

Note that the "quadratic approximation", of MBAW, as described, for example, by Hull (1996), is nothing other than the quasi-stationary approximation after the application of a further change of variables to make the method more accurate. The change of variable developed is given by

$$\tau = T - t \quad h(\tau) = 1 - e^{-r\tau} \quad k_1 = \frac{2r}{\sigma^2} \quad k_2 = \frac{2(r-q)}{\sigma^2} \tag{13}$$

and setting

$$\rho = h(\tau) \Psi(S, h) \tag{14}$$

leads to

$$S^2 \frac{\partial^2 \Psi}{\partial S^2} + k_2 S \frac{\partial \Psi}{\partial S} - \frac{k_1 \Psi}{h} - (1-h) k_1 \frac{\partial \Psi}{\partial h} = 0 \tag{15}$$

with quasi-stationary approximation

$$S^2 \frac{\partial^2 \Psi}{\partial S^2} + k_2 S \frac{\partial \Psi}{\partial S} - \frac{k_1 \Psi}{h} = 0 \tag{16}$$

This is an equation of homogeneous type, which is easily solved in terms of a power of S. Only one solution satisfies the requirement that the result tends to zero for large S. The answer may be written as

$$\Psi = A_1 \left(\frac{S}{B_t} \right)^{\gamma_1} \tag{17}$$

where

$$\gamma_1 = \frac{1}{2} \left(1 - k_2 - \sqrt{(1-k_2)^2 + \frac{4 k_1}{h}} \right) \tag{18}$$

The values of A_1 and the critical price B_t are now determined by imposing the boundary conditions at the critical price. A_1 is eliminated first using continuity of the derivative at the critical price. Then the other boundary condition is imposed. This leads to an implicit non-linear equation for B_t. To do the first step, let Δ be the Delta for the ordinary European Put. Then continuity of the derivative establishes that

$$A_1 = -\frac{B_t\,(\Delta + 1)}{\gamma_1} \tag{19}$$

The continuity of the value at the free boundary is then given by the equation

$$K - P(B_t, t) - B_t = -\frac{B_t\,(\Delta + 1)}{\gamma_1} \tag{20}$$

This can now be solved in *Mathematica*.

11.5 Mathematica Implementation of the MBAW Method

Once the standard European option Package has been loaded, we can make use of some standard Put functions. Let's remind ourselves of their operation.

? BlackScholesPut

```
    BlackScholesPut[price, strike, vol, riskfree, divyield, expiry]
        returns the Black-Scholes value of a vanilla European Put.
```

? BlackScholesPutDelta

```
    BlackScholesPutDelta[price, strike, vol, riskfree, divyield,
        expiry]  returns the delta of a vanilla European Put.
```

Now we define the functions in the MBAW analysis. We begin with a clean-up:

```
Clear[r, q, t, σ, K];
h[r_, t_] = 1 - Exp[-r t];
k₁[r_, σ_] := 2 r / σ²;
k₂[r_, σ_, q_] := 2 (r - q) / σ²;
γ₁[r_, σ_, t_, q_] :=
  1 / 2 (1 - k₂[r, σ, q] - Sqrt[(1 - k₂[r, σ, q])^2 + 4 k₁[r, σ] / h[r, t]])
```

Just to check, we enter these expressions and ask for γ_1 in ordinary mathematical notation:

```
TraditionalForm[γ₁[r, s, t, q]]
```

$$\frac{1}{2}\left(-\frac{2\,(r-q)}{s^2}-\sqrt{\left(1-\frac{2\,(r-q)}{s^2}\right)^2+\frac{8\,r}{(1-e^{-r\,t})\,s^2}}+1\right)$$

The boundary condition governing the location of the critical price is equivalent to setting the following function to zero, where we use Sf to denote the critical price:

```
Msolver[Sf_, σ_, r_, t_, K_, q_] =
  Simplify[K - Sf - BlackScholesPut[Sf, K, σ, r, q, t] +
    Sf (BlackScholesPutDelta[Sf, K, σ, r, q, t] + 1)
    ─────────────────────────────────────────────────
                    γ₁[r, σ, t, q]                    ]
```

$$K - Sf - \frac{1}{2}\,E^{-r\,t}\,K\left(1+\mathrm{Erf}\left[\frac{t\,(2\,q-2\,r+\sigma^2)-2\,\mathrm{Log}\left[\frac{Sf}{K}\right]}{2\sqrt{2}\,\sqrt{t}\,\sigma}\right]\right) -$$

$$\frac{1}{2}\,E^{-q\,t}\,Sf\left(-1+\mathrm{Erf}\left[\frac{t\,(-2\,q+2\,r+\sigma^2)+2\,\mathrm{Log}\left[\frac{Sf}{K}\right]}{2\sqrt{2}\,\sqrt{t}\,\sigma}\right]\right) +$$

$$\frac{E^{-q\,t}\,Sf\left(-1+2\,E^{q\,t}+\mathrm{Erf}\left[\frac{t\,(-2\,q+2\,r+\sigma^2)+2\,\mathrm{Log}\left[\frac{Sf}{K}\right]}{2\sqrt{2}\,\sqrt{t}\,\sigma}\right]\right)}{1-\sqrt{\left(1+\frac{2\,(q-r)}{\sigma^2}\right)^2+\frac{8\,r}{(1-E^{-r\,t})\,\sigma^2}}+\frac{2\,(q-r)}{\sigma^2}}$$

Now we can define two functions, one for the critical price, the other for the value of the option. Let's take a look at the front first of all:

```
MMFront[K_, t_, σ_, r_, q_] :=
  Module[{Sf, v},
    Sf = v /. FindRoot[Msolver[v, σ, r, t, K, q] == 0,
      {v, K - 1}, MaxIterations -> 40][[1]];
    Sf]
```

We can make a table of values, and plot it, for zero dividend yield:

```
mmfrontdata =
  Table[{t, MMFront[50, t, 0.4, 0.1, 0]}, {t, 0.0001, 20, 0.2}];
```

```
mbawplot =
  ListPlot[mmfrontdata, PlotJoined -> True, PlotRange -> {20, 50}];
```

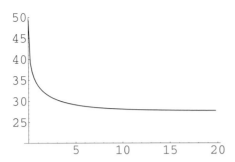

The function appears to level off. If we analyse the function **Msolver** for very late times, and for q zero, it is easy to find the late-time asymptote. It is given by (see Section 4.2 for another derivation)

$$B_\infty = \frac{K}{1 + \frac{\sigma^2}{2r}} \tag{21}$$

or, in *Mathematica*, by

```
FrontLate[K_, r_, σ_, t_] := K / (1 + σ²/(2 r))
```

We can overlay a plot of the late-time asymptote with the front, as follows:

```
asymptote = Plot[FrontLate[50, 0.1, 0.4, t], {t, 0, 20},
    PlotRange -> {20, 50}, DisplayFunction -> Identity];
```

```
Show[mbawplot, asymptote];
```

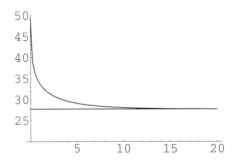

For the valuation itself, it is convenient to first introduce A_1 as an explicit function:

```
aone[Sf_, r_, q_, σ_, t_, K_] =
    - Sf (BlackScholesPutDelta[Sf, K, σ, r, q, t] + 1) / γ₁[r, σ, t, q];
```

Now we can give the valuation of the American Put in approximate closed form:

```
AmericanPutMBAW[S_, K_, σ_, r_, q_, t_] :=
  Module[{Sf, v},
    Sf = v /. FindRoot[Msolver[v, σ, r, t, K, q] == 0,
      {v, K - 1}, MaxIterations -> 40][[1]];
    If[S <= Sf, K - S, aone[Sf, r, q, σ, t, K] * (S / Sf)^γ₁[r, σ, t, q] +
      BlackScholesPut[S, K, σ, r, q, t]]]
```

We can now both calculate and plot the European and approximate American values, showing the rough level of the early exercise premium.

```
AmericanPutMBAW[10, 10, 0.4, 0.1, 0, 1]
```

    1.20244

```
BlackScholesPut[10, 10, 0.4, 0.1, 0, 1]
```

    1.08022

```
Plot[{AmericanPutMBAW[S, 10, 0.4, 0.1, 0, 1],
  BlackScholesPut[S, 10, 0.4, 0.1, 0, 1]}, {S, 2, 20}];
```

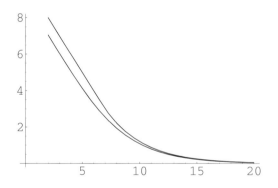

11.6 Extrapolation of the Quasi-stationary Method?

The MBAW method can be shown, using a binomial, or finite-difference, model, to be a rather inaccurate estimator. We shall discuss a more accurate analytical model shortly. However, it is worth remarking on how one might, at least in principle, refine the method, bearing in mind corresponding results in heat and chemical diffusion. The MBAW method involves a change of variables, before invoking the quasi-stationary method. In heat problems, with Stefan boundary conditions (such as arise for melting ice), there are well-known methods for taking a quasi-stationary approximation and improving it. Crank (1984) gives the basic idea, which amounts to taking the quasi-stationary solution and adding a correction which satisfies the quasi-stationary heat equation (QSHE) with a source that is the time-derivative of the quasi-stationary solution. Once this PDE is solved, a third correction is added, which satisfies the QSHE with a source that is the time-derivative of the first correction, and so on. So the solution gets refined, at the price of introducing higher and higher order differential equations for the critical point.

This method can sometimes be added up exactly, by the use of Laplace transforms, to get an integral equation for the critical point. I have studied this myself for chemical diffusion in cylindrical domains, as discussed in Shaw (1989) (the relevant appendix is available direct from the author). Similar results are given by Hill and Dewynne (1985a, 1985b). Sometimes such integral equations can be derived directly, as discussed by Ockendon (1974) without recourse to an iterative improvement of the quasi-stationary solution.

It is noteworthy that this is actually rather difficult to implement for the American Put problem, and I have so far not succeeded. I encourage others to explore the issue. However, two things have put me off investigating the matter further, at least for now. The full specification of the problem, in terms of an integral equation for the Laplace transform of the critical point curve, cannot in general be solved exactly, but does sometimes lead to formulae for the small-time and late-time asymptotes. Intermediate times typically remain awkward. The second point is that an integral equation for the critical point curve for the American Put has already been found directly, by Carr, Jarrow and Myneni (1992), and I shall instead pursue a discussion of their method, and an accurate approximate solution scheme.

11.7 The Carr-Jarrow-Myneni Formulae

The main achievement of Carr, Jarrow and Myneni (CJM) in their 1992 paper was to give the early exercise premium as an *exact* functional of the critical price curve. It involves integration of a non-linear function of the critical price for all times between evaluation and maturity, rather than a non-linear function of just the critical price curve at evaluation. To write down the result (you should refer to CJM, 1992, for a derivation), we introduce the variables

$$\rho_1 = r + \frac{\sigma^2}{2} \tag{22}$$

$$\rho_2 = r - \frac{\sigma^2}{2} = \rho_1 - \sigma^2 \tag{23}$$

and let the critical stock price be $S_0(t)$. We work with zero dividend yield, as with CJM. The CJM formula for the early exercise premium ρ is then just

$$\rho = rK \int_0^T e^{-rt} \, \text{Norm}\left(\frac{\log(\frac{S_0[T-t]}{S}) - \rho_2 \, t}{\sigma \sqrt{t}} \right) dt \tag{24}$$

The great advantage of this representation is that we can get the Delta and Gamma by symbolic differentiation. For the premium component, we have

$$\Delta_\rho = -\frac{rK}{S\sigma} \int_0^T \frac{e^{-rt} \, e^{-\frac{b^2}{2}}}{\sqrt{2\pi t}} \, dt \tag{25}$$

where

$$b = \frac{\log(\frac{S_0[T-t]}{S}) - \rho_2 \, t}{\sigma \sqrt{t}} \tag{26}$$

The Gamma of the premium component is, by a further differentiation,

$$\Gamma_\rho = \frac{rK}{S^2 \, \sigma \, \sqrt{2\pi}} \int_0^T e^{-rt} \, e^{-\frac{b^2}{2}} \Big[\frac{1}{\sqrt{t}} - \frac{b}{\sigma t} \Big] dt \tag{27}$$

$$\tag{28}$$

One can write down two equivalent integral equations by applying either of the boundary conditions at the critical price. The one involving continuity of Delta is the simplest, and will be used for our analysis. To express the boundary condition concisely, we write the critical price in terms of the strike K, other known parameters, and another unknown k_{1T}, as follows:

$$S_0[T] = K \, e^{-\rho_1 T - \sigma \sqrt{T} \, k_{1T}} \tag{29}$$

Then we define b_{2t}, following CJM, by the equation

$$b_{2t} = \frac{k_{1T} \sqrt{T} - k_{1T-t} \sqrt{T-t}}{\sqrt{t}} + \sigma \sqrt{t} \tag{30}$$

The integral equation we have to solve, arising from continuity of Delta at the critical price, is given by

$$\frac{rK}{\sqrt{2\pi}} \int_0^T \frac{e^{-rt} \, e^{-\frac{b_{2t}^2}{2}}}{\sqrt{t}} \, dt = \sigma \, S_0[T] \, \mathrm{Norm}(-k_{1T}) \tag{31}$$

11.8 An Accurate Semi-analytical Treatment of CJM

The integral equation for the critical price could be solved by numerical methods, but we wish to try to give an accurate semi-analytical alternative. The key to this is the square root of t in the denominator. The integral is therefore dominated by the behaviour of its integrand near the origin. The unknown under the integral is b_{2t}, given by

$$b_{2t} = \frac{k_{1T} \sqrt{T} - k_{1T-t} \sqrt{T-t}}{\sqrt{t}} + \sigma \sqrt{t} = \beta(T, t) \sqrt{t} \tag{32}$$

which now serves to define $\beta(T, t)$.

$$\beta(T, t) = \sigma + \frac{k_{1T} \sqrt{T} - k_{1T-t} \sqrt{T-t}}{t} \tag{33}$$

To evaluate the integral approximately we only need to know the behaviour of this function near $t = 0$. We therefore set

$$\beta(T) = \lim_{t \to 0} \beta(T, t) \tag{34}$$

and we can evaluate this as just

$$\beta(T) = \sigma + \lim_{t \to 0} \frac{k_{1T} \sqrt{T} - k_{1T-t} \sqrt{T-t}}{t} = \sigma + \frac{k_{1T}}{2\sqrt{T}} - \sqrt{T} \, \frac{\partial k_{T-t}}{\partial t} \Big|_{t=0} \tag{35}$$

Let us defer for a moment the business of estimating $\beta(T)$, and consider the integral equation for the critical point. The left side of this equation is now just

$$\frac{rK}{\sqrt{2\pi}} \int_0^T \frac{e^{-\left(\frac{\beta(T)^2}{2}+r\right)t}}{\sqrt{t}}\, dt \tag{36}$$

which can now be evaluated exactly in terms of the error function, to give a non-linear equation for k_{1T} and hence for the critical price:

$$\frac{(rK)\,\mathrm{erf}\left(\sqrt{\tfrac{1}{2}(\beta(T)^2+2r)}\right)}{\sqrt{\beta(T)^2+2r}} = \sigma K\, e^{-\rho_1 T + -\sigma\sqrt{T}\, k_{1T}}\, \mathrm{Norm}(-k_{1T}) \tag{37}$$

Given a model of $\beta(T)$, this can be solved by Newton-Raphson methods, just as in the case of the MBAW model.

A Model of β

All we know is that

$$\beta(T) = \frac{k_{1T}}{2\sqrt{T}} - \sqrt{T}\, \frac{\partial k_{T-t}}{\partial t}\bigg|_{t=0} \tag{38}$$

and the second term on the right side is unknown - we solve the resulting equation for k_{1T}, but we need an independent estimate of the derivative. It turns out that we can learn a great deal about the structure of this term by using the asymptotic behaviour of the critical price for large times, or by postulating a simple model for the time-dependence of the front consistent with this asymptotic behaviour. We have already worked out the late-time asymptote in the MBAW framework, but in order to re-derive it within the CJM model, we shall just suppose it has a value of S_∞. We also know that

$$\frac{S_0[T]}{K} = e^{-\sqrt{T}\,\sigma k_{1T} - \rho_1 T} \tag{39}$$

Now letting T become large, and inverting this relationship, we must have that, asymptotically,

$$k_{1T} \sim \frac{\log(\frac{K}{S_\infty})}{\sigma\sqrt{T}} - \frac{\rho_1\sqrt{T}}{\sigma} \tag{40}$$

This allows us to estimate the unknown derivative and hence $\beta(T)$. A few lines of algebra reveal that, as T becomes infinite, *irrespective of the value of S_∞*,

$$\beta(T) \sim \sigma - \frac{\rho_1}{\sigma} \tag{41}$$

for large values of T. Note, for consistency, that since then

$$\mathrm{Norm}(-k_{1T}) \sim 1 \tag{42}$$

our integral equation for the critical price gives, as before,

$$S_\infty = \frac{r\,K}{\sigma\sqrt{\beta^2 + 2\,r}} = \frac{K}{1 + \frac{\sigma^2}{2\,r}} \tag{43}$$

More generally

$$k_{1\,T} = -\frac{\rho_1\sqrt{T}}{\sigma} + \frac{\log\left(\frac{K}{S(T)}\right)}{\sigma\sqrt{T}} \tag{44}$$

We have shown that any choice of "constant" $S(T)$ in the second term gets wiped out in the calculation of β. So one family of estimates for β is given by just using the first term and making the replacement

$$-\sqrt{T}\,\frac{\partial k_{T-t}}{\partial t}\Big|_{t=0} \longrightarrow -\frac{\rho_1}{2\,\sigma} \tag{45}$$

which leads to

$$\beta(T) = \sigma - \frac{\rho_1}{2\,\sigma} + \frac{k_{1\,T}}{2\sqrt{T}} \tag{46}$$

and the non-linear equation for solution is

$$\frac{(r\,K)\,\mathrm{erf}\left(\sqrt{\tfrac{1}{2}\,(\beta(T)^2 + 2\,r)}\right)}{\sqrt{\beta(T)^2 + 2\,r}} = \sigma\,K\,e^{-\rho_1\,T + -\sigma\sqrt{T}\,k_{1\,T}}\,\mathrm{Norm}(-k_{1\,T}) \tag{47}$$

It is possible to consider more sophisticated models for β based, for example, on the Van Moerbeke (1976) model for the front, but this involves a free parameter which must be fixed. The use of such a scheme can lead to extremely accurate results if the free parameter is suitably fiddled, but the method is somewhat unsatisfactory in requiring such tuning. Other models can be developed based on the observation that $\beta(T,\,T) = \sigma + k_{1\,T}/\sqrt{T}$ exactly, but this result only relates to the integral away from $t = 0$, which is the dominating region. The author would be pleased to hear of anyone's experiments for more suitable models of β. In particular, better independent estimates of the small-time or late-time asymptotes could be used to refine the model presented here.

11.9 *Mathematica* Implementation of Approximate CJM

We can now give a very concise representation of the CJM model. We define a function for the critical price, or front, then three functions for value, Delta and Gamma:

```
Front[strike_, r_, sigma_, t_] :=
If[t==0,strike,Module[{l,m},
l= m/. FindRoot[Exp[-m*sigma*Sqrt[t] - (r+sigma^2/2)*t]*(1+Erf[-
m/Sqrt[2]]) ==
r*Sqrt[2/(r+(sigma - (r/(2*sigma)+sigma/4) +m/(2*Sqrt[t]))^2/2)]*
Erf[Sqrt[(r+(sigma-(r/(2*sigma)+sigma/4)+m/(2*Sqrt[t]))^2/2)*t]]/sigma,
{m,-(sigma/2+r/sigma)*Sqrt[t]},MaxIterations -> 30];
strike*Exp[-l*sigma*Sqrt[t] - (r+sigma^2/2)*t]]]
```

```
AmericanPutCJM[price_,strike_,sigma_,r_,t_]  :=
If[price <= Front[strike, r, sigma, t], strike - price,
BlackScholesPut[price,strike,sigma,r,0,t]+
r*strike*NIntegrate[Exp[-r*u]*Norm[(Log[
Front[strike,r,sigma,t-u]/price]-(r-sigma^2/2)*u)/(sigma*Sqrt[u])],
{u,0,t}]]

AmericanPutDeltaCJM[price_,strike_,sigma_,r_,t_]  :=
If[price <= Front[strike, r, sigma, t],   -1,
N[BlackScholesPutDelta[price,strike,sigma,r,0,t]  -
(r*strike/(price*sigma*Sqrt[2.*Pi]))*NIntegrate[Exp[-r*u]*Exp[-((Log[
Front[strike,r,sigma,t-u]/price]-(r-sigma^2/2)*u)/(sigma*-
Sqrt[u]))^2/2]/Sqrt[u],
{u, 0, t}]]]

AmericanPutGammaCJM[price_,strike_,sigma_,r_,t_]  :=
If[price <= Front[strike, r, sigma, t], 0,
N[
BlackScholesPutGamma[price,strike,sigma,r,0,t]+
(r*strike/(price^2*sigma*Sqrt[2.*Pi]))*NIntegrate[Exp[-r*u]*Exp[-((Log[
Front[strike,r,sigma,t-u]/price]-(r-sigma^2/2)*u)/(sigma*-
Sqrt[u]))^2/2]*
(1/Sqrt[u] -
(Log[Front[strike,r,sigma,t-u]/price]-(r-sigma^2/2)*u)/(sigma*-
Sqrt[u])/(sigma*u)),
{u, 0, t}]]]
```

Let's take a look at the critical price frontier, for the same parameters as before:

```
cjmplot = Plot[{Front[50, 0.1, 0.4, t], FrontLate[50, 0.1, 0.4, t]},
  {t, 0, 20}, PlotRange -> {25, 50},
  PlotStyle -> {{Thickness[0.01], Dashing[{0.025, 0.025}]},
    Dashing[{0.025, 0.025}]}];
```

We can overlay this with the MBAW frontier and see the approximate consistency:

```
Show[cjmplot, mbawplot];
```

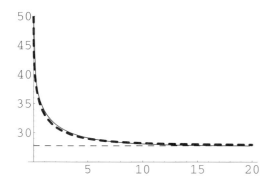

Let's make a few comparisons for a strike at 50, volatility 40%, interest rate 10% (cc) with one year to expiry:

At-the-Money Comparisons

European

```
BlackScholesPut[50, 50, 0.4, 0.1, 0, 1]
```

    5.40111

American MBAW

```
AmericanPutMBAW[50, 50, 0.4, 0.1, 0, 1]
```

    6.01219

American CJM

```
AmericanPutCJM[50, 50, 0.4, 0.1, 1]
```

    5.97688

In-the-Money Comparisons

European

```
BlackScholesPut[45, 50, 0.4, 0.1, 0, 1]
```

```
    7.27411
```

American MBAW

```
AmericanPutMBAW[45, 50, 0.4, 0.1, 0, 1]
```

```
    8.18143
```

American CJM

```
AmericanPutCJM[45, 50, 0.4, 0.1, 1]
```

```
    8.18064
```

Out-of-the-Money Comparisons

European

```
BlackScholesPut[55, 50, 0.4, 0.1, 0, 1]
```

```
    3.97888
```

American MBAW

```
AmericanPutMBAW[55, 50, 0.4, 0.1, 0, 1]
```

```
    4.40626
```

American CJM

```
AmericanPutCJM[55, 50, 0.4, 0.1, 1]
```

```
    4.34906
```

It is an interesting question how one decides which result is best. We shall revisit this question after developing binomial and finite-difference models, where it will turn out that the CJM approximation not only is substantially better, but will typically outperform binomial models unless several hundred time-steps are taken.

11.10 The Roll-Geske-Whaley Model for American Calls

When we turn our attention to American Calls, if the dividend payments are large enough early exercise may be desirable on the last dividend date. (Here we confine attention to discrete dividends.) This can be accounted for analytically by the Roll (1977), Geske (1979, 1981) and Whaley (1981) formulae, as described also by Hull (1996). Their formula reduces to the "jump-adjusted" Black-Scholes formula for low enough dividends, where there is no early exercise (see Chapter 24 for further discussion of the treatment of discrete dividends). The evaluation of this formula requires various special functions. First we need to define the cumulative normal distribution in two variables - we get it by loading a Package in *Mathematica*.

```
Needs["Statistics`MultinormalDistribution`"]
```

The objects loaded are associated with the following distribution list.

```
? Statistics`MultinormalDistribution`*
```

HotellingTSquareDistribution MultivariateTDistribution   WishartDistri-
bution
MultinormalDistribution   QuadraticFormDistribution

We want the two-variable case of the multinormal object:

```
? MultinormalDistribution
```

> MultinormalDistribution[mu, sigma] represents the
>   multivariate normal (Gaussian) distribution with mean
>   vector mu and covariance matrix sigma. For a p-variate
>   random vector to be distributed MultinormalDistribution[
>   mu, sigma], mu must be a p-variate vector, and sigma
>   must be a p x p symmetric positive definite matrix.

The RGW M-function, as described by Hull, is defined as follows:

```
mu = {0, 0};
sigma[rho_] := {{1, rho}, {rho, 1}}
```

Here is a plot for zero correlation:

```
Mfunc[a_, b_, rho_] :=
  CDF[MultinormalDistribution[mu, sigma[rho]], {a, b}]
```

```
Plot3D[Mfunc[a, b, 0], {a, -3, 3}, {b, -3, 3}];
```

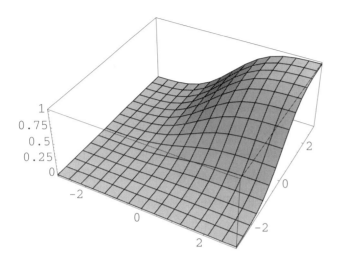

Critical Dividend and Price Computations

As discussed by Hull, there is a critical dividend level below which early exercise at the last dividend payment date is not optimal. If the dividend occurs at a point t before expiry, the critical dividend level is given by the formula

$$\text{CriticalDiv}(K, r, t) = K (1 - e^{-rt}) \tag{48}$$

```
CriticalDiv[K_, r_, t_] := K * (1 - Exp[-r * t])

CriticalDiv[50, 0.1, 0.5]
```

```
    2.43853
```

For dividends above this value, the critical price governing exercise is given by the solution of the equation:

```
CriticalPrice[ strike_, vol_, r_, q_, div_, divtime_] :=
price /.
  FindRoot[BlackScholesCall[price , strike, vol, r, q, divtime] ==
    price + div - strike, {price, strike}, MaxIterations -> 30]

CriticalPrice[50, 0.25, 0.1, 0, 3, 0.5]
```

```
    58.4246
```

```
CriticalPrice[50, 0.25, 0.1, 0, 6, 0.5]
```

    47.1163


```
CriticalPrice[50, 0.25, 0.1, 0, 2.4385287, 0.5]
```

    116.445

The algorithm for finding the critical price falls in a heap if we lower the dividend value below the critical dividend level:

```
CriticalPrice[50, 0.25, 0.1, 0, 2.43, 0.5]
```

    FindRoot::jsing :
     Encountered a singular Jacobian at the point price =
        1310.95393224636502`. Try perturbing the initial point(s).

    price /. FindRoot[
      BlackScholesCall[price, 50, 0.25, 0.1, 0, 0.5] == price + 2.43 - 50,
      {price, 50}, MaxIterations → 30]

This is because the equation is being solved for the zero of a function that does not attain the value zero - for example, if we plot it with a value of the dividend of 2.4, we can see this explicitly:

```
checkplot[strike_, vol_, r_, q_, div_, divtime_]  :=
 Plot[BlackScholesCall[price , strike, vol, r, q, divtime] -
   (price + div - strike),
  {price, 40, 100}, PlotRange -> {0, 0.5}]
```

```
checkplot[50, 0.25, 0.1, 0, 2.40, 0.5];
```

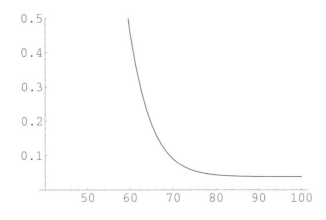

The Roll-Geske-Whaley Result

We write a formula that uses Black-Scholes if we are past the final dividend date, else it checks to see if the dividend is below the critical dividend for early exercise. In that case the jump-adjusted Black-Scholes is used. If the dividend is above the critical value for early exercise, the RGW result is used. Note that formulae are given by both Hull (1996), and Jarrow and Rudd (1983) - these are contradictory. I have chosen to use Hull's result - there were several papers written by Roll, Geske and Whaley (see the Bibliography) and it took several years to get the right result - Jarrow and Rudd's book was written before their analysis was finished.

```
RGWH[price_, strike_, vol_, r_, div_, τ_, tdiv_] :=
Module[{criticalprice, adjprice, aone, atwo, bone, btwo},
criticalprice = CriticalPrice[ strike, vol, r, 0, div, τ - tdiv];
adjprice = price - div * Exp[-r * tdiv];
 aone = (Log[adjprice / strike] + (r + vol^2 / 2) * τ) / (vol * Sqrt[τ]);
 atwo = aone - vol * Sqrt[τ];
 bone = (Log[adjprice / criticalprice] + (r + vol^2 / 2) * tdiv) /
   (vol * Sqrt[tdiv]);
 btwo = bone - vol * Sqrt[tdiv];
Chop[adjprice * Norm[bone] +
   adjprice * Mfunc[aone, -bone, -Sqrt[tdiv / τ]] -
   strike * Exp[-r * τ] * Mfunc[atwo, -btwo, -Sqrt[tdiv / τ]] -
   (strike - div) * Exp[-r * tdiv] * Norm[btwo], 10 ^ (-4)]]
```

Here it is wrapped up for an American Call with one discrete dividend:

```
AmCallH[price_, strike_, vol_, r_, τ_, div_, tdiv_] :=
 Module[{divtime = τ - tdiv},
 If[divtime <= 0, BlackScholes[price, strike, vol, r, τ],
   If[div <= CriticalDiv[strike , r, divtime],
     BlackScholesCall[price - div * Exp[-r * tdiv], strike, vol, r, 0, τ],
     RGWH[price, strike, vol, r, div, τ, tdiv]]]]
```

Here is a sample evaluation:

```
AmCallH[100, 100, 0.2, Log[1.05], 1, 20, 184 / 365]

   5.86669
```

Packages Re-visited

We shall need to refer to the results of the approximate form of the CJM model again, when we consider models of American options based on finite differences and binomial or trinomial trees. It is therefore desirable to put these particular functions into a Package. This is useful for other reasons, in that it is a good opportunity to show how to build a Package that uses another Package - the CJM model assumes that we already have a model for the European Put in place. This is very simply accomplished by adding the name of the existing required Package into the **BeginPackage** statement. In the following listing we have kept things to a minimum in order to focus on just this point and defining just the functions we need, but this gives us a prototype that could be used to build Packages out of most of the functions in this book.

We remind the reader that *this particular code is both experimental and approximate* - it should not be used in isolation - our purpose here is mainly to make it easy to re-load the CJM functions when we need them. Do not execute the following listing.

```
(* :Title:
    Approximate American Options *)

(* :Context:
    Derivatives`AnalyticAmerican` *)

(* :Author:
    Dr William T. Shaw *)

(* :Summary:
    *)

(* :Package Version:
    1.0 *)

(* :Mathematica Version:
    3.0 *)

(* :History:
    V.1.0, January 1997, by William T. Shaw *)


BeginPackage["Derivatives`AnalyticAmerican`", "Derivatives`BlackScholes`"]

(* Usage Information *)

Front::usage =
"Front[strike, interest, vol, t] returns the critical price frontier for a
vanilla zero-dividend American Put according to an approximation of the Carr-
Jarrow-Myneni integral equation."

AmericanPutCJM::usage =
"AmericanPutCJM[price,strike,vol,r,t]  returns an APPROXIMATE value for a
vanilla zero-dividend American Put according to an approximation of the Carr-
Jarrow-Myneni integral equation."


AmericanPutDeltaCJM::usage =
"AmericanPutDeltaCJM[price,strike,vol,r,t]  returns an APPROXIMATE value for
the delta of a vanilla zero-dividend American Put according to an
approximation of the Carr-Jarrow-Myneni integral equation."


AmericanPutGammaCJM::usage =
"AmericanPutGammaCJM[price,strike,vol,r,t]  returns an APPROXIMATE value for
the gamma of a vanilla zero-dividend American Put according to an
approximation of the Carr-Jarrow-Myneni integral equation."


(* END OF USAGE INFORMATION *)
```

```
Begin["`Private`"]

Front[strike_, r_, sigma_, t_] :=
If[t==0,strike,Module[{l,m},
l= m/. FindRoot[Exp[-m*sigma*Sqrt[t] - (r+sigma^2/2)*t]*(1+Erf[-m/Sqrt[2]]) ==
 r*Sqrt[2/(r+(sigma - (r/(2*sigma)+sigma/4) +m/(2*Sqrt[t]))^2/2)]*Erf[Sqrt[(
r+(sigma-(r/(2*sigma)+sigma/4)+m/(2*Sqrt[t]))^2/2)*t]]/sigma,
{m,-(sigma/2+r/sigma)*Sqrt[t]},MaxIterations -> 30];
strike*Exp[-l*sigma*Sqrt[t] - (r+sigma^2/2)*t]]]

AmericanPutCJM[price_,strike_,sigma_,r_,t_] :=
If[price <= Front[strike, r, sigma, t], strike - price,
BlackScholesPut[price,strike,sigma,r,0,t]+r*strike*NIntegrate[Exp[-r*u]*Norm[(
Log[Front[strike,r,sigma,t-u]/price]-(r-sigma^2/2)*u)/(sigma*Sqrt[u])],
{u,0,t}]]

AmericanPutDeltaCJM[price_,strike_,sigma_,r_,t_] :=
If[price <= Front[strike, r, sigma, t],  -1,
N[BlackScholesPutDelta[price,strike,sigma,r,0,t] - (r*strike/(price*sigma*
Sqrt[2.*Pi]))*NIntegrate[Exp[-r*u]*Exp[-((Log[Front[strike,r,sigma,t-u]/
price]-(r-sigma^2/2)*u)/(sigma*Sqrt[u]))^2/2]/Sqrt[u],
{u, 0, t}]]]

AmericanPutGammaCJM[price_,strike_,sigma_,r_,t_] :=
If[price <= Front[strike, r, sigma, t], 0,
N[
BlackScholesPutGamma[price,strike,sigma,r,0,t]+(r*strike/(price^2*sigma*Sqrt[
2.*Pi]))*NIntegrate[Exp[-r*u]*Exp[-((Log[Front[strike,r,sigma,t-u]/price]-(r-
sigma^2/2)*u)/(sigma*Sqrt[u]))^2/2]*(1/Sqrt[u] - (Log[Front[strike,r,sigma,t-
u]/price]-(r-sigma^2/2)*u)/(sigma*Sqrt[u])/(sigma*u)),
{u, 0, t}]]]


End[]

EndPackage[]
```

The contents of this file, called **AnalyticAmerican.m**, should be placed in the Derivatives folder or directory with the **BlackScholes.m** Package. We now test it. First we quit the kernel, and restart a fresh session by loading the package.

```
Quit[]
```

```
Needs["Derivatives`AnalyticAmerican`"]
```

```
?Derivatives`AnalyticAmerican`*
```

```
AmericanPutCJM       AmericanPutDeltaCJM AmericanPutGammaCJM Front
```

```
?AmericanPutCJM
```

```
AmericanPutCJM[price,strike,vol,r,t]   returns an APPROXIMATE
   value for a vanilla zero-dividend American Put according to
   an approximation of the Carr-Jarrow-Myneni integral equation.
```

Note that the European formulae are now also loaded:

? BlackScholesPut

```
BlackScholesPut[price, strike, vol, riskfree, divyield, expiry]
   returns the Black-Scholes value of a vanilla European Put.
```

We can check that the Package functions work as before:

```
{AmericanPutCJM[50, 50, 0.4, 0.1, 1],
 AmericanPutDeltaCJM[50, 50, 0.4, 0.1, 1],
 AmericanPutGammaCJM[50, 50, 0.4, 0.1, 1]}
```

```
{5.97688, -0.377896, 0.0229194}
```

Chapter 11 Bibliography

Barone-Adesi, G. and Whaley, R., 1987, Efficient analytic approximation of American option values, *Journal of Finance*, 42, p. 301.

Carr, P., Jarrow, R. and Myneni, R., 1992, Alternative characterization of American put options, *Mathematical Finance*, 2, p. 87.

Crank, J., 1984, *Free and Moving Boundary Value Problems*, Clarendon Press.

Geske, R., 1979, A note on an analytic valuation formula for unprotected American call options on stocks with known dividends, *Journal of Financial Economics*, 7, p. 375.

Geske, R., 1981, Comments on Whaley's note, *Journal of Financial Economics*, 9, p. 213.

Geske, R. and Johnson, H.E., 1984, The American put option valued analytically, *Journal of Finance*, 39, p. 1511.

Hill, J.M. and Dewynne, J.N., 1985a, A note on Langford's cylinder functions, *Quarterly Journal of Applied Mathematics*, 43, p. 179.

Hill, J.M. and Dewynne, J.N., 1985b, Bounds for moving boundary problems with two chemical reactions, *Nonlinear Analysis, Theory, Methods and Applications*, 9, p. 1293.

Hull, J.C., 1996, *Options, Futures, and Other Derivatives*, 3rd edition, Prentice-Hall.

Jarrow, R.A. and Rudd, A., 1983, *Option Pricing*, Irwin.

MacMillan, L., 1986, Analytic approximation for the American put option, *Advances in Futures and*

Options Research, 1, p. 119.

McKean, Jr, H.P., 1965, A free boundary problem for the heat equation arising from a problem in mathematical economics, *Ind. Management Rev.* 6, p. 32.

Ockendon, J.R., 1974, Techniques of analysis, in *Moving Boundary Problems in Heat Flow and Diffusion*, ed. Ockendon and Hodgkins, Oxford University Press.

Roll, R., 1977, An analytic formula for unprotected American call options on stocks with known dividends, *Journal of Financial Economics,* 5, p. 251.

Shaw, W. T., 1989, Redox front motion in a system of bentonite and fractured rock, *Swedish Nuclear Power Inspectorate Technical Report*, SKI TR 89:7 (Appendix on moving boundary problems available from WTS).

Van Moerbeke, P., 1976, On optimal stopping and free boundary problems, *Arch. Rational Mech. Anal.*, 60, p. 101.

Whaley, R., 1981, On the valuation of American call options on stocks with known dividends, *Journal of Financial Economics,* 9, p. 207.

Chapter 12.
Double-Barrier, Compound, Quanto Options and Other Exotics

A Bundle of Exotics and Diverse Methodologies

12.1 Introduction

In this chapter we consider several different types of "exotic" option, including:

double-barrier options;
compound options;
Quanto options;
warrants;
chooser options;
exchange options.

These illustrate various types of problem. Double-barriers can be solved both by series methods and by Laplace transform methods - we present an analysis of the latter based on methods similar to those used for arithmetic options. Compound and Quanto options have in common the fact that there are two correlated random processes present. In the case of compound options these are treated through the bivariate normal distribution, whereas Quantos can be reduced to a single-factor approach by using the arguments developed already for linked asset-FX options. We also consider some other commonplace exotics, such as chooser and exchange options, and briefly explore the effects of dilution in warrant pricing.

12.2 Double-Barrier   Options

Introduction

A double-barrier option is characterized by a strike, K, and two barriers, L, the lower barrier, and U, the upper barrier, with

$$L < K < U \tag{1}$$

The option knocks out, i.e., becomes worthless, if the stock price hits either barrier (rebates can be added). Otherwise the payoff at maturity is a standard type. In this section we shall consider a double-barrier Call option. The naive approach to this problem is to do a Fourier series type of solution - indeed, the solution of the corresponding heat equation, where the temperature is fixed at zero at both ends, and subject to an initial condition, is a familiar exercise in undergraduate mathematics. Just as familiar are the associated phenomena of the Gibbs effect, arising from a discontinuity between the initial condition and the boundary condition. This plays havoc with the series solution, at least for smaller times to maturity, in the neighbourhood of the offending barrier. The derivatives of the answer are particularly vulnerable to corruption in this respect. Such a discontinuity in initial-boundary data can also cause trouble for schemes such as simple implicit finite-difference schemes. Monte Carlo and tree methods also have their own problems when it comes to barrier options. In the case of Monte Carlo methods, an extremely large number of samples may be required to model effectively the impact of the barrier, and tree models are easily corrupted by nodes not being placed on the barrier. Not all series-type methods suffer from the kinds of pathology we have described. Work by Kunitomo and Ikeda (1992) describes another method which they argue converges very rapidly.

It is highly desirable therefore to have a non-series analytical approach. Remarkably, one such method was published very recently while this text was in preparation. Geman and Yor (1996) have shown how to employ the Laplace transform method to double-barrier options, using methods similar to those deployed for arithmetic Asian options, as described in Chapter 10. They demonstrated good agreement with Kunitomo and Ikeda. We shall explore and implement this method here.

The Geman-Yor Model for Double-Barrier Options

We shall work with a double knock-out Call option whose value is given by C. It is convenient to write

$$C = C_{BS} - e^{-rt} S \phi \tag{2}$$

where C_{BS} is the standard Black-Scholes value and ϕ is a normalized measure of the diminution in the value due to the double barrier. As usual, the risk-free continuously compounded interest rate is r and the stock price at initiation of the option is S. The knock-out term ϕ can be written as a function of the time to expiry t and the scaled variables:

$$h = \frac{K}{S} \quad m = \frac{L}{S} \quad M = \frac{U}{S} \tag{3}$$

The knock-out term is Laplace-transformed to define

$$\psi(\lambda) = \int_0^\infty e^{-\lambda \tau} \phi(\tau, h, m, M) \, d\tau \tag{4}$$

Further scaling arguments can be used to write

$$\psi(\lambda) = \frac{\Phi(\frac{\lambda}{\sigma^2})}{\sigma^2} \tag{5}$$

What is remarkable about the Geman-Yor analysis is that they have found an exact closed-form solution for $\Phi(\theta)$. First of all, we need to define (positive) quantities a and b by

$$m = e^{-a} \quad M = e^b \tag{6}$$

We also introduce a variable v as follows:

$$v = \frac{r - q - \frac{\sigma^2}{2}}{\sigma^2} \tag{7}$$

Finally we need a variable μ given in terms of the argument of the Laplace transform by

$$\mu = \sqrt{v^2 + 2\theta} \tag{8}$$

To build the Laplace transform, we now introduce some auxiliary functions f and g given by

$$f(a, h, \mu, v) = \frac{h^{-\mu+v+1} \, e^{-\mu a}}{\mu \, (\mu - v) \, (\mu - v - 1)} \tag{9}$$

$$g(b, h, \mu, v) = \frac{2 \, e^{b \, (v+1)}}{\mu^2 - (v + 1)^2} - \frac{2 \, h \, e^{b \, v}}{\mu^2 - v^2} + \frac{h^{\mu+v+1} \, e^{-\mu b}}{\mu \, (\mu + v) \, (\mu + v + 1)} \tag{10}$$

Then the Laplace transform Φ is given by

$$\Phi(\theta) = \frac{g(b, h, \mu, v) \sinh(\mu a)}{\sinh(\mu \, (a + b))} + \frac{f(a, h, \mu, v) \sinh(\mu b)}{\sinh(\mu \, (a + b))} \tag{11}$$

At the time of writing, a closed-form inversion of this is not known, but one can use a variety of numerical methods to invert it.

Mathematica Implementation of the Geman-Yor Model

First we just enter the basic changes of variable employed:

```
h[S_, K_]  := K/S;
a[S_, L_]  := Log[S/L];
b[S_, U_]  := Log[U/S];
v[σ_, r_, q_]  := (r - q - σ^2/2)/σ^2;
μ[v_, θ_]  := Sqrt[v^2 + 2*θ]
```

We now define the auxiliary functions f and g given by

```
f[a_, h_, μ_, v_]  :=  h^{-μ+v+1} Exp[-μ a]
                       ─────────────────────
                       μ (μ - v) (μ - v - 1)
```

```
g[b_, h_, μ_, v_]  :=  2 Exp[b (v + 1)]     2 h Exp[b v]    h^{μ+v+1} Exp[-μ b]
                       ────────────────  -  ────────────  +  ───────────────────
                         μ² - (v + 1)²         μ² - v²         μ (μ + v) (μ + v + 1)
```

Then the Laplace transform Φ is given by

```
Φ[S_, K_, L_, U_, σ_, r_, q_, θ_] :=
  Module[{A = a[S, L], B = b[S, U], H = h[S, K], n = v[σ, r, q], m},

  m = μ[n, θ];   g[B, H, m, n] Sinh[m A] + f[A, H, m, n] Sinh[m B]
                 ──────────────────────────────────────────────── ]
                                 Sinh[m (A + B)]
```

Just to check that all this is working, and to see the type of function we are dealing with, we plot the absolute value of the transform in complex θ-space:

```
Plot3D[Abs[Φ[20, 15, 10, 30, 0.2, 0.1, 0.5, s + I t]],
  {s, -200, 0}, {t, -10, 10}, PlotRange -> All, PlotPoints -> 80];
```

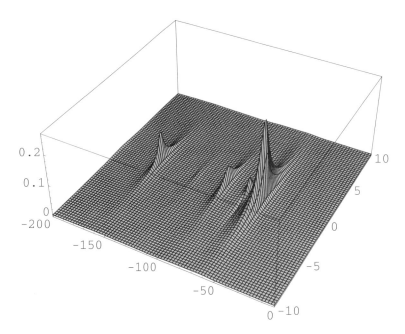

We now need to invert this transform, evaluated at $\sigma^2\,T$, and subtract its discounted and normalized value from the Black-Scholes value. First we load the standard Package:

```
Needs["Derivatives`BlackScholes`"]
```

Now we define the Laplace inversion by integration along a vertical line in the complex plane (the Bromwich integral), with real part given by the parameter "contour". The integration works, but *Mathematica* may warn us about the oscillatory nature of the integration. If you get irritating messages, you might like to turn it off, using

```
Off[NIntegrate::slwcon]
```

However, if you choose parameters that result in the integration yielding strange answers, you should not turn off such messages, but possibly explore the options to **NIntegrate**, as discussed in Section 2.6. Now we define the valuation by the use of the Bromwich integral. In the following function we compute the Black-Scholes value, the correction for a unit stock price, and the full value.

```
DoubleBarrierCall[S_, K_, L_, U_, σ_, r_, q_, t_, contour_] :=
Module[{bs, dim},
  bs = BlackScholesCall[S, K, σ, r, q, t];
  dim = Re[ Exp[-r*t] / (2 Pi) *NIntegrate[
      Φ[S, K, L, U, σ, r, q, contour + I p] * Exp[(contour + I p) *σ^2 * t],
  {p, -1500, 1500}, MaxRecursion -> 10]];  bs - S *dim]
```

Note that I have not explored the various approximate schemes for transform inversion - here we are concerned with getting an accurate answer. You may wish to explore the stability under changes in the contour parameter. The range $\{\{contour - 1500\,I\}, \{contour + 1500\,I\}\}$ has been found to be useful for several practical cases. The truncation at 1500 may need to be modified for other situations.

Examples

The following numerical examples were considered by Geman and Yor (1996) - we see that our model agrees with their numerical simulations:

```
DoubleBarrierCall[2, 2, 1.5, 2.5, 0.2, 0.02, 0, 1, 10]
```

```
    0.0410889
```

```
DoubleBarrierCall[2, 2, 1.5, 3.0, 0.5, 0.05, 0, 1, 10]
```

```
    0.0178568
```

```
DoubleBarrierCall[2, 1.75, 1.0, 3.0, 0.5, 0.05, 0, 1, 10]
```

```
    0.0761723
```

Hedging the Double-Barrier Call

It is now just a matter of differentiating the analytical result to obtain Delta and the other Greeks. Geman and Yor worked out a formula for the Laplace transform of Delta, then inverted this. We can just as easily get *Mathematica* to do the differentiations for us. Let's first take a look at Delta. We need to bear in mind that we have put a factor involving the current stock price outside the integration. So the differentiation will involve the original diminution function, and its derivative.

```
DeltaΦ[S_, K_, L_, U_, σ_, r_, q_, z_] =
  Evaluate[D[Φ[S, K, L, U, σ, r, q, z], S]];

DoubleBarrierCallDelta[S_, K_, L_, U_, σ_, r_, q_, t_, contour_] :=
Module[{bsd, dim, deldim},
  bsd = BlackScholesCallDelta[S, K, σ, r, q, t];
  deldim = Re[ Exp[-r * t] / (2 Pi) * NIntegrate[
    (Φ[S, K, L, U, σ, r, q, contour + I p] + S * DeltaΦ[S, K, L, U,
         σ, r, q, contour + I p]) * Exp[(contour + I p) * σ^2 * t],
    {p, -1500, 1500}, MaxRecursion -> 10]]; bsd - deldim]
```

```
DoubleBarrierCallDelta[2, 2, 1.5, 2.5, 0.2, 0.02, 0, 1, 10]

    0.0118013
```

The Other Greeks

For all the other Greeks except Theta, we just differentiate the Laplace transform, and add some corrections. The following set of functions serve as the necessary building blocks - the function **AuxΦ** plays a role in the computation of both Theta and Vega, and Theta, the time-derivative, is computed using the appropriate Laplace transform identity. In what follows we define each Greek and perform a sample evaluation.

```
GammaΦ[S_, K_, L_, U_, σ_, r_, q_, z_] =
  Evaluate[D[Φ[S, K, L, U, σ, r, q, z], {S, 2}]];
RhoΦ[S_, K_, L_, U_, σ_, r_, q_, z_] =
  Evaluate[D[Φ[S, K, L, U, σ, r, q, z], r]];
VegaΦ[S_, K_, L_, U_, σ_, r_, q_, z_] =
  Evaluate[D[Φ[S, K, L, U, σ, r, q, z], σ]];
AuxΦ[S_, K_, L_, U_, σ_, r_, q_, z_] = z * Φ[S, K, L, U, σ, r, q, z];


DoubleBarrierCallGamma[S_, K_, L_, U_, σ_, r_, q_, t_, contour_] :=
Module[{bsd, dim, deldim, gamdim},
  bsg = BlackScholesCallGamma[S, K, σ, r, q, t];
  gamdim = Re[ Exp[-r * t] / (2 Pi) * NIntegrate[
    (2 * DeltaΦ[S, K, L, U, σ, r, q, contour + I p] + S * GammaΦ[S, K, L, U,
             σ, r, q, contour + I p]) * Exp[(contour + I p) * σ^2 * t],
    {p, -1500, 1500}, MaxRecursion -> 10]]; bsg - gamdim]


DoubleBarrierCallGamma[2, 2, 1.5, 2.5, 0.2, 0.02, 0, 1, 10]

    -0.454404


DoubleBarrierCallRho[S_, K_, L_, U_, σ_, r_, q_, t_, contour_] :=
Module[{bsr, dim, rhodim},
  bsr = BlackScholesCallRho[S, K, σ, r, q, t];
  rhodim = Re[ Exp[-r * t] / (2 Pi) * NIntegrate[
    (t * Φ[S, K, L, U, σ, r, q, contour + I p] - RhoΦ[S, K, L, U,
            σ, r, q, contour + I p]) * Exp[(contour + I p) * σ^2 * t],
  {p, -1500, 1500}, MaxRecursion -> 10]]; bsr + S * rhodim]
```

```
DoubleBarrierCallRho[2, 2, 1.5, 2.5, 0.2, 0.02, 0, 1, 10]
```

```
0.0791006
```

```
DoubleBarrierCallVega[S_, K_, L_, U_, σ_, r_, q_, t_, contour_] :=
Module[{bsr, dim, rhodim},
  bsv = BlackScholesCallVega[S, K, σ, r, q, t];
  vegdim = Re[ Exp[-r * t] / (2 Pi) * NIntegrate[
      (VegaΦ[S, K, L, U, σ, r, q, contour + I p] + 2 * σ * t * AuxΦ[S, K, L,
          U, σ, r, q, contour + I p]) * Exp[(contour + I p) * σ^2 * t],
  {p, -1500, 1500}, MaxRecursion -> 10]]; bsv - S * vegdim]
```

```
DoubleBarrierCallVega[2, 2, 1.5, 2.5, 0.2, 0.02, 0, 1, 10]
```

```
-0.38284
```

```
DoubleBarrierCallTheta[S_, K_, L_, U_, σ_, r_, q_, t_, contour_] :=
Module[{bsr, dim, rhodim},
  bst = BlackScholesCallTheta[S, K, σ, r, q, t];
  thetdim = Re[ Exp[-r * t] / (2 Pi) * NIntegrate[
      (σ^2 * AuxΦ[S, K, U, σ, r, q, contour + I p] - r * Φ[S, K, L, U,
          σ, r, q, contour + I p]) * Exp[(contour + I p) * σ^2 * t],
  {p, -1500, 1500}, MaxRecursion -> 10]]; bst + S * thetdim]
```

```
DoubleBarrierCallTheta[2, 2, 1.5, 2.5, 0.2, 0.02, 0, 1, 10]
```

```
0.036702
```

All of these results can be confirmed by the use of an approximate differencing technique, but they illustrate the point that the use of symbolic differentiation is not just confined to where there is a simple "nice formula".

12.3 Compound Options

In this section we present a brief summary of the valuation of compound options as described by Hull (1996) together with some sample valuations. This model makes use of the standard **BlackScholes** package for determining critical prices at the first exercise date, and also uses the Package (new in version 3.0) for the multivariate normal distribution. This avoids the use of any approximations, but at the time of writing I can only recommend its use for two dimensions. We have not considered the computation of Greeks here due to some issues it raises differentiating the new Package functions - they can of course be estimated using central difference methods.

Basic Package Functions

First we load some routines to model the multivariate normal distribution and vanilla options.

```
Needs["Statistics`MultinormalDistribution`"]
```

```
Needs["Derivatives`BlackScholes`"]
```

We want the two-variable case of the multi-normal object:

```
? MultinormalDistribution
```

> MultinormalDistribution[mu, sigma] represents the
> multivariate normal (Gaussian) distribution with mean
> vector mu and covariance matrix sigma. For a p-variate
> random vector to be distributed MultinormalDistribution[
> mu, sigma], mu must be a p-variate vector, and sigma
> must be a p x p symmetric positive definite matrix.

The cumulative distribution function (CDF) or M-function, as described by Hull (1996), is defined as follows:

```
mu = {0, 0};
sigma[rho_] := {{1, rho}, {rho, 1}}

Mfunc[a_, b_, rho_] :=
  Re[CDF[MultinormalDistribution[mu, sigma[rho]], {a, b}]]
```

Here is a plot for zero correlation:

```
Plot3D[Mfunc[a, b, 0], {a, -3, 3}, {b, -3, 3}];
```

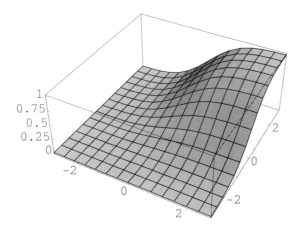

Critical Price Computations

First a point of notation: The first exercise date is T_1 and the second exercise date is T_2. The second phase of option has time to expiry $T_2 - T_1$ and a strike X_2. First we get the critical prices where the value of the underlying option equals the strike on the first date.

```
CriticalCallPrice[Xone_, Xtwo_, σ_, r_, q_, Tone_, Ttwo_] :=
  P /. FindRoot [
    Xone == BlackScholesCall[P, Xtwo, σ, r, q, Ttwo - Tone], {P, Xtwo}]
```

```
CriticalCallPrice[3, 50, 0.2, 0.05, 0, 1, 2]
```

      46.0059

```
CriticalPutPrice[Xone_, Xtwo_, σ_, r_, q_, Tone_, Ttwo_] :=
  P /. FindRoot [
    Xone == BlackScholesPut[P, Xtwo, σ, r, q, Ttwo - Tone], {P, Xtwo}]
```

```
CriticalPutPrice[3, 50, 0.2, 0.05, 0, 1, 2]
```

      49.4298

Auxiliary Functions

To define the compound valuations we need the following functions:

```
aone[S_, Sc_, σ_, r_, q_, Tone_] :=
  (Log[S / Sc] + (r - q + σ^2 / 2) * Tone) / (σ * Sqrt[Tone])
```

```
atwo[S_, Sc_, σ_, r_, q_, Tone_] :=
  (Log[S / Sc] + (r - q - σ^2 / 2) * Tone) / (σ * Sqrt[Tone])
```

```
bone[S_, Xtwo_, σ_, r_, q_, Ttwo_] :=
  (Log[S / Xtwo] + (r - q + σ^2 / 2) * Ttwo) / (σ * Sqrt[Ttwo])
```

```
btwo[S_, Xtwo_, σ_, r_, q_, Ttwo_] :=
  (Log[S / Xtwo] + (r - q - σ^2 / 2) * Ttwo) / (σ * Sqrt[Ttwo])
```

Call on a Call

We are now able to write down the pricing formula for the Call on a Call.

```
CallCall[S_, Xo_, Xt_, σ_, r_, q_, To_, Tt_] :=
Module[{Sc, ao, at, bo, bt, rho = Sqrt[To / Tt]},
  Sc = CriticalCallPrice[Xo, Xt, σ, r, q, To, Tt];
  ao = aone[S, Sc, σ, r, q, To];
  at = atwo[S, Sc, σ, r, q, To];
  bo = bone[S, Xt, σ, r, q, Tt];
  bt = btwo[S, Xt, σ, r, q, Tt];
  S * Exp[-q * Tt] * Mfunc[ao, bo, rho] -
    Xt * Exp[-r * Tt] * Mfunc[at, bt, rho] - Xo * Exp[-r * To] * Norm[at]
]
```

If you prefer to work with interest rates and yields in annual terms rather than the continuously compounded ones employed above we make a simple change of variables:

```
CallCallAnn[S_, Xo_, Xt_, σ_, r_, q_, To_, Tt_] :=
  CallCall[S, Xo, Xt, σ, Log[1 + r] , Log[1 + q], To, Tt]
```

Sample Values

Let us consider the following sample parameters:

index: 100;
strike on underlying: 100;
volatility: 20%
risk-free: 10% *annual*;
dividend-yield: 5% *annual*;
total time to final expiry (Hull's T_2): 1 year.

The strikes on the intermediate date are given by the following list:

```
klist = {5, 7.5, 10, 12.5, 15};
```

The times to the intermediate date, T_1, are given by the following list, giving fractions of a year:

```
tlist = {0.25, 0.375, 0.50, 0.625, 0.75};
```

We make a list of values as a matrix, with each row corresponding to a fixed strike, each column corresponding to one time.

```
TableForm[Table[CallCallAnn[100, klist[[i]], 100, 0.2,
   0.10, 0.05, tlist[[j]], 1.0], {i, 1, 5}, {j, 1, 5}]]
```

5.30918	5.66388	6.01016	6.35066	6.69356
3.6633	4.20004	4.68152	5.13252	5.57095
2.44416	3.07077	3.62277	4.13383	4.62502
1.58336	2.21509	2.78303	3.31375	3.82489
0.998576	1.57701	2.12127	2.64174	3.14854

These results compare well with those published by Rubenstein in his *RISK Magazine* article (1991) and elsewhere.

Put on a Call

```
PutCall[S_, Xo_, Xt_, σ_, r_, q_, To_, Tt_] :=
Module[{Sc, ao, at, bo, bt, rho = Sqrt[To / Tt]},
  Sc = CriticalCallPrice[Xo, Xt, σ, r, q, To, Tt];
  ao = aone[S, Sc, σ, r, q, To];
  at = atwo[S, Sc, σ, r, q, To];
  bo = bone[S, Xt, σ, r, q, Tt];
  bt = btwo[S, Xt, σ, r, q, Tt];
  -S * Exp[-q * Tt] * Mfunc[-ao, bo, -rho] +
    Xt * Exp[-r * Tt] * Mfunc[-at, bt, -rho] + Xo * Exp[-r * To] * Norm[-at]
]
```

```
PutCallAnn[S_, Xo_, Xt_, σ_, r_, q_, To_, Tt_] :=
  PutCall[S, Xo, Xt, σ, Log[1 + r], Log[1 + q], To, Tt]
```

Sample Values

For the sample parameters as considered previously, we have

```
TableForm[Table[PutCallAnn[100, klist[[i]], 100, 0.2,
   0.10, 0.05, tlist[[j]], 1.0], {i, 1, 5}, {j, 1, 5}]]
```

0.414163	0.711039	1.00019	1.28423	1.57133
1.20941	1.65943	2.05521	2.42151	2.77625
2.43141	2.94238	3.38011	3.77825	4.15786
4.01175	4.49893	4.92402	5.3136	5.68525
5.8681	6.27307	6.64592	6.99701	7.33643

Put on a Put

```
PutPut[S_, Xo_, Xt_, σ_, r_, q_, To_, Tt_] :=
Module[{Sc, ao, at, bo, bt, rho = Sqrt[To / Tt]},
 Sc = CriticalPutPrice[Xo, Xt, σ, r, q, To, Tt];
 ao = aone[S, Sc, σ, r, q, To];
 at = atwo[S, Sc, σ, r, q, To];
 bo = bone[S, Xt, σ, r, q, Tt];
 bt = btwo[S, Xt, σ, r, q, Tt];
 S * Exp[-q * Tt] * Mfunc[ao, -bo, -rho] -
   Xt * Exp[-r * Tt] * Mfunc[at, -bt, -rho]  + Xo * Exp[-r * To] * Norm[at]
]
```

```
PutPutAnn[S_, Xo_, Xt_, σ_, r_, q_, To_, Tt_] :=
 PutPut[S, Xo, Xt, σ, Log[1 + r] , Log[1 + q], To, Tt]
```

Sample Values

For the sample parameters as considered previously, we have

```
TableForm[Table[PutPutAnn[100, klist[[i]], 100, 0.2,
   0.10, 0.05, tlist[[j]], 1.0], {i, 1, 5}, {j, 1, 5}]]
```

1.12555	1.42676	1.68969	1.93407	2.1739
2.69241	2.97936	3.23818	3.48304	3.72561
4.68042	4.86422	5.05548	5.25099	5.45405
6.90662	6.96542	7.06211	7.18218	7.32128
9.25438	9.20243	9.2016	9.23689	9.30039

Call on a Put

```
CallPut[S_, Xo_, Xt_, σ_, r_, q_, To_, Tt_] :=
Module[{Sc, ao, at, bo, bt, rho = Sqrt[To / Tt]},
 Sc = CriticalPutPrice[Xo, Xt, σ, r, q, To, Tt];
 ao = aone[S, Sc, σ, r, q, To];
 at = atwo[S, Sc, σ, r, q, To];
 bo = bone[S, Xt, σ, r, q, Tt];
 bt = btwo[S, Xt, σ, r, q, Tt];
 -S * Exp[-q * Tt] * Mfunc[-ao, -bo, rho] +
   Xt * Exp[-r * Tt] * Mfunc[-at, -bt, rho]  - Xo * Exp[-r * To] * Norm[-at]
]
```

```
CallPutAnn[S_, Xo_, Xt_, σ_, r_, q_, To_, Tt_] :=
  CallPut[S, Xo, Xt, σ, Log[1 + r], Log[1 + q], To, Tt]
```

Sample Values

For the sample parameters as considered previously, we have

```
TableForm[Table[CallPutAnn[100, klist[[i]], 100, 0.2,
   0.10, 0.05, tlist[[j]], 1.0], {i, 1, 5}, {j, 1, 5}]]
```

1.69156	2.05059	2.37066	2.6715	2.96712
0.817287	1.19096	1.5355	1.86504	2.1913
0.364159	0.663606	0.969139	1.27757	1.59221
0.149225	0.35258	0.592112	0.853327	1.1319
0.0558511	0.177365	0.347942	0.552615	0.783484

12.4 Quanto Options

In Chapter 3 we analysed the various options that can be built out of an asset and an FX rate. We identified a certain class of options, known as "Quantos", that could be reduced to a standard form by various parameter substitutions. One can also approach Quantos through the methods given by Reiner (1992). Reiner's model is based on the *foreign-market* valuation model. The Quanto option may be valued in much the same way as was done with the geometric Asian, by making a suitable substitution for the parameters.

Quanto Effective Yield

We define a function **QuantoQ** (the x_3 as defined by Reiner), which is the effective yield parameter.

```
QuantoQ[q_, rd_, rf_, sdstock_, sdforex_, corr_] :=
  q + rd - rf + corr * sdstock * sdforex
```

Note that we are using continuously compounded variables for **rf**, **q**, **r**, whereas Reiner uses annual measures.

Quanto Call Exact Symbolic Model

The Black-Scholes formula for the Quanto valuation is

```
QuantoCall[pricef_, strikef_,
  sdstock_, sdforex_, rf_, rd_, q_, corr_, t_, fixed_, spot_] :=
fixed * BlackScholesCall[pricef, strikef, sdstock,
   rd, QuantoQ[q, rd, rf, sdstock, sdforex, corr], t]
```

Now we compute the sensitivities by symbolic differentiation. Note that the Delta is given w.r.t. domestic currency, so that it does indeed give the required amount of the equity in a hedge portfolio. The hedge portfolio should also contain some domestic and foreign cash. Note also that we define a pair of Rhos and a pair of Vegas.

```
QuantoCallDelta[pricef_, strikef_, sdstock_, sdforex_, rf_, rd_, q_,
corr_, t_, fixed_, spot_] =
D[QuantoCall[pricef, strikef, sdstock, sdforex, rf, rd,q, corr, t,
fixed, spot], pricef]/spot;

QuantoCallGamma[pricef_, strikef_, sdstock_, sdforex_, rf_, rd_, q_,
corr_, t_, fixed_, spot_] =
D[QuantoCall[pricef, strikef, sdstock, sdforex, rf, rd, q,
corr, t, fixed, spot], {pricef,2}]/spot^2;

QuantoCallTheta[pricef_, strikef_, sdstock_, sdforex_, rf_, rd_, q_,
corr_, t_, fixed_, spot_] =
D[-QuantoCall[pricef, strikef, sdstock, sdforex, rf, rd, q,
corr, t, fixed, spot], t];

QuantoCallVega[pricef_, strikef_, sdstock_, sdforex_, rf_, rd_, q_,
corr_, t_, fixed_,spot_] =
D[QuantoCall[pricef, strikef, sdstock, sdforex, rf, rd, q,
corr, t, fixed, spot], sdstock];

QuantoCallVegaFX[pricef_, strikef_, sdstock_, sdforex_, rf_, rd_, q_,
corr_, t_, fixed_,spot_] =
D[QuantoCall[pricef, strikef, sdstock, sdforex, rf, rd, q,
corr, t, fixed, spot], sdforex];

QuantoCallRhoF[pricef_, strikef_, sdstock_, sdforex_, rf_, rd_, q_,
corr_, t_, fixed_, spot_] =
D[QuantoCall[pricef, strikef, sdstock, sdforex, rf, rd, q,
corr, t, fixed, spot], rf];

QuantoCallRhoD[pricef_, strikef_, sdstock_, sdforex_, rf_, rd_, q_,
corr_, t_, fixed_, spot_] =
D[QuantoCall[pricef, strikef, sdstock, sdforex, rf, rd, q,
corr, t, fixed, spot], rd];
```

Quanto Put Exact Symbolic Model

First we introduce the value function:

```
QuantoPut[pricef_, strikef_, sdstock_,
sdforex_, rf_, rd_, q_, corr_, t_, fixed_, spot_] :=
fixed * BlackScholesPut[pricef, strikef, sdstock,
    rd, QuantoQ[q, rd, rf, sdstock, sdforex, corr], t]
```

Now we compute the sensitivities by differentiation:

```
QuantoPutDelta[pricef_, strikef_, sdstock_, sdforex_, rf_, rd_, q_,
corr_, t_, fixed_, spot_] =
D[QuantoPut[pricef, strikef, sdstock, sdforex, rf, rd, q,
corr, t, fixed, spot], pricef]/spot;

QuantoPutGamma[pricef_, strikef_, sdstock_, sdforex_, rf_, rd_, q_,
corr_, t_, fixed_, spot_] =
D[QuantoPut[pricef, strikef, sdstock, sdforex, rf, rd, q,
corr, t, fixed, spot], {pricef,2}]/spot^2;

QuantoPutTheta[pricef_, strikef_, sdstock_, sdforex_, rf_, rd_, q_,
corr_, t_, fixed_, spot_] =
D[-QuantoPut[pricef, strikef, sdstock, sdforex, rf, rd, q,
corr, t, fixed, spot], t];

QuantoPutVega[pricef_, strikef_, sdstock_, sdforex_, rf_, rd_, q_,
corr_, t_, fixed_,spot_] =
D[QuantoPut[pricef, strikef, sdstock, sdforex, rf, rd, q,
corr, t, fixed, spot], sdstock];

QuantoPutVegaFX[pricef_, strikef_, sdstock_, sdforex_, rf_, rd_, q_,
corr_, t_, fixed_, spot_] =
D[QuantoPut[pricef, strikef, sdstock, sdforex, rf, rd, q,
corr, t, fixed, spot], sdforex];

QuantoPutRhoF[pricef_, strikef_, sdstock_, sdforex_, rf_, rd_, q_,
corr_, t_, fixed_, spot_] =
D[QuantoPut[pricef, strikef, sdstock, sdforex, rf, rd, q,
corr, t, fixed, spot], rf];

QuantoPutRhoD[pricef_, strikef_, sdstock_, sdforex_, rf_, rd_, q_,
corr_, t_, fixed_, spot_] =
D[QuantoPut[pricef, strikef, sdstock, sdforex, rf, rd, q,
corr, t, fixed, spot], rd];
```

12.5 Warrants

The Lauterbach-Schultz Analysis

In their 1990 paper, Lauterbach and Schultz (LS) looked at two types of models for valuing warrants. The Black-Scholes model may be modified to value European warrants issued by a company on its own stock, by allowing for the effects of dilution. This dilution adjustment was discussed in the original Black-Scholes paper and so will be referred to as the Black-Scholes warrant-pricing model. This clearly ignores the following: early exercise; extension; volatility and interest-rate changes over the warrant life; variation of volatility with stock price.

Cox, in a 1975 Stanford working paper, introduced constant elasticity of variance (CEV) models where the volatility is inversely proportional to a fractional power of the stock price. The special case where the volatility varied as the inverse square root of the stock price is called the SRCEV model. This family of models is discussed in detail in Chapter 28.

LS also investigated other modifications to BS that allow for early exercise, and stochastic interest rates and volatilities. They concluded that these modifications did not generate significant improvements in the pricing model. They concluded, however, that the CEV model outperforms the dilution-adjusted Black-Scholes model given a good estimate of the elasticity, but that it is also true that the SRCEV model also outperforms the dilution-adjusted BS model. These comparisons were based on a data analysis study that explored how well the models explained market price data - the BS model was found to be markedly inferior at explaining prices across the market. Nevertheless, the dilution-adjusted Black-Scholes model has found widespread popularity, and it is the one developed here.

The reasons for focusing on this model are pedagogical, since the main focus of this book is on explaining how to do things in *Mathematica*. The CEV models can be coded up using the functions built in to *Mathematica*. However, the dilution adjustment to the Black-Scholes model requires the solution of non-linear equations (somewhat analogously to the computation of implied volatility), so it is perhaps more instructive to review this here. As noted, the CEV family is discussed in Chapter 28, and can also be made the basis of a dilution-adjusted CEV calculation of value.

The Dilution Equation for Warrants

Let us suppose that the unknown warrant price is "warprice", and that the number of outstanding warrants is "warrants", the number of shares per warrant is "shperwar" and the number of outstanding shares is "shares".

```
WarrantEqn[p_,k_,sd_,r_,q_,t_,warprice_, shares_, warrants_,
shperwar_] :=
(shares*shperwar/(shares+shperwar*warrants))*
BlackScholesCall[p*Exp[-q*t]+warrants*warprice/shares,k,sd,r,0,t]
```

```
WarrantValue[p_,k_,sd_,r_,q_,t_, shares_, warrants_, shperwar_] :=
warprice /. FindRoot[
warprice == WarrantEqn[p,k,sd,r,q,t,warprice,shares, warrants, shper-
war],
{warprice, 10}]
```

```
WarrantValue[100,90,0.1,0.04,0.04,1, 100, 10, 1]
```

```
    10.1664
```

As remarked in the introduction to this section, this calculation is meant to be illustrative, as there are other real-world complications in warrant pricing that are mostly beyond the scope of this text, but see Chapter 28 for a discussion of CEV models. One irritating point of convention is that, depending on the market, warrant values may be reported in raw currency terms or as a percentage of their denomination! Also, the formulae we have given relate to purely domestic warrants - see Section 3.6 for a discussion of FX management.

Warrant Hedge Parameters

These are slightly tricky as the dilution effect rears its head in several places simultaneously. Here is an illustrative computation of Delta with the dilution effect incorporated - along the way the warrant value is reported.

```
Clear[wdone];
```

```
wdone[price_, strike_, sd_, r_, t_] :=
(Log[price/strike] + (r + sd^2/2)*t)/(sd*Sqrt[t])
```

```
di[shares_, warrants_, shperwar_] :=
(shares*shperwar)/(shares+warrants*shperwar)
```

```
WarrantDelta[price_, strike_, vol_, rate_, q_, time_,shares_,
warrants_, shperwar_] :=
Module[{w, delbs, dil},
w = WarrantValue[price,strike,vol,rate,q, time,shares, warrants,
shperwar];
Print[w];
dil = di[shares,warrants,shperwar];
delbs = Norm[wdone[price*Exp[-
q*time]+w*warrants/shares,strike,vol,rate,time]];
dil*delbs*Exp[-q*time]/(1 - dil*delbs*warrants/shares)]
```

```
WarrantDelta[780,815,0.238120,4*Log[1+1.351875/400],0.03,3.89589,
29135984,9550,513.619632]
```

```
    47542.1
```

```
    220.202
```

It may be useful to express Delta on a "per share" basis - here it is divided by the shares per warrant (some people also prefer to multiply by 100 to get Delta on a percentage basis):

```
%/513.619632
```

```
    0.428725
```

12.6 Chooser Options

The option known as a "chooser option", also called by Hull (1996) an "as you like it" option, contains the feature that the holder can decide after a specified period of time whether the option is a Call or a Put. The basic European chooser option can be regarded as a package consisting of a simple Call and a simple Put with different weights and times to expiry. As usual, we need the functions used in Black-Scholes, and load the Package to treat this first.

```
Needs["Derivatives`BlackScholes`"]
```

Additional Functions for Use in Choosers

We now use the characterization of a chooser as a package consisting of a Call and a Put. Note that **t** is time until expiry, **tchoose** is time until choice.

```
EuroChooser[p_, k_, sd_, r_, q_, t_, tchoose_] :=
BlackScholesCall[p,k,sd,r,q,t] +
Exp[-q*(t - tchoose)]*BlackScholesPut[p,k*Exp[-(r-q)*(t -
tchoose)],sd,r,q,tchoose]
```

```
EuroChooserDelta[p_, k_, sd_, r_, q_, t_, tchoose_] =
D[EuroChooser[p, k, sd, r, q, t, tchoose], p];
```

```
EuroChooserGamma[p_, k_, sd_, r_, q_, t_, tchoose_] =
D[EuroChooser[p, k, sd, r, q, t, tchoose], {p,2}];
```

In the following, note that we want the rate of change with respect to real time - we therefore decrement BOTH the time to expiry and the time till choice.

```
EuroChooserTheta[p_, k_, sd_, r_, q_, t_, tchoose_] =
D[-EuroChooser[p, k, sd, r, q, t, tchoose], t] -
D[EuroChooser[p, k, sd, r, q, t, tchoose], tchoose];


EuroChooserVega[p_, k_, sd_, r_, q_, t_, tchoose_] =
D[EuroChooser[p, k, sd, r, q, t, tchoose], sd];


EuroChooserRho[p_, k_, sd_, r_, q_, t_, tchoose_] =
D[EuroChooser[p, k, sd, r, q, t, tchoose], r];
```

12.7 Exchange Options

These options are in one sense rather like Quantos, in that although there are two correlated random processes, the combination can be reduced to a one-dimensional system. The matter was sorted out by Margrabe (1978).

Additional Functions for Use in Exchange Options

Our notation is as used by Hull (1996), Section 18.1. First we check that our standard Package has been loaded, by confirming that the CDF for the normal distribution is available:

```
? Norm
```

```
    Norm[x] returns the cumulative
       normal distribution function evaluated at x.
```

We just need the following two functions:

```
done[priceone_, pricetwo_, sdone_, sdtwo_, qone_, qtwo_, corr_, t_] :=
(Log[pricetwo/priceone] +
(qone - qtwo + (sdone^2+sdtwo^2 - 2 corr sdone sdtwo)/2)*t)/
Sqrt[(sdone^2 + sdtwo^2 - 2 corr sdone sdtwo)*t]


dtwo[priceone_, pricetwo_, sdone_, sdtwo_, qone_, qtwo_, corr_, t_] :=
(Log[pricetwo/priceone] +
(qone - qtwo - (sdone^2+sdtwo^2 - 2 corr sdone sdtwo)/2)*t)/
Sqrt[(sdone^2 + sdtwo^2 - 2 corr sdone sdtwo)*t]
```

Exchange Value and Sensitivity Functions

The relevant variables are **priceone** and **pricetwo**, the values of the two underlying assets, and their volatilities **sdone**, **sdtwo** and continuous yields **qone**, **qtwo**. The correlation between the assets is **corr** and the time to expiry is **t**.

First we introduce the value function:

```
Exchange[priceone_, pricetwo_, sdone_, sdtwo_, qone_, qtwo_, corr_,
t_]:=
pricetwo*Exp[-qtwo*t]*
Norm[done[priceone, pricetwo, sdone, sdtwo, qone, qtwo, corr, t]] -
priceone*Exp[-qone*t]*
Norm[dtwo[priceone, pricetwo, sdone, sdtwo, qone, qtwo, corr, t]]
```

We check that it evaluates correctly:

```
Exchange[100, 100, 0.25, 0.25, 0, 0, 0, 366/365]
```

```
    14.0506
```

Now we compute the sensitivities by differentiation. The formulae are independent of the risk-free rate so Rho is always zero.

```
ExchangeDeltaOne[pone_, ptwo_, sdone_, sdtwo_, qone_, qtwo_, corr_,
t_] =
D[Exchange[pone, ptwo, sdone, sdtwo, qone, qtwo, corr, t], pone];

ExchangeDeltaTwo[pone_, ptwo_, sdone_, sdtwo_, qone_, qtwo_, corr_,
t_] =
D[Exchange[pone, ptwo, sdone, sdtwo, qone, qtwo, corr, t], ptwo];

ExchangeGammaOne[pone_, ptwo_, sdone_, sdtwo_, qone_, qtwo_, corr_,
t_] =
D[Exchange[pone, ptwo, sdone, sdtwo, qone, qtwo, corr, t], {pone, 2}];

ExchangeGammaTwo[pone_, ptwo_, sdone_, sdtwo_, qone_, qtwo_, corr_,
t_] =
D[Exchange[pone, ptwo, sdone, sdtwo, qone, qtwo, corr, t], {ptwo, 2}];

ExchangeVegaOne[pone_, ptwo_, sdone_, sdtwo_, qone_, qtwo_, corr_, t_]
=
D[Exchange[pone, ptwo, sdone, sdtwo, qone, qtwo, corr, t], sdone];

ExchangeVegaTwo[pone_, ptwo_, sdone_, sdtwo_, qone_, qtwo_, corr_, t_]
=
D[Exchange[pone, ptwo, sdone, sdtwo, qone, qtwo, corr, t], sdtwo];
```

```
ExchangeTheta[pone_, ptwo_, sdone_, sdtwo_, qone_, qtwo_, corr_, t_] =
-D[Exchange[pone, ptwo, sdone, sdtwo, qone, qtwo, corr, t], t];
```

Chapter 12 Bibliography

Geman, H. and Yor, M., 1996, Pricing and hedging double-barrier options: a probabilistic approach, *Mathematical Finance*, 6, p. 365.

Hull, J.C., 1996, *Options, Futures, and Other Derivatives*, 3rd edition, Prentice-Hall.

Kunitomo, N. and Ikeda, M., 1992, Pricing options with curved boundaries, *Mathematical Finance,* 2, p. 275.

Lauterbach, B. and Schultz, P., 1990, Pricing warrants: an empirical study of the Black-Scholes model and its alternatives, *Journal of Finance*, 45, p. 1181.

Margrabe, W., 1978, The value of an option to exchange one asset for another, *Journal of Finance*, 33, p. 177.

Reiner, E., 1992, Quanto Mechanics, in *From Black-Scholes to Black Holes* (based on *RISK Magazine*, Mar. 92), RISK-FINEX Publications.

Rubenstein, M., 1991, Double trouble, in *RISK Magazine*, Dec. 1991-Jan.1992.

Chapter 13.
The Discipline of the Greeks and Overview of Finite-Difference Schemes

Error Magnification and an Overview of Simple Difference Schemes

13.1 A Reminder from Basic Analysis

When we value derivatives, we usually need to establish the value of not just a function of many variables, but also several of its first and second derivatives. The complications and additional discipline that the computation of such "Greeks" imposes seem not to be widely appreciated. To appreciate what is happening it is a good idea to temporarily forget all about financial derivatives and remind ourselves of some results from basic analysis.

Non-uniform Convergence 101

Suppose we have a sequence of functions $f_n(x)$, $n = 1, 2, 3, \ldots$, with the property that for each x,

$$f_n(x) \to 0 \tag{1}$$

as

$$n \to \infty \tag{2}$$

In this case a (mathematical!) analyst would say that f *converges pointwise to zero*. The question arises what happens to the derivative of f_n as n becomes large. Naively one might expect that also becomes small as the function becomes small. Indeed, many functions satisfy this requirement, such as, for x in some finite interval,

$$\frac{x^m}{n} \qquad \frac{\sin(x)}{n} \tag{3}$$

Unfortunately, it is not always true. The following classic example makes this clear. We consider the function

$$f(x, n) = \frac{\sin(n\,x)}{n} \tag{4}$$

Its first derivative, or "Delta", is then

$$\frac{\partial f(x, n)}{\partial x} = \cos(n\,x) \tag{5}$$

Its second derivative, or "gamma", is then

$$\frac{\partial^2 f(x, n)}{\partial x^2} = -n \sin(n\,x) \tag{6}$$

So we have the situation where although the function goes to zero, its Delta remains of $O(1)$, and its Gamma is $O(n)$ and tends to infinity. This is important to us when we regard f as the error arising in some numerical scheme. In other words, the following situation is perfectly possible:

Small error in function $\Rightarrow$ possible moderate error in first derivative $\delta \Rightarrow$ possible huge error in second derivative Γ.

There are other strange things that can happen with sequences of functions, although the issue with differentiation is of the most importance to derivatives modelling. There is a type of convergence called uniform convergence, which places stringent limits on the behaviour of functions which ensure that derivatives are as controlled as the function value itself.

For our purposes it is more important to appreciate when we consider the function f to represent the errors in some numerical scheme, where n is related to some grid or tree parameter, there may be substantial errors in our Greeks even when a close inspection of the valuation suggests all is well.

The Key Utility of *Mathematica*

The power of *Mathematica* as an investigative tool becomes clear in this context. We can define a series of test problems and use the special-function capabilities of *Mathematica* to characterize the solution exactly, using *Mathematica*'s infinite-precision arithmetic to get the results with as much accuracy as we desire. Next we use *Mathematica*'s symbolic calculus capabilities to define the Greeks by differentiation. Then we make a comparison, within *Mathematica*, between these analytical forms and the numerical solution. The latter is obtained using compiled forms of tridiagonal, SOR and PSOR solvers, where the enhancements of *Mathematica* version 3.0 are used to optimize the numerics.

13.2 Overview of Difference Schemes

Two-Time-Level Schemes for the Diffusion Equation

Our initial goal is to derive a family of difference schemes for investigation. A discussion of difference schemes is given in many good books on numerical PDEs. For option-pricing practitioners perhaps the most accessible texts are the excellent pair by Wilmott *et al* (1993, 1995), but they are also described in the classic survey by Richtmyer and Morton (1957). The more recent and more widely available Oxford text by Smith (1985) also gives an excellent discussion. Perhaps the cleanest derivation of an entire family of schemes involving two time-levels is given by Mitchell and Griffiths (1980), and in what follows we present an argument that borrows heavily from their approach.

We assume that one way or the other, our option-pricing problem has been reduced to the diffusion equation:

$$\frac{\partial u}{\partial \tau} = \frac{\partial^2 u}{\partial x^2} \tag{7}$$

We shall describe in detail how this may be done for various types of option later - for now it is much more useful to study this problem in isolation, and we shall also do so numerically for a test problem in Chapter 14. We introduce a discrete grid with steps $\Delta \tau$, Δx, where Δx is the grid step for the (log) stock price, and $\Delta \tau$ is the grid step for the time, and set

$$u_n^m = u(m \, \Delta \tau, n \, \Delta x) \tag{8}$$

All the difference schemes involve a parameter α that is given by

$$\alpha = \frac{\Delta \tau}{\Delta x^2} \tag{9}$$

The Operator Approach

We introduce the operators L, D, given by

$$L f = \frac{\partial f}{\partial t} \qquad D f = \frac{\partial f}{\partial x} \tag{10}$$

So the diffusion equation is just $L f = D^2 f$. Assuming that the Taylor series expansion holds, we can write

$$u(\tau + \Delta \tau, x) = e^{\Delta \tau L} u(\tau, x) \tag{11}$$

In other words

$$u_n^{m+1} = e^{\Delta \tau L} u_n^m = e^{\Delta \tau D^2} u_n^m \tag{12}$$

More generally, if we consider the value, u_θ of u at $x = n \, \Delta x$, and $\tau = \theta n \, \Delta \tau + (1 - \theta)(n + 1) \, \Delta \tau$, we can write it in two ways. First, by using a forwards Taylor expansion, we have

$$u_\theta = e^{\Delta \tau (1-\theta) L} u_n^m = e^{\Delta \tau (1-\theta) D^2} u_n^m \tag{13}$$

By considering a Taylor series backwards from the next time-level, we can also say that

$$u_\theta = e^{-\Delta \tau \theta L} u_n^{m+1} = e^{-\Delta \tau \theta D^2} u_n^{m+1} \tag{14}$$

So on the assumption that we have such Taylor series, we can equate the two to obtain

$$e^{-\Delta \tau \theta D^2} u_n^{m+1} = e^{\Delta \tau (1-\theta) D^2} u_n^m \tag{15}$$

Note that no approximations have been made.

The Difference Operators

Now we define the difference operator δ_x by

$$\delta_x u_n^m = u_{n+\frac{1}{2}}^m - u_{n-\frac{1}{2}}^m \tag{16}$$

Its square is

$$\delta_x^2 u_n^m = u_{n+1}^m + u_{n-1}^m - 2 u_n^m \tag{17}$$

It can be shown that there is an *exact* relationship (see e.g. Hildebrand, 1956) of the form

$$D = \frac{2 \sinh^{-1}(\frac{\delta_x}{2})}{\Delta x} \tag{18}$$

Now our diffusion equation (15) involves $\Delta\tau D^2$, which, after some algebra, we can expand out as

$$\alpha\left(\delta_x^2 - \frac{\delta_x^4}{12} + \frac{\delta_x^6}{90} + ...\right) \tag{19}$$

This can be confirmed and extended using *Mathematica*:

```
g[x_] := 2 ArcSinh[x / 2];

Series[(g[x])^2, {x, 0, 10}]
```

$$x^2 - \frac{x^4}{12} + \frac{x^6}{90} - \frac{x^8}{560} + \frac{x^{10}}{3150} + O[x]^{11}$$

General High Order Difference Versions of the Diffusion Equation

We can combine our exact diffusion equation (15) with the series expansion of the operators contained within it (19) to obtain a description of the problem to any desired order. Keeping all terms up to order δ_x^6, and performing some tedious simplifications, the combination of (15) and (19) becomes, neglecting eighth and higher order differences

$$u_n^{m+1} - \alpha\,\theta\,\delta_x^2 u_n^{m+1} + \frac{1}{2}\,\theta\,\alpha\left(\alpha\,\theta + \frac{1}{6}\right)\delta_x^4 u_n^{m+1} - \alpha\,\theta\left(\frac{\theta^2\,\alpha^2}{6} + \frac{\theta\,\alpha}{12} + \frac{1}{90}\right)\delta_x^6 u_n^{m+1} =$$

$$u_n^m + \alpha\,(1-\theta)\,\delta_x^2 u_n^m + \left(\frac{1}{2}\,(1-\theta)^2\,\alpha^2 - \frac{1}{12}\,(1-\theta)\,\alpha\right)\delta_x^4 u_n^m \tag{20}$$

$$+ \left(\frac{1}{6}\,\alpha^2(1-\theta)^2 - \frac{1}{12}\,\alpha\,(1-\theta) + \frac{1}{90}\right)\alpha\,(1-\theta)\,\delta_x^6 u_n^m$$

This in general is a matrix equation, and can be represented in terms of a difference matrix, A, that governs the mapping from one time level to the next. All such schemes can be written in the form $u^{m+1} = A\,u^m$ for a suitable difference matrix A.

Explicit Schemes

These are obtained by setting $\theta = 0$, thereby obtaining, to sixth order,

$$u_n^{m+1} = u_n^m + \alpha\,\delta_x^2 u_n^m + \frac{\alpha}{2}\left(\alpha - \frac{1}{6}\right)\delta_x^4 u_n^m + \frac{\alpha}{6}\left(\alpha^2 - \frac{1}{2}\,\alpha + \frac{1}{15}\right)\delta_x^6 u_n^m \tag{21}$$

Second Order Explicit and Binomial Schemes

If we keep terms to second order we obtain

$$u_n^{m+1} = u_n^m + \alpha \; \delta_x^2 \, u_n^m \tag{22}$$

The choice $\alpha = 1/2$ gives the scheme embodied by the binomial model (if a tree-shaped grid is used instead of a rectangular grid).

Fourth Order Explicit and Pentanomial/Trinomial Schemes

If we keep terms to fourth order we obtain the family of pentanomial schemes

$$u_n^{m+1} = u_n^m + \alpha \; \delta_x^2 \, u_n^m + \frac{\alpha}{2} \left(\alpha - \frac{1}{6} \right) \delta_x^4 \, u_n^m \tag{23}$$

The choice $\alpha = 1/6$ gives the scheme embodied by the trinomial model (if a tree-shaped grid is used instead of a rectangular grid) - the fourth order terms then vanish identically, and we have a simple scheme but with high order accuracy.

Implicit Crank-Nicholson and Douglas Schemes

Working to second order, the choice $\theta = 1/2$ gives the scheme known as the Crank-Nicholson scheme. If we write it out it becomes

$$u_n^{m+1} - \frac{1}{2} \alpha \, (u_{n-1}^{m+1} + u_{n+1}^{m+1} - 2 \, u_n^{m+1}) = u_n^m + \frac{1}{2} \alpha \, (u_{n-1}^m + u_{n+1}^m - 2 \, u_n^m) \tag{24}$$

Another interesting set of schemes can be derived by taking the general high order scheme with $\theta = 1/2$, and multiplying both sides by

$$1 + \mu \, \delta_x^2 - \lambda \, \delta_x^4 \tag{25}$$

The choice $\mu = 1/12, \; \lambda = \alpha^2 / 8$ leads to a very interesting equation where the fourth order terms disappear:

$$u_n^{m+1} - (1/12 - \alpha/2) \, \delta_x^2 \, u_n^{m+1} + O(\delta_x^6 \, u_n^{m+1}) = u_n^m + (1/12 + \alpha/2) \, \delta_x^2 \, u_n^m + O(\delta_x^6 \, u_n^m) \tag{26}$$

The second order truncated form, i.e.

$$u_n^{m+1} - (1/12 - \alpha/2) \, \delta_x^2 \, u_n^{m+1} = u_n^m + (1/12 + \alpha/2) \, \delta_x^2 \, u_n^m \tag{27}$$

is called the *Douglas* scheme. It is very important due to the fact that it is exact to order δ_x^4, even though it contains terms only of order δ_x^2.

The θ-Method Family

The θ-method schemes are given by just keeping the second order terms in the general scheme. Writing it out, we have

$$u_n^{m+1} - \theta\alpha \, (u_{n-1}^{m+1} + u_{n+1}^{m+1} - 2 \, u_n^{m+1}) = u_n^m + \alpha \, (1 - \theta) \, (u_{n-1}^m + u_{n+1}^m - 2 \, u_n^m) \tag{28}$$

The explicit scheme is given by $\theta = 0$, and fully implicit is given by $\theta = 1$. Crank-Nicholson is given by the choice $\theta = 1/2$. A point of interest is that Wilmott *et al* remark that it is not particularly sensible to use the θ method for values of $\theta < 1/2$, since "this involves all the complications of solving implicit sets of equations without any great improvement in stability over the explicit method." This unfortunately conceals a powerful improvement over ordinary Crank-Nicholson schemes. The *Douglas* scheme can now be seen to correspond to a special value of θ given by

$$\theta = \frac{1}{2} - \frac{1}{12\,\alpha} \tag{29}$$

and is in fact unconditionally stable (see the remarks later on stability). It is therefore highly desirable to use this particular value of θ, even though it is less than $1/2$. This will be seen from our discussion of a test problem in Chapter 14, but there are more fundamental reasons. These can be seen by writing down the truncation errors for each scheme. These have been defined and given by Richtmyer and Morton.

Truncation errors:

Explicit : $\qquad O(\Delta\tau) + O((\Delta x)^2)$ $\qquad$ (30)

Fully implicit : $\quad O(\Delta\tau) + O((\Delta x)^2)$ $\qquad$ (31)

CN : $\qquad\quad O(\Delta\tau^2) + O((\Delta x)^2)$ $\qquad$ (32)

Douglas : $\qquad O(\Delta\tau^2) + O((\Delta x)^4)$ $\qquad$ (33)

Although we eventually want to use the implicit schemes with high α, we note that the Douglas scheme has an optimal α where the truncation error is minimized. This is

$$\alpha = \frac{1}{\sqrt{20}} \tag{34}$$

but this is rather small to be of practical use. It can, however, produce amazing error reduction - see Section 14.10.

13.3 Reminder on Binomial and Trinomial Trees

The following links may be helpful in appreciating the importance of the Douglas scheme, and its relationship to the explicit schemes that are common to FD models and binomial/trinomial tree models.

(1) When $\theta = 0$ and $\alpha = 1/2$, our schemes, as noted already, become the same difference rule as in a binomial tree, but on a rectangular grid in standard coordinates.

(2) When we use a Douglas scheme with $\alpha = 1/6$, so that equation (29) gives $\theta = 0$, we obtain a highly accurate explicit scheme. This is none other than the trinomial tree model, but on a rectangular grid. So the Douglas scheme is the natural implicit form of the trinomial model. We have already seen that this case also corresponds to a special case of a high order explicit scheme based on five points (the pentanomial scheme).

The relationship to standard tree models may be made clearer if we state up front the simplest form of the change of variables used to reduce an option-pricing problem to the diffusion equation. For a problem with a flat term-structure parametrized by a variable K (e.g. strike or barrier) the change of coordinates being used here is, for an underlying S, time variable T, volatility σ,

$$x = \log\left(\frac{S}{K}\right) \tag{35}$$

$$\tau = \frac{\sigma^2 T}{2} \tag{36}$$

so that

$$\Delta S = S\sigma \sqrt{\frac{\Delta T}{2\alpha}} \tag{37}$$

In these coordinates the relationship between FD schemes with $\alpha = 1/2$, $1/6$ and binomial, trinomial trees becomes clear. To practitioners, we emphasize the following:

The two-time-level Douglas scheme is the natural implicit generalization of the trinomial tree.

13.4 Stability and Non-smooth Initial Conditions

Unfortunately this is far from the end to the story as far as option pricing is concerned. Matters are more complicated when the initial condition ("payoff") is an option-like payoff function. In Chapter 15 we will consider first a simple Put payoff. When we use smaller time-steps, the Douglas method gives slightly more accurate results than Crank-Nicholson, and both are more accurate than explicit or fully implicit. However, as the time-step is increased, both the Crank-Nicholson method and the Douglas scheme start to exhibit bad behaviour in a neighbourhood of the strike, and the accuracy of both is compromised.

This effect is actually quite well known and is understood in theoretical terms. Both Richtmyer and Morton, and the book by Smith, point out that non-smooth initial data, or other discontinuities between boundary conditions and initial conditions, can induce slowly decaying oscillations in otherwise "stable" models. The Crank-Nicholson scheme has long been understood to be vulnerable to this. Smith explains how the diffusion equation can be discretized using the notion of Padé approximations. This gives a family of schemes labelled $P(s, t)$, for s and t integers, and an associated notion of *strong stability*, which is satisfied if $s > t$. The Crank-Nicholson scheme fails this test as it is a $P(1, 1)$ scheme. In fact, fully implicit is a $P(1, 0)$ scheme, so does not suffer from the oscillation problem, but it is not generally accurate enough because the truncation error is only $O(\Delta t)$.

This leaves us with a problem. We can either look at implementations of Padé $(2, 0)$ or $(2, 1)$ schemes, or at extrapolations of the Padé $(1, 0)$ scheme, or try something else. Richtmyer and Morton have a family of suggestions for combining stability with large time step, with the elimination of oscillatory terms while retaining reasonable accuracy. These are based on three-time-level schemes. Both Richtmyer and Morton, and Smith give three-time-level analogues of Crank-Nicholson and Douglas, and state they are the best schemes to use when the initial data are non-smooth. We shall look at these in Chapter 15, where we shall aim to give our final view on a "benchmark" finite-difference scheme.

The choice of a good scheme is closely related to various notions of stability. It is beyond the scope of this text to supply a critique of the various notions of stability and supply a catalogue of what schemes are stable in each sense. Each of the texts in the bibliography contains some form of discussion, and the reader is referred to these, and Smith in particular, for a theoretical discussion. For our purposes, it would take a substantial amount of theory to get to our final conclusions. We shall reach our view instead by doing some numerical experiments with *Mathematica* that reveal the very practical issues that arise in the option-pricing context. These suggest that given our desire for accurate estimates of both a function and its derivatives, only the strongest notion of stability, as discussed by Smith, is actually useful. Other notions of stability, in particular, those associated with merely requiring that the eigenvalues of the difference matrix are less than unity in magnitude, are not sufficient, though they are certainly necessary.

Chapter 13 Bibliography

Hildebrand, F.B., 1956, *Introduction to Numerical Analysis*, McGraw-Hill.

Mitchell, A.R. and Griffiths, D.F., 1980, *The Finite Difference Method in Partial Differential Equations*, John Wiley (Corrected reprinted edition, 1994).

Richtmyer, R.D. and Morton, K.W., 1957, *Difference Methods for Initial Value Problems*, Krieger reprint, 1994, reprinted from the Wiley 1957/1967 original.

Smith, G.D., 1985, *Numerical Solution of Partial Differential Equations: Finite Difference Methods*, Oxford University Press.

Wilmott, P., Dewynne, J. and Howison, S., 1993, *Option Pricing - Mathematical Models and Computation*, Oxford Financial Press.

Wilmott, P., Dewynne, J. and Howison, S., 1995, *The Mathematics of Financial Derivatives*, Cambridge University Press.

Chapter 14.
Finite-Difference Schemes for the Diffusion Equation with Smooth Initial Conditions

Explicit, Implicit, Crank-Nicholson and Douglas
Finite-Difference Schemes Compared

14.1 Introduction

In binomial models there is a link between the time-step and the price-step in the binomial tree. Accuracy requires a small price-step, and this implies a small time-step. For long times to expiry many thousands of time-steps may be needed to get an accurate result. In finite-difference (FD) schemes, explicit schemes have the same properties in this respect as binomial models, and there are finite-difference analogues of both binomial and trinomial schemes. Much more interesting are the implicit schemes. These allow much larger time-steps with no appreciable loss in accuracy, and therefore in principle offer much more efficient computation.

Furthermore, the nature of these schemes is such that they allow rather more accurate and straightforward computations of Delta, Gamma and Theta to be given than with (a standard implementation of) a binomial model. Also, there is an increasing understanding of how to map many types of option-pricing problems into finite-difference form.

14.2 Schemes Investigated

In this chapter we compare the accuracy of various difference schemes for solving the diffusion equation. This is the equation that arises when the Black-Scholes differential equation is transformed into a form suitable for treatment by finite-difference methods. We compare

(a) explicit finite-difference, with 400 time-steps (equivalent to the use of a binomial model, but on a grid rather than a tree);
(b) fully implicit, also with 400 time-steps;
(c) Crank-Nicholson, with 40 time-steps;
(d) Douglas, with 40 time-steps.

The solution method for type (a) is a simple updating rule, while (b), (c), (d) require the solution of tridiagonal systems of equations.

14.3 *Mathematica* Solution Issues

We will be able to write down a closed-form solution and implement it in *Mathematica*. Then we shall describe two types of solution method that work for FD schemes applied to the ordinary diffusion equation and European-style options. The first is an explicit and compiled solver, while the second uses a compiled solver for tridiagonal systems. We shall see later how to generalize these to compiled SOR and PSOR solvers. Another use of *Mathematica* is to automatically interpolate the solution using various types of interpolation scheme.

14.4 A Simple Test Problem with Smooth Initial Conditions

We consider the diffusion equation

$$\frac{\partial u}{\partial \tau} = \frac{\partial^2 u}{\partial x^2} \tag{1}$$

on the region defined by

$$-2 \leq x \leq 2 \qquad \tau \geq 0 \tag{2}$$

The initial condition is

$$u(x, 0) = \sin\left(\frac{\pi x}{2}\right) \tag{3}$$

and the boundary conditions are

$$u(2, \tau) = u(-2, \tau) = 0 \tag{4}$$

This has the exact solution

$$u(x, \tau) = \sin\left(\frac{\pi x}{2}\right) e^{-\frac{\pi^2 \tau}{4}} \tag{5}$$

So it is a simple matter to test various difference schemes by comparing with this known exact solution. It should be emphasized that this type of smooth initial data, which also joins continuously onto the boundary conditions, is rather atypical of option-pricing problems. Our purpose here is to simplify matters to get a general feel for the relative merits of explicit and implicit schemes, and to introduce the idea of compiling within *Mathematica*.

14.5 Explicit Scheme

Here are the price- and time-steps used, together with an output that is the value of the parameter $\alpha = \frac{\Delta\tau}{\Delta x^2}$.

```
dx = 0.025; dtau = 0.00025; alpha = dtau/dx^2
```

```
0.4
```

The number of time-steps is 400, and there are 160 space-steps.

```
M=400; nminus = 80; nplus = 80;
```

Here we set initial and boundary conditions, and plot the latter:

```
initial =
  Table[N[Sin[Pi * (k - 1 - nminus) / nminus] ], {k, 1, nminus + nplus + 1}];
lower = Table[0, {m, 1, M + 1}];
upper = Table[0, {m, 1, M + 1}];
```

```
ListPlot[initial];
```

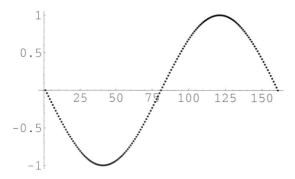

Here we define a function to solve the problem on the given grid, the output is a list of values of *u* for the last three time points (for now we shall just use the final value). Note that this function makes use of the *Mathematica* 3.0 vector compilation methods. The initial conditions, and upper and lower boundary conditions, are supplied as vectors, **initial**, **lower**, **upper**, as rank-1 objects that are real - note the syntax {**initial**, **_Real**, **1**} to denote this. For further details see section 2.5.14 of the *Mathematica* book.

```
ExplicitSolver =
  Compile[
   {{initial, _Real, 1}, {lower, _Real, 1}, {upper, _Real, 1}, alpha},
   Module[{wold = initial, wnew = initial, wvold, m,
     k, tsize = Length[lower], xsize = Length[initial]},
    For[m = 2, m <= tsize, m++,
     (wvold = wold; wold = wnew;
      For[k = 2, k < xsize, k++,
       (wnew[[k]] =
          alpha (wold[[k - 1]] + wold[[k + 1]]) + (1 - 2 alpha) wold[[k]])];
       wnew[[1]] = lower[[m]];
       wnew[[xsize]] = upper[[m]])];
     {wvold, wold, wnew}]
   ];
```

Now we apply this to get the solution at a time $\tau = 0.1$ (the units are irrelevant for our analysis):

```
soln = ExplicitSolver[initial, lower, upper, alpha];
```

Now we do the interpolation to supply a continuous function:

```
interpoldata =
Table[{(k-nminus-1)*dx ,soln[[3,k]]}, {k, 1, nminus+nplus+1}];
```

```
ufunc = Interpolation[interpoldata, InterpolationOrder -> 3]
```

```
    InterpolatingFunction[{{-2., 2.}}, <>]
```

Now we plot the error in the answer:

```
Plot[ufunc[x] - Sin[Pi*x/2]*Exp[-Pi^2 0.1/4], {x, -2, 2},
PlotPoints -> 50];
```

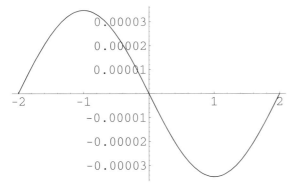

Finally we tabulate the numerical result, the exact result and the error in the numerical scheme. With 400 time-steps the error is manageably small.

```
samples = TableForm[Join[{{"x", "Exact", "Explicit FD", "Error"}},
Table[Map[PaddedForm[N[Chop[#1]],{5,6}]&,
N[{x,
ufunc[x],
Sin[Pi*x/2]*Exp[-(Pi^2 0.1)/4],
ufunc[x]- Sin[Pi*x/2]*Exp[-(Pi^2 0.1)/4]},5]],
{x, -2, 2, 0.25}]]]
```

x	Exact	Explicit FD	Error
-2.000000	0.000000	0.000000	0.000000
-1.750000	-0.298990	-0.299010	0.000013
-1.500000	-0.552470	-0.552490	0.000025
-1.250000	-0.721840	-0.721870	0.000032
-1.000000	-0.781310	-0.781340	0.000035
-0.750000	-0.721840	-0.721870	0.000032
-0.500000	-0.552470	-0.552490	0.000025
-0.250000	-0.298990	-0.299010	0.000013
0.000000	0.000000	0.000000	0.000000
0.250000	0.298990	0.299010	-0.000013
0.500000	0.552470	0.552490	-0.000025
0.750000	0.721840	0.721870	-0.000032
1.000000	0.781310	0.781340	-0.000035
1.250000	0.721840	0.721870	-0.000032
1.500000	0.552470	0.552490	-0.000025
1.750000	0.298990	0.299010	-0.000013
2.000000	0.000000	0.000000	0.000000

14.6 Compiled Tridiagonal Solver for Implicit Schemes

We introduce the following compiled *Mathematica* function to solve tridiagonal systems of equations. It mimics the Package function originally written by J. Keiper (see the *Mathematica* Packages documentation) with some minor adjustments to get it through the compiler.

```
CompTridiagSolve =
 Compile[{{a, _Real, 1}, {b, _Real, 1}, {c, _Real, 1}, {r, _Real, 1}},
 Module[{len = Length[r], solution = r,
    aux = 1 / (b[[1]]), aux1 = r, a1 = Prepend[a, 0.0], iter},
  solution[[1]] = aux * r[[1]];
  Do[aux1[[iter]] = c[[iter - 1]] aux;
    aux = 1 / (b[[iter]] - a1[[iter]] * aux1[[iter]]);
    solution[[iter]] =
    (r[[iter]] - a1[[iter]] solution[[iter - 1]]) aux,
   {iter, 2, len}];
  Do[solution[[iter]] -= aux1[[iter + 1]] solution[[iter + 1]],
   {iter, len - 1, 1, -1}];
 solution]];
```

14.7 Fully Implicit Scheme

Here is the matrix involved in the computations:

```
FullyImpCMatrix[alpha_, nminus_, nplus_] :=
Sequence[Table[-alpha, {nplus+nminus-2}],
Table[1+2*alpha, {nplus+nminus-1}],
Table[-alpha, {nplus+nminus-2}]]
```

Here are the parameters of this particular model:

```
M=400;
nminus = 80;
nplus = 80;
dx = 0.025;
dtau = 0.00025;
alpha = dtau/dx^2
```

    0.4

Initial and boundary conditions, and problem initialization:

```
initial =
 Table[N[Sin[Pi * (k - 1 - nminus) / nminus] ], {k, 1, nminus + nplus + 1}];
lower = Table[0, {m, 1, M + 1}];
upper = Table[0, {m, 1, M + 1}];
wold = initial;
wvold = wold;
wnew = wold;
```

The solution:

```
CMat = FullyImpCMatrix[alpha,nminus,nplus];

For[m=2, m<=M+1, m++,
(wvold = wold; wold = wnew;
rhs = Take[wold, {2, -2}]+
Table[Which[k==1, alpha*lower[[m]],
k== nplus + nminus-1,  alpha*upper[[m]], True, 0],
{k, 1, nplus + nminus-1}];
temp = CompTridiagSolve[CMat, rhs];
wnew = Join[{lower[[m]]}, temp, {upper[[m]]}])]
```

Interpolating the answer:

```
interpoldata =
Table[{(k-nminus-1)*dx ,wnew[[k]]}, {k, 1, nminus+nplus+1}];
ufunc = Interpolation[interpoldata, InterpolationOrder -> 3];
```

Now we plot the error:

```
Plot[ufunc[x] - Sin[Pi*x/2]*Exp[-Pi^2 0.1/4], {x, -2, 2},
PlotPoints -> 50];
```

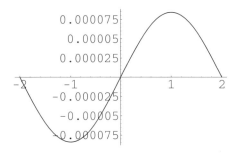

Finally we tabulate the numerical result, the exact result and the error in the numerical scheme.

```
samples = TableForm[Join[{{"x", "Exact", "Implicit FD", "Error"}},
Table[Map[PaddedForm[N[Chop[#1]],{5,6}]&,
N[{x, ufunc[x], Sin[Pi*x/2]*Exp[-(Pi^2 0.1)/4],
ufunc[x]- Sin[Pi*x/2]*Exp[-(Pi^2 0.1)/4]},5]],
{x, -2, 2, 0.25}]]]
```

x	Exact	Implicit FD	Error
-2.000000	0.000000	0.000000	0.000000
-1.750000	-0.299040	-0.299010	-0.000032
-1.500000	-0.552550	-0.552490	-0.000060
-1.250000	-0.721950	-0.721870	-0.000078
-1.000000	-0.781430	-0.781340	-0.000084
-0.750000	-0.721950	-0.721870	-0.000078
-0.500000	-0.552550	-0.552490	-0.000060
-0.250000	-0.299040	-0.299010	-0.000032
0.000000	0.000000	0.000000	0.000000
0.250000	0.299040	0.299010	0.000032
0.500000	0.552550	0.552490	0.000060
0.750000	0.721950	0.721870	0.000078
1.000000	0.781430	0.781340	0.000084
1.250000	0.721950	0.721870	0.000078
1.500000	0.552550	0.552490	0.000060
1.750000	0.299040	0.299010	0.000032
2.000000	0.000000	0.000000	0.000000

Note that we obtain no improvement in accuracy over the explicit scheme. The only advantage of the fully implicit scheme over the explicit scheme is the fact that we can increase α, that is, the time-step for a given price-step, without the system going unstable. You might like to explore this, and the accuracy of the solution, by adjusting the parameters given above and re-running.

14.8 Crank-Nicholson

The fully implicit analysis can be repeated with a Crank-Nicholson scheme by making some minor changes to the difference algorithm. This time we also implement two different interpolation schemes - linear and cubic.

```
CNCMatrix[alpha_, nminus_, nplus_] :=
Sequence[Table[-alpha/2, {nplus+nminus-2}],
Table[1+alpha, {nplus+nminus-1}],
Table[-alpha/2, {nplus+nminus-2}]];

CNDMatrix[alpha_, vec_List] := Module[{temp},
temp = (1 - alpha)*vec + (alpha/2)*(RotateRight[vec] + Rotate-
Left[vec]);
temp[[1]] = Simplify[First[temp] - alpha*Last[vec]/2];
temp[[-1]] = Simplify[Last[temp] - alpha*First[vec]/2];
temp];

M=40; nminus = 80; nplus = 80;
dx = 0.025; dtau = 0.0025; alpha = dtau/dx^2;
initial=Table[N[Sin[Pi*(k-1-nminus)/nminus] ], {k,1, nminus+nplus+1}];
lower=Table[0, {m, 1, M+1}]; upper=Table[0, {m, 1, M+1}];
wold = initial; wvold = wold; wnew = wold;
CMat = CNCMatrix[alpha,nminus,nplus];

For[m=2, m<=M+1, m++,
(wvold = wold; wold = wnew;
rhs = CNDMatrix[alpha, Take[wold, {2, -2}]]+
Table[Which[k==1, alpha*(lower[[m-1]] + lower[[m]])/2,
k== nplus + nminus-1,  alpha*(upper[[m-1]] + upper[[m]])/2,
True, 0],
{k, 1, nplus + nminus-1}];
temp = CompTridiagSolve[CMat, rhs];
wnew = Join[{lower[[m]]}, temp, {upper[[m]]}])
];

interpoldata =
Table[{(k-nminus-1)*dx ,wnew[[k]]}, {k, 1, nminus+nplus+1}];

ufunca = Interpolation[interpoldata, InterpolationOrder -> 1];
ufuncb = Interpolation[interpoldata, InterpolationOrder -> 3];
```

Now we plot the error, with both linear and cubic interpolation:

```
Plot[ufunca[x] - Sin[Pi*x/2]*Exp[-Pi^2 0.1/4], {x, -2, 2},
PlotPoints -> 50];
```

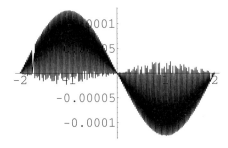

```
Plot[ufuncb[x] - Sin[Pi*x/2]*Exp[-Pi^2 0.1/4], {x, -2, 2},
PlotPoints -> 50];
```

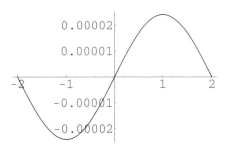

It is clear that linear is hopeless, especially if we wish to differentiate. Finally we tabulate the numerical result with cubic interpolation, the exact result and the error in the numerical scheme.

```
samples = TableForm[Join[{{"x", "Exact", "Crank-Nich", "Error"}},
Table[Map[PaddedForm[N[Chop[#1]], {5,6}]&,
N[{x, ufuncb[x], Sin[Pi*x/2]*Exp[-(Pi^2 0.1)/4],
ufuncb[x] - Sin[Pi*x/2]*Exp[-(Pi^2 0.1)/4]}, 5]], {x, -2, 2, 0.25}]]]
```

x	Exact	Crank-Nich	Error
-2.000000	0.000000	0.000000	0.000000
-1.750000	-0.299020	-0.299010	-9.246900×10^{-6}
-1.500000	-0.552510	-0.552490	-0.000017
-1.250000	-0.721890	-0.721870	-0.000022
-1.000000	-0.781370	-0.781340	-0.000024
-0.750000	-0.721890	-0.721870	-0.000022
-0.500000	-0.552510	-0.552490	-0.000017
-0.250000	-0.299020	-0.299010	-9.246900×10^{-6}
0.000000	0.000000	0.000000	0.000000
0.250000	0.299020	0.299010	9.246900×10^{-6}
0.500000	0.552510	0.552490	0.000017
0.750000	0.721890	0.721870	0.000022
1.000000	0.781370	0.781340	0.000024
1.250000	0.721890	0.721870	0.000022
1.500000	0.552510	0.552490	0.000017
1.750000	0.299020	0.299010	9.246900×10^{-6}
2.000000	0.000000	0.000000	0.000000

Note that we obtain comparable accuracy or better with a tenth the number of time-steps that where used for the explicit case. This example nicely illustrates the power of the implicit approach and the Crank-Nicholson scheme in particular. As we shall see, we can do rather better.

14.9 Douglas

A further small change to the matrices takes us to the Douglas algorithm. This time we just stick with cubic interpolation.

Douglas Algorithm

```
DougCMatrix[alpha_, nminus_, nplus_] :=
Sequence[Table[1-6*alpha, {nplus+nminus-2}], Table[10+12*alpha,
{nplus+nminus-1}],
Table[1-6*alpha, {nplus+nminus-2}]];

DougDMatrix[alpha_, vec_List] := Module[{temp},
temp = (10 - 12*alpha)*vec + (1+6*alpha)*(RotateRight[vec] + Rotate-
Left[vec]);
temp[[1]] = Simplify[First[temp] -  (1+6*alpha)*Last[vec]];
temp[[-1]] = Simplify[Last[temp] - (1 + 6*alpha)*First[vec]];
temp];

M=40;nminus = 80;nplus = 80;dx = 0.025;
dtau = 0.0025;alpha = dtau/dx^2;
CMat = DougCMatrix[alpha,nminus,nplus];
initial = Table[N[Sin[Pi*(k-1-nminus)/nminus] ], {k,1,
nminus+nplus+1}];
lower=Table[0, {m, 1, M+1}];
upper=Table[0, {m, 1, M+1}];
wold = initial; wvold = wold; wnew = wold;

For[m=2, m<=M+1, m++,
(wvold = wold;
wold = wnew;
rhs = DougDMatrix[alpha, Take[wold, {2, -2}]]+
Table[
Which[
k==1, (6*alpha+1)*lower[[m-1]] +(6 alpha - 1)*lower[[m]],
k== nplus + nminus-1,
(6*alpha+1)*upper[[m-1]] + (6 alpha - 1)*upper[[m]],
True, 0],
{k, 1, nplus + nminus-1}];
temp = CompTridiagSolve[CMat, rhs];
wnew = Join[{lower[[m]]}, temp, {upper[[m]]}])];
interpoldata =
Table[{(k-nminus-1)*dx ,wnew[[k]]}, {k, 1, nminus+nplus+1}];
ufunc = Interpolation[interpoldata, InterpolationOrder -> 3];
```

Now we plot the error - note the vertical scale!

```
Plot[ufunc[x] - Sin[Pi*x/2]*Exp[-Pi^2 0.1/4], {x, -2, 2},
PlotPoints -> 50];
```

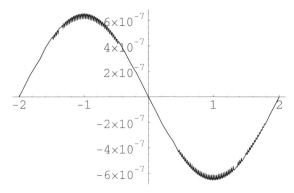

Finally we tabulate the *x*-value, the numerical result, the exact result and the error in the numerical scheme.

```
samples = TableForm[Join[{{"x", "Exact", "Douglas", "Error"}},
Table[Map[PaddedForm[N[Chop[#1]],{5,6}]&,
N[{x, ufunc[x], Sin[Pi*x/2]*Exp[-(Pi^2 0.1)/4],
ufunc[x]- Sin[Pi*x/2]*Exp[-(Pi^2 0.1)/4]},5]],
{x, -2, 2, 0.25}]]]
```

x	Exact	Douglas	Error
-2.000000	0.000000	0.000000	0.000000
-1.750000	-0.299010	-0.299010	2.332100×10^{-7}
-1.500000	-0.552490	-0.552490	4.309100×10^{-7}
-1.250000	-0.721870	-0.721870	5.630100×10^{-7}
-1.000000	-0.781340	-0.781340	6.094000×10^{-7}
-0.750000	-0.721870	-0.721870	5.630100×10^{-7}
-0.500000	-0.552490	-0.552490	4.309100×10^{-7}
-0.250000	-0.299010	-0.299010	2.332100×10^{-7}
0.000000	0.000000	0.000000	0.000000
0.250000	0.299010	0.299010	-2.332100×10^{-7}
0.500000	0.552490	0.552490	-4.309100×10^{-7}
0.750000	0.721870	0.721870	-5.630100×10^{-7}
1.000000	0.781340	0.781340	-6.094000×10^{-7}
1.250000	0.721870	0.721870	-5.630100×10^{-7}
1.500000	0.552490	0.552490	-4.309100×10^{-7}
1.750000	0.299010	0.299010	-2.332100×10^{-7}
2.000000	0.000000	0.000000	0.000000

Note that we obtain errors of about 1/40 *of those obtained with Crank-Nicholson, with identical time-step parameters*. Some oscillations have started to appear in our error plot, though it should be appreciated that our plot is now given on a much finer scale than was the case for CN - if we plot the Douglas errors on the same scale as the CN error plot the difference is clearer:

```
Plot[ufunc[x] - Sin[Pi*x/2]*Exp[-Pi^2 0.1/4], {x, -2, 2},
PlotPoints -> 50, PlotRange -> {-0.000025, 0.000025}];
```

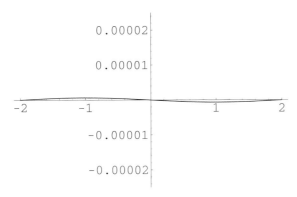

It is interesting to re-run the Douglas scheme with $\alpha = 32$ - do it yourself with the code - here are the results. The errors are now similar in scale to the CN results, but with an α that is eight times as large - we have only used five time-steps!

```
samples = TableForm[Join[{{"x", "Exact", "Douglas", "Error"}},
Table[Map[PaddedForm[N[Chop[#1]], {5,6}]&,
N[{x, ufunc[x], Sin[Pi*x/2]*Exp[-(Pi^2 0.1)/4],
ufunc[x]- Sin[Pi*x/2]*Exp[-(Pi^2 0.1)/4]},5]],
{x, -2, 2, 0.25}]]]
```

x	Exact	Douglas	Error
-2.000000	0.000000	0.000000	0.000000
-1.750000	-0.298990	-0.299010	0.000015
-1.500000	-0.552470	-0.552490	0.000028
-1.250000	-0.721830	-0.721870	0.000036
-1.000000	-0.781300	-0.781340	0.000039
-0.750000	-0.721830	-0.721870	0.000036
-0.500000	-0.552470	-0.552490	0.000028
-0.250000	-0.298990	-0.299010	0.000015
0.000000	0.000000	0.000000	0.000000
0.250000	0.298990	0.299010	-0.000015
0.500000	0.552470	0.552490	-0.000028
0.750000	0.721830	0.721870	-0.000036
1.000000	0.781300	0.781340	-0.000039
1.250000	0.721830	0.721870	-0.000036
1.500000	0.552470	0.552490	-0.000028
1.750000	0.298990	0.299010	-0.000015
2.000000	0.000000	0.000000	0.000000

14.10 Summary

We have shown how to use *Mathematica*'s compiler to build efficient solvers for explicit, fully implicit, Crank-Nicholson and Douglas two-time-level difference schemes. Using *Mathematica*'s built-in functions, we have compared the results with the exact solution of the diffusion equation using a simple smooth initial condition. These comparisons reveal the advantages of implicit schemes over explicit schemes, and the advantage in accuracy of the Douglas scheme over the Crank-Nicholson scheme. Within a general θ-method approach, the Douglas scheme is given by setting

$$\theta = \frac{1}{2} - \frac{1}{12\,\alpha} \tag{6}$$

and this method is the natural implicit generalization of the trinomial scheme, to which it reduces when $\alpha = 1/6$. On problems with smooth initial data, it is frequently the method of choice, and can give results nearly two orders of magnitude more accurate than a Crank-Nicholson scheme on such problems, for similar numerical parameters, or results of comparable accuracy with much larger time-steps.

We have not tried to optimize the Douglas scheme for accuracy by making the optimal choice of the parameter α. If the choice $\alpha = 1/\sqrt{20}$ is made the accuracy gains over Crank-Nicholson can be quite spectacular. Mitchell and Griffiths (1980) give an example where the errors in Douglas are about 1/400000 of those in Crank-Nicholson with otherwise identical numerical parameters.

We emphasize that this is *not* the end of the story. There are additional complications in the presence of initial and boundary conditions that are not smooth, which is almost always the case in option-pricing problems. These will be considered in the next chapter.

Chapter 14 Bibliography

Mathematica 3.0 *Standard Add-on Packages*, 1996, Wolfram Media-Cambridge University Press.

Mitchell, A.R. and Griffiths, D.F., 1980, *The Finite Difference Method in Partial Differential Equations*, John Wiley (Corrected reprinted edition, 1994).

Wolfram, S., 1996, *The Mathematica Book* (for version 3.0), Wolfram Media-Cambridge University Press.

Chapter 15.
Finite-Difference Schemes for the Black-Scholes Equation with Non-smooth Payoff Initial Conditions

Errors in Greeks in Crank-Nicholson Schemes, and Resolution by Three-Time-Level Douglas Finite-Difference Scheme

15.1 Introduction

In Chapter 14 I gave some examples which show how various commonly used finite-difference schemes behave with simple smooth "payoffs". With smooth payoffs the Douglas finite-difference scheme gives much better results than Crank-Nicholson. In this chapter we will look at how these methods behave when applied to real-world option-pricing problems. This will highlight the potentially nasty behaviour of both Crank-Nicholson and Douglas finite-difference schemes when applied to simple option-pricing problems. The discontinuous nature of the payoff or its derivative in the neighbourhood of the strike induces slowly decaying oscillations into the solution of the finite-difference equations, when we use a larger time-step (which is the main point of introducing FD schemes in the first place.) These introduce small errors into the valuation itself, and undermine attempts to compute δ, Γ and other "derivative" quantities in the neighbourhood of the strike. Note that, even with the oscillatory components small in the valuation error, the effect on the slope of the function is larger (generating larger errors in δ), and there are even larger errors in Γ.

In this chapter I also look at the three-time-level version of the Douglas scheme. This chapter shows how the problem of larger errors in the Greeks can be cured, in that the oscillations can be removed, at least for our test payoffs, and the method ("three-time-level Douglas") will be used as the basis of our programme to define benchmark numerical algorithms.

The problems with the Crank-Nicholson scheme, and its resolution, are exemplified by a test problem with $\alpha = 8$. Although the error in the valuation is small, we get *an error in Γ near the strike of 12% of its exact value, and an error in θ of 200%*. In the corresponding three-time-level Douglas solution, the error in Γ near the strike is reduced to 0.2%, and that in θ to 2.6%.

Users of implicit finite-difference models may wish to evaluate their choices of difference scheme in the light of these results, and at least to look at the effect of changing the difference scheme to a three-time-level one such as that presented here.

15.2 Finite-Difference Schemes with Three Time-Levels

Various schemes have been proposed for treating problems introduced by considering discontinuous or non-smooth boundary/initial conditions. These schemes aim to damp fast oscillations more effectively, by adjusting the spectrum of eigenvalues of the difference matrix. Richtmyer and Morton (1957) give a list of 14 difference schemes for the diffusion equation, and recommend schemes (their numbering) 9, 11 and 13 for non-smooth initial data. Schemes 9 and 13 are also recommended for these purposes by Smith (1985). These two schemes may be regarded as the three-time-level versions of the Crank-Nicholson scheme and the Douglas scheme, since they have truncation errors of a similar character to their two-time-level counterparts. We already know that the truncation error characteristics of the Douglas scheme make it preferable, so we shall use this. As in Chapter 14, we let m denote the time-step, and n denote the x-step. The three time-level Douglas scheme is then given by

$$
\begin{aligned}
&\left(\frac{1}{8} - \alpha\right)(u_{n-1}^{m+1} + u_{n+1}^{m+1}) + \left(\frac{5}{4} + 2\alpha\right)u_n^{m+1} \\
&= \frac{1}{6}(u_{n-1}^m + u_{n+1}^m + 10\,u_n^m) - \frac{1}{24}(u_{n-1}^{m-1} + u_{n+1}^{m-1} + 10\,u_n^{m-1})
\end{aligned}
\tag{1}
$$

This type of process requires a kick-off procedure, since initially we only know u^1. We use the ordinary Douglas two-time-level scheme:

$$
(1 - 6\alpha)(u_{n-1}^{m+1} + u_{n+1}^{m+1}) + (10 + 12\alpha)u_n^{m+1} = (1 + 6\alpha)(u_{n-1}^m + u_{n+1}^m) + (10 - 12\alpha)u_n^m
\tag{2}
$$

once with $\alpha \to \alpha/4$, then the three-time-level scheme once with $\alpha/4$ and then again with $\alpha/2$.

This gives us our vector u^2 to allow the three-time-level iteration to proceed normally thereafter. We shall develop SOR and PSOR variations of this scheme in Chapter 16 to treat American-style options.

Some Required Functions from Chapter 14

```
ExplicitSolver =
 Compile[
  {{initial, _Real, 1}, {lower, _Real, 1}, {upper, _Real, 1}, alpha},
  Module[{wold = initial, wnew = initial, wvold, m,
    k, tsize = Length[lower], xsize = Length[initial]},
   For[m = 2, m <= tsize, m++,
    (wvold = wold; wold = wnew;
     For[k = 2, k < xsize, k++,
     (wnew[[k]] =
        alpha (wold[[k - 1]] + wold[[k + 1]]) + (1 - 2 alpha) wold[[k]])];
     wnew[[1]] = lower[[m]];
     wnew[[xsize]] = upper[[m]])];
   {wvold, wold, wnew}]
  ];
```

```
CompTridiagSolve =
 Compile[{{a, _Real, 1}, {b, _Real, 1}, {c, _Real, 1}, {r, _Real, 1}},
  Module[{len = Length[r], solution = r, aux = 1 / (b[[1]]), aux1 = r,
    a1 = Prepend[a, 0.0], iter},  solution[[1]] = aux * r[[1]];
   Do[aux1[[iter]] = c[[iter - 1]] aux;
    aux = 1 / (b[[iter]] - a1[[iter]] * aux1[[iter]]);
    solution[[iter]] = (r[[iter]] - a1[[iter]] solution[[iter - 1]]) aux,
    {iter, 2, len}];
   Do[solution[[iter]] -= aux1[[iter + 1]] solution[[iter + 1]],
    {iter, len - 1, 1, -1}];
      solution]];

FullyImpCMatrix[alpha_, nminus_, nplus_] :=
Sequence[Table[-alpha, {nplus+nminus-2}],
Table[1+2*alpha, {nplus+nminus-1}],
Table[-alpha, {nplus+nminus-2}]]
```

15.3 Case Study 1: the Vanilla European Put Option

We now begin a detailed study of the first of two examples that we have picked for detailed investigation. The reader might wonder why we are considering such a trivial case for which there is a known analytic solution. Our goal here is to try out various difference schemes and find out what works well, by testing them on a case for which the solution is known and where the errors can be precisely described. In this way we can see what is happening without all the complications of other real-world effects.

Factors Common to All Our FD Schemes

The Black-Scholes differential equation

$$\frac{1}{2} S^2 \frac{\partial^2 V}{\partial S^2} \sigma^2 - r V + (r - q) S \frac{\partial V}{\partial S} + \frac{\partial V}{\partial t} = 0 \tag{3}$$

with constant coefficients (in particular r is a constant) can be transformed into the diffusion equation (1) by a standard change of variables, as given by Wilmott *et al* (Chapter 17, equations 17.1, 17.2). We can therefore implement our four standard schemes on the problem, and use solutions of (3) with known payoffs. In the following example we use the vanilla European Put.

Standardization of Variables

With a strike K and constant parameters r, q, σ, we make the changes of variables

$$\tau = \frac{\sigma^2 (T - t)}{2} \qquad k_1 = \frac{2 r}{\sigma^2} \qquad k_2 = \frac{2 (r - q)}{\sigma^2} \tag{4}$$

$$V(S, t) = K\, e^{-\frac{1}{2}(k_2 - 1)x - \left(\frac{1}{4}(k_2 - 1)^2 + k_1\right)\tau}\, u(x, \tau)$$

The *Mathematica* implementation of this requires the following functions.

$$\text{NonDimExpiry}[T\_, \ \sigma\_] := \frac{\sigma^2 \ T}{2};$$

$$\text{kone}[r\_, \ \sigma\_] := \frac{2 \ r}{\sigma^2}; \quad \text{ktwo}[r\_, \ q\_, \ sd\_] := \frac{2 \ (r - q)}{sd^2};$$

$$\text{ValuationMultiplier}[\text{strike}\_, \ r\_, \ q\_, \ x\_, \ tau\_, \ sd\_] := \text{strike}$$
$$\text{Exp}\left[-\frac{1}{2} \ (\text{ktwo}[r, \ q, \ sd] - 1) \ x - \left(\frac{1}{4} \ (\text{ktwo}[r, \ q, \ sd] - 1)^2 + \text{kone}[r, \ sd]\right) \ tau\right]$$

Initial (Expiry) Conditions

The initial conditions for a Call and a Put, in the transformed coordinates, are

$$\text{CallExercise}[x\_, \ r\_, \ q\_, \ sd\_] :=$$
$$\text{Max}\left[\text{Exp}\left[\frac{1}{2} \ (\text{ktwo}[r, \ q, \ sd] - 1) \ x\right] \ (\text{Exp}[x] - 1), \ 0\right];$$
$$\text{PutExercise}[x\_, \ r\_, \ q\_, \ sd\_] :=$$
$$\text{Max}\left[\text{Exp}\left[\frac{1}{2} \ (\text{ktwo}[r, \ q, \ sd] - 1) \ x\right] \ (1 - \text{Exp}[x]), \ 0\right];$$

Black-Scholes Model for Verification

We load our standard Package for vanilla European options:

```
Needs["Derivatives`BlackScholes`"]
```

We test the loading of the Package by executing the function we shall use in our test problem:

```
BlackScholesPut[8, 10, 0.2, 0.05, 0, 3]
```

    1.47045

Put Boundary Conditions

For our boundary conditions the upper boundary condition is to set the function to zero, while the lower takes the asymptotic Put value as the stock price approaches zero.

```
g[x_, tau_, r_, q_, sd_] := 0;
f[x_, tau_, r_, q_, sd_] := Exp[1/2 (ktwo[r, q, sd] - 1) x +
    1/4 ((ktwo[r, q, sd] - 1)² + 4 kone[r, sd]) tau] - Exp[
  1/2 (ktwo[r, q, sd] + 1) x + 1/4 ((ktwo[r, q, sd] - 1)² + 4 kone[r, sd]) tau];
```

15.4 Explicit Scheme for Put

```
dx = 0.025; dtau = 0.00025; alpha = dtau/dx²;
```

```
M = 400; nminus = 160; nplus = 160;
```

Setting the Initial (Expiry) Condition and Boundary Conditions

```
initial = Table[
  PutExercise[(k - 1 - nminus) dx, 0.05, 0, 0.2], {k, nminus + nplus + 1}];
lower = Table[f[-nminus dx, (m - 1) dtau, 0.05, 0, 0.2], {m, 1, M + 1}];
upper = Table[g[+nplus dx,  (m - 1) dtau, 0.05, 0, 0.2], {m, 1, M + 1}];
```

Note that the payoff looks slightly odd in these coordinates - the main point to notice is the discontinuity in slope at the strike.

```
ListPlot[initial];
```

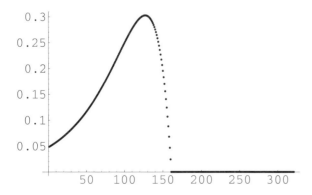

Solving the PDE (Explicit Method)

We use the same functions as we used for our smooth test problem in Chapter 14 - load the function **ExplicitSolver** now if you are using the electronic form:

```
soln = ExplicitSolver[initial, lower, upper, alpha ];
```

Interpolating to Supply a Continuous Function

```
interpoldata =
  Table[{(k - nminus - 1) dx, soln[[3, k]]}, {k, 1, nminus + nplus + 1}];
```

```
ufunc = Interpolation[interpoldata, InterpolationOrder → 3];
```

$$\text{Valuation}[strike\_, r\_, q\_, S\_, T\_, sd\_] :=$$
$$\text{ValuationMultiplier}\left[strike, r, q, \text{Log}\left[\frac{S}{strike}\right], \frac{sd^2\ T}{2}, sd\right]$$
$$\text{ufunc}\left[\text{Log}\left[\frac{S}{strike}\right]\right]$$

Error Plot

```
Plot[Valuation[10, 0.05, 0, S, 5, 0.2] -
  BlackScholesPut[S, 10, 0.2, 0.05, 0, 5],
 {S, 1, 20}, PlotPoints → 50, PlotRange → All];
```

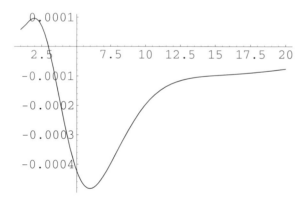

```
samples = TableForm[Join[{{"S", "Exact", "Explicit FD", "Error"}},
 Table[(PaddedForm[N[#1], {5, 5}] &) /@
   {S, Valuation[10, 0.05, 0, S, 5, 0.2],
      BlackScholesPut[S, 10, 0.2, 0.05, 0, 5],
      Valuation[10, 0.05, 0, S, 5, 0.2] -
       BlackScholesPut[S, 10, 0.2, 0.05, 0, 5]},
     {S, 2, 16, 1}
   ]]]
```

S	Exact	Explicit FD	Error
2.00000	5.78870	5.78860	0.00010
3.00000	4.80050	4.80050	-2.16410×10^{-6}
4.00000	3.86130	3.86150	-0.00023
5.00000	3.02040	3.02090	-0.00042
6.00000	2.31040	2.31080	-0.00048
7.00000	1.73820	1.73870	-0.00044
8.00000	1.29290	1.29320	-0.00035
9.00000	0.95451	0.95478	-0.00027
10.00000	0.70167	0.70187	-0.00020
11.00000	0.51477	0.51492	-0.00015
12.00000	0.37754	0.37766	-0.00012
13.00000	0.27715	0.27726	-0.00011
14.00000	0.20384	0.20394	-0.00010
15.00000	0.15030	0.15040	-0.00010
16.00000	0.11116	0.11125	-0.00010

15.5 Fully Implicit Scheme for Put

These algorithms are by now self-explanatory - first the initialization:

$$dx = 0.025; \quad dtau = 0.00025; \quad alpha = \frac{dtau}{dx^2};$$

```
M = 400; nminus = 160; nplus = 160;

initial = Table[
  PutExercise[(k - 1 - nminus) dx, 0.05, 0, 0.2], {k, nminus + nplus + 1}];
lower = Table[f[-nminus dx, (m - 1) dtau, 0.05, 0, 0.2], {m, 1, M + 1}];
upper = Table[g[+nplus dx,  (m - 1) dtau, 0.05, 0, 0.2], {m, 1, M + 1}];
wold = initial;
wvold = wold; wnew = wold;

CMat = FullyImpCMatrix[alpha, nminus, nplus];
```

Evolving the solution (this may take some time):

```
For[m=2, m<=M+1, m++,
(wvold = wold;
wold = wnew;
rhs = Take[wold, {2, -2}]+
Table[
Which[
k==1, alpha*lower[[m]],
k== nplus + nminus-1,  alpha*upper[[m]],
True, 0],
{k, 1, nplus + nminus-1}];
temp = CompTridiagSolve[CMat, rhs];
wnew = Join[{lower[[m]]}, temp, {upper[[m]]}])
]
```

Interpolation and construction of valuation function:

```
interpoldata = Table[{(k - nminus - 1)*dx, wnew[[k]]},
    {k, 1, nminus + nplus + 1}];
```

```
ufuncb = Interpolation[interpoldata, InterpolationOrder -> 3];
```

```
Valuation[strike_, r_, q_, S_, T_, sd_] :=
  ValuationMultiplier[strike, r, q, Log[S/strike], (sd^2*T)/2, sd]*
  ufuncb[Log[S/strike]]
```

Error Plot

```
Plot[Valuation[10, 0.05, 0, S, 5, 0.2] -
  BlackScholesPut[S, 10, 0.2, 0.05, 0, 5],
 {S, 1, 20}, PlotPoints -> 50, PlotRange -> All];
```

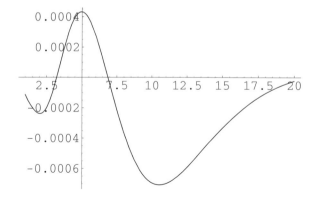

```
samples = TableForm[Join[{{"S", "Exact", "Implicit FD", "Error"}},
 Table[(PaddedForm[N[#1], {5, 5}] &) /@
   {S, Valuation[10, 0.05, 0, S, 5, 0.2],
      BlackScholesPut[S, 10, 0.2, 0.05, 0, 5],
      Valuation[10, 0.05, 0, S, 5, 0.2] -
       BlackScholesPut[S, 10, 0.2, 0.05, 0, 5]},   {S, 2, 16, 1}]]]
```

S	Exact	Implicit FD	Error
2.00000	5.78830	5.78860	-0.00024
3.00000	4.80040	4.80050	-0.00007
4.00000	3.86180	3.86150	0.00028
5.00000	3.02130	3.02090	0.00043
6.00000	2.31110	2.31080	0.00028
7.00000	1.73860	1.73870	-0.00005
8.00000	1.29290	1.29320	-0.00037
9.00000	0.95419	0.95478	-0.00059
10.00000	0.70118	0.70187	-0.00069
11.00000	0.51422	0.51492	-0.00070
12.00000	0.37702	0.37766	-0.00064
13.00000	0.27672	0.27726	-0.00054
14.00000	0.20351	0.20394	-0.00044
15.00000	0.15007	0.15040	-0.00033
16.00000	0.11101	0.11125	-0.00024

Note that we obtain no improvement in accuracy over the explicit scheme. The only advantage of the fully implicit scheme over the explicit scheme is the fact that we can, if we wish, increase α, that is, the time-step for a given price-step, without the system going unstable.

15.6 Crank-Nicholson

We now increase the time-step by a factor of 20, so that $\alpha = 8$. Otherwise all this proceeds as before:

Initialization:

```
M=20; nminus = 160; nplus = 160;
dx = 0.025; dtau = 0.005; alpha = dtau/dx^2
```

    8.

```
initial = Table[
  PutExercise[(k - 1 - nminus) dx, 0.05, 0, 0.2], {k, nminus + nplus + 1}];
lower = Table[f[-nminus dx, (m - 1) dtau, 0.05, 0, 0.2], {m, 1, M + 1}];
upper = Table[g[+nplus dx,  (m - 1) dtau, 0.05, 0, 0.2], {m, 1, M + 1}];
wold = initial;
wvold = wold;
wnew = wold;


CNCMatrix[alpha_, nminus_, nplus_] :=
Sequence[Table[-alpha / 2, {nplus + nminus - 2}],
Table[1 + alpha, {nplus + nminus - 1}],
Table[-alpha / 2, {nplus + nminus - 2}]];


CNDMatrix[alpha_, vec_List] := Module[{temp},
temp =
    (1 - alpha) * vec + (alpha / 2) * (RotateRight[vec] + RotateLeft[vec]);
temp[[1]] = Simplify[First[temp] -
alpha * Last[vec] / 2];
temp[[-1]] = Simplify[Last[temp] - alpha * First[vec] / 2];
temp];


CMat = CNCMatrix[alpha, nminus, nplus];
```

Evolution:

```
For[m=2, m<=M+1, m++,
(wvold = wold;
wold = wnew;
rhs = CNDMatrix[alpha, Take[wold, {2, -2}]]+
Table[
Which[
k==1, alpha*(lower[[m-1]] + lower[[m]])/2,
k== nplus + nminus-1,  alpha*(upper[[m-1]] + upper[[m]])/2,
True, 0],
{k, 1, nplus + nminus-1}];
temp = CompTridiagSolve[CMat, rhs];
wnew = Join[{lower[[m]]}, temp, {upper[[m]]}])
]
```

Interpolation

```
interpoldatab =
Table[{(k - nminus - 1)*dx, wnew[[k]]}, {k, 1, nminus+nplus+1}];

ufuncb = Interpolation[interpoldatab, InterpolationOrder -> 3];
```

```
Valuation[strike_, r_, q_, S_, T_, sd_] :=
ValuationMultiplier[strike, r, q, Log[S/strike], (sd^2*T)/2,
sd]*ufuncb[Log[S/strike]]
```

Error Plot

```
Plot[Valuation[10, 0.05, 0, S, 5, 0.2] -
  BlackScholesPut[S, 10, 0.2, 0.05, 0, 5],
 {S, 1, 20}, PlotPoints → 50, PlotRange → All];
```

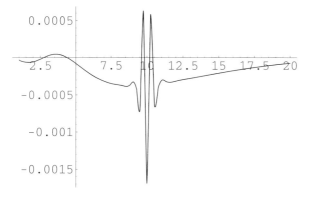

```
samples = TableForm[Join[{{"S", "Exact", "Crank-Nich", "Error"}},
 Table[(PaddedForm[N[#1], {5, 5}] &) /@
  {S, Valuation[10, 0.05, 0, S, 5, 0.2],
    BlackScholesPut[S, 10, 0.2, 0.05, 0, 5],
    Valuation[10, 0.05, 0, S, 5, 0.2] -
     BlackScholesPut[S, 10, 0.2, 0.05, 0, 5]}, {S, 2, 16, 1}]]]
```

S	Exact	Crank-Nich	Error
2.00000	5.78850	5.78860	-0.00006
3.00000	4.80050	4.80050	0.00002
4.00000	3.86160	3.86150	0.00003
5.00000	3.02080	3.02090	-0.00008
6.00000	2.31060	2.31080	-0.00022
7.00000	1.73830	1.73870	-0.00032
8.00000	1.29290	1.29320	-0.00036
9.00000	0.95445	0.95478	-0.00033
10.00000	0.70016	0.70187	-0.00171
11.00000	0.51461	0.51492	-0.00031
12.00000	0.37735	0.37766	-0.00032
13.00000	0.27698	0.27726	-0.00028
14.00000	0.20370	0.20394	-0.00024
15.00000	0.15019	0.15040	-0.00021
16.00000	0.11108	0.11125	-0.00018

Note that we obtain comparable accuracy to the explicit method with 1/20 the number of time-steps, but near the strike things are becoming "interesting". We obtain a sharply oscillatory error in the neighbourhood of the strike. Note that the size of the error is quite small, but it is steeply sloped. Bearing in mind the warnings given in Chapter 14 this is now the time to build the Greeks.

Construction and Verification of Interpolated Valuation and Greeks

We begin by writing down a function that does numerical differentiation of a list. This function uses a simple central difference algorithm for points in the interior of the list. The end points are treated using a special difference algorithm.

```
listd[data_, step_] :=
Module[{dleft,dright,len},
len = Length[data];
dleft = (4*data[[2]]-3*data[[1]]-data[[3]])/(2*step);
dright = (3*data[[len]]-4*data[[len-1]]+data[[len-2]])/(2*step);
Join[{dleft}, Take[RotateLeft[data]-RotateRight[data], {2,
-2}]/(2*step), {dright}]
]
```

Next we fix the values of the *k*-parameters:

```
kt = ktwo[0.05, 0, 0.2];
ko = kone[0.05, 0.2];
```

$$\text{deltadata} = \text{listd[wnew, dx]} - \frac{1}{2} \, (\text{kt} - 1) \, \text{wnew};$$

$$\text{gammadata} = \text{listd[deltadata, dx]} - \frac{1}{2} \, (\text{kt} + 1) \, \text{deltadata};$$

$$\text{thetadata} = \frac{3 \, \text{wnew} - 4 \, \text{wold} + \text{wvold}}{2 \, \text{dtau}} - \left(\frac{1}{4} \, (\text{kt} - 1)^2 + \text{ko} \right) \text{wnew};$$

```
points = Table[(k - nminus - 1)*dx, {k, 1, nminus + nplus + 1}];

deltainterpoldata = Transpose[{points, deltadata}];
gammainterpoldata = Transpose[{points, gammadata}];
thetainterpoldata = Transpose[{points, thetadata}];

dfunc = Interpolation[deltainterpoldata, InterpolationOrder → 3];
gfunc = Interpolation[gammainterpoldata, InterpolationOrder → 3];
tfunc = Interpolation[thetainterpoldata, InterpolationOrder → 3];
```

```
CNDelta[strike_, r_, q_, S_, T_, sd_] :=
```

$$\frac{1}{S}\left(\text{ValuationMultiplier}\left[\text{strike}, r, q, \text{Log}\left[\frac{S}{\text{strike}}\right], \frac{sd^2\,T}{2}, sd\right]\right.$$

$$\left. dfunc\left[\text{Log}\left[\frac{S}{\text{strike}}\right]\right]\right)$$

```
CNGamma[strike_, r_, q_, S_, T_, sd_] :=
```

$$\frac{1}{S^2}\left(\text{ValuationMultiplier}\left[\text{strike}, r, q, \text{Log}\left[\frac{S}{\text{strike}}\right], \frac{sd^2\,T}{2}, sd\right]\right.$$

$$\left. gfunc\left[\text{Log}\left[\frac{S}{\text{strike}}\right]\right]\right)$$

```
CNTheta[strike_, r_, q_, S_, T_, sd_] :=
```

$$-\frac{1}{2}\,sd^2\,\text{ValuationMultiplier}\left[\text{strike}, r, q, \text{Log}\left[\frac{S}{\text{strike}}\right], \frac{sd^2\,T}{2}, sd\right]$$

$$tfunc\left[\text{Log}\left[\frac{S}{\text{strike}}\right]\right]$$

Delta Analysis

To investigate the errors in Delta we make a comparison with the exact solution, which we have already loaded as a function:

```
? BlackScholesPutDelta
```

    BlackScholesPutDelta[price, strike, vol, riskfree, divyield,
        expiry] returns the delta of a vanilla European Put.

We make a table of values of the percentage errors in delta, given in the last column of the following table:

```
deltasamples = TableForm[
Table[Map[PaddedForm[N[#], {5, 4}]&,
{S, CNDelta[10, 0.05, 0, S, 5, 0.2],
BlackScholesPutDelta[S,10,0.2,0.05,0, 5],
100*(CNDelta[10,0.05,0,S,5,0.2]/BlackScholesPutDelta[S,10,0.2,0.05,0,
5]-1)}],
{S, 2, 16, 1}]]
```

2.0000	-0.9979	-0.9976	0.0324
3.0000	-0.9721	-0.9719	0.0234
4.0000	-0.8975	-0.8973	0.0205
5.0000	-0.7786	-0.7785	0.0130
6.0000	-0.6404	-0.6404	-0.0004
7.0000	-0.5059	-0.5060	-0.0147
8.0000	-0.3882	-0.3883	-0.0256
9.0000	-0.2921	-0.2922	-0.0166
10.0000	-0.2168	-0.2169	-0.0631
11.0000	-0.1594	-0.1597	-0.2056
12.0000	-0.1169	-0.1170	-0.0505
13.0000	-0.0854	-0.0855	-0.0394
14.0000	-0.0624	-0.0624	-0.0436
15.0000	-0.0456	-0.0456	-0.0467
16.0000	-0.0333	-0.0334	-0.0518

The error in Delta is at most 0.2 % of the exact value, near the strike. This is still not too bad, and we can plot the error in Delta.

Plot of Computed and Exact Delta

If we overlay the computed and exact Delta, the wobbles are starting to show:

```
Plot[{CNDelta[10, 0.05, 0, S, 5, 0.2],
BlackScholesPutDelta[S, 10, 0.2, 0.05, 0, 5]}, {S, 8, 12},
  PlotPoints -> 50, PlotRange -> All];
```

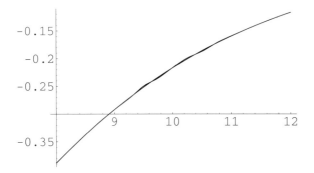

Plotting the difference makes the problem clear:

```
Plot[CNDelta[10, 0.05, 0, S, 5, 0.2] -
  BlackScholesPutDelta[S, 10, 0.2, 0.05, 0, 5],
 {S, 8, 12}, PlotPoints -> 50,
 PlotRange -> All];
```

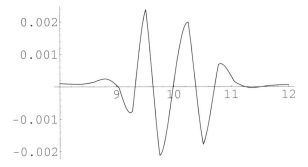

Gamma Analysis

Here we tabulate the exact and numerical solution, with the percentage error in Gamma in the last column:

```
gammasamples = TableForm[Table[Map[PaddedForm[N[#], {5, 4}]&,
{S, CNGamma[10, 0.05, 0, S, 5, 0.2],
BlackScholesPutGamma[S,10,0.2,0.05,0,  5],
100*(CNGamma[10,0.05,0,S,5,0.2]/BlackScholesPutGamma[S,10,0.2,0.05,0,5]
-1)}],{S, 2, 16, 1}]]
```

2.0000	0.0084	0.0085	-1.2291
3.0000	0.0481	0.0480	0.2056
4.0000	0.1002	0.1000	0.2123
5.0000	0.1332	0.1329	0.1812
6.0000	0.1395	0.1394	0.1186
7.0000	0.1275	0.1274	0.0496
8.0000	0.1070	0.1071	-0.0589
9.0000	0.0830	0.0853	-2.7845
10.0000	0.0738	0.0657	12.4110
11.0000	0.0490	0.0494	-0.8800
12.0000	0.0366	0.0366	-0.0030
13.0000	0.0269	0.0269	-0.0717
14.0000	0.0196	0.0196	-0.0693
15.0000	0.0143	0.0143	-0.0595
16.0000	0.0104	0.0104	-0.0500

The error in Γ is 12.4% of the exact value at the strike.

Gamma Plot

The wobbles in Γ are now manifest in a plot of the computed and exact version - the error is of a similar scale to the value of Γ:

```
Plot[{CNGamma[10, 0.05, 0, S, 5, 0.2],
  BlackScholesPutGamma[S, 10, 0.2, 0.05, 0, 5]},
 {S, 8, 12}, PlotPoints -> 50, PlotRange -> All];
```

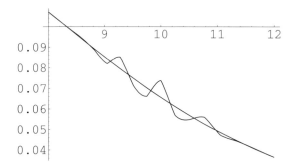

Theta Analysis

Now we make a table of percentage errors with the Crank-Nicholson scheme:

```
thetasamples = TableForm[
Table[Map[PaddedForm[N[#], {5, 4}]&,
{S, CNTheta[10, 0.05, 0, S, 5, 0.2],
BlackScholesPutTheta[S,10,0.2,0.05,0, 5],
100*(CNTheta[10,0.05,0,S,5,0.2]/BlackScholesPutTheta[S,10,0.2,0.05,0,5]
-1)}], {S, 2, 16, 1}]]
```

2.0000	0.3885	0.3885	0.0008
3.0000	0.3772	0.3772	0.0125
4.0000	0.3406	0.3405	0.0134
5.0000	0.2792	0.2792	-0.0157
6.0000	0.2072	0.2073	-0.0541
7.0000	0.1390	0.1391	-0.0780
8.0000	0.0828	0.0829	-0.1037
9.0000	0.0401	0.0410	-2.2123
10.0000	0.0362	0.0122	196.8900
11.0000	-0.0065	-0.0060	8.7857
12.0000	-0.0163	-0.0164	-0.5050
13.0000	-0.0214	-0.0214	-0.0244
14.0000	-0.0230	-0.0230	0.0136
15.0000	-0.0226	-0.0225	0.0668
16.0000	-0.0209	-0.0209	0.1011

Near the strike the error in Theta peaks at 197% of the exact value.

Theta Plots

```
Plot[{CNTheta[10, 0.05, 0, S, 5, 0.2],
BlackScholesPutTheta[S, 10, 0.2, 0.05, 0, 5]}, {S, 8, 12}, PlotPoints -> 5
  PlotRange -> All];
```

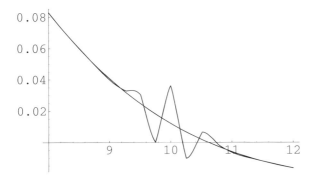

Remark on Two-Time-Level Douglas

This time, there is no benefit in going to the simple Douglas scheme as described in Chapter 14. The oscillations in the neighbourhood of the strike are just as bad. For example, here is the error in valuation with $\alpha = 8$, for comparison. There are corresponding problems in the Greeks.

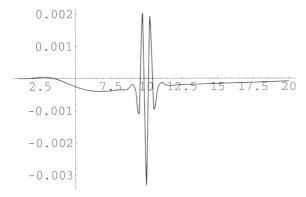

15.7 Douglas Three-Time-Level Solution

We have to work a little harder to develop a scheme that eliminates the oscillation problem. What we do is to implement the three-time-level extension of the Douglas scheme discussed in Section 15.2. We shall not go into a detailed theoretical discussion of this algorithm - we shall content ourselves with an explicit demonstration that it works.

Necessary Functions

```
DougCMatrix[alpha_, nminus_, nplus_] :=
Sequence[Table[1-6*alpha, {nplus+nminus-2}], Table[10+12*alpha,
{nplus+nminus-1}],
Table[1-6*alpha, {nplus+nminus-2}]]


DougCCMatrix[alpha_, nminus_, nplus_] :=
Sequence[Table[1/8-alpha, {nplus+nminus-2}],
Table[5/4+2*alpha, {nplus+nminus-1}],
Table[1/8-alpha, {nplus+nminus-2}]]


DougDMatrix[alpha_, vec_List] := Module[{temp},
temp = (10 - 12*alpha)*vec + (1+6*alpha)*(RotateRight[vec] + Rotate-
Left[vec]);
temp[[1]] = Simplify[First[temp] -  (1+6*alpha)*Last[vec]];
temp[[-1]] = Simplify[Last[temp] - (1 + 6*alpha)*First[vec]];
temp]


DougDDMatrix[vec_List] := Module[{temp},
temp = (10*vec + RotateRight[vec] + RotateLeft[vec]);
temp[[1]] = Simplify[First[temp] - Last[vec]];
temp[[-1]] = Simplify[Last[temp] - First[vec]];
temp/6]
```

Douglas Three Time-Level Solution Evolution

The initialization consists of defining the grid parameters, setting boundary and initial conditions and defining the various matrices.

```
dx = 0.025;
dtau = 0.005;
alpha = dtau/dx^2;
M=20;
nminus = 160;
nplus = 160;
alpha
```

    8.


```
initial = Table[
  PutExercise[(k - 1 - nminus) dx, 0.05, 0, 0.2], {k, nminus + nplus + 1}];
lower = Table[f[-nminus dx, (m - 1) dtau, 0.05, 0, 0.2], {m, 1, M + 1}];
upper = Table[g[+nplus dx,  (m - 1) dtau, 0.05, 0, 0.2], {m, 1, M + 1}];

w = Table[0, {m, 1, 3},    {k, 1, nminus+nplus+1}];
```

```
vold = Take[initial, {2, -2}];

w[[1,1]] = f[-nminus*dx, dtau/4, 0.05, 0, 0.2];
w[[1, nminus+nplus+1]] = g[nplus*dx, dtau/4, 0.05, 0, 0.2];
w[[2,1]] = f[-nminus*dx, dtau/2, 0.05, 0, 0.2];
w[[2, nminus+nplus+1]] = g[nplus*dx, dtau/2, 0.05, 0, 0.2];
w[[3,1]] = f[-nminus*dx, dtau, 0.05, 0, 0.2];
w[[3, nminus+nplus+1]] = g[nplus*dx, dtau, 0.05, 0, 0.2];

CMat = DougCMatrix[alpha,nminus,nplus];
CCMat = DougCCMatrix[alpha,nminus,nplus];

CMatQ = DougCMatrix[alpha/4,nminus,nplus];
CMatH = DougCMatrix[alpha/2,nminus,nplus];

CCMatQ = DougCCMatrix[alpha/4,nminus,nplus];
CCMatH = DougCCMatrix[alpha/2,nminus,nplus];
```

Kick-Off Phase

This begins with the simple Douglas scheme with two time-levels and 1/4 the basic time-step.

```
rhs = DougDMatrix[alpha/4, vold]+
Table[
Which[
k==1, (6*alpha/4+1)*lower[[1]] + (6*alpha/4-1)*w[[1, 1]],
k== nplus + nminus-1,   (6*alpha/4+1)*upper[[1]] +
(6*alpha/4-1)*w[[1, nplus+nminus+1]],
True, 0],
{k, 1, nplus + nminus-1}];
vnew = CompTridiagSolve[CMatQ, rhs];
w[[1]] = Join[{w[[1,1]]}, vnew, {w[[1,nplus+nminus+1]]}];
vvold = vold;
vold = vnew;
```

Now we have two iterations of the three-time-level Douglas system, doubling the time-step at each stage:

```
rhs = DougDDMatrix[vold] - DougDDMatrix[vvold]/4 +
Table[
Which[
k==1, (alpha/4-1/8)*w[[2, 1]] + w[[1,1]]/6 - lower[[1]]/24,
k==nplus + nminus-1,   (alpha/4-1/8)*w[[2, nplus+nminus+1]] +
w[[1,nplus+nminus+1]]/6 - upper[[1]]/24,
True, 0],
{k, 1, nplus + nminus-1}];
vnew = CompTridiagSolve[CCMatQ, rhs];
w[[2]] = Join[{w[[2,1]]}, vnew, {w[[2,nplus+nminus+1]]}];
vold = vnew;
```

```
rhs = DougDDMatrix[vold] - DougDDMatrix[vvold]/4 +
Table[
Which[
k==1, (alpha/2-1/8)*w[[3, 1]] + w[[2,1]]/6 - lower[[1]]/24,
k==nplus + nminus-1,  (alpha/2-1/8)*w[[3, nplus+nminus+1]] +
w[[2,nplus+nminus+1]]/6 - upper[[1]]/24,
True, 0],
{k, 1, nplus + nminus-1}];
vnew = CompTridiagSolve[CCMatH, rhs];
w[[3]] = Join[{w[[3,1]]}, vnew, {w[[3,nplus+nminus+1]]}];
wold = initial;
wvold = initial;
wnew = w[[3]];
```

Main Evolution Phase and Interpolation of Solution

```
For[m=3, m<=M+1, m++,
(wvold = wold;
wold = wnew;
rhs = DougDDMatrix[Take[wold, {2, -2}]] -
DougDDMatrix[Take[wvold, {2, -2}]]/4 +
Table[
Which[
k==1,
(alpha-1/8)*lower[[m]] + lower[[m-1]]/6 - lower[[m-2]]/24,
k==nplus + nminus-1,
(alpha-1/8)*upper[[m]] + upper[[m-1]]/6 - upper[[m-2]]/24,
True, 0],
{k, 1, nplus + nminus-1}];
temp = CompTridiagSolve[CCMat, rhs];
wnew = Join[{lower[[m]]}, temp, {upper[[m]]}])
]

points = Table[(k - nminus - 1) * dx,
{k, 1, nplus + nminus + 1}];

finalstep = wnew;

prevstep = wold;

pprevstep = wvold;

interpoldata = Transpose[{points, finalstep}];

ufunc = Interpolation[interpoldata, InterpolationOrder -> 3];
```

```
Valuation[strike_, r_, q_, S_, T_, sd_] :=
ValuationMultiplier[strike, r, q, Log[S/strike], sd^2*T/2, sd]*
ufunc[Log[S/strike]]
```

Valuation Analysis

```
Plot[Valuation[10, 0.05, 0, S, 5, 0.2]-
BlackScholesPut[S,10,0.2,0.05,0,5], {S, 1, 20},
PlotPoints -> 50, PlotRange -> All];
```

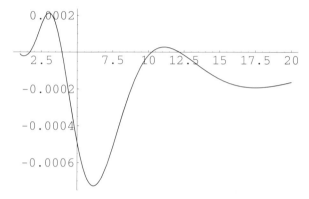

```
samples = TableForm[
Table[Map[PaddedForm[N[#], {5, 4}]&, {S, Valuation[10, 0.05, 0, S, 5,
0.2],
BlackScholesPut[S,10,0.2,0.05,0,5],
Valuation[10, 0.05, 0, S, 5, 0.2]-
BlackScholesPut[S,10,0.2,0.05,0,5]}],
{S, 2, 16, 1}]]
```

2.0000	5.7886	5.7886	0.0000
3.0000	4.8007	4.8005	0.0002
4.0000	3.8615	3.8615	-0.0000
5.0000	3.0204	3.0209	-0.0005
6.0000	2.3101	2.3108	-0.0007
7.0000	1.7380	1.7387	-0.0006
8.0000	1.2928	1.2932	-0.0004
9.0000	0.9546	0.9548	-0.0002
10.0000	0.7019	0.7019	-0.0000
11.0000	0.5149	0.5149	0.0000
12.0000	0.3777	0.3777	7.4967×10^{-6}
13.0000	0.2772	0.2773	-0.0000
14.0000	0.2038	0.2039	-0.0001
15.0000	0.1503	0.1504	-0.0001
16.0000	0.1111	0.1113	-0.0002

Construction and Verification of Interpolated Valuation and Greeks

In this section we define some functions to compute Delta, Gamma and Theta directly from the finite-differencegrid.

```
listd[data_, step_] :=
Module[{dleft,dright,len},
len = Length[data];
dleft = (4*data[[2]]-3*data[[1]]-data[[3]])/(2*step);
dright = (3*data[[len]]-4*data[[len-1]]+data[[len-2]])/(2*step);
Join[{dleft}, Take[RotateLeft[data]-RotateRight[data], {2,
-2}]/(2*step), {dright}]
]


kt = ktwo[0.05, 0, 0.2];
ko = kone[0.05, 0.2];
```

$$\text{deltadata} = \text{listd[finalstep, dx]} - \frac{1}{2}\ (kt - 1)\ \text{finalstep};$$

$$\text{gammadata} = \text{listd[deltadata, dx]} - \frac{1}{2}\ (kt + 1)\ \text{deltadata};$$

$$\text{thetadata} =$$
$$\frac{3\ \text{finalstep} - 4\ \text{prevstep} + \text{pprevstep}}{2\ \text{dtau}} - \left(\frac{1}{4}\ (kt - 1)^2 + ko\right)\ \text{finalstep};$$

```
deltainterpoldata = Transpose[{points, deltadata}];
gammainterpoldata = Transpose[{points, gammadata}];
thetainterpoldata = Transpose[{points, thetadata}];

dfunc = Interpolation[deltainterpoldata, InterpolationOrder -> 3];
gfunc = Interpolation[gammainterpoldata, InterpolationOrder -> 3];
tfunc = Interpolation[thetainterpoldata, InterpolationOrder -> 3];

DougDelta[strike_, r_, q_, S_, T_, sd_] :=
```

$$\frac{1}{S}\left(\text{ValuationMultiplier}\left[\text{strike, r, q, Log}\left[\frac{S}{\text{strike}}\right], \frac{sd^2\ T}{2}, sd\right]\right.$$
$$\left. \text{dfunc}\left[\text{Log}\left[\frac{S}{\text{strike}}\right]\right]\right)$$

```
DougGamma[strike_, r_, q_, S_, T_, sd_] :=
```

$$\frac{1}{S^2} \left(\text{ValuationMultiplier}\left[\text{strike}, r, q, \text{Log}\left[\frac{S}{\text{strike}}\right], \frac{sd^2\, T}{2}, sd\right] \right.$$

$$\left. \text{gfunc}\left[\text{Log}\left[\frac{S}{\text{strike}}\right]\right]\right)$$

```
DougTheta[strike_, r_, q_, S_, T_, sd_] :=
```

$$-\frac{1}{2}\, sd^2\, \text{ValuationMultiplier}\left[\text{strike}, r, q, \text{Log}\left[\frac{S}{\text{strike}}\right], \frac{sd^2\, T}{2}, sd\right]$$

$$\text{tfunc}\left[\text{Log}\left[\frac{S}{\text{strike}}\right]\right]$$

Delta Analysis

The percentage error in Delta is given in the last column of the following table.

```
deltasamples = TableForm[
Table[Map[PaddedForm[N[#], {5, 4}]&,
{S, DougDelta[10, 0.05, 0, S, 5, 0.2],
BlackScholesPutDelta[S,10,0.2,0.05,0,5],
100*(DougDelta[10,0.05,0,S,5,0.2]/BlackScholesPutDelta[S,10,0.2,0.05,0,
5]-1)}],
{S, 2, 16, 1}]]
```

2.0000	-0.9978	-0.9976	0.0186
3.0000	-0.9722	-0.9719	0.0288
4.0000	-0.8979	-0.8973	0.0634
5.0000	-0.7789	-0.7785	0.0464
6.0000	-0.6404	-0.6404	-0.0119
7.0000	-0.5056	-0.5060	-0.0693
8.0000	-0.3879	-0.3883	-0.0997
9.0000	-0.2919	-0.2922	-0.0978
10.0000	-0.2168	-0.2169	-0.0695
11.0000	-0.1596	-0.1597	-0.0258
12.0000	-0.1170	-0.1170	0.0225
13.0000	-0.0855	-0.0855	0.0660
14.0000	-0.0625	-0.0624	0.0975
15.0000	-0.0456	-0.0456	0.1126
16.0000	-0.0334	-0.0334	0.1088

The error in Delta is at most 0.1% of the exact value.

Delta Plots

```
Plot[{DougDelta[10, 0.05, 0, S, 5, 0.2],
BlackScholesPutDelta[S, 10, 0.2, 0.05, 0, 5]}, {S, 8, 12}, PlotPoints -> 5
 PlotRange -> All];
```

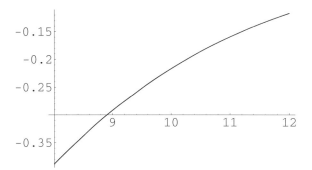

Gamma Analysis

The percentage error in Gamma is given by the last column of the following table:

```
gammasamples = TableForm[Table[Map[PaddedForm[N[#], {5, 4}]&,
{S, DougGamma[10, 0.05, 0, S, 5, 0.2],
BlackScholesPutGamma[S,10,0.2,0.05,0,5],
100*(DougGamma[10,0.05,0,S,5,0.2]/BlackScholesPutGamma[S,10,0.2,0.05,0,
5]-1)}],
{S,2,16,1}]]
```

2.0000	0.0084	0.0085	-0.3285
3.0000	0.0477	0.0480	-0.6616
4.0000	0.1001	0.1000	0.0932
5.0000	0.1335	0.1329	0.4113
6.0000	0.1398	0.1394	0.3334
7.0000	0.1276	0.1274	0.1276
8.0000	0.1070	0.1071	-0.0616
9.0000	0.0852	0.0853	-0.1835
10.0000	0.0655	0.0657	-0.2327
11.0000	0.0493	0.0494	-0.2225
12.0000	0.0365	0.0366	-0.1710
13.0000	0.0268	0.0269	-0.0961
14.0000	0.0196	0.0196	-0.0127
15.0000	0.0143	0.0143	0.0671
16.0000	0.0104	0.0104	0.1348

The error is at most 0.7% of the exact value - near the strike it is now only 0.2%.

Plot of Computed and Exact Gamma

```
Plot[{DougGamma[10, 0.05, 0, S, 5, 0.2],
BlackScholesPutGamma[S, 10, 0.2, 0.05, 0, 5]}, {S, 8, 12}, PlotPoints -> 5
 PlotRange -> All];
```

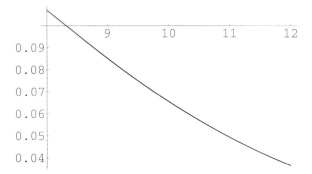

Theta Analysis

The percentage error in Theta is given by the last column of the following table:

```
thetasamples = TableForm[
Table[Map[PaddedForm[N[#], {5, 4}]&,
{S, DougTheta[10, 0.05, 0, S, 5, 0.2],
BlackScholesPutTheta[S,10,0.2,0.05,0,5],
100*(DougTheta[10,0.05,0,S,5,0.2]/BlackScholesPutTheta[S,10,0.2,0.05,0,
5]-1)}],
{S, 2, 16, 1}]]
```

2.0000	0.3885	0.3885	-0.0085
3.0000	0.3773	0.3772	0.0260
4.0000	0.3407	0.3405	0.0498
5.0000	0.2792	0.2792	-0.0232
6.0000	0.2071	0.2073	-0.1296
7.0000	0.1389	0.1391	-0.1795
8.0000	0.0828	0.0829	-0.1116
9.0000	0.0410	0.0410	0.1566
10.0000	0.0124	0.0122	1.2132
11.0000	-0.0058	-0.0060	-2.5962
12.0000	-0.0163	-0.0164	-0.6827
13.0000	-0.0214	-0.0214	-0.2323
14.0000	-0.0230	-0.0230	0.0372
15.0000	-0.0226	-0.0225	0.2296
16.0000	-0.0210	-0.0209	0.3672

Near the strike the error in θ peaks at 2.6% of the exact value. Note that with the two-time-level Crank-Nicholson scheme the error was almost 200%.

Theta Plots

```
Plot[{DougTheta[10, 0.05, 0, S, 5, 0.2],
BlackScholesPutTheta[S, 10, 0.2, 0.05, 0, 5]}, {S, 8, 12}, PlotPoints -> 5
 PlotRange -> All];
```

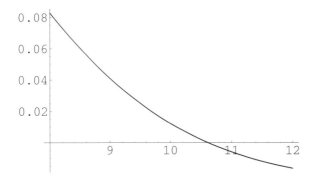

As a reminder, here is the corresponding plot for the Crank-Nicholson scheme.

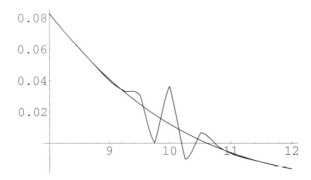

15.8 Summary

The introduction of implicit schemes is motivated by the desire to get accurate and stable solutions for larger values of α. When we need to compute the values of both a function and its first and second derivatives, popular schemes such as the ordinary two-time-level Crank-Nicholson scheme are unsuitable for option-pricing problems, unless small to moderate values of α are used. This is because of the fact that non-smoothness in initial data (almost always present in option payoffs) propagates through the solution causing small oscillatory errors. When amplified by the process of differentiation, these may cause substantial errors in the Greeks. The errors in the valuation may be deceptively small - to quote Wilmott *et al* (1993) when discussing a verification example with the Crank-Nicholson scheme, they remark that "Even with $\alpha = 10$, the numerical and exact results differ only marginally." This is absolutely right, at least when we just look at the valuation. When we inspect the Greeks a rather more disturbing picture emerges. We have given an example where estimates of θ from a computation with $\alpha = 8$ are in error by about 200%.

The oscillation problem and the resulting corruption of the Greeks can be cured by the use of a more suitable difference scheme. When the initial data were smooth, we saw in Chapter 14 that the two-time-level Douglas scheme was sufficient, but in the presence of non-smooth data the corresponding three-time-level Douglas scheme cures the problems quite dramatically.

The transition to such a scheme for practitioners should not be a great leap. We have already seen that the two-time-level Douglas scheme is the natural implicit generalization of the trinomial model, when carried out on a grid rather than a tree. The next step to a three-time-level scheme allows larger time steps still to be taken without a loss of accuracy or corruption of the Greeks.

Chapter 15 Bibliography

Richtmyer, R.D. and Morton, K.W., 1957, *Difference Methods for Initial Value Problems*, Krieger reprint, 1994, reprinted from the Wiley 1957/1967 original.

Smith, G.D., 1985, *Numerical Solution of Partial Differential Equations: Finite Difference Methods*, Oxford University Press.

Wilmott, P., Dewynne, J. and Howison, S., 1993, *Option Pricing - Mathematical Models and Computation*, Oxford Financial Press.

Chapter 16.
SOR and PSOR Schemes for the Three-Time-Level Douglas Scheme and Application to American Options

16.1 Introduction

The purpose of our finite-difference analysis is to define, implement, test and apply an accurate and efficient implicit finite-difference scheme for the valuation of derivatives with a long time to expiry. We consider, as our next example, American options with a long time to expiry. Other related instruments include convertible bonds (CBs).

In the previous two chapters I explained why Crank-Nicholson and other simple two-time-level schemes are unsuitable for the accurate modelling of option-pricing problems, due to their poor management of non-smooth payoffs, and the resulting errors in the Greeks arising from the propagation of errors from non-smooth points in the payoff. An example was given to justify the replacement of Crank-Nicholson methods with three-time-level Douglas finite-difference schemes (3TLD). This is justified for two reasons:

(a) a better truncation error (arising from the use of a Douglas-type scheme);
(b) a better decay of oscillations arising from non-smoothness of the payoff at the strike.

We have also made the point that schemes of the Douglas type are actually the natural implicit generalization of the trinomial model. So we shall use 3TLD from here on. As a reminder, we wish to use implicit finite-difference models for long-expiry options because, compared to explicit FD or binomial methods, they

(a) are just as accurate, but with a larger time-step;
(b) allow more straightforward and accurate computation of some of the Greeks.

Here we get this scheme going in a framework designed to allow the valuation of American options and CBs and other options with a long time to expiry. We test it by re-valuing European options, and we apply it by testing American option models with long times to expiry. Exactly the same methods can be used to look at simple convertible bonds. These are obtained from the model described here by making simple changes to expiry/boundary conditions and to payoff constraints. Otherwise the maths of the solution scheme is identical, though one needs to combine the management of discrete dividends (Chapter 24) with the form of the model suitable for modelling coupons and time-dependent interest rates, as discussed in Chapter 3.

In this chapter we shall implement a solution method for European options (SOR, for successive over-relaxation) that is generalizable to American options (PSOR, or projected SOR). When this generalization is made some new issues arise regarding computation of Greeks in the neighbourhood of the moving boundary (this is just the critical stock price that gives the boundary between regions where early exercise is desirable and where it is not). The use of standard differencing formulae gives rise to problems, particularly in Γ, arising from the fact that "standard" formulae essentially smooth the Greeks over the boundary region. In fact, Γ has a discontinuity at the critical price, and special techniques are needed to capture this. We expect this issue to arise whenever there is some type of boundary or barrier/trigger within a problem. We give some suggestions for fixing the problem and show how they work. Additional modifications to PSOR can also be used to treat embedded Puts and Calls in a CB model.

We compare the results of the PSOR-3TLD model with the analytical approximation for the Carr-Jarrow-Myneni scheme given in Chapter 11, and demonstrate good consistency between our analytical and numerical methods for American options. Later in the book we shall also explore the corresponding binomial tree results. We emphasize that we regard the 3TLD model as vastly superior to binomial.

We wish to point out that the considerations of this chapter are intensively numerical. A fast machine is recommended!

16.2 Mathematical Reminders

Transforming to the Diffusion Equation

The Black-Scholes differential equation

$$\frac{\sigma^2}{2} S^2 \frac{\partial^2 V}{\partial S^2} - r V + (r - q) S \frac{\partial V}{\partial S} + \frac{\partial V}{\partial t} = 0 \tag{1}$$

can be transformed into the diffusion equation by a standard change of variables, as given in Chapter 15. When all this is done, one obtains

$$\frac{\partial^2 u}{\partial x^2} = \frac{\partial u}{\partial \tau} \tag{2}$$

We now wish to solve this with appropriate modifications for early exercise.

3TLD Reminder

The diffusion equation (2) is solved by introducing a discrete grid with steps Δx and $\Delta \tau$, and setting

$$u_n^m = u(n\Delta x, m\Delta \tau) \tag{3}$$

A key role is played by the parameter α, given by

$$\alpha = \frac{\Delta \tau}{(\Delta x)^2} \qquad\qquad \Delta \log S = \sigma \sqrt{\frac{\Delta T}{2 \alpha}} \tag{4}$$

In the previous chapter, we explored various schemes for solving for u, and decided that the three-time-level Douglas scheme best satisfied the goals of accuracy, particularly around the strike, and eliminated problems such as oscillations that corrupted the Greeks. As in the previous chapter, the scheme we use is 3TLD:

$$\left(\frac{1}{8} - \alpha\right)(u_{n-1}^{m+1} + u_{n+1}^{m+1}) + \left(\frac{5}{4} + 2\alpha\right)u_n^{m+1}$$

$$= \frac{1}{6}(u_{n-1}^m + u_{n+1}^m + 10\,u_n^m) - \frac{1}{24}(u_{n-1}^{m-1} + u_{n+1}^{m-1} + 10\,u_n^{m-1}) \tag{5}$$

This type of process requires a kick-off procedure. We use the ordinary Douglas two-time-level scheme to kick off the process. This is given by

$$(1 - 6\alpha)(u_{n-1}^{m+1} + u_{n+1}^{m+1}) + (10 + 12\alpha)\,u_n^{m+1} = (1 + 6\alpha)(u_{n-1}^m + u_{n+1}^m) + (10 - 12\alpha)\,u_n^m \tag{6}$$

and is used once with α replaced by $\alpha/4$, followed by the three-time-level scheme with $\alpha/4$ and then again with $\alpha/2$.

16.3 Detailed Modelling Studies

Equation Solvers

The solution of equations such as (5) and (6) requires a solver. Furthermore, when treating American options or CBs, we wish to solve differential inequalities related to equation (1), based on an associated constraint surface (the payoff). We therefore need a solution mechanism for (5) and (6) that allows this. The tridiagonal solver used in the previous chapter does not allow for early exercise problems, so we replace it by the successive over-relaxation (SOR) method, and for early exercise problems, or those with other constraints, the projected form (PSOR). For details of the SOR and PSOR methods in general, you should see Wilmott *et al* (1993). Although we are using rather novel difference schemes here, the solution methods of SOR and PSOR are otherwise identical to those of Wilmott *et al*.

Compiled SOR and PSOR Solvers

The matrix equations resulting from (5) and (6) are symmetric and also have constant values down the diagonal and off-diagonal lines, so we can simplify matters in constructing the SOR/PSOR solvers:

```
CompSORSolve = Compile[
    {{diag, _Real, 0}, {offdiag, _Real, 0}, {r, _Real, 1},
    {eps, _Real, 0}, {relax, _Real, 0}, {kickoff, _Real, 1}},
    Module[{len = Length[r], yu = r,
      yold = kickoff, ynew = r, error = 100.0},
        While[error > eps^2,
  (Do[(yu[[i]] = r[[i]] / diag - (offdiag / diag) *
            (If[i == 1, 0, ynew[[i - 1]]]] + If[i < len, yold[[i + 1]], 0]);
  ynew[[i]] = yold[[i]] + relax * (yu[[i]] - yold[[i]])), {i,
        1, len}];
  error = (ynew - yold) . (ynew - yold);
  yold = ynew)]; yold]];
```

We make one change to this routine to define the corresponding projected form suitable for valuing American options:

```
CompPSORSolve = Compile[
   {{diag, _Real, 0}, {offdiag, _Real, 0}, {r, _Real, 1},
   {eps, _Real, 0}, {relax, _Real, 0}, {kickoff, _Real, 1}},
   Module[{len = Length[r], yu = r,
    yold = kickoff, ynew = r, error = 100.0},
     While[error > eps^2,
 (Do[(yu[[i]] = r[[i]] / diag - (offdiag / diag) *
         (If[i == 1, 0, ynew[[i - 1]]] + If[i < len, yold[[i + 1]], 0]);
  ynew[[i]] = Max[kickoff[[i]], yold[[i]] +
        relax * (yu[[i]] - yold[[i]])] ), {i, 1, len}];
 error = (ynew - yold) . (ynew - yold);
 yold = ynew)]; yold]];
```

Standardization of Variables

Here we implement the standard transformations used to obtain the diffusion equation:

$$\text{NonDimExpiry}[T\_, sd\_] := \frac{sd^2\ T}{2};$$

$$\text{kone}[r\_, sd\_] := \frac{2\ r}{sd^2};$$

$$\text{ktwo}[r\_, q\_, sd\_] := \frac{2\ (r - q)}{sd^2};$$

$$\text{ValuationMultiplier}[strike\_, r\_, q\_, x\_, tau\_, sd\_] := strike$$
$$\text{Exp}\left[-\frac{1}{2}\ (\text{ktwo}[r, q, sd] - 1)\ x - \left(\frac{1}{4}\ (\text{ktwo}[r, q, sd] - 1)^2 + \text{kone}[r, sd]\right) tau\right]$$

Initial (Expiry) Conditions

$$\text{CallExercise}[x\_, r\_, q\_, sd\_] :=$$
$$\text{Max}\left[\text{Exp}\left[\frac{1}{2}\ (\text{ktwo}[r, q, sd] - 1)\ x\right]\ (\text{Exp}[x] - 1), 0\right];$$

$$\text{PutExercise}[x\_, r\_, q\_, sd\_] :=$$
$$\text{Max}\left[\text{Exp}\left[\frac{1}{2}\ (\text{ktwo}[r, q, sd] - 1)\ x\right]\ (1 - \text{Exp}[x]), 0\right];$$

Black-Scholes Model for Verification

We load a Package built with the approximate analytical model described in Chapter 11. This model describes the American Put by an accurate analytical approximation.

```
Needs["Derivatives`AnalyticAmerican`"]
```

```
? Derivatives`AnalyticAmerican`*
```

```
AmericanPutCJM          AmericanPutGammaCJM  Front
AmericanPutDeltaCJM
```

```
? AmericanPutCJM
```

> AmericanPutCJM[price,strike,vol,r,t] returns an APPROXIMATE
>   value for a vanilla zero-dividend American Put according to
>   an approximation of the Carr-Jarrow-Myneni integral equation.

Note that this Package automatically loads the ordinary vanilla European options package:

```
? BlackScholesPut
```

> BlackScholesPut[price, strike, vol, riskfree, divyield, expiry]
>   returns the Black-Scholes value of a vanilla European Put.

Boundary Conditions and Payoff Constraints

Since the most interesting basic early-exercise problem arises for the Put, with zero dividends, we focus on that problem. We define both the boundary conditions and the payoff constraint function. Note that these are expressed in terms of the variables of the diffusion equation.

Put (q=0) Boundary Conditions

```
g[x_, tau_, r_, q_, sd_] := 0;

f[x_, tau_, r_, q_, sd_] := Exp[ 1/2 (ktwo[r, q, sd] - 1) x +

    1/4 ((ktwo[r, q, sd] - 1)^2 + 4 kone[r, sd]) tau] - Exp[

    1/2 (ktwo[r, q, sd] + 1) x + 1/4 ((ktwo[r, q, sd] - 1)^2 + 4 kone[r, sd]) tau];
```

Payoff Constraint Function

```
PutPayoff[x_, tau_, r_, q_, sd_] :=
  Exp[((ktwo[r, q, sd] - 1)^2 / 4 + kone[r, sd]) * tau] *
Max[Exp[(ktwo[r, q, sd] - 1) * x / 2] (1 - Exp[x]), 0];
```

16.4 Douglas Three-Time-Level SOR Solution Functions

Necessary Functions

The matrix used in solving the two-time-level Douglas system is characterized by its diagonal and off-diagonal entries:

```
DougDiag[alpha_] := 10 + 12 alpha;
DougOffDiag[alpha_] := 1 - 6 alpha;
```

The matrix used in solving the three-time-level Douglas system is characterized by its own diagonal and off-diagonal entries:

```
ThreeDougDiag[alpha_] := 5 / 4 + 2 alpha;
ThreeDougOffDiag[alpha_] := 1 / 8 - alpha;
```

The right-hand side of our implicit difference scheme is constructed using the same matrices as before:

```
DougDMatrix[alpha_, vec_List] := Module[{temp},
temp = (10 - 12*alpha)*vec + (1+6*alpha)*(RotateRight[vec] + Rotate-
Left[vec]);
temp[[1]] = Simplify[First[temp] -  (1+6*alpha)*Last[vec]];
temp[[-1]] = Simplify[Last[temp] - (1 + 6*alpha)*First[vec]];
temp]


DougDDMatrix[vec_List] := Module[{temp},
temp = (10*vec + RotateRight[vec] + RotateLeft[vec]);
temp[[1]] = Simplify[First[temp] - Last[vec]];
temp[[-1]] = Simplify[Last[temp] - First[vec]];
temp/6]
```

16.5 Verification of 3TLD-SOR Algorithm on European Options

We quickly check that the replacement of the tridiagonal solver (used in the previous chapter) by the SOR solver makes no significant difference provided the tolerance and relaxation parameters are suitably set. *Readers who wish to take this on trust can skip to Section 16.6 where we treat the American option, but should note that here we will show that we can use just 20 time-steps to deal accurately with a problem with five years to expiry.*

```
dx = 0.0125; dtau = 0.005; alpha = dtau/dx^2;
M=20; nminus = 320; nplus = 160; alpha
```

    32.

So we have a very large value of $\alpha = 32$. Next we introduce the additional parameters for the SOR solver. A variety of experiments, not reported here, reveal that a good value for the relaxation parameter is about 1.1. The choice of tolerance parameter affects the accuracy and speed of solution. Here we are being fairly demanding on accuracy, and have set a value of 10^{-6} at the price of speed. You are encouraged to explore the impact of changing this parameter for yourselves.

```
relax = 1.1;
tolerance = 0.000001;

initial = Table[
   PutExercise[(k - 1 - nminus) dx, 0.05, 0, 0.2], {k, nminus + nplus + 1}];
lower = Table[f[-nminus dx, (m - 1) dtau, 0.05, 0, 0.2], {m, 1, M + 1}];
upper = Table[g[+nplus dx,  (m - 1) dtau, 0.05, 0, 0.2], {m, 1, M + 1}];

w = Table[0, {m, 1, 3},    {k, 1, nminus+nplus+1}];

vold = Take[initial, {2, -2}];
payoff = vold;

w[[1,1]] = f[-nminus*dx, dtau/4, 0.05, 0, 0.2];
w[[1, nminus+nplus+1]] = g[nplus*dx, dtau/4, 0.05, 0, 0.2];
w[[2,1]] = f[-nminus*dx, dtau/2, 0.05, 0, 0.2];
w[[2, nminus+nplus+1]] = g[nplus*dx, dtau/2, 0.05, 0, 0.2];
w[[3,1]] = f[-nminus*dx, dtau, 0.05, 0, 0.2];
w[[3, nminus+nplus+1]] = g[nplus*dx, dtau, 0.05, 0, 0.2];
```

First Kick-Off Component

Simple Douglas with two time-levels and $dt/4$ time-step.

```
rhs = DougDMatrix[alpha/4, vold]+
Table[
Which[
k==1, (6*alpha/4+1)*lower[[1]] + (6*alpha/4-1)*w[[1, 1]],
k== nplus + nminus-1,  (6*alpha/4+1)*upper[[1]] +
(6*alpha/4-1)*w[[1, nplus+nminus+1]],
True, 0],
{k, 1, nplus + nminus-1}];
vnew = CompSORSolve[DougDiag[alpha/4], DougOffDiag[alpha/4], rhs, toler-
ance, relax, payoff];
w[[1]] = Join[{w[[1,1]]}, vnew, {w[[1,nplus+nminus+1]]}];
vvold = vold;
vold = vnew;
```

Now we have two iterations of the three-time-level Douglas system, doubling the time-step at each stage:

```
rhs = DougDDMatrix[vold] - DougDDMatrix[vvold]/4 +
Table[
Which[
k==1, (alpha/4-1/8)*w[[2, 1]] + w[[1,1]]/6 - lower[[1]]/24,
k==nplus + nminus-1, (alpha/4-1/8)*w[[2, nplus+nminus+1]] +
w[[1,nplus+nminus+1]]/6 - upper[[1]]/24,
True, 0],
{k, 1, nplus + nminus-1}];
vnew = CompSORSolve[ThreeDougDiag[alpha/4], ThreeDougOffDiag[alpha/4],
rhs, tolerance, relax, payoff];
w[[2]] = Join[{w[[2,1]]}, vnew, {w[[2,nplus+nminus+1]]}];
vold = vnew;


rhs = DougDDMatrix[vold] - DougDDMatrix[vvold]/4 +
Table[
Which[
k==1, (alpha/2-1/8)*w[[3, 1]] + w[[2,1]]/6 - lower[[1]]/24,
k==nplus + nminus-1, (alpha/2-1/8)*w[[3, nplus+nminus+1]] +
w[[2,nplus+nminus+1]]/6 - upper[[1]]/24,
True, 0],
{k, 1, nplus + nminus-1}];
vnew = CompSORSolve[ThreeDougDiag[alpha/2], ThreeDougOffDiag[alpha/2],
rhs, tolerance, relax, payoff];
w[[3]] = Join[{w[[3,1]]}, vnew, {w[[3,nplus+nminus+1]]}];
wold = initial;
wvold = initial;
wnew = w[[3]];
```

Main Block

We evolve this using the SOR solver. You are warned that this method takes a lot longer than the tridiagonal solution method. It is not recommended as a working model for European options. Here we are carrying out a verification of the SOR solution method, before proceeding to the PSOR approach for American options. (In the following we have suppressed the output of **Print** that shows progress.)

```
For[m=3, m<=M+1, m++,
(wvold = wold;
wold = wnew;
Print[m];
rhs = DougDDMatrix[Take[wold, {2, -2}]] -
DougDDMatrix[Take[wvold, {2, -2}]]/4 +
Table[
Which[
k==1,
(alpha-1/8)*lower[[m]] + lower[[m-1]]/6 - lower[[m-2]]/24,
k==nplus + nminus-1,
(alpha-1/8)*upper[[m]] + upper[[m-1]]/6 - upper[[m-2]]/24,
True, 0],
{k, 1, nplus + nminus-1}];
temp = CompSORSolve[ThreeDougDiag[alpha], ThreeDougOffDiag[alpha],
```

```
rhs, tolerance, relax, payoff];
wnew = Join[{lower[[m]]}, temp, {upper[[m]]}])
]


points = Table[(k-nminus-1)*dx,    {k, 1, nplus+nminus+1}];
finalstep = wnew;
prevstep = wold;
pprevstep = wvold;
interpoldata = Transpose[{points,finalstep}];
ufunc = Interpolation[interpoldata, InterpolationOrder -> 3];
Valuation[strike_, r_, q_, S_, T_, sd_] :=
ValuationMultiplier[strike, r, q, Log[S/strike], sd^2*T/2, sd]*
ufunc[Log[S/strike]]
```

Sample Errors in Valuation

We now tabulate, for verification purposes, the computed and exact values and the absolute error. Note that this is a five-year option and we have only used 20 time-steps ($\alpha = 32$).

```
samples = TableForm[Join[{{"S", "SOR", "Exact", "Absolute Error"}},
Table[Map[PaddedForm[N[#], {5, 4}]&, {S, Valuation[10, 0.05, 0, S, 5,
0.2],
BlackScholesPut[S,10,0.2,0.05,0,5],
Valuation[10, 0.05, 0, S, 5, 0.2]-BlackScholesPut[S,10,0.2,0.05,0,5]}],
{S, 2, 16, 1}]]]
```

S	SOR	Exact	Absolute Error
2.0000	5.7889	5.7886	0.0003
3.0000	4.8012	4.8005	0.0007
4.0000	3.8621	3.8615	0.0006
5.0000	3.0209	3.0209	0.0000
6.0000	2.3105	2.3108	-0.0003
7.0000	1.7383	1.7387	-0.0004
8.0000	1.2930	1.2932	-0.0002
9.0000	0.9547	0.9548	-0.0001
10.0000	0.7019	0.7019	0.0000
11.0000	0.5150	0.5149	0.0000
12.0000	0.3777	0.3777	0.0000
13.0000	0.2772	0.2773	-0.0000
14.0000	0.2039	0.2039	-0.0001
15.0000	0.1503	0.1504	-0.0001
16.0000	0.1111	0.1113	-0.0002

So even with a large α we have obtained excellent results.

Construction and Verification of Interpolated Valuation and Greeks

In this section we define some functions to compute Delta, Gamma and Theta directly from the finite-differencegrid. *These are done with ordinary differencing methods.*

```
listd[data_, step_] :=
Module[{dleft,dright,len},
len = Length[data];
dleft = (4*data[[2]]-3*data[[1]]-data[[3]])/(2*step);
dright = (3*data[[len]]-4*data[[len-1]]+data[[len-2]])/(2*step);
Join[{dleft}, Take[RotateLeft[data]-RotateRight[data], {2,
-2}]/(2*step), {dright}]
]


kt = ktwo[0.05, 0, 0.2];
ko = kone[0.05, 0.2];
```

$$\text{deltadata} = \text{listd[finalstep, dx]} - \frac{1}{2} \, (kt - 1) \, \text{finalstep};$$

$$\text{gammadata} = \text{listd[deltadata, dx]} - \frac{1}{2} \, (kt + 1) \, \text{deltadata};$$

thetadata =
$$\frac{3\,\text{finalstep} - 4\,\text{prevstep} + \text{pprevstep}}{2\,\text{dtau}} - \left(\frac{1}{4} \, (kt - 1)^2 + ko\right) \text{finalstep};$$

```
deltainterpoldata = Transpose[{points, deltadata}];
gammainterpoldata = Transpose[{points, gammadata}];
thetainterpoldata = Transpose[{points, thetadata}];

dfunc = Interpolation[deltainterpoldata, InterpolationOrder → 3];
gfunc = Interpolation[gammainterpoldata, InterpolationOrder → 3];
tfunc = Interpolation[thetainterpoldata, InterpolationOrder → 3];
```

SORDelta[strike_, r_, q_, S_, T_, sd_] :=
$$\frac{1}{S} \left(\text{ValuationMultiplier}\left[\text{strike, r, q, Log}\left[\frac{S}{\text{strike}}\right], \frac{sd^2 \, T}{2}, sd\right] \right.$$

$$\left. \text{dfunc}\left[\text{Log}\left[\frac{S}{\text{strike}}\right]\right]\right)$$

```
SORGamma[strike_, r_, q_, S_, T_, sd_] :=
```

$$\frac{1}{S^2}\left(\text{ValuationMultiplier}\left[\text{strike, r, q, Log}\left[\frac{S}{\text{strike}}\right], \frac{sd^2\,T}{2}, sd\right]\right.$$

$$\left.\text{gfunc}\left[\text{Log}\left[\frac{S}{\text{strike}}\right]\right]\right)$$

```
SORTheta[strike_, r_, q_, S_, T_, sd_] :=
```

$$-\frac{1}{2}\,sd^2\,\text{ValuationMultiplier}\left[\text{strike, r, q, Log}\left[\frac{S}{\text{strike}}\right], \frac{sd^2\,T}{2}, sd\right]$$

$$\text{tfunc}\left[\text{Log}\left[\frac{S}{\text{strike}}\right]\right]$$

Table of Percentage Errors in Delta

```
deltasamples = TableForm[Join[{{"S", "SOR Delta", "Exact Delta", "%
Error"}},
Table[Map[PaddedForm[N[#], {5, 4}]&,
{S, SORDelta[10, 0.05, 0, S, 5, 0.2],
BlackScholesPutDelta[S,10,0.2,0.05,0,5],
100*(SORDelta[10,0.05,0,S,5,0.2]/BlackScholesPutDelta[S,10,0.2,0.05,0,5
]-1)}],
{S, 2, 16, 1}]]]
```

S	SOR Delta	Exact Delta	% Error
2.0000	-0.9972	-0.9976	-0.0415
3.0000	-0.9718	-0.9719	-0.0127
4.0000	-0.8978	-0.8973	0.0533
5.0000	-0.7791	-0.7785	0.0660
6.0000	-0.6406	-0.6404	0.0273
7.0000	-0.5058	-0.5060	-0.0217
8.0000	-0.3881	-0.3883	-0.0526
9.0000	-0.2920	-0.2922	-0.0566
10.0000	-0.2168	-0.2169	-0.0373
11.0000	-0.1597	-0.1597	-0.0036
12.0000	-0.1170	-0.1170	0.0343
13.0000	-0.0855	-0.0855	0.0672
14.0000	-0.0624	-0.0624	0.0884
15.0000	-0.0456	-0.0456	0.0932
16.0000	-0.0334	-0.0334	0.0792

The error in Delta peaks at about 0.1%.

Table of Percentage Errors in Gamma

The corresponding table of values of Gamma reveals a maximum error of about 1.3% of the true value.

```
gammasamples = TableForm[Join[{{"S", "SOR Gamma", "Exact Gamma", "%
Error"}},
Table[Map[PaddedForm[N[#], {5, 4}]&,
{S, SORGamma[10, 0.05, 0, S, 5, 0.2],
BlackScholesPutGamma[S,10,0.2,0.05,0,5],
100*(SORGamma[10,0.05,0,S,5,0.2]/BlackScholesPutGamma[S,10,0.2,0.05,0,5
]-1)}],
{S, 2, 16, 1}]]]
```

S	SOR Gamma	Exact Gamma	% Error
2.0000	0.0085	0.0085	0.6715
3.0000	0.0474	0.0480	-1.3355
4.0000	0.0997	0.1000	-0.3275
5.0000	0.1332	0.1329	0.2057
6.0000	0.1397	0.1394	0.2711
7.0000	0.1276	0.1274	0.1518
8.0000	0.1071	0.1071	0.0082
9.0000	0.0853	0.0853	-0.0944
10.0000	0.0656	0.0657	-0.1401
11.0000	0.0493	0.0494	-0.1351
12.0000	0.0366	0.0366	-0.0937
13.0000	0.0269	0.0269	-0.0318
14.0000	0.0196	0.0196	0.0367
15.0000	0.0143	0.0143	0.1004
16.0000	0.0104	0.0104	0.1510

Table of Percentage Errors in Theta

```
thetasamples = TableForm[Join[{{"S", "SOR Theta", "Exact Theta", "%
Error"}},
Table[Map[PaddedForm[N[#], {5, 4}]&,
{S, SORTheta[10, 0.05, 0, S, 5, 0.2],
BlackScholesPutTheta[S,10,0.2,0.05,0,5],
100*(SORTheta[10,0.05,0,S,5,0.2]/BlackScholesPutTheta[S,10,0.2,0.05,0,5
]-1)}],
{S, 2, 16, 1}]]]
```

S	SOR Theta	Exact Theta	% Error
2.0000	0.3884	0.3885	-0.0272
3.0000	0.3772	0.3772	-0.0002
4.0000	0.3406	0.3405	0.0272
5.0000	0.2791	0.2792	-0.0369
6.0000	0.2071	0.2073	-0.1312
7.0000	0.1389	0.1391	-0.1633
8.0000	0.0828	0.0829	-0.0643
9.0000	0.0411	0.0410	0.2754
10.0000	0.0124	0.0122	1.6405
11.0000	-0.0058	-0.0060	-3.4529
12.0000	-0.0162	-0.0164	-0.9706
13.0000	-0.0213	-0.0214	-0.4259
14.0000	-0.0230	-0.0230	-0.1158
15.0000	-0.0226	-0.0225	0.1007
16.0000	-0.0210	-0.0209	0.2553

The maximum error in Theta is about 3.5% of the true value. This has been obtained with $\alpha = 32$, and we remind the reader that a corresponding treatment with a Crank-Nicholson scheme with $\alpha = 8$ led to errors of about 200%. So the use of the SOR solver preserves the good characteristics of the 3TLD system. Note that we have used just 20 time-steps to deal accurately with a problem with five years to expiry. The price of the use of the SOR method is a decrease in speed compared to the tridiagonal solver. Note that we are not recommending the use of such a method for vanilla European options. Its appearance here is purely to act as a test of the method - to demonstrate that we have correctly implemented successive over-relaxation as a method for solving the difference problem.

16.6 3TLD-PSOR Analysis of American Options

Now we proceed to the first "real" application of these methods, and look at the valuation of American Puts using a PSOR scheme based on 3TLD. We shall keep most parameters to be identical with the previous European case, but consider expiry times of one, five and ten years, and also we shall turn on "early exercise" within the solution scheme by turning on the projection onto the payoff constraint surface. We shall fix the number of time-steps at 20, so in the case of one year to expiry, $\alpha = 6.4$. This case has been contrived to reveal an important issue regarding the calculation of Greeks near the critical price.

```
dx = 0.0125; dtau = 0.001; alpha = dtau / dx^2;
M = 20; nminus = 320; nplus = 160; alpha
```

```
    6.4
```

```
relax = 1.1;
tolerance = 0.000001;
```

```
initial = Table[
  PutExercise[(k - 1 - nminus) dx, 0.05, 0, 0.2], {k, nminus + nplus + 1}];
lower = Table[f[-nminus dx, (m - 1) dtau, 0.05, 0, 0.2], {m, 1, M + 1}];
upper = Table[g[+nplus dx,  (m - 1) dtau, 0.05, 0, 0.2], {m, 1, M + 1}];

w = Table[0, {m, 1, 3},    {k, 1, nminus+nplus+1}];


vold = Take[initial, {2, -2}];


w[[1,1]] = f[-nminus*dx, dtau/4, 0.05, 0, 0.2];
w[[1, nminus+nplus+1]] = g[nplus*dx, dtau/4, 0.05, 0, 0.2];
w[[2,1]] = f[-nminus*dx, dtau/2, 0.05, 0, 0.2];
w[[2, nminus+nplus+1]] = g[nplus*dx, dtau/2, 0.05, 0, 0.2];
w[[3,1]] = f[-nminus*dx, dtau, 0.05, 0, 0.2];
w[[3, nminus+nplus+1]] = g[nplus*dx, dtau, 0.05, 0, 0.2];
```

First Kick-Off Component

Simple Douglas with two time levels and $dt/4$ time-step. We treat the American case by replacing SOR with PSOR based on the payoff function. You should note that there are only two differences between this analysis and the previous European case:

(a) the payoff constraint function is re-computed for each time step;
(b) the solver is the **CompPSORSolve** function, which projects the solution onto the payoff constraint surface.

```
payoff = Table[PutPayoff[(k - 1 - nminus) * dx, dtau / 4, 0.05, 0, 0.2],
  {k, 2, nminus + nplus}];


rhs = DougDMatrix[alpha/4, vold]+
Table[
Which[
k==1, (6*alpha/4+1)*lower[[1]] + (6*alpha/4-1)*w[[1, 1]],
k== nplus + nminus-1,  (6*alpha/4+1)*upper[[1]] +
(6*alpha/4-1)*w[[1, nplus+nminus+1]],
True, 0],
{k, 1, nplus + nminus-1}];
vnew = CompPSORSolve[DougDiag[alpha/4], DougOffDiag[alpha/4], rhs,
tolerance, relax, payoff];
w[[1]] = Join[{w[[1,1]]}, vnew, {w[[1,nplus+nminus+1]]}];
vvold = vold;
vold = vnew;
```

Now we have two iterations of the three-time-level Douglas system, doubling the time-step at each stage, and re-computing the payoff.

```
payoff = Table[PutPayoff[(k - 1 - nminus) * dx, dtau / 2, 0.05, 0, 0.2],
  {k, 2, nminus + nplus}];

rhs = DougDDMatrix[vold] - DougDDMatrix[vvold]/4 +
Table[
Which[
k==1, (alpha/4-1/8)*w[[2, 1]] + w[[1,1]]/6 - lower[[1]]/24,
k==nplus + nminus-1,  (alpha/4-1/8)*w[[2, nplus+nminus+1]] +
w[[1,nplus+nminus+1]]/6 - upper[[1]]/24,
True, 0],
{k, 1, nplus + nminus-1}];
vnew = CompPSORSolve[ThreeDougDiag[alpha/4], ThreeDougOff-
Diag[alpha/4], rhs, tolerance, relax, payoff];
w[[2]] = Join[{w[[2,1]]}, vnew, {w[[2,nplus+nminus+1]]}];
vold = vnew;

payoff = Table[PutPayoff[(k - 1 - nminus) * dx, dtau, 0.05, 0, 0.2],
  {k, 2, nminus + nplus}];

rhs = DougDDMatrix[vold] - DougDDMatrix[vvold]/4 +
Table[
Which[
k==1, (alpha/2-1/8)*w[[3, 1]] + w[[2,1]]/6 - lower[[1]]/24,
k==nplus + nminus-1,  (alpha/2-1/8)*w[[3, nplus+nminus+1]] +
w[[2,nplus+nminus+1]]/6 - upper[[1]]/24,
True, 0],
{k, 1, nplus + nminus-1}];
vnew = CompPSORSolve[ThreeDougDiag[alpha/2], ThreeDougOff-
Diag[alpha/2], rhs, tolerance, relax, payoff];
w[[3]] = Join[{w[[3,1]]}, vnew, {w[[3,nplus+nminus+1]]}];
wold = initial;
wvold = initial;
wnew = w[[3]];
```

Main Block

Once the kick-off phase has been completed, we go into the main block. Again, the only difference between this case and European is re-computation of the payoff constraint and the use of the PSOR solver. The output of the **Print** statement is not shown.

```
For[m=3, m<=M+1, m++,
(wvold = wold;
wold = wnew;
Print[m];
rhs = DougDDMatrix[Take[wold, {2, -2}]] -
DougDDMatrix[Take[wvold, {2, -2}]]/4 +
Table[
Which[
k==1,
```

```
(alpha-1/8)*lower[[m]] + lower[[m-1]]/6 - lower[[m-2]]/24,
k==nplus + nminus-1,
(alpha-1/8)*upper[[m]] + upper[[m-1]]/6 - upper[[m-2]]/24,
True, 0],
{k, 1, nplus + nminus-1}];
payoff = Table[PutPayoff[(k-1-nminus)*dx, (m-1)*dtau, 0.05, 0, 0.2],
{k, 2, nminus+nplus}];
temp = CompPSORSolve[ThreeDougDiag[alpha], ThreeDougOffDiag[alpha],
rhs, tolerance, relax, payoff];
wnew = Join[{lower[[m]]}, temp, {upper[[m]]}])
]
```

```
points = Table[(k - nminus - 1) * dx, {k, 1, nplus + nminus + 1}];
finalstep = wnew;
prevstep = wold;
pprevstep = wvold;
interpoldata = Transpose[{points, finalstep}];
ufunc = Interpolation[interpoldata, InterpolationOrder -> 3];
Valuation[strike_, r_, q_, S_, T_, sd_] :=
ValuationMultiplier[strike, r, q, Log[S / strike], sd^2 * T / 2, sd] *
ufunc[Log[S / strike]]
```

Plot Sample Valuation

```
Plot[Valuation[10, 0.05, 0, S, 1, 0.2], {S, 2, 20},
PlotPoints -> 50, PlotRange -> All];
```

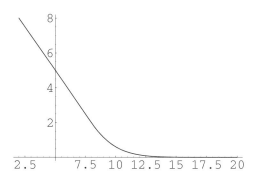

Construction and Verification of Interpolated Valuation and Greeks - First Simple Method

In this subsection we define some functions to compute Delta, Gamma and Theta directly from the finite-difference grid. Note that these are based on the same simple differencing and interpolation methods as were used for the European case, but we should anticipate that these methods introduce some inappropriate smoothing in the neighbourhood of the critical price frontier, where Γ is actually discontinuous.

```
listd[data_, step_] :=
Module[{dleft,dright,len},
len = Length[data];
dleft = (4*data[[2]]-3*data[[1]]-data[[3]])/(2*step);
dright = (3*data[[len]]-4*data[[len-1]]+data[[len-2]])/(2*step);
Join[{dleft}, Take[RotateLeft[data]-RotateRight[data], {2,
-2}]/(2*step), {dright}]
]

kt = ktwo[0.05, 0, 0.2];
ko = kone[0.05, 0.2];
```

$$\text{deltadata} = \text{listd[finalstep, dx]} - \frac{1}{2}\,(kt - 1)\,\text{finalstep};$$

$$\text{gammadata} = \text{listd[deltadata, dx]} - \frac{1}{2}\,(kt + 1)\,\text{deltadata};$$

$$\text{thetadata} = \frac{3\,\text{finalstep} - 4\,\text{prevstep} + \text{pprevstep}}{2\,\text{dtau}} - \left(\frac{1}{4}\,(kt - 1)^2 + ko\right)\text{finalstep};$$

```
deltainterpoldata = Transpose[{points, deltadata}];
gammainterpoldata = Transpose[{points, gammadata}];
thetainterpoldata = Transpose[{points, thetadata}];

dfunc = Interpolation[deltainterpoldata, InterpolationOrder → 3];
gfunc = Interpolation[gammainterpoldata, InterpolationOrder → 3];
tfunc = Interpolation[thetainterpoldata, InterpolationOrder → 3];
```

SORDelta[strike_, r_, q_, S_, T_, sd_] :=

$$\frac{1}{S}\left(\text{ValuationMultiplier}\left[\text{strike, r, q, Log}\left[\frac{S}{\text{strike}}\right], \frac{sd^2\,T}{2}, sd\right]\right.$$

$$\left.\text{dfunc}\left[\text{Log}\left[\frac{S}{\text{strike}}\right]\right]\right)$$

SORGamma[strike_, r_, q_, S_, T_, sd_] :=

$$\frac{1}{S^2}\left(\text{ValuationMultiplier}\left[\text{strike, r, q, Log}\left[\frac{S}{\text{strike}}\right], \frac{sd^2\,T}{2}, sd\right]\right.$$

$$\left.\text{gfunc}\left[\text{Log}\left[\frac{S}{\text{strike}}\right]\right]\right)$$

```
SORTheta[strike_, r_, q_, S_, T_, sd_] :=
```

$$-\frac{1}{2} sd^2 \, \text{ValuationMultiplier}\left[strike, r, q, \text{Log}\left[\frac{S}{strike}\right], \frac{sd^2 \, T}{2}, sd\right]$$

$$tfunc\left[\text{Log}\left[\frac{S}{strike}\right]\right]\Big]$$

16.7 Summary of Finite-Difference Results - One Year Expiry

In the following table we give a table of underlying values, option valuations, Deltas, Gammas and Thetas for a range of underlying prices with the strike at 10. (1 year to expiry, 20% volatility, 5% interest rate, 20 time-steps, 480 price-steps)

```
samples = TableForm[Join[{{"S", "SOR Value", "SOR Delta", "SOR Gamma",
"SOR Theta"}},
Table[Map[PaddedForm[N[#], {5, 3}]&, {S, Valuation[10, 0.05, 0, S, 1,
0.2],
SORDelta[10, 0.05, 0, S, 1, 0.2],SORGamma[10, 0.05, 0, S, 1, 0.2],
SORTheta[10, 0.05, 0, S, 1, 0.2]}],
{S, 2, 16, 1}]]]
```

S	SOR Value	SOR Delta	SOR Gamma	SOR Theta
2.000	8.000	-1.000	-0.000	1.528×10^{-6}
3.000	7.000	-1.000	-0.000	1.337×10^{-6}
4.000	6.000	-1.000	-0.000	1.146×10^{-6}
5.000	5.000	-1.000	-0.000	9.552×10^{-7}
6.000	4.000	-1.000	-0.000	7.642×10^{-7}
7.000	3.000	-1.000	-0.000	5.731×10^{-7}
8.000	2.000	-1.000	0.065	0.000
9.000	1.149	-0.683	0.312	-0.141
10.000	0.609	-0.411	0.229	-0.223
11.000	0.299	-0.224	0.147	-0.217
12.000	0.137	-0.111	0.082	-0.164
13.000	0.059	-0.051	0.042	-0.104
14.000	0.024	-0.022	0.019	-0.058
15.000	0.010	-0.009	0.008	-0.030
16.000	0.004	-0.004	0.003	-0.014

Note the value of Gamma at a stock price of 8 - the option value and Delta would suggest exercise has occurred, but $\Gamma \neq 0$.

16.8 Review of the Analytical Approximation

We can apply the CJM approximation scheme to our test problem under consideration. For values of the underlying beyond the critical price, it returns exact results:

```
{AmericanPutCJM[7, 10, 0.2, 1, 0.05],
 AmericanPutDeltaCJM[7, 10, 0.2, 1, 0.05],
 AmericanPutGammaCJM[7, 10, 0.2, 1, 0.05]}
```

```
    {3, -1, 0}
```

We have obtained the following results by applying this to several values in the neighbourhood of the strike. *Mathematica* may produce some complaints about underflow error or slow convergence of integrals when you run this model. This calculation takes some time, and this in some ways emphasizes the power of a good finite-difference scheme - once the FD solution has been computed on a rectangular grid, it is then very little effort to extract the valuation, Delta, Gamma and Theta for various values of the underlying. In contrast, in our analytical model, an integral has to be computed for each result. This could, in principle, be optimized further by precomputing a large number of values of the critical price and storing these for use in an approximate integration scheme. Here we shall proceed just using the Package functions already defined.

```
CJMsamples = TableForm[Join[{{"S", "CJM Value", "CJM Delta", "CJM
Gamma"}},
Table[Map[PaddedForm[N[#], {5, 3}]&, {S,
AmericanPutCJM[S, 10, 0.2, 0.05, 1], AmericanPutDeltaCJM[S, 10, 0.2,
0.05, 1],
AmericanPutGammaCJM[S, 10, 0.2, 0.05, 1]}], {S, 2, 16, 1}]]]
```

S	CJM Value	CJM Delta	CJM Gamma
2.000	8.000	-1.000	0.000
3.000	7.000	-1.000	0.000
4.000	6.000	-1.000	0.000
5.000	5.000	-1.000	0.000
6.000	4.000	-1.000	0.000
7.000	3.000	-1.000	0.000
8.000	2.000	-1.000	0.000
9.000	1.149	-0.683	0.312
10.000	0.609	-0.411	0.230
11.000	0.299	-0.224	0.147
12.000	0.137	-0.111	0.082
13.000	0.059	-0.051	0.041
14.000	0.024	-0.022	0.019
15.000	0.010	-0.009	0.008
16.000	0.004	-0.004	0.003

This gives very good agreement with the finite-difference model - at this level of precision, the level of agreement is almost perfect, giving confidence in both the analytical approximation and the FD scheme. We can't rule out some form of common-mode failure affecting both approaches, but the agreement is reassuring. We have already noted that the FD model has a glitch in Γ at $S = 8$. We can ask the analytical model where the front is, approximately:

```
Front[10, 0.05, 0.2, 1]
```

8.08038

So Γ is going to be discontinuous around $S = 8.08$, and the FD estimate is smoothed somewhat, as is revealed by the following plot of the numerical Γ.

```
Plot[SORGamma[10, 0.05, 0, S, 1, 0.2], {S, 5, 15}];
```

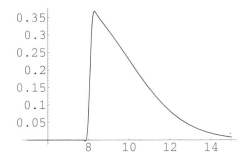

Note that the interpolation scheme has also produced a spurious negative value for Γ near $S = 8$. We need to make a modification of the means by which Γ is estimated in a neighbourhood of the front. The corresponding plot for Δ is

```
Plot[SORDelta[10, 0.05, 0, S, 1, 0.2], {S, 5, 15}];
```

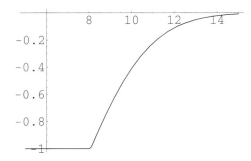

Revised Method for Greeks

In this sub-section we define some new functions to compute Delta, Gamma and Theta directly from the finite-difference grid. Note that these are based on the same simple differencing and interpolation methods as were used for the European case, but we should expect these to introduce some inappropriate smoothing in the neighbourhood of the critical price frontier, where Γ is actually discontinuous.

```
listd[data_, step_] := Module[{dleft,dright,len},
len = Length[data];
dleft = (4*data[[2]]-3*data[[1]]-data[[3]])/(2*step);
dright = (3*data[[len]]-4*data[[len-1]]+data[[len-2]])/(2*step);
Join[{dleft}, Take[RotateLeft[data]-RotateRight[data], {2,
-2}]/(2*step), {dright}]]
```

```
kt = ktwo[0.05, 0, 0.2]; ko = kone[0.05, 0.2];
```

$$rawdeltadata = listd[finalstep, dx] - \frac{1}{2}(kt - 1) finalstep;$$

$$rawgammadata = listd[rawdeltadata, dx] - \frac{1}{2}(kt + 1) rawdeltadata;$$

At this point we interrupt the process, and work out the derivative of Γ with respect to x - this has a spike at the frontier:

```
thirddata = listd[rawgammadata, dx];
thirdinterpoldata = Transpose[{points, thirddata}];
ListPlot[thirdinterpoldata, PlotJoined -> True, PlotRange -> All];
```

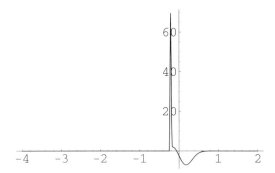

We can locate the maximum easily:

```
sortthird = Transpose[{Abs[thirddata], points}];
```

```
Last[Sort[sortthird]]
```

```
        {69.4731, -0.2125}
```

```
criticalx = -0.2125;
```

So the maximum is attained at the value of $x = -0.2125$. This is a simple estimate of the location of the critical price. We now split the data into two blocks, each side of this region.

```
rawdata = Transpose[{points, finalstep}];
finala = Select[rawdata, (#[[1]] >= criticalx &)];
finalb = Select[rawdata, (#[[1]] <= criticalx &)];

xa = Transpose[finala][[1]]; xb = Transpose[finalb][[1]];
dataa = Transpose[finala][[2]]; datab = Transpose[finalb][[2]];
```

```
deltadataa = listd[dataa, dx] - 1/2 (kt - 1) dataa;

deltadatab = listd[datab, dx] - 1/2 (kt - 1) datab;


gammadataa = listd[deltadataa, dx] - 1/2 (kt + 1) deltadataa;

gammadatab = listd[deltadatab, dx] - 1/2 (kt + 1) deltadatab;


deltainterpoldataa = Transpose[{xa, deltadataa}];
deltainterpoldatab = Transpose[{xb, deltadatab}];
gammainterpoldataa = Transpose[{xa, gammadataa}];
gammainterpoldatab = Transpose[{xb, gammadatab}];


dfunca = Interpolation[deltainterpoldataa, InterpolationOrder -> 3];
dfuncb = Interpolation[deltainterpoldatab, InterpolationOrder -> 3];
gfunca = Interpolation[gammainterpoldataa, InterpolationOrder -> 3];
gfuncb = Interpolation[gammainterpoldatab, InterpolationOrder -> 3];
```

We now give piecewise definitions of Δ and Γ:

```
SORDelta[strike_, r_, q_, S_, T_, sd_] :=
  If[Log[S/strike] >= criticalx, 1/S*ValuationMultiplier[strike, r, q,
Log[S/strike], (sd^2*T)/2, sd]*dfunca[Log[S/strike]],-1]

SORGamma[strike_, r_, q_, S_, T_, sd_] :=
  1/S^2*ValuationMultiplier[strike, r, q, Log[S/strike], (sd^2*T)/2,
sd]*If[Log[S/strike] >= criticalx, gfunca[Log[S/strike]], 0]
```

The value at 8 is now precisely zero:

```
SORGamma[10, 0.05, 0, 8, 1, 0.2]
```

```
    0
```

The plot now has a sharp discontinuity and no negative regions:

```
Plot[SORGamma[10, 0.05, 0, S, 1, 0.2], {S, 5, 15}];
```

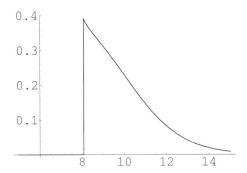

16.9 American Options - Five-Year Expiry

Now we proceed to a second test application of these methods, and look at the valuation of American Puts
with a longer time to expiry, using a PSOR scheme based on 3TLD. We shall keep most parameters to be
identical with the previous one-year case, but multiply the time-step by five. Since we are using implicit
methods, we can do so without adjusting any other parameters. Previously α was 6.4 - it is now 32 (64
times the size in a binomial model with the same price-steps).

Summary of Finite-Difference Results - Five-Year Expiry

In the following table we give a table of valuations, Deltas, Gammas and Thetas for a range of underlying
prices with the strike at 10. (5 years to expiry, 20% volatility, 5% interest rate, 20 time-steps, 480 price-
steps) Note that this has been re-computed without showing the details.

```
samples = TableForm[Table[Map[PaddedForm[N[#], {6, 4}]&, {S,
Valuation[10, 0.05, 0, S, 5, 0.2],
SORDelta[10, 0.05, 0, S, 5, 0.2],SORGamma[10, 0.05, 0, S, 5, 0.2],
SORTheta[10, 0.05, 0, S, 5, 0.2]}],
{S, 2, 16, 1}]]
```

2.0000	8.0000	-1.0000	0.0000	0.0000
3.0000	7.0000	-1.0000	0.0000	0.0000
4.0000	6.0000	-1.0000	0.0000	0.0000
5.0000	5.0000	-1.0000	0.0000	0.0000
6.0000	4.0000	-1.0000	0.0000	0.0000
7.0000	3.0000	-1.0000	0.0000	0.0000
8.0000	2.0614	-0.7856	0.3367	-0.0136
9.0000	1.4196	-0.5194	0.2083	-0.0326
10.0000	0.9904	-0.3512	0.1343	-0.0434
11.0000	0.6980	-0.2413	0.0891	-0.0479
12.0000	0.4959	-0.1677	0.0602	-0.0479
13.0000	0.3549	-0.1176	0.0412	-0.0452
14.0000	0.2556	-0.0831	0.0286	-0.0409
15.0000	0.1852	-0.0591	0.0199	-0.0361
16.0000	0.1349	-0.0423	0.0140	-0.0312

```
CJMsamples = TableForm[Join[{{"S", "CJM Value", "CJM Delta", "CJM
Gamma"}},Table[Map[PaddedForm[N[#], {5, 3}]&, {S, AmericanPutCJM[S,
10, 0.2, 0.05, 5],AmericanPutDeltaCJM[S, 10, 0.2, 0.05, 5],AmericanPut-
GammaCJM[S, 10, 0.2, 0.05, 5]}], {S, 5, 15, 2}]]]
```

S	CJM Value	CJM Delta	CJM Gamma
5.000	5.000	-1.000	0.000
7.000	3.000	-1.000	0.000
9.000	1.419	-0.520	0.209
11.000	0.697	-0.241	0.089
13.000	0.354	-0.117	0.041
15.000	0.185	-0.059	0.020

Again we can see good agreement between our analytical model and the finite-difference scheme.

16.10 American Options - Ten-Year Expiry

Summary of Finite-Difference Results

In the following table we give a table of valuations, Deltas, Gammas and Thetas for a range of underlying prices with the strike at 10. (10 years to expiry, 20% volatility, 5% interest rate, 40 time-steps, 320 price-steps, $\alpha = 32$)

```
samples = TableForm[Table[Map[PaddedForm[N[#], {6, 4}]&, {S,
Valuation[10, 0.05, 0, S, 10, 0.2],
SORDelta[10, 0.05, 0, S, 10, 0.2],SORGamma[10, 0.05, 0, S, 10, 0.2],
SORTheta[10, 0.05, 0, S, 10, 0.2]}], {S, 2, 16, 1}]]
```

2.0000	8.0000	-1.0000	0.0000	0.0000
3.0000	7.0000	-1.0000	0.0000	0.0000
4.0000	6.0000	-1.0000	0.0000	0.0000
5.0000	5.0000	-1.0000	0.0000	0.0000
6.0000	4.0000	-1.0000	0.0000	0.0000
7.0000	3.0000	-1.0000	0.0000	0.0000
8.0000	2.1077	-0.7209	0.3121	-0.0057
9.0000	1.5187	-0.4779	0.1869	-0.0116
10.0000	1.1211	-0.3288	0.1179	-0.0153
11.0000	0.8437	-0.2327	0.0775	-0.0173
12.0000	0.6451	-0.1686	0.0526	-0.0181
13.0000	0.4998	-0.1245	0.0367	-0.0181
14.0000	0.3918	-0.0934	0.0262	-0.0176
15.0000	0.3101	-0.0710	0.0190	-0.0167
16.0000	0.2477	-0.0547	0.0140	-0.0156

For comparison, here are the corresponding results from the CJM approximation. We still get good agreement, even with 10-year maturity and only 40 time-steps in our numerical scheme.

```
CJMsamples = TableForm[Join[{{"S", "CJM Value", "CJM Delta", "CJM
Gamma"}},
Table[Map[PaddedForm[N[#], {5, 3}]&, {S, AmericanPutCJM[S, 10, 0.2,
0.05, 10],AmericanPutDeltaCJM[S, 10, 0.2, 0.05, 10],AmericanPutGamma-
CJM[S, 10, 0.2, 0.05, 10]}],
{S, 5, 15, 2}]]]
```

S	CJM Value	CJM Delta	CJM Gamma
5.000	5.000	-1.000	0.000
7.000	3.000	-1.000	0.000
9.000	1.527	-0.483	0.190
11.000	0.846	-0.234	0.078
13.000	0.501	-0.125	0.037
15.000	0.310	-0.071	0.019

We have shown that the 3TLD scheme gives results in agreement with the CJM approximation using very few time-steps, and that the agreement extends to the basic Greeks. The computation of Rho and Vega for American options, so far as this author is aware, requires recomputation with shifted values for the interest rate and volatility. A generalization to American options of the identities relating Rho to Delta and Vega to Gamma would be of considerable benefit!

The 3TLD scheme was recommended for non-smooth initial data at least forty years ago, but has been widely ignored in option-pricing. Given the universality of non-smooth initial data in option-pricing problems, it is perhaps appropriate to suggest that this should change. Of course, practitioners use other devices to get around the problem, such as smoothing the initial data, or performing some analytical smoothing for small times to expiry. The fully-implicit scheme is also quite good, but requires smaller steps for comparable accuracy. No other scheme combines the properties of smoothing and high order truncation error that are the characteristics of 3TLD. Its only real drawback is that the scheme must be managed carefully in the neighbourhood of discrete events such as dividends and coupons, or time-discontinuous shifts in payoffs - under such situations the scheme should either be re-started, or numerical data adjusted with care.

Chapter 16 Bibliography

Wilmott, P., Dewynne, J. and Howison, S., 1993, *Option Pricing - Mathematical Models and Computation*, Oxford Financial Press.

Chapter 17.
Linear Programming Alternatives to PSOR and Regression

The Dempster-Hutton LP Method, and Robust Regression via LP

17.1 Introduction

The PSOR method described in the previous chapters works, but raises questions of efficiency, even when employed with a cunning difference scheme. The number of time-steps is small, but there is quite a lot of effort per time-step to solve the problem. I am not aware of a complete answer to the question of what the most efficient solution method is - the standard *explicit* methods employing trees will be discussed in the next three chapters - but an interesting alternative for dealing with an implicit approach, with considerable potential, is the linear programming (LP) approach introduced recently by M. Dempster and J. Hutton (1996) - henceforth abbreviated LPDH. The idea is to map the constrained matrix equations into a linear programming problem, and then to solve them using standard, or, preferably, carefully optimized, LP techniques.

There are several different ways of solving LP problems, and it is possible to consider:

(a) exploitation of the dual problem;
(b) use of interior point methods instead of the simplex method;
(c) exploitation of the special structure of the problem due to the tridiagonal matrix equations.

This is very much research in progress, and the purpose of this chapter is really to perform the simplest task of implementing one version of the algorithm using *Mathematica*'s built-in LP solver, in order to illustrate the basic idea, and, hopefully, to stimulate efforts into further algorithm optimization. The author would be delighted to hear of much more efficient implementations.

Once we get familiar with the LP solver in *Mathematica*, we shall also show quickly how to use it to do robust regression analysis. Although not directly related to derivative modelling, fitting lines to data is commonplace in finance, and frequently the data are very noisy. LP methods allow one to implement one form of "robust" regression algorithm.

17.2 Naive Summary of LPDH Approach

What we want to do is to obtain an alternative to the PSOR routine

```
CompPSORSolve = Compile[
   {{diag, _Real, 0}, {offdiag, _Real, 0}, {r, _Real, 1},
   {eps, _Real, 0}, {relax, _Real, 0}, {kickoff, _Real, 1}},
   Module[{len = Length[r], yu = r,
    yold = kickoff, ynew = r, error = 100.0},
     While[error > eps^2,
 (Do[(yu[[i]] = r[[i]] / diag - (offdiag / diag) *
           (If[i == 1, 0, ynew[[i - 1]]]) + If[i < len, yold[[i + 1]], 0]);
  ynew[[i]] = Max[kickoff[[i]], yold[[i]] +
           relax * (yu[[i]] - yold[[i]])] ), {i, 1, len}];
  error = (ynew - yold) . (ynew - yold);
  yold = ynew)]; yold]];
```

for solving a constrained tridiagonal matrix problem. The matrix problem, that is the discretization of the diffusion equation, can be written formally in the form

$$A . y = b \tag{1}$$

where the diffusion equation is satisfied, and the condition

$$y \geq \text{payoff} \tag{2}$$

expresses the early exercise constraint. In this text we have not concerned ourselves with the formal manner in which these equations can be represented and combined mathematically. If we wish to extend (1) to cover both the possibility of early exercise and smooth diffusion evolution, we have to write it in the form

$$A . y \geq b \tag{3}$$

What this inequality means financially is that the return on a risk-neutral portfolio is at least that given by a bank deposit, leading to the Black-Scholes partial differential *inequality*, and formula (3) once discretized. We also have the condition that either the payoff inequality, or the (discretized) PDE inequality, must be satisfied as equality. Putting it all together, we get the conditions

$$y \geq \text{payoff}$$
$$A . y \geq b \tag{4}$$
$$\text{Min}[A . y - b, \ y - \text{payoff}] = 0$$

where the minimization applies component by component. This type of representation is known as an order complementarity problem, and a detailed discussion of the associated mathematics is given by Borwein and Dempster (1989). This type of formulation has been popularized by Wilmott *et al* (1993) who call it a linear complementarity problem.

Under certain circumstances, which amount to requiring that the log-price discrete step is sufficiently small, the class of schemes represented by the θ method are shown by Dempster and Hutton to be equivalent to an LP problem:

$$\text{Min}[c^T . y]$$
$$y \geq \text{payoff} \tag{5}$$
$$A . y \geq b$$

where c is a positive vector with the same length as the desired solution. This can be standardized further to the problem

$$\text{Min}[c^T . u]$$
$$u \geq 0 \qquad\qquad\qquad\qquad\qquad\qquad\qquad\qquad\qquad\qquad\qquad\qquad\qquad (6)$$
$$A . u \geq b - A . \text{payoff}$$

It will be evident that this is now an absolutely standard LP problem, which can be passed to a generic or suitably optimized solver. Clearly our next task is to understand the generic LP solver provided by *Mathematica*.

17.3 Linear Programming in *Mathematica*

There are three basic functions in *Mathematica* that deal with the task of finding the minimum or maximum of a linear function with a linear constraint set. Here is the first pair (**ConstrainedMax** is obvious from this one):

```
?ConstrainedMin
```

```
    ConstrainedMin[f, {inequalities}, {x, y, ... }] finds the global
       minimum of f in the domain specified by the inequalities.
       The variables x, y, ...  are all assumed to be non-negative.
```

These two are symbolic in their operation, and require the introduction of named symbols to operate. The third, just called **LinearProgramming**, just takes the numerical matrix entries:

```
? LinearProgramming
```

```
    LinearProgramming[c, m, b] finds the vector x which minimizes the
       quantity c.x subject to the constraints m.x >= b and x >= 0.
```

They can be used in equivalent ways - for example:

```
ConstrainedMin[x + 2 y - z, {x - y - z > 2 }, {x, y, z}]
```

```
    {2, {x → 2, y → 0, z → 0}}
```

```
LinearProgramming[{1, 2, -1}, {{1, -1, -1}}, {2}]
```

```
    {2, 0, 0}
```

We shall use **LinearProgramming** to implement the LPDH method. Later we shall see how to use **ConstrainedMin** to solve a regression problem.

17.4 Basic Implementation of LPDH

We now have all we need to implement a basic LPDH solution of the American option problem, and will use the Put as our working example. We shall do so on the basis of the two-time-level Douglas algorithm, where the A matrix is defined by the difference equation

$$(1 - 6\alpha)(u_{n-1}^{m+1} + u_{n+1}^{m+1}) + (10 + 12\alpha)u_n^{m+1} = (1 + 6\alpha)(u_{n-1}^m + u_{n+1}^m) + (10 - 12\alpha)u_n^m \qquad (7)$$

The right side of this form of the discretized diffusion equation is given by relations involving the matrix:

```
DougDMatrix[alpha_, vec_List] := Module[{temp},
temp = (10 - 12*alpha)*vec + (1+6*alpha)*(RotateRight[vec] + Rotate-
Left[vec]);
temp[[1]] = Simplify[First[temp] - (1+6*alpha)*Last[vec]];
temp[[-1]] = Simplify[Last[temp] - (1 + 6*alpha)*First[vec]];
temp]
```

Previously, for the left side of the discrete equation, we just wrote down the sequence of tridiagonal entries

```
DougCMatrix[alpha_, nminus_, nplus_] :=
Sequence[Table[1-6*alpha, {nplus+nminus-2}], Table[10+12*alpha,
{nplus+nminus-1}],
Table[1-6*alpha, {nplus+nminus-2}]]
```

or even just the elements of the diagonal and off-diagonal entries:

```
DougDiag[alpha_] := 10 + 12 alpha;
DougOffDiag[alpha_] := 1 - 6 alpha;
```

The first of these, for example, could be fed to the tridiagonal solver, to treat European-style options:

```
CompTridiagSolve =     Compile[
  {{a, _Real, 1}, {b, _Real, 1},     {c, _Real, 1}, {r, _Real, 1}},
  Module[{len = Length[r], solution = r,
    aux = 1 / (b[[1]]),     aux1 = r, a1 = Prepend[a, 0.0], iter},
    solution[[1]] = aux * r[[1]];    Do[aux1[[iter]] = c[[iter - 1]] aux;
      aux = 1 / (b[[iter]] - a1[[iter]] * aux1[[iter]]);
        solution[[iter]] = (r[[iter]] - a1[[iter]] solution[[iter - 1]]) aux
  {iter, 2, len}];
  Do[solution[[iter]] -= aux1[[iter + 1]] solution[[iter + 1]],
  {iter, len - 1, 1, -1}]; solution]];
```

and the second simpler form could be fed to the PSOR solver given earlier, for American-style implementations. In our implementation of the LPDH alternative for this case, we need a full matrix to supply to the LP solver, at least in its default implementation:

```
FullDougCMatrix[alpha_, nminus_, nplus_] :=
Table[Which[i == j, 10 + 12 * alpha,
   i == j - 1 || i == j + 1,   1 - 6 * alpha, True, 0],
  {i, 1, nplus + nminus - 1},
  {j, 1, nplus + nminus - 1}]
```

For example:

```
MatrixForm[FullDougCMatrix[1 / 2, 3, 3]]
```

$$\begin{pmatrix} 16 & -2 & 0 & 0 & 0 \\ -2 & 16 & -2 & 0 & 0 \\ 0 & -2 & 16 & -2 & 0 \\ 0 & 0 & -2 & 16 & -2 \\ 0 & 0 & 0 & -2 & 16 \end{pmatrix}$$

It is clear that for realistically sized problems, this particular implementation is not suitable, as we waste memory and time creating and filling a large array consisting mostly of zeros. However, here we are using it for illustration, and will now proceed to solve the Put equation. As a reminder, the changes of variables used to obtain the diffusion equation are, as before,

```
NonDimExpiry[T_, sd_] := sd² T / 2;
```
$$\text{NonDimExpiry}[T\_, sd\_] := \frac{sd^2\, T}{2};$$

```
kone[r_, sd_] := 2 r / sd²;   ktwo[r_, q_, sd_] := 2 (r - q) / sd²;
```
$$\text{kone}[r\_, sd\_] := \frac{2\,r}{sd^2};\quad \text{ktwo}[r\_, q\_, sd\_] := \frac{2\,(r - q)}{sd^2};$$

```
ValuationMultiplier[strike_, r_, q_, x_, tau_, sd_] := strike
```
$$\text{Exp}\left[-\frac{1}{2}\,(\text{ktwo}[r, q, sd] - 1)\, x - \left(\frac{1}{4}\,(\text{ktwo}[r, q, sd] - 1)^2 + \text{kone}[r, sd]\right)\, tau\right]$$

The boundary conditions for a Put are expressed by

```
g[x_, tau_, r_, q_, sd_] := 0;
```
$$f[x\_, tau\_, r\_, q\_, sd\_] := \text{Exp}\left[\frac{1}{2}\,(\text{ktwo}[r, q, sd] - 1)\, x + \right.$$
$$\left. \frac{1}{4}\,((\text{ktwo}[r, q, sd] - 1)^2 + 4\,\text{kone}[r, sd])\, tau\right] - \text{Exp}\left[\right.$$
$$\left. \frac{1}{2}\,(\text{ktwo}[r, q, sd] + 1)\, x + \frac{1}{4}\,((\text{ktwo}[r, q, sd] - 1)^2 + 4\,\text{kone}[r, sd])\, tau\right];$$

Initial conditions are defined by

```
PutExercise[x_, r_, q_, sd_] :=
```
$$\mathrm{Max}\Big[\mathrm{Exp}\Big[\frac{1}{2}\ (\mathrm{ktwo}[r,\ q,\ sd]\ -\ 1)\ x\Big]\ (1\ -\ \mathrm{Exp}[x]),\ 0\Big];$$

This is a special case of the general payoff constraint function, which is given by

```
PutPayoff[x_, tau_, r_, q_, sd_] :=
  Exp[((ktwo[r, q, sd] - 1)^2/4 + kone[r, sd])*tau]*
Max[Exp[(ktwo[r, q, sd] - 1)*x/2] (1 - Exp[x]), 0];
```

Now let's build a sample solution, with $\alpha = 8$, first using the old PSOR routine in its standard implementation. The initialization is essentially the same whether we use PSOR or LPDH, and is set up for a one-year-expiry American Put option with zero continuous dividend yield.

```
M = 20; nminus = 60; nplus = 60; dx = 0.05; dtau = 0.001;
alpha = dtau / dx^2;
initial = Table[PutExercise[(k - 1 - nminus) dx, 0.05, 0, 0.2],
   {k, nminus + nplus + 1}];
lower = Table[f[-nminus dx, (m - 1) dtau, 0.05, 0, 0.2], {m, 1, M + 1}];
upper = Table[g[+nplus dx, (m - 1) dtau, 0.05, 0, 0.2], {m, 1, M + 1}];
wold = initial; wvold = wold; wnew = wold;
CMat = DougCMatrix[alpha, nminus, nplus];
relax = 1.1;
tolerance = 0.000001;
```

So first we do the evolution with PSOR, and time it (266 MHz G3 results);

```
Timing[For[m=2, m<=M+1, m++,
(wvold = wold;
wold = wnew;
rhs = DougDMatrix[alpha, Take[wold, {2, -2}]]+
Table[
Which[
k==1, (6*alpha+1)*lower[[m-1]] + (6*alpha-1)*lower[[m]],
k== nplus + nminus-1, (6*alpha+1)*upper[[m-1]] + (6*alpha--
1)*upper[[m]],
True, 0],
{k, 1, nplus + nminus-1}];
payoff = Table[PutPayoff[(k-1-nminus)*dx, (m-1)*dtau, 0.05, 0, 0.2],
{k, 2, nminus+nplus}];
temp = CompPSORSolve[DougDiag[alpha], DougOffDiag[alpha], rhs, toler-
ance, relax, payoff];
wnew = Join[{lower[[m]]}, temp, {upper[[m]]}])
]]
```

```
{2.41667 Second, Null}
```

Next we do an interpolation to supply a continuous function:

```
interpoldatab =
   Table[{(k - nminus - 1)*dx, wnew[[k]]}, {k, 1, nminus + nplus +
1}];
ufuncb = Interpolation[interpoldatab, InterpolationOrder -> 3];
Valuation[strike_, r_, q_, S_, T_, sd_] :=
   ValuationMultiplier[strike, r, q, Log[S/strike], (sd^2*T)/2, sd]*
      ufuncb[Log[S/strike]]
```

Next we do a valuation plot and tabulation of sample values:

```
Plot[Valuation[10, 0.05, 0, S, 1, 0.2],
  {S, 2, 18}, PlotPoints → 50, PlotRange → All];
```

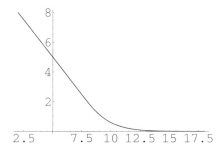

```
samples = TableForm[Join[{{"S", "Douglas PSOR"}},
  Table[(PaddedForm[N[#1], {5, 4}] &) /@
    {S, Valuation[10, 0.05, 0, S, 1, 0.2]},
  {S, 2, 16, 1}]]]
```

S	Douglas PSOR
2.0000	8.0000
3.0000	7.0000
4.0000	6.0000
5.0000	5.0000
6.0000	4.0000
7.0000	3.0000
8.0000	1.9968
9.0000	1.1462
10.0000	0.6051
11.0000	0.2953
12.0000	0.1344
13.0000	0.0576
14.0000	0.0235
15.0000	0.0093
16.0000	0.0035

These values are approximately consistent with those given in Section 16.7 - it should be noted that here, for our experiments, we are using rather fewer stock-price-steps.

Evolution with LPDH - Simple Version

The initialization is almost identical, except that we compute and store the matrix, and the *c*-vector to be used in the objective function - here this is taken to be a vector of unit elements. The initialization takes longer as we are setting up a big square array for the constraint matrix - this and the definition of the default objective function are highlighted by a comment.

```
M = 20; nminus = 60; nplus = 60; dx = 0.05; dtau = 0.001;
alpha = dtau / dx^2;
initial = Table[PutExercise[(k - 1 - nminus) dx, 0.05, 0, 0.2],
  {k, nminus + nplus + 1}];
lower = Table[f[-nminus dx, (m - 1) dtau, 0.05, 0, 0.2], {m, 1, M + 1}];
upper = Table[g[+nplus dx,  (m - 1) dtau, 0.05, 0, 0.2], {m, 1, M + 1}];
wold = initial; wvold = wold;
wnew = wold;
(* Creation of the constraint matrix and objective function for LP *)
AMat = FullDougCMatrix[alpha, nminus, nplus];
cvec = Table[1, {k, 1, nminus + nplus - 1}];
```

Next we do the evolution with a simple implementation of LPDH. Note that only two lines have changed - these are highlighted by comments:

```
Timing[For[m=2, m<=M+1, m++,
(wvold = wold;
wold = wnew;
rhs = DougDMatrix[alpha, Take[wold, {2, -2}]]+
Table[
Which[
k==1, (6*alpha+1)*lower[[m-1]] + (6*alpha-1)*lower[[m]],
k== nplus + nminus-1, (6*alpha+1)*upper[[m-1]] + (6*alpha--
1)*upper[[m]],
True, 0],
{k, 1, nplus + nminus-1}];
payoff = Table[PutPayoff[(k-1-nminus)*dx, (m-1)*dtau, 0.05, 0, 0.2],
{k, 2, nminus+nplus}];
(* The next two lines replace the PSOR with LPDH *)
fullrhs = rhs - AMat.payoff;
temp = payoff + LinearProgramming[cvec, AMat, fullrhs];
wnew = Join[{lower[[m]]}, temp, {upper[[m]]}])
]]
```

```
    {56.2833 Second, Null}
```

We shall discuss the efficiency presently - first we need to check the result. We do an interpolation to supply a continuous function:

```
interpoldatab =
   Table[{(k - nminus - 1)*dx, wnew[[k]]}, {k, 1, nminus + nplus +
1}];
ufuncb = Interpolation[interpoldatab, InterpolationOrder -> 3];
Valuation[strike_, r_, q_, S_, T_, sd_] :=
   ValuationMultiplier[strike, r, q, Log[S/strike], (sd^2*T)/2, sd]*
   ufuncb[Log[S/strike]]
```

As before, we next do a valuation plot and tabulation of sample values:

```
Plot[Valuation[10, 0.05, 0, S, 1, 0.2],
  {S, 2, 18}, PlotPoints → 50, PlotRange → All];
```

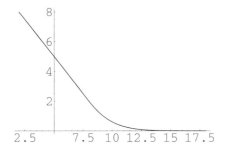

```
samples = TableForm[Join[{{"S", "Douglas LPDH"}},
 Table[(PaddedForm[N[#1], {5, 4}] &) /@
  {S, Valuation[10, 0.05, 0, S, 1, 0.2]},
 {S, 2, 16, 1}]]]
```

S	Douglas LPDH
2.0000	8.0000
3.0000	7.0000
4.0000	6.0000
5.0000	5.0000
6.0000	4.0000
7.0000	3.0000
8.0000	1.9968
9.0000	1.1462
10.0000	0.6051
11.0000	0.2953
12.0000	0.1344
13.0000	0.0576
14.0000	0.0235
15.0000	0.0093
16.0000	0.0035

We note that the results are *identical*, to the precision given, with those from the PSOR approach, which constitutes a crude numerical verification of the LPDH algorithm.

Towards Efficient Implementations of LPDH

The example above illustrates the ease with which we can try out a new algorithm with *Mathematica*, bearing in mind the rich supply of built-in functions. However, this particular implementation of LPDH and the timing comparison with PSOR are maximally idiotic and unfair respectively! Our PSOR routine was carefully coded and compiled to exploit the constant and tridiagonal nature of the constraint matrix, whereas we have built a huge square matrix, mostly of zeros, to feed to the standard built-in LP function, which expects a matrix. The important point is that the two methods give identical answers, and no particular significance should be attached to this particular timing comparison.

In particular, we have not used the tridiagonal nature of the matrix A at all. It was very explicitly exploited in the PSOR solver we have described. The efficient implementation of LPDH relies on a form of the simplex method tailored to constraint matrices that are tridiagonal. A discussion of this is beyond the scope of this text, as is a proper treatment of the application of duality and interior point methods, but work in progress suggests that a massive improvement in speed can be obtained by a carefully optimized LP solver (Dempster *et al* 1997).

17.5 LP and Robust Regression

As we are exploring *Mathematica*'s LP functions, it is a good point at which to introduce the subject of robust regression. One form of "robust" technique is L^1 regression, and it is easy to use the **ConstrainedMin** function to implement it symbolically.

Conventional regression is based on the method of least squares. This is highly prone to the influence of outliers. An alternative is to use the method of "least absolute deviation", where the fit is determined by minimizing the sum of the absolute deviations of the fitted and actual values. It is well known how to transform this problem into an LP form. Given the data, we work out the number ("len") of observations. Then we isolate the x-values and the y-values, by transposition if we are given a list of pairs, and invent 2 len variables $n[i]$, $p[i]$ to plug into the LP solver, where we minimize the sum of the $n[i]$ and $p[i]$ subject to the constraints that the observed $y[i]$ differs from the fitted $y[i]$ by $n[i] - p[i]$. The slope and intercept are free to take positive or negative values, but the $n[i]$ and $p[i]$ are non-negative. The LP solver requires that all variables are non-negative, so we write the intercept, for example, as $a = \text{ap} - \text{an}$, with $\text{ap} \geq 0$ and $\text{an} \leq 0$. The following module takes care of all this, and nicely illustrates the symbolic LP solver, **ConstrainedMin**, in action:

```
LADLPFit[data_?MatrixQ, x_] :=
Module[{tr=Transpose[data],len, nvar,n,pvar,p,ap,an,bp,bn, vbls, min-
fun, ineq, lhs}, len = Length[tr[[1]]];
(* Invent all the LP variables *)
nvar = Array[n, len]; pvar = Array[p, len];
vbls = Join[nvar,pvar,{ap,an,bp,bn}];
(* Write objective function as sum of absolute deviations *)
minfun = Apply[Plus, Join[nvar,pvar]];
lhs = (ap-an)*Table[1,{len}]+(bp-bn)*tr[[1]]+nvar-pvar;
ineq = MapThread[#1==#2&,{lhs,tr[[2]]}];
ap-an+(bp-bn)*x /. ConstrainedMin[minfun,ineq,vbls][[2]]]
```

To visualize the answers, we introduce the following function to plot the data and the fitted function (of x).

```
plotfit[fn_, data_] :=
Module[{minind, maxind, mindep, maxdep},
minind = Min[Transpose[data][[1]]];
maxind = Max[Transpose[data][[1]]];
mindep = Min[Transpose[data][[2]]];
maxdep = Max[Transpose[data][[2]]];
Plot[fn, {x, Floor[minind], Ceiling[maxind]},
AxesOrigin -> {Floor[minind], 0},
PlotRange -> {Min[0, Floor[mindep]], Ceiling[maxdep] + 1},
Evaluate[Epilog -> {PointSize[0.02], Map[Point, data]}]]]
```

The Pilot Plant Example

The book by Rousseeuw and Leroy (1987) gives some interesting examples that we can use to check our approach and illustrate the method. Here are the basic clean "Pilot Plant" data described by Daniel and Wood (1971):

```
rdata = {{123, 76}, {109, 70}, {62, 55}, {104, 71}, {57, 55}, {37,
48},
{44, 50}, {100, 66}, {16, 41}, {28, 43}, {138, 82}, {105, 68}, {159,
88}, {75, 58}, {88, 64}, {164, 88}, {169, 89}, {167, 88}, {149, 84},
{167, 88}};
```

First we do a standard least-squares fit, and look at it.

```
lsqfit = Fit[rdata, {1, x}, x]
```

   35.4583 + 0.321608 x

```
plotfit[lsqfit, rdata];
```

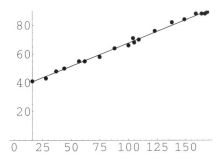

In this case the LP solution is given in exact form, and is very close to the least squares solution.

```
LADLPFit[rdata, x]
```

$$\frac{1329}{37} + \frac{47\,x}{148}$$

```
%//N
```

$$35.9189 + 0.317568\,x$$

Making an Outlier

We move one of the points to create an outlier:

```
rdata[[6]] = {370, 48};
```

The least squares approach now fails dramatically:

```
lsqfit = Fit[rdata, {1, x}, x]
```

$$58.9388 + 0.0807115\,x$$

```
plotfit[lsqfit, rdata];
```

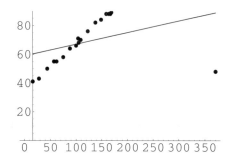

The LAD approach still tracks the main trend.

```
pilotladfit = LADLPFit[rdata, x]//N
```

$$36.4065 + 0.308943\,x$$

```
plotfit[pilotladfit, rdata];
```

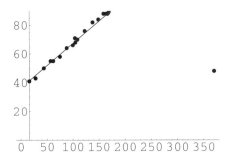

This is one illustration of the power of robust methods - for others see Rousseeuw and Leroy (1987), who present many techniques. The approach given above is easily generalized to many variables (Shaw and Wilson, 1998).

Chapter 17 Bibliography

Borwein, J.M. and Dempster, M.A.H., 1989, The linear order complementarity problem, *Mathmeatics of OR*, 14, p. 534.

Daniel, C. and Wood, F.S., 1971, *Fitting Equations to Data*, John Wiley and Sons.

Demspter, M.A.H. and Hutton, J.P., 1996, Pricing American options by linear programming, submitted to *Mathematical Finance*.

Demspter, M.A.H., Richards, D.G. and Hutton, J.P., 1997, LP valuations of exotic American options exploiting structure, Judge Institute, Cambridge, working paper.

Rousseeuw, P.J. and Leroy, A.M., 1987, *Robust Regression and Outlier Detection*, John Wiley and Sons.

Shaw, W.T. and Wilson, R., 1998, Robust regression in *Mathematica* - submitted to the *Mathematical Journal*.

Wilmott, P., Dewynne, J. and Howison, S., 1993, *Option Pricing - Mathematical Models and Computation*, Oxford Financial Press.

Chapter 18.
Traditional and Supersymmetric Trees

A Re-specification of Trees by Symmetry Methods,
and Lifting the Curse of "Negative Probability"

18.1 Introduction

The discussion of trees is an awkward topic within a text employing *Mathematica,* and taking a largely continuum-PDE-oriented approach. For a large class of derivatives that are specifiable by a partial differential equation, there are usually more efficient implicit finite-difference methods that can be used. On the other hand, tree systems may sometimes be a good way of defining a new model, for which a continuum PDE model may or may not emerge. Secondly, it is very easy and elegant to write down brief and concise code to model trees within *Mathematica*, but these particular models are rather inefficient in execution. One can define uglier code that executes rather more quickly, and we shall do so in Chapter 20.

There are two standard families of models of (binomial) trees in common use. The first family is based on the original Cox, Ross and Rubenstein model (Cox, Ross and Rubenstein, 1979), involving a binomial model with certain up and down states and probabilities associated with them. A variant family developed by Jarrow and Rudd (1983) is also in common use. We shall refer to these families as the CRR and JR families respectively, using obvious abbreviations. One can construct trinomial variants of either of these families. I use the term "family" rather than "model" as each family contains at least two models, though that is perhaps not quite how their originators might see it! The two basic variants are, in each case, obtained by considering an exact and an approximate solution of the constraints laid down for each family. Both the CRR and JR "approximate" binomial models are described by Hull (1996), who in a footnote makes reference also to an "exact" form of CRR. The exact forms of both families are those described by Wilmott *et al* (1995). A significant difference between the two models is that the "up" and "down" "probabilities" in the CRR model can become negative if the time-step is large enough, whereas the JR probabilities are always 0.5. The abscissas of the exact JR model can also become negative. The CRR model is, if anything, used more frequently, but neither family is entirely satisfactory - Wilmott *et al* (1995) note that either method can fail if the time-steps are such that the weights or abscissas become negative. A notable exception is the "approximate" Jarrow-Rudd model - we shall see how this model emerges as the *exact* solution of a new set of constraints based on symmetry considerations, along with other new models with non-negative weights and abscissas.

Our discussion of trees is also complicated by the desire of the author to attempt to develop a more systematic approach to tree models. I hope by now that the reader will have learnt to forgive the author for criticizing almost every numerical method known (gripes about Monte Carlo forthcoming), but the literature on trees appears dominated by a certain probabilistic approach. It turns out to be neither necessary nor consistent to have a probabilistic view. That the probabilistic approach is inconsistent is manifest in the original models, as the "probabilities" can become negative, as has been said. Here we shall demonstrate that the probabilistic methodology is also completely unnecessary. We will present a new approach, based on symmetry relations associated with simple solutions of the Black-Scholes equation, which contains as special cases all the well-known models, but which also allows us to construct new models. The most interesting new models are supersymmetric in a sense which we shall define, and do not suffer from any problems with negative weights or abscissas. Our approach does allow the derivation of the traditional models, but it also makes it transparent that these are sub-optimal.

In order to give a complete presentation I will devote three chapters to tree models, covering the following three topics:

(a) (this chapter) A general and rigorous re-specification of tree models, including what we argue is a more systematic approach;
(b) the elegant recursive tree models developed by Benninga *et al* (1993) and generalizations, and overview of what can go wrong (Chapter 19);
(c) how to optimize tree models at the price of some elegance (Chapter 20).

I can thoroughly recommend the article in *The Mathematica Journal* by Simon Benninga, Raz Steinmetz and John Stroughair (1993) for an introduction to this type of scheme and to Monte Carlo schemes. Several of my own studies were in part motivated by that article, and by their elegant approach to tree modelling in particular. The approach taken in the next chapter will be based in part on the methods introduced by Benninga *et al*. However, as well as specifying the tree theory rather more carefully than is traditional, we shall also be seeking generalizations that allow us to

(a) compute hedge parameters;
(b) deal with trinomial or other systems;
(c) proceed more efficiently.

The first two of these topics will be addressed in Chapter 19, while (c) is deferred to Chapter 20. Also, and in the spirit of this text, we shall first investigate how binomial and trinomial systems emerge from a solution-symmetry approach. The paper by Benninga *et al* motivates matters from a binomial-probabilistic viewpoint, which is discussed by many authors - this is the view we argue is somewhat incomplete. Here we shall take a complementary approach.

18.2 A General Definition of a Tree Rule

A "tree rule" is a backwards evolution scheme of the form

$$V(S, t) = e^{-r \Delta t} \sum_{i=1}^{n} w_i \, V(S \, u_i, t + \Delta t) \tag{1}$$

The variables w_i are called the "weights", and the points $S \, u_i$ are called the "abscissas". If $n = 2$ we have a general binomial scheme, whereas if $n = 3$ we have a general trinomial scheme. Some authors may use the term "binary" rather than binomial. Explicitly, for a binomial scheme

$$V(S, t) = e^{-r \Delta t} \, [w_1 \, V(S \, u_1, t + \Delta t) + w_2 \, V(S \, u_2, t + \Delta t)] \tag{2}$$

Basic Solutions of the Black-Scholes Equation

What we shall do now is to write down some basic solutions of the Black-Scholes equation that we wish to be exactly consistent with the tree rule. Clearly the choice is somewhat arbitrary. One question is, how many solutions to use? For a binomial scheme, we have two weights, two abscissas, and the parameter Δt, making five in total. For a trinomial scheme, there are three weights, three abscissas and the time-step, totalling seven. It is traditional to allow the time-step to remain free - we shall do so for the present, though we shall see in the next chapter that there are all kinds of good reasons for picking the time-step very carefully. That leaves us with four unknowns for a binomial scheme, and six for trinomial.

What we shall demand is that the tree rule is such that it is consistent with certain power solutions of the Black-Scholes equation. Recall, from Chapter 4, Section 4.2, that the general power solution is of the form

$$V = Z \left(\frac{S}{K}\right)^n e^{-r(T-t)} e^{\left((r-q)n + \frac{n(n-1)\sigma^2}{2}\right)(T-t)} \tag{3}$$

The special cases for $n = 0, 1, 2$ are

$$V = Z e^{-r(T-t)} \tag{4}$$

$$V = Z \left(\frac{S}{K}\right) e^{-r(T-t)} e^{(r-q)(T-t)} \tag{5}$$

$$V = Z \left(\frac{S}{K}\right)^2 e^{-r(T-t)} e^{(2(r-q)+\sigma^2)(T-t)} \tag{6}$$

Considering the first negative powers, for $n = -1, -2$ we have

$$V = Z \left(\frac{K}{S}\right) e^{-r(T-t)} e^{(-(r-q)+\sigma^2)(T-t)} \tag{7}$$

$$V = Z \left(\frac{K}{S}\right)^2 e^{-r(T-t)} e^{(-2(r-q)+3\sigma^2)(T-t)} \tag{8}$$

Let's demand that the binomial rule be consistent with some of the power solutions. We call a tree satisfying some subset of these consistency conditions a "symmetric tree", because the mapping defined by the tree rule preserves, or leaves invariant, the solution. In general, making the substitution and simplifying, this gives us

$$w_1 u_1^n + w_2 u_2^n = e^{\left(\frac{1}{2}\sigma^2 n(n-1)+n(r-q)\right)\Delta t} \tag{9}$$

and for $n = 0, 1, 2$ this gives

$$w_1 + w_2 = 1 \tag{10}$$

$$w_1 u_1 + w_2 u_2 = e^{(r-q)\Delta t} \tag{11}$$

$$w_1 u_1^2 + w_2 u_2^2 = e^{(2(r-q)+\sigma^2)\Delta t} \tag{12}$$

The first pair of negative powers give us the equations, for $-1, -2$ respectively,

$$\frac{w_1}{u_1} + \frac{w_2}{u_2} = e^{(-(r-q)+\sigma^2)\Delta t} \tag{13}$$

$$\frac{w_1}{u_1^2} + \frac{w_2}{u_2^2} = e^{(-2(r-q)+3\sigma^2)\Delta t} \tag{14}$$

The condition that the weights sum to unity is traditionally regarded as arising from the fact that these weights are "up" and "down" probabilities. We see that such a view is unnecessary - it suffices to demand that "cash is conserved", i.e., that the tree preserves the simple cash ($n = 0$) solution to the Black-Scholes PDE.

It is clear that we can generate weights and abscissas consistent with up to four simple solutions of the Black-Scholes equation. A scheme preserving a full *four* of the solutions will be called "supersymmetric". First, however, we need to see how some standard models arise, and to note that these are only partially symmetric.

Schemes of Jarrow-Rudd Type

We impose the condition

$$w_1 = w_2 = \frac{1}{2} \tag{15}$$

which solves the cash conservation condition, leaving us with two unknowns and two conditions based on the first and second power solutions:

$$u_1 + u_2 = 2\,e^{(r-q)\Delta t} \tag{16}$$

$$u_1^2 + u_2^2 = 2\,e^{(2(r-q)+\sigma^2)(T-t)} \tag{17}$$

These can be solved with pen and paper, by solving a quadratic equation, but we can be lazy and get *Mathematica* to do the work.

```
soln = Solve[
    {u₁ + u₂ == 2 Exp[(r - q) Δt], u₁² + u₂² == 2 Exp[(2 (r - q) + σ²) Δt]}, {u₁, u₂}];
```

The larger of the two solutions (here u_2) is now called u, as is conventional. (In the following the output has been converted to Traditional Form and set as a numbered equation - this is not quite WYSIWMAMA)

```
u = Simplify[PowerExpand[Simplify[u₂ /. soln〖1〗]]]
```

$$e^{(r-q)\Delta t}\left(1 + \sqrt{e^{\Delta t \sigma^2} - 1}\right) \tag{18}$$

```
d = Simplify[PowerExpand[Simplify[u₁ /. soln[[1]]]]]
```

$$e^{(r-q)\Delta t}\left(1 - \sqrt{e^{\Delta t \sigma^2} - 1}\right) \tag{19}$$

These give us the "exact" forms of the up and down states in the Jarrow-Rudd model. Note that the d variable can become zero if the time-step is large enough. We need

$$\sigma^2 \, \Delta t << \log(2)$$ (20)

for this particular scheme to be sensible. However, there is an interesting approximation, which is the version described by Hull (1996), and appears to be the form originally given by Jarrow and Rudd. This is based on comparing the expansions for

```
1+Series[Sqrt[Exp[x]-1], {x, 0, 1}]
```

$$1 + \sqrt{x} + O[x]^{3/2}$$

and for

```
1-Series[Sqrt[Exp[x]-1], {x, 0, 1}]
```

$$1 - \sqrt{x} + O[x]^{3/2}$$

with those for the pair

```
Series[Exp[Sqrt[x]-x/2], {x, 0, 1}]
```

$$1 + \sqrt{x} + O[x]^{3/2}$$

```
Series[Exp[-Sqrt[x]-x/2], {x, 0, 1}]
```

$$1 - \sqrt{x} + O[x]^{3/2}$$

In the approximate (original) form of the Jarrow-Rudd model, the up and down scale factors are written as

$$u \; = \; e^{(r-q-\sigma^2/2)\Delta t} \, e^{\sigma \sqrt{\Delta t}} \qquad d \; = \; e^{(r-q-\sigma^2/2)\Delta t} \, e^{-\sigma \sqrt{\Delta t}}$$ (21)

This then conveniently avoids the down scale factor becoming negative, but this approximation violates the matching of both the solutions in S and S^2. We shall, nevertheless, see later how this "approximation" can be justified exactly from the symmetry method, even though it does not quite fit our considerations so far.

Schemes of Cox-Ross-Rubenstein Type

To get schemes similar to those of CRR, we impose the conditions

$$u_1 \; = \; 1/u_2$$ (22)

leaving us with two unknowns and conditions:

$$w_1 \, u_1 + (1 - w_1) \, 1/u_1 \; = \; e^{(r-q)\Delta t}$$ (23)

$$w_1 \, u_1^2 + (1 - w_1) \, 1 / u_1^2 \;=\; e^{(2\,(r-q)+\sigma^2)\Delta t} \tag{24}$$

Clearly the first equation allows us to write w_1 as an explicit function of u_1:

$$w_1 \;=\; \frac{e^{(r-q)\Delta t} - u_2}{u_1 - u_2} \;=\; \frac{e^{(r-q)\Delta t} - \frac{1}{u_1}}{u_1 - \frac{1}{u_1}} \tag{25}$$

The second equation allows us to equate this also to

$$w_1 \;=\; \frac{e^{(2\,(r-q)+\sigma^2)\Delta t} - u_2^2}{u_1^2 - u_2^2} \;=\; \frac{e^{(2\,(r-q)+\sigma^2)\Delta t} - \frac{1}{u_1^2}}{u_1^2 - \frac{1}{u_1^2}} \tag{26}$$

If we re-organize these two expressions for w_1 a little, we obtain the quadratic equation

$$u_1^2 - \left(e^{(\sigma^2 + (r-q))\Delta t} + e^{-(r-q)\Delta t}\right) u_1 + 1 = 0 \tag{27}$$

```
soln = u₁ /.
    Solve[u₁² - (Exp[((r - q) + σ^2) Δt] + Exp[-(r - q) Δt]) u₁ + 1 == 0, u₁];

{d, u} = soln;

u
```

$$\frac{1}{2}\left(E^{(q-r)\,\Delta t} + E^{\Delta t\,(-q+r+\sigma^2)} + \sqrt{-4 + \left(-E^{(q-r)\,\Delta t} - E^{\Delta t\,(-q+r+\sigma^2)}\right)^2}\,\right)$$

```
d
```

$$\frac{1}{2}\left(E^{(q-r)\,\Delta t} + E^{\Delta t\,(-q+r+\sigma^2)} - \sqrt{-4 + \left(-E^{(q-r)\,\Delta t} - E^{\Delta t\,(-q+r+\sigma^2)}\right)^2}\,\right)$$

So far we have described the "exact" form of the CRR model. However, the form of the up and down factors for small time-intervals is interesting:

```
PowerExpand[Normal[Series[{u, d}, {Δt, 0, 1}]]]
```

$$\left\{ 1 + \sqrt{\Delta t}\,\sigma + \frac{\Delta t\,\sigma^2}{2}, \; 1 - \sqrt{\Delta t}\,\sigma + \frac{\Delta t\,\sigma^2}{2} \right\}$$

They are identical, to this order, to those for $e^{\sigma\sqrt{\Delta t}}$ and its inverse:

```
PowerExpand[
 Normal[Series[{Exp[σ Sqrt[Δt]], Exp[-σ Sqrt[Δt]]}, {Δt, 0, 1}]]]
```

$$\left\{1 + \sqrt{\Delta t}\ \sigma + \frac{\Delta t\ \sigma^2}{2},\ 1 - \sqrt{\Delta t}\ \sigma + \frac{\Delta t\ \sigma^2}{2}\right\}$$

This gives an approximate form of the CRR model that is in common use. Note however that if we work with the exact forms, the constant, asset and asset-squared solutions are matched exactly. Jarrow and Rudd (1983) commented that alternative approximate forms of the CRR model do not satisfy the quadratic conditions (in their approach, expressed statistically as the second moment). This is only a feature of the approximation - there is no fundamental problem with a solution of CRR type matching the second moment, as we have shown. A drawback of the CRR form is that the weight factor can become zero - at least it is a drawback if a probabilistic interpretation is used. This is not avoided by the use of the approximate form. Note that Hull (1996) describes the original approximate form in detail, and makes references to the exact form in a footnote. The exact forms of both the CRR and the JR models are the ones described by Wilmott *et al* (1995).

18.3 Supersymmetric Schemes?

The question arises whether we can build other schemes that are supersymmetric. We can pick one other solution to the PDE, and arrange both the weights and abscissas so that this solution is preserved. The question arises which one to take. We can only rely on common sense here, and some experimentation with different possibilities is helpful. We note that in some situations, for example Put options, we wish to model cases where the option value tends to zero for large values of the underlying. This suggests that it might be useful to consider setting the tree parameters so that the solution for a negative power of S is preserved exactly. If we use the first negative power, this creates a supersymmetric tree rule, and will turn out to have an interesting impact on the behaviour of the weights and abscissas.

With this choice, we have the four unknowns satisfying the four equations (the last one is the new constraint that replaces the *ansatz* used by CRR or JR)

$$w_1 + w_2 = 1 \tag{28}$$

$$w_1\, u_1 + w_2\, u_2 = e^{(r-q)\Delta t} \tag{29}$$

$$w_1\, u_1^2 + w_2\, u_2^2 = e^{(2(r-q)+\sigma^2)\Delta t} \tag{30}$$

$$\frac{w_1}{u_1} + \frac{w_2}{u_2} = e^{(-(r-q)+\sigma^2)\Delta t} \tag{31}$$

Eliminating w_2 leads to

$$w_1\, u_1 + (1-w_1)\,u_2 = e^{(r-q)\Delta t} \tag{32}$$

$$w_1\, u_1^2 + (1-w_1)\,u_2^2 = e^{(2(r-q)+\sigma^2)\Delta t} \tag{33}$$

$$\frac{w_1}{u_1} + \frac{(1-w_1)}{u_2} = e^{(-(r-q)+\sigma^2)\Delta t} \tag{34}$$

Each of these can be solved for w_1, leading to the relations

$$w_1 = \frac{e^{(r-q)\Delta t} - u_2}{u_1 - u_2} = \frac{e^{(2(r-q)+\sigma^2)\Delta t} - u_2^2}{u_1^2 - u_2^2} = \frac{e^{(-(r-q)+\sigma^2)\Delta t} - 1/u_2}{1/u_1 - 1/u_2} \tag{35}$$

The last of these three can be re-organized, leading to

$$\frac{e^{(r-q)\Delta t} - u_2}{u_1 - u_2} = \frac{e^{(2(r-q)+\sigma^2)\Delta t} - u_2^2}{u_1^2 - u_2^2} = \frac{-(u_1 \, u_2 \, e^{(-(r-q)+\sigma^2)\Delta t} - u_1)}{u_1 - u_2} \tag{36}$$

Now multiplying through by the common denominator gives us

$$e^{(r-q)\Delta t} - u_2 = \frac{e^{(2(r-q)+\sigma^2)\Delta t} - u_2^2}{u_1 + u_2} = u_1 - u_1 \, u_2 \, e^{(-(r-q)+\sigma^2)\Delta t} \tag{37}$$

Now we use the equality of the left and right sides of (37) to get the condition

$$(u_1 + u_2) - u_1 \, u_2 \, e^{(-(r-q)+\sigma^2)\Delta t} = e^{(r-q)\Delta t} \tag{38}$$

Equality of the left and middle terms of (37) gives the condition

$$(u_1 + u_2) \, e^{(r-q)\Delta t} - u_1 \, u_2 = e^{(2(r-q)+\sigma^2)\Delta t} \tag{39}$$

This is just a pair of simultaneous linear equations for the sum and the product. The solution for the sum and product is

$$u_1 + u_2 = e^{(r-q)\Delta t}\left(1 + e^{\sigma^2 \Delta t}\right) \tag{40}$$

$$u_1 \, u_2 = e^{2(r-q)\Delta t} \tag{41}$$

It is evident that we can write

$$u_i = e^{(r-q)\Delta t} \, v_i \tag{42}$$

where the v_i satisfy

$$v_1 + v_2 = 1 + e^{\sigma^2 \Delta t} \tag{43}$$

$$v_1 \, v_2 = 1 \tag{44}$$

So these abscissas satisfy a reciprocity condition similar to that in CRR, but the linear drift factor has been incorporated into the abscissas in the same spirit as JR. The solution for the v_i is now just a matter of solving a quadratic equation of the form

$$v^2 - \left(e^{\sigma^2 \Delta t} + 1\right) v + 1 = 0 \tag{45}$$

This has two roots - we extract them in the usual way, multiply back by the drift factor, and finally get the "up" and "down" scale factors - up is defined by taking the larger of the two roots.

$$u = \frac{1}{2} \, e^{(r-q)\Delta t} \left(e^{\sigma^2 \Delta t} + 1 + \sqrt{(e^{\sigma^2 \Delta t} + 1)^2 - 4}\right) = e^{(r-q)\Delta t} \, \alpha \tag{46}$$

$$d = \frac{1}{2}\, e^{(r-q)\Delta t}\left(e^{\sigma^2\,\Delta t} + 1 - \sqrt{(e^{\sigma^2\,\Delta t} + 1)^2 - 4}\,\right) = e^{(r-q)\Delta t}\,\beta \qquad (47)$$

These relations serve to define α, β. The up and down weights are given by

$$p = w_1 = \frac{e^{(r-q)\Delta t} - d}{u - d} = \frac{1 - \beta}{\alpha - \beta}$$

$$q = 1 - p = w_2 = \frac{\alpha - 1}{\alpha - \beta} \qquad (48)$$

Properties of the New Model

This new model has some interesting theoretical features:

(1) The weights/probabilities are independent of the interest rate and yield - they only depend on the volatility and time-step.

(2) Consider the down scale function term β. Let's look at it with *Mathematica*, as a function of $x = \sigma^2\,\Delta t$:

```
β[x_] := 1/2(Exp[x] + 1 - Sqrt[(Exp[x]+1)^2-4])
```

```
Plot[Log[β[x]], {x, 0, 10}];
```

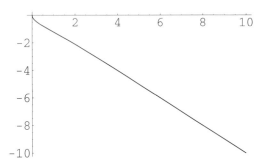

Although this function can be made arbitrarily small, it never becomes zero or negative, no matter how large the time-step. *So we see that unlike the exact JR form, the down scale factor cannot become zero or negative.*

(3) It is evident from this plot that it is also true that $\beta < 1$ for $\Delta t > 0$, hence, $p > 0$ always. Similarly, if we look at α

```
α[x_] := 1/2(Exp[x] + 1 + Sqrt[(Exp[x]+1)^2-4])
```

This function is strictly greater than unity for $\Delta t > 0$. Hence $q > 0$ always. *Putting this together, we see that, unlike the CRR form, the up and down weights can never be negative.*

(4) For small time-steps, the up and down weights are given by

```
p[x_] := (1 - β[x])/(α[x] - β[x]);
q[x_] := (α[x]-1)/(α[x] - β[x]);
```

in terms of $x = \sigma^2 \Delta t$, and

```
Series[p[x], {x, 0, 3}]
```

$$\frac{1}{2} - \frac{\sqrt{x}}{4} - \frac{x^{3/2}}{32} + \frac{7\,x^{5/2}}{1536} + O[x]^{7/2}$$

```
Series[q[x], {x, 0, 3}]
```

$$\frac{1}{2} + \frac{\sqrt{x}}{4} + \frac{x^{3/2}}{32} - \frac{7\,x^{5/2}}{1536} + O[x]^{7/2}$$

The small time-step behaviour of the up-down scale factors is given by multiplying the drift exponential by the following series:

```
Series[α[x], {x, 0, 1}]
```

$$1 + \sqrt{x} + \frac{x}{2} + O[x]^{3/2}$$

```
Series[β[x], {x, 0, 1}]
```

$$1 - \sqrt{x} + \frac{x}{2} + O[x]^{3/2}$$

Note that to this order the α series is identical to those for $e^{-\sqrt{x}}$:

```
Series[Exp[-Sqrt[x]], {x, 0, 1}]
```

$$1 - \sqrt{x} + \frac{x}{2} + O[x]^{3/2}$$

The small-time behaviour is seen to be a hybrid of the CRR and JR behaviour. This model does, however, avoid the negative probability problem of the former, and the negative down-scale problem of the (exact form of the) latter is also avoided. It therefore merits serious consideration as an alternative scheme. That is not to say it is not already in use, for this scheme is essentially the result of applying the CRR model, but to the forward price containing the drift correction. We now have six schemes to play with: CRR, JR and the new form, which we label SS, for supersymmetric; these can be used in either their exact or their approximate forms. These schemes will be considered in Chapter 19, where they are implemented using methods similar to those introduced by Benninga *et al* (1993). Before doing so, we shall explore how the supersymmetric scheme may be re-derived in *Mathematica*, and hence extended to a trinomial system.

18.4 Creating Supersymmetric Schemes with *Mathematica*

In the binomial case, we wish to solve four of the relations,

$$w_1 \, u_1^n + u_2^n \, w_2 = e^{\left(\frac{1}{2} n(n-1)\sigma^2 + n(r-q)\right)\Delta t} = A^n \, B^{\frac{1}{2} n(n-1)}$$
$$A = e^{(r-q)\Delta t} \quad B = e^{\sigma^2 \Delta t}$$

$$(49)$$

which serve to define *A*, *B* also. We give the set with $n = -1, 0, 1, 2$ to *Mathematica* as the system of equations (you may find it helpful to quit and re-start the kernel before carrying out the following):

```
symmetries =
```
$$\left\{w_1 + w_2 == 1, \; w_1 \, u_1 + w_2 \, u_2 == A, \; w_1 \, u_1^2 + w_2 \, u_2^2 == A^2 \, B, \; \frac{w_1}{u_1} + \frac{w_2}{u_2} == B/A\right\};$$

When *Mathematica* is asked to solve this system, two solutions are obtained - we just extract the first, and simplify the result:

```
soln = Simplify[Solve[symmetries, {w₁, w₂, u₁, u₂}]〚1〛]
```

$$\left\{w_1 \rightarrow \frac{3 + B - \sqrt{-3 + 2\,B + B^2}}{6 + 2\,B}, \; w_2 \rightarrow \frac{3 + B + \sqrt{-3 + 2\,B + B^2}}{6 + 2\,B}, \right.$$
$$\left. u_2 \rightarrow \frac{1}{2}\,A\left(1 + B - \sqrt{-3 + 2\,B + B^2}\right), \; u_1 \rightarrow \frac{1}{2}\,A\left(1 + B + \sqrt{-3 + 2\,B + B^2}\right)\right\}$$

These are the same ones we found "manually" - as can be checked, for example, by examining the following:

```
Simplify[u₁ u₂ /. soln]
```

$$A^2$$

```
Simplify[u₁ * u₁ / A^2 - (B + 1) u₁ / A + 1 /. soln]
```

0

One can also check the extent to which neighbouring symmetries apply:

```
CheckSymmetry[u_, d_, p_, q_, n_, a_, b_, solution_] :=
 Factor[Simplify[p u^n + q d^n - a^n b^((n-1) n/2) /. solution]]
```

```
CheckSymmetry[u₁, u₂, w₁, w₂, 3, A, B, soln]
```

$$-A^3 (-1+B)^2 (1+B)$$

```
CheckSymmetry[u₁, u₂, w₁, w₂, 4, A, B, soln]
```

$$-A^4 (-1+B)^2 (1+B) (1+2B+B^2+B^3)$$

```
CheckSymmetry[u₁, u₂, w₁, w₂, -2, A, B, soln]
```

$$-\frac{(-1+B)^2 (1+B)}{A^2}$$

They are approximately satisfied, all having a factor $(B-1)^2$, which is $O((\Delta t)^2)$. We can look at a whole range of such variables, looking at the series expansion in the form $B = 1 + x$. You should think of x as $\sigma^2 \Delta t$.

```
MatrixForm[Table[
  Series[(CheckSymmetry[u₁, u₂, w₁, w₂, n, A, B, soln] /. B -> 1 + x),
  {x, 0, 2}], {n, -6, 6}]]
```

$$
\begin{pmatrix}
-\frac{140 x^2}{A^6} + O[x]^3 \\
-\frac{70 x^2}{A^5} + O[x]^3 \\
-\frac{30 x^2}{A^4} + O[x]^3 \\
-\frac{10 x^2}{A^3} + O[x]^3 \\
-\frac{2 x^2}{A^2} + O[x]^3 \\
0 \\
0 \\
0 \\
0 \\
-2 A^3 x^2 + O[x]^3 \\
-10 A^4 x^2 + O[x]^3 \\
-30 A^5 x^2 + O[x]^3 \\
-70 A^6 x^2 + O[x]^3
\end{pmatrix}
$$

We see that all the positive and negative solutions are preserved with an error that is of order $(\Delta t)^2$, though the size of the error increases as the power of S increases away from zero. I conjecture that this is the best that can be done with a binomial scheme. With a trinomial scheme, it turns out one can do quite a lot better. One exercise that I have not yet considered is the extent to which the supersymmetric binomial tree specification can be tightened further by either considering a different family of power solutions, or including symmetries associated with other elementary solutions. There is one very important case that should be noted, however.

"Approximate" Jarrow-Rudd Re-visited

Let's go back to the fundamental binomial constraint. We set the yield to zero (we can replace r by $r - q$ later - for the present this prevents any confusion between yield and the down weight) and consider the constraint in the standard notation, obtained by setting

```
f[u_, d_, p_, q_, n_, r_, σ_] :=
  p u^n + q d^n - Exp[(1/2 n (n - 1) σ^2 + n r) Δt]
```

to zero. Previously we have considered this equation for just integer values of n, and $-1, 0, 1, 2$ in particular. We can instead consider it for real n, and demand instead that we set as many derivatives as possible to zero, evaluated at $n = 0$. This is very illuminating. The zeroth derivative gives us

```
ea = f[u, d, p, q, 0, r, σ]
```

$$-1 + p + q$$

The first derivative generates a constraint based on setting the following to zero:

```
eb = D[f[u, d, p, q, n, r, σ], n] /. n -> 0
```

$$-\Delta t \left(r - \frac{\sigma^2}{2}\right) + q\, \text{Log}[d] + p\, \text{Log}[u]$$

Astute readers will note that this is the same constraint as would arise from demanding that the tree preserve the log solution. The second and third derivatives give us the conditions that the following are also zero:

```
· ec = D[f[u, d, p, q, n, r, σ], {n, 2}] /. n -> 0
```

$$-\Delta t\, \sigma^2 - \Delta t^2 \left(r - \frac{\sigma^2}{2}\right)^2 + q\, \text{Log}[d]^2 + p\, \text{Log}[u]^2$$

```
ed = D[f[u, d, p, q, n, r, σ], {n, 3}] /. n -> 0
```

$$-3 \, \Delta t^2 \, \sigma^2 \left(r - \frac{\sigma^2}{2} \right) - \Delta t^3 \left(r - \frac{\sigma^2}{2} \right)^3 + q \, \text{Log}[d]^3 + p \, \text{Log}[u]^3$$

Now let's solve these four conditions, and take the second solution:

```
Simplify[Solve[{ea == 0, eb == 0, ec == 0, ed == 0}, {u, d, p, q}][[2]]]
```

$$\left\{ p \rightarrow \frac{1}{2}, \ q \rightarrow \frac{1}{2}, \ u \rightarrow E^{r \, \Delta t + \sqrt{\Delta t} \ \sigma - \frac{\Delta t \, \sigma^2}{2}}, \ d \rightarrow E^{r \, \Delta t - \sqrt{\Delta t} \ \sigma - \frac{\Delta t \, \sigma^2}{2}} \right\}$$

This is precisely the "approximate" Jarrow-Rudd form, which is now seen to be an approximate supersymmetric system, based on the four equations associated with the zero to third derivatives of the symmetry constraint. Like the one we have derived based on four different integer powers, its weights and factors remain non-negative.

18.5 A Supersymmetric Trinomial Scheme?

In the trinomial case we have three abscissa ratios, which we call u, d, s, and associated weights p, q, r. They solve the symmetry conditions

```
{p + q + r == 1,
p*u + q*d + r*s == A,
p*u^2 + q*d^2 + r*s^2 == A^2 B,
p*u^3 + q*d^3 + r*s^3 == A^3 B^3,
p/u + q/d+ r/s == B/A,
p/u^2 + q/d^2+ r/s^2 == B^3/A^2};
```

We may as well scale A out of the problem immediately, by writing

```
u = A U;
d = A D;
s = A S;
```

This leads to the set

```
teqns = {p + q + r == 1,
p*U + q*D + r*S == 1,
p*U^2 + q*D^2 + r*S^2 == B,
p*U^3 + q*D^3 + r*S^3 == B^3,
p/U + q/D+ r/S == B,
p/U^2 + q/D^2+ r/S^2 == B^3};
```

This family is solved most easily by a little manual intervention coupled with *Mathematica*. We first eliminate *r* by using the first condition, and substitute in the remaining five equations:

```
teqnsb = Rest[teqns /. r -> (1-p-q)]
```

$$\left\{ D q + (1-p-q) S + p U == 1, \right.$$
$$D^2 q + (1-p-q) S^2 + p U^2 == B, \quad D^3 q + (1-p-q) S^3 + p U^3 == B^3,$$
$$\frac{q}{D} + \frac{1-p-q}{S} + \frac{p}{U} == B, \quad \frac{q}{D^2} + \frac{1-p-q}{S^2} + \frac{p}{U^2} == B^3 \left. \right\}$$

The first two of these five serve to define *p*, *q* in terms of *U*, *D*, *S*:

```
teqnsc = teqnsb[[{1, 2}]]
```

$$\{ D q + (1-p-q) S + p U == 1, \quad D^2 q + (1-p-q) S^2 + p U^2 == B \}$$

So we solve them explicitly:

```
tsolnc = Simplify[Solve[teqnsc, {p, q}][[1]]]
```

$$\left\{ p \to \frac{B + D (-1+S) - S}{(D-U)(S-U)}, \quad q \to \frac{B + S (-1+U) - U}{(D-S)(D-U)} \right\}$$

Note that the weights are again independent of the drift. Now we substitute these relations into the remaining three and simplify:

```
teqnsd = Simplify[teqnsb[[{3,4,5}]] /. tsolnc]
```

$$\left\{ -S U + B (D+S+U) + D (-S-U+S U) == B^3, \quad \frac{B - S - U + S U + D (-1+S+U)}{D S U} == B, \right.$$
$$\frac{1}{D^2 S^2 U^2} (S U (-S-U+S U) + D^2 (-S+S^2-U+S U+U^2) +$$
$$D (-S^2 - 2 S U + S^2 U - U^2 + S U^2) + B (S U + D (S+U))) == B^3 \left. \right\}$$

This is still quite hard to solve. In the binomial case we found a solution with $U = 1/D$ in the current notation - let's see if this still works:

```
teqnse = Simplify[teqnsd /. D->1/U]
```

$$\Big\{-1 + S\left(1 + B - \frac{1}{U} - U\right) + B\left(\frac{1}{U} + U\right) == B^3, \quad \frac{B + \frac{(-1+S)\ (1-U+U^2)}{U}}{S} == B,$$

$$\frac{U\ (-1 + U + B\,U - U^2) - S\ (1 + U^2)\ (1 - (1 + B)\ U + U^2) + S^2\ (1 - U + U^2 - U^3 + U^4)}{S^2\ U^2}$$

$$== B^3\Big\}$$

Does this set have a solution?

```
provisionalsoln = Solve[teqnse, {U, S}]
```

    Solve::svars :
      Equations may not give solutions for all "solve" variables.

$$\Big\{\Big\{S \to 1,\ U \to \frac{1}{2}\left(-(-1-B)\,B - \sqrt{-4 + B^2 + 2\,B^3 + B^4}\right)\Big\},$$

$$\Big\{S \to 1,\ U \to \frac{1}{2}\left(-(-1-B)\,B + \sqrt{-4 + B^2 + 2\,B^3 + B^4}\right)\Big\},$$

$$\{U \to -I\},\ \{U \to I\},\ \{U \to 1\},\ \{U \to 1\}\Big\}$$

The second of this lot looks promising:

```
trialsoln = provisionalsoln[[2]]
```

$$\Big\{S \to 1,\ U \to \frac{1}{2}\left(-(-1-B)\,B + \sqrt{-4 + B^2 + 2\,B^3 + B^4}\right)\Big\}$$

Let's see if it really works:

```
Simplify[teqnse /. trialsoln]
```

    {True, True, True}

It does! We have found a supersymmetric trinomial tree satisfying all six symmetry conditions. We write the complete answer as follows:

```
U[ b_] := 1 / 2 (b * (b + 1) + Sqrt[b^4 + 2 b^3 + b^2 - 4]);
up[a_, b_] := a U[b]; down[a_, b_] := a / U[b];
stay[a_, b_] := a;
p[a_, b_] := (b - 1) / (1 - U[b]) / (1 / U[b] - U[b]);
q[a_, b_] := (b - 1) / (1 - 1 / U[b]) / (U[b] - 1 / U[b]);
r[a_, b_] := 1 - p[1, b] - q[1, b]
```

Here is the full form of the up and down and "stay" multipliers - note all three have a drift factor:

```
TraditionalForm[up[Exp[(r - q) Δt], Exp[σ^2 Δt]]]
```

$$\frac{1}{2} e^{(r-q)\,\Delta t} \left(e^{\Delta t\,\sigma^2} \left(1 + e^{\Delta t\,\sigma^2}\right) + \sqrt{-4 + e^{2\,\Delta t\,\sigma^2} + 2\,e^{3\,\Delta t\,\sigma^2} + e^{4\,\Delta t\,\sigma^2}} \right)$$

```
TraditionalForm[down[Exp[(r - q) Δt], Exp[σ^2 Δt]]]
```

$$\frac{2\,e^{(r-q)\,\Delta t}}{e^{\Delta t\,\sigma^2}\left(1 + e^{\Delta t\,\sigma^2}\right) + \sqrt{-4 + e^{2\,\Delta t\,\sigma^2} + 2\,e^{3\,\Delta t\,\sigma^2} + e^{4\,\Delta t\,\sigma^2}}}$$

```
TraditionalForm[stay[Exp[(r - q) Δt], Exp[σ^2 Δt]]]
```

$$e^{(r-q)\,\Delta t}$$

As in the binomial case, none of the weights/abscissas can become zero or negative. The small-time behaviour of these functions is instructive - the weights do tend to (1/6, 2/3, 1/6) as the time-step tends to zero, and the up and down scale factors do look like those corresponding to $e^{\sigma\sqrt{3\,\Delta t}}$, and its inverse, multiplied by a drift correction.

```
PowerExpand[Series[up [Exp[(r - q) t], Exp[σ^2 t]], {t, 0, 1}]]
```

$$1 + \sqrt{3}\,\sigma\sqrt{t} + \frac{1}{2}\,(2\,(-q + r) + 3\,\sigma^2)\,t + O[t]^{3/2}$$

```
PowerExpand[Series[down [Exp[(r - q) t], Exp[σ^2 t]], {t, 0, 1}]]
```

$$1 - \sqrt{3}\,\sigma\sqrt{t} + 2\left(\frac{1}{2}\,(-q + r) + \frac{3\,\sigma^2}{4}\right)t + O[t]^{3/2}$$

```
PowerExpand[Series[stay [Exp[(r - q) t], Exp[σ^2 t]], {t, 0, 1}]]
```

$$1 + (-q + r)\,t + O[t]^2$$

```
PowerExpand[Series[p [Exp[(r - q) t], Exp[σ^2 t]], {t, 0, 1}]]
```

$$\frac{1}{6} - \frac{\sigma \sqrt{t}}{4 \sqrt{3}} - \frac{\sigma^2 t}{18} + \frac{7 \sigma^3 t^{3/2}}{96 \sqrt{3}} + O[t]^2$$

```
PowerExpand[Series[q [Exp[(r - q) t], Exp[σ^2 t]], {t, 0, 1}]]
```

$$\frac{1}{6} + \frac{\sigma \sqrt{t}}{4 \sqrt{3}} - \frac{\sigma^2 t}{18} - \frac{7 \sigma^3 t^{3/2}}{96 \sqrt{3}} + O[t]^2$$

```
PowerExpand[Series[r [Exp[(r - q) t], Exp[σ^2 t]], {t, 0, 1}]]
```

$$\frac{2}{3} + \frac{\sigma^2 t}{9} + O[t]^2$$

Satisfaction of Neighbouring Symmetry Constraints

As in the binomial case, we can explore the extent to which other neighbouring symmetries apply.

```
CheckSymmetry[u_, d_, s_, p_, q_, r_, n_, a_, b_] :=
  Factor[Simplify[p u^n + q d^n + r s^n - a^n b^((n - 1) n / 2)]]
```

We leave it as an exercise for the reader to confirm that this function yields zero for the cases $n = -2, -1, 0, 1, 2, 3$. As to neighbouring values:

```
Series[(CheckSymmetry[up[a, b], down[a, b], stay[a, b],
    p[a, b], q[a, b], r[a, b], 4, a, b] /. b -> 1 + x), {x, 0, 3}]
```

$$-6 a^4 x^3 + O[x]^4$$

```
Series[
  (CheckSymmetry[up[a, b], down[a, b], stay[a, b], p[a, b], q[a, b],
    r[a, b], -3, a, b] /. b -> 1 + x), {x, 0, 3}]
```

$$-\frac{6 x^3}{a^3} + O[x]^4$$

```
Series[
  (CheckSymmetry[up[a, b], down[a, b], stay[a, b], p[a, b], q[a, b],
    r[a, b], -4, a, b] /. b -> 1 + x), {x, 0, 3}]
```

$$-\frac{42\,x^3}{a^4}+O[x]^4$$

So the neighbouring symmetries are satisfied with an error that is of order Δt^3. I conjecture that this is the best that can be done with a trinomial system - six symmetry relations are preserved exactly and all others are preserved with an error that is now *cubic* in the time-step. The reader may wish to explore the solution of the constraints arising from differentiation, rather than from integer powers.

Chapter 18 Bibliography

Benninga, S., Steinmetz, R., and Stroughair, J., 1993, Implementing numerical option pricing models, *Mathematica Journal*, 3, issue 4, p. 66, 1993.

Cox, J., Ross, S. and Rubenstein, M., 1979, Option pricing: a simplified approach. *Journal of Financial Economics*, 7. p. 229.

Hull, J. , 1996, *Options, Futures and Other Derivatives*, 3rd edition, Prentice-Hall.

Jarrow, R.A., and Rudd, A., 1983, *Option Pricing*, Irwin.

Wilmott, P., Dewynne, J. and Howison, S., 1995, *The Mathematics of Financial Derivatives*, Cambridge University Press.

Chapter 19.
Tree Implementation in *Mathematica* and Basic Tree Pathology

Recursive Tree Modelling, Extracting Greeks, and Problems with Barriers

19.1 Introduction

The discussion in Chapter 18 established the theoretical forms of the CRR, JR and supersymmetric tree models that we wish to consider. We have two forms of each of CRR and JR, based on exact and approximate matching of the second moments. We shall now turn to their implementation in *Mathematica*, and explore some basic problems that arise. We proceed first on the basis of a recursive definition of the option values, based directly on the results of Chapter 18, using the work of Benninga *et al* (1993) as a basic prototype, but extending their approach to include the Greeks. In Chapter 20 we shall see how these models can be optimized using the *Mathematica* compiler, at the expense, perhaps, of a little elegance.

Before reading this chapter you are encouraged to review Section 2.7 of Chapter 2, and the sub-subsection on "Defining a Function Recursively", which looks at a prototype recursive algorithm using the Fibonacci series as a model case. It is important that you are aware of the syntax for getting *Mathematica* to remember previously computed values.

Our plan is to first look at the definition of European options, making the usual comparisons with the exact solutions to get a feel for the accuracy of the various models. Then we shall consider the modification necessary for American-style options and explore some valuations. Finally we shall look at the nasty problems which arise when trees are naively applied to problems involving Barriers.

Throughout this chapter we shall be using recursive structures rather larger than is standard - to this end we re-set the following *Mathematica* system parameter:

```
$RecursionLimit=5000;
```

19.2 *Mathematica* Tree Models 1 - CRR Forms

We proceed to define the various functions to explore the CRR form of a tree, following the general plan of Benninga *et al* (1993), but generalizing to allow non-zero continuous dividend yield, and also implementing the slightly more accurate form that matches the second moment exactly. First we have the universal functions giving the up and down steps for a CRR-like binomial tree, and the discounting factor, for the case where the second moments are matched exactly.

```
CRRupBin[n_, r_, q_, σ_, T_] := Module[{f = Exp[-(r-q)*T/n],
g = Exp[(r-q+σ^2)*T/n]},
1/2*(f+g + Sqrt[(f+g)^2-4])];
CRRdownBin[n_, r_, q_, σ_, T_] := Module[{f = Exp[-(r-q)*T/n],
g = Exp[(r-q+σ^2)*T/n]},
1/2*(f+g - Sqrt[(f+g)^2-4])];
Discount[n_, r_, T_] := N[Exp[-r T/n]];
```

When the simplified form of the up and down step is used (cf. the description of the approximate model by Hull, 1996):

```
CRRupBinApp[n_, r_, q_, σ_, T_] := Exp[σ Sqrt[T/n]];
CRRdownBinApp[n_, r_, q_, σ_, T_] := Exp[-σ Sqrt[T/n]];
```

We also need the function

```
afunc[n_, r_, q_, T_] := N[Exp[(r - q) T / n]];
```

Now in the implementation of the original CRR form, we only need the probabilities of up and down, and they are always multiplied by the discount factor - these are called *P* and *Q*:

```
Pfunc[up_,down_, a_, disc_] := N[disc*(a-down)/(up-down)];
Qfunc[up_, down_, a_, disc_] := N[disc-Pfunc[up,down,a,disc]];
```

Now for verification purposes, we do want to evaluate these functions rather more quickly than the general recursive tree model permits, so we define two functions appropriate only to the European case, where we just extract the valuation by just summing over final values:

```
FastCRREuropeanOption[S_, n_, σ_, T_, r_, q_, exercise_Function] :=
Module[{u = CRRupBin[n,r,q,σ,T], d = CRRdownBin[n,r,q,σ,T],
a = afunc[n,r,q,T],disc = Discount[n,r,T],P,Q},
P = Pfunc[u,d,a,disc];
Q = Qfunc[u,d,a,disc];
Sum[exercise[S u^j d^(n-j)] Binomial[n,j] P^j Q^(n-j), {j,0,n}]];

FastCRREuropeanOptionApp[S_, n_, σ_, T_, r_, q_, exercise_Function] :=
Module[{u = CRRupBinApp[n,r,q,σ,T], d = CRRdownBinApp[n,r,q,σ,T],
a = afunc[n,r,q,T],disc = Discount[n,r,T],P,Q},
P = Pfunc[u,d,a,disc];
Q = Qfunc[u,d,a,disc];
Sum[exercise[S u^j d^(n-j)] Binomial[n,j] P^j Q^(n-j), {j,0,n}]];
```

Next we make functions specific to Calls and Puts, for each of the two versions:

```
FastCRREuropeanPut[S_, K_, n_, r_, q_, σ_, T_] :=FastCRREuropeanOption[-
S,n,σ,T,r,q,Max[K-#,0]&];
FastCRREuropeanCall[S_, K_, n_, r_, q_, σ_, T_] :=FastCRREuropean-
Option[S,n,σ,T,r,q,Max[#-K,0]&];
FastCRREuropeanPutApp[S_, K_, n_, r_, q_, σ_, T_] :=FastCRREuropeanOp-
tionApp[S,n,σ,T,r,q,Max[K-#,0]&];
FastCRREuropeanCallApp[S_, K_, n_, r_, q_, σ_, T_] :=FastCRREuropeanOp-
tionApp[S,n,σ,T,r,q,Max[#-K,0]&];
```

For the American case, the same approach is again used whatever the exercise function, and given the availability of information about all nodes in the tree, we also define those Greeks that can be computed by one call to the tree. Note carefully the definitions of Delta, Gamma and Theta. They are in fact estimates of Delta and Theta at time Δt, and of Gamma at time $2\Delta t$. Vega and Rho can be calculated in two ways. The first is potentially both more accurate but numerically precarious - one changes the volatility and interest rate and re-computes, using a one-sided or central difference. A less accurate but robust method is to use the Vega-Gamma identity and the Rho-Delta identity for a robust estimate.

```
CRRAmericanOption[S_, n_, σ_, T_, r_, q_, exercise_Function] :=
Module[{u = CRRupBin[n,r,q,σ,T], d = CRRdownBin[n,r,q,σ,T],
a = afunc[n,r,q,T],
disc = Discount[n,r,T],
P,Q,TreeOp,value,delta,gamma,theta},
P = Pfunc[u,d,a,disc];
Q = Qfunc[u,d,a,disc];
TreeOp[node_,level_] := TreeOp[node,level] =
If[level==n,exercise[S d^node u^(level-node)],
Max[{P,Q}.{TreeOp[node,level+1],TreeOp[node+1,level+1]},
exercise[S d^node u^(level-node)]]];
value=TreeOp[0,0];
delta = (TreeOp[0,1] - TreeOp[1,1])/((u-d)*S);
gamma = ((TreeOp[0,2] - TreeOp[1,2])/(u^2-1) -
(TreeOp[1,2]-TreeOp[2,2])/(1-d^2))*2/(S^2 (u^2-d^2));
theta = (TreeOp[1,2] - TreeOp[0,0])/(2*T/n);
Clear[TreeOp];
{value, delta, gamma, theta}];
```

This can be done with the approximate weights also:

```
CRRAmericanOptionApp[S_, n_, σ_, T_, r_, q_, exercise_Function] :=
Module[{u = CRRupBinApp[n,r,q,σ,T], d = CRRdownBinApp[n,r,q,σ,T],
a = afunc[n,r,q,T],
disc = Discount[n,r,T],
P,Q,TreeOp,value,delta,gamma,theta},
P = Pfunc[u,d,a,disc];
Q = Qfunc[u,d,a,disc];
TreeOp[node_,level_] := TreeOp[node,level] =
If[level==n,exercise[S d^node u^(level-node)],
Max[{P,Q}.{TreeOp[node,level+1],TreeOp[node+1,level+1]},
exercise[S d^node u^(level-node)]]];
value=TreeOp[0,0];
delta = (TreeOp[0,1] - TreeOp[1,1])/((u-d)*S);
```

```
gamma = ((TreeOp[0,2] - TreeOp[1,2])/(u^2-1) -
(TreeOp[1,2]-TreeOp[2,2])/(1-d^2))*2/(S^2 (u^2-d^2));
theta = (TreeOp[1,2] - TreeOp[0,0])/(2*T/n);
Clear[TreeOp];
{value, delta, gamma, theta}];
```

Next we give functions specific to a Call and a Put:

```
CRRAmericanPut[S_, K_, n_, r_, q_, σ_, T_] :=CRRAmericanOption[S,n,σ,-
T,r,q,Max[K-#,0]&];
CRRAmericanCall[S_, K_, n_, r_, q_, σ_, T_] :=CRRAmericanOption[S,n,σ,-
T,r,q,Max[#-K,0]&];
CRRAmericanPutApp[S_, K_, n_, r_, q_, σ_, T_] :=CRRAmericanOptionApp[-
S,n,σ,T,r,q,Max[K-#,0]&];
CRRAmericanCallApp[S_, K_, n_, r_, q_, σ_, T_] :=CRRAmericanOptionApp[-
S,n,σ,T,r,q,Max[#-K,0]&];
```

Evaluation and Verification of CRR-Style Models

We load the standard European options Package.

```
Needs["Derivatives`BlackScholes`"]
```

```
{BlackScholesCall[9, 10, 0.2, 0.1, 0, 1],
 BlackScholesCallDelta[9, 10, 0.2, 0.1, 0, 1],
 BlackScholesCallGamma[9, 10, 0.2, 0.1, 0, 1],
 BlackScholesCallTheta[9, 10, 0.2, 0.1, 0, 1]}
```

```
    {0.694898, 0.529175, 0.221042, -0.764856}
```

What we shall do is to compute sample values for a Call and a Put in both forms of the CRR model. We shall look at a Call and a Put and will sample their values for trees with between 15 and 300 time-steps, and explore the convergence. The expiry will be fairly short, at just one year.

```
ExactData =
  Table[{k, BlackScholesCall[9, 10, 0.2, 0.1, 0, 1]}, {k, 15, 300}];
ExactDataP =
  Table[{k, BlackScholesPut[9, 10, 0.2, 0.1, 0, 1]}, {k, 15, 300}];
explot = ListPlot[ExactData, PlotJoined -> True,
   DisplayFunction -> Identity];
explotP = ListPlot[ExactDataP, PlotJoined -> True,
   DisplayFunction -> Identity];
```

In the following we work with the "fast" versions, so that we can compute a large number of different trees.

```
CRRdataC = Table[
  {k, FastCRREuropeanCall[9, 10, k, 0.1, 0, 0.2, 1]}, {k, 15, 300}];
CRRdataP = Table[{k, FastCRREuropeanPut[9, 10, k, 0.1, 0, 0.2, 1]},
  {k, 15, 300}];
CRRdataCApp = Table[{k, FastCRREuropeanCallApp[
    9, 10, k, 0.1, 0, 0.2, 1]}, {k, 15, 300}];
CRRdataPApp = Table[{k, FastCRREuropeanPutApp[9,
    10, k, 0.1, 0, 0.2, 1]}, {k, 15, 300}];


CRRPlotC =
 ListPlot[CRRdataC, PlotJoined -> True, DisplayFunction -> Identity];
CRRPlotP = ListPlot[CRRdataP, PlotJoined -> True,
  DisplayFunction -> Identity];
CRRPlotCApp = ListPlot[CRRdataCApp, PlotJoined -> True,
  DisplayFunction -> Identity];
CRRPlotPApp = ListPlot[CRRdataPApp, PlotJoined -> True,
  DisplayFunction -> Identity];
```

We can now view the convergence properties. First the Call with the "exact" form of the tree rule:

```
Show[explot, CRRPlotC, PlotRange -> {0.69, 0.70},
  DisplayFunction -> $DisplayFunction];
```

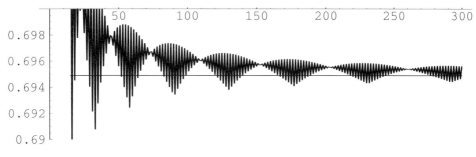

Next the Call with the approximate tree rule:

```
Show[explot, CRRPlotCApp, PlotRange -> {0.69, 0.70},
  DisplayFunction -> $DisplayFunction];
```

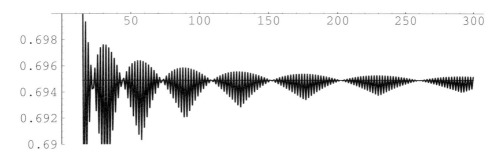

Next the Put with the "exact" form of the tree rule:

```
Show[explotP, CRRPlotP, PlotRange -> {0.74, 0.75},
  DisplayFunction -> $DisplayFunction];
```

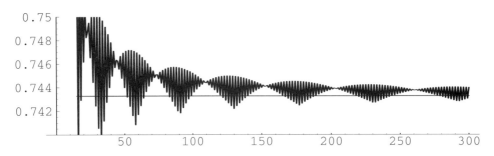

Finally the Put with the approximate rule:

```
Show[explotP, CRRPlotPApp, PlotRange -> {0.74, 0.75},
  DisplayFunction -> $DisplayFunction];
```

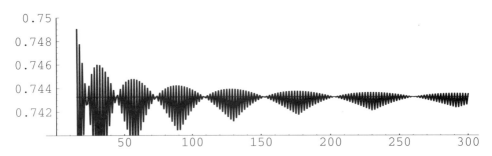

These results dramatically illustrate the fact that the "exact" form of the model is not necessarily better than the approximate form. The convergence properties of both are rather odd. This interesting state of affairs is somewhat perplexing, but gives weight to the idea that verification is a good idea!

However, the convergence of both the exact and the approximate scheme is problematic, being highly oscillatory. Even the envelope of the fast oscillations does not decrease monotonically. The envelope is also asymmetric, so that the common practice of averaging the odd-even calculations is not particularly reliable either. The average is shown, roughly, by the interference pattern in the plots. In the exact and approximate cases it bounces off the exact solution from different sides.

Magic Tree Sizes 1 - Approximate CRR Models

There are clearly values of n where the approximate CRR model is particularly good. We can locate these values by hypothesizing that they may have something to do with the location of tree nodes at expiry in relation to the strike. Two possibilities stand out (and the author has to admit to being slightly surprised by the outcome) - one where a node sits right on the strike, and the other where the strike is between two nodes. A quick calculation shows that the strike sits on a node when

$$n = \frac{m^2 \, \sigma^2 \, T}{\log^2(\frac{K}{S})} \tag{1}$$

and the strike is placed half-way between two nodes when

$$n = \frac{(m + \frac{1}{2})^2 \, \sigma^2 \, T}{\log^2(\frac{K}{S})} \tag{2}$$

Mathematica functions for these cases are

```
sitCRR[m_, σ_, T_, S_, K_] := m^2 σ^2 T / Log[K / S]^2;
straddleCRR[m_, σ_, T_, S_, K_] := (m + 1 / 2)^2 σ^2 T / Log[K / S]^2
```

Let's evaluate these values (in real form - you would take the nearest integer in practice) for the parameters used above:

```
Table[sitCRR[m, 0.2, 1, 9, 10], {m, 1, 10}]

    {3.60333, 14.4133, 32.43, 57.6533, 90.0833,
    129.72, 176.563, 230.613, 291.87, 360.333}


Table[straddleCRR[m, 0.2, 1, 9, 10], {m, 1, 10}]

    {8.1075, 22.5208, 44.1408, 72.9675, 109.001,
    152.241, 202.687, 260.341, 325.201, 397.267}
```

Inspection of the plots makes the point - the latter gives the locations where the model is best. This illustrates the point that the time-step of the tree should be adapted to the option. It is also a first example of what we call a magic tree size. These magic tree sizes are not universal - sometimes the size has to be adapted to meet other considerations. For an approximate CRR tree they are given by setting n to be the integer closest to

$$\frac{(m + \frac{1}{2})^2 \, \sigma^2 \, T}{\log^2(\frac{K}{S})} \tag{3}$$

for some integer m. Let's look at the convergence on the straddling cases:

```
CRRdataCMagic = Table[{k, FastCRREuropeanCallApp[9, 10,
    Floor[straddleCRR[k, 0.2, 1, 9, 10]], 0.1, 0, 0.2, 1]}, {k, 3, 20}];
```

CRRdataCMagic

```
{{3, 0.694859}, {4, 0.694687}, {5, 0.694861},
 {6, 0.694859}, {7, 0.694902}, {8, 0.694875}, {9, 0.694882},
 {10, 0.694892}, {11, 0.694895}, {12, 0.694891},
 {13, 0.694896}, {14, 0.694896}, {15, 0.694891}, {16, 0.694894},
 {17, 0.694893}, {18, 0.694895}, {19, 0.694895}, {20, 0.694895}}
```

Here is a reminder of the exact value:

```
BlackScholesCall[9, 10, 0.2, 0.1, 0, 1]
```

```
    0.694898
```

This is clearly much more satisfactory. However, it is evident from the magic tree size formula that things go wrong with this notion when the option is at the money:

```
CRRdataATM = Table[
   {k, FastCRREuropeanCall[10, 10, k, 0.1, 0, 0.2, 1]}, {k, 15, 300}];
```

```
ListPlot[CRRdataATM, PlotJoined -> True];
```

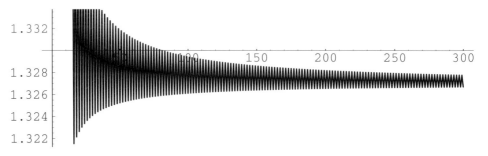

As now expected, the magic tree sizes no longer exist - we have an oscillatory convergence that takes a very long time to attain the exact value.

```
BlackScholesCall[10, 10, 0.2, 0.1, 0, 1]
```

```
    1.32697
```

19.3 *Mathematica* Tree Models 2 - JR Forms

We proceed to define the various functions to explore the JR form of a trees. The up and down weights are always given by

```
JRUpWeight[disc_]  := N[disc/2];
JRDownWeight[disc_] :=N[disc/2];
```

In the exact form of the model, where all terms are matched exactly, the up and down scale factors are given by

```
JRupBin[n_, r_, q_, σ_, T_] := Exp[(r-q)*T/n]*(1 + Sqrt[Exp[σ^2*-
T/n]-1]);
JRdownBin[n_, r_, q_, σ_, T_] := Exp[(r-q)*T/n]*(1 - Sqrt[Exp[σ^2*-
T/n]-1]);
```

In the approximate form of the model, the scale factors are given instead by the functions

```
JRupBinApp[n_, r_, q_, σ_, T_] := Exp[(r-q-σ^2/2)*T/n + σ Sqrt[T/n]];
JRdownBinApp[n_, r_, q_, σ_, T_] := Exp[(r-q-σ^2/2)*T/n - σ Sqrt[T/n]];
```

The same approach is used whatever the exercise function, first for results of a European-style option. The approximate form just uses the approximate up and down states (recall that this does have the advantage that the latter does not vanish):

```
FastJREuropeanOption[S_, n_, σ_, T_, r_, q_, exercise_Function] :=
Module[{u = JRupBin[n,r,q,σ,T], d = JRdownBin[n,r,q,σ,T],disc =
Discount[n,r,T]},
(disc/2)^n Sum[exercise[S u^j d^(n-j)] Binomial[n,j], {j,0,n}]];

FastJREuropeanOptionApp[S_, n_, σ_, T_, r_, q_, exercise_Function] :=
Module[{u = JRupBinApp[n,r,q,σ,T], d = JRdownBinApp[n,r,q,σ,T],disc =
Discount[n,r,T],P,Q},
(disc/2)^n Sum[exercise[S u^j d^(n-j)] Binomial[n,j], {j,0,n}]];
```

Next we give functions specific to a Call and a Put:

```
FastJREuropeanPut[S_, K_, n_, r_, q_, σ_, T_] :=FastJREuropeanOption[-
S,n,σ,T,r,q,Max[K-#,0]&];
FastJREuropeanCall[S_, K_, n_, r_, q_, σ_, T_] := FastJREuropean-
Option[S, n, σ, T, r, q, Max[# - K, 0]&];
FastJREuropeanPutApp[S_, K_, n_, r_, q_, σ_, T_] :=FastJREuropeanOption-
App[S,n,σ,T,r,q,Max[K-#,0]&];
FastJREuropeanCallApp[S_, K_, n_, r_, q_, σ_, T_] := FastJREuropeanOp-
tionApp[S, n, σ, T, r, q, Max[# - K, 0]&];
```

Next we give the American versions. These follow the same pattern as for CRR, but we have to make some careful adjustments to the Greeks. Delta can be done the same way, and Gamma needs a little more care with the differencing. Both are evaluated at not quite the right time and not exactly the right stock price. Theta has to be evaluated by first correcting the [1,2] node for Delta effects. There is a common misconception that it is somehow more difficult to extract the basic Greeks from one call to a tree in the Jarrow-Rudd model. This is nonsense. One just has to remember that $u\,d$ is not unity!

```
JRAmericanOption[S_, n_, σ_, T_, r_, q_, exercise_Function] :=
Module[{u,d,disc = Discount[n,r,T],P,Q,TreeOp,value,delta,gamma,theta},
u = JRupBin[n,r,q,σ,T];
d = JRdownBin[n,r,q,σ,T];
P = JRUpWeight[disc];
Q = P;
TreeOp[node_,level_] := TreeOp[node,level] =
If[level==n,exercise[S d^node u^(level-node)],
Max[{P,Q}.{TreeOp[node,level+1],TreeOp[node+1,level+1]},
exercise[S d^node u^(level-node)]]];
value=TreeOp[0,0];
delta = (TreeOp[0,1] - TreeOp[1,1])/((u-d)*S);
gamma = ((TreeOp[0,2] -TreeOp[1,2])/(u^2-u*d) -
(TreeOp[1,2]-TreeOp[2,2])/(u*d-d^2))*2/(S^2(u^2-d^2));
theta = (TreeOp[1,2]-(u*d-1)*S*delta - TreeOp[0,0])/(2*T/n);
Clear[TreeOp];
{value, delta, gamma, theta}];


JRAmericanOptionApp[S_, n_, σ_, T_, r_, q_, exercise_Function] :=
Module[{u,d,disc = Discount[n,r,T],P,Q,TreeOp,value,delta,gamma,theta},
u = JRupBinApp[n,r,q,σ,T];
d = JRdownBinApp[n,r,q,σ,T];
P = JRUpWeight[disc];
Q = P;
TreeOp[node_,level_] := TreeOp[node,level] =
If[level==n,exercise[S d^node u^(level-node)],
Max[{P,Q}.{TreeOp[node,level+1],TreeOp[node+1,level+1]},
exercise[S d^node u^(level-node)]]];
value=TreeOp[0,0];
delta = (TreeOp[0,1] - TreeOp[1,1])/((u-d)*S);
gamma = ((TreeOp[0,2] -TreeOp[1,2])/(u^2-u*d) -
(TreeOp[1,2]-TreeOp[2,2])/(u*d-d^2))*2/(S^2(u^2-d^2));
theta = (TreeOp[1,2]-(u*d-1)*S*delta - TreeOp[0,0])/(2*T/n);
Clear[TreeOp];
{value, delta, gamma, theta}];
```

Finally functions specific to an American Call and a Put:

```
JRAmericanPut[S_, K_, n_, r_, q_, σ_, T_] :=JRAmericanOption[S,n,σ,-
T,r,q,Max[K-#,0]&];
JRAmericanCall[S_, K_, n_, r_, q_, σ_, T_] :=JRAmericanOption[S,n,σ,-
T,r,q,Max[#-K,0]&];
JRAmericanPutApp[S_, K_, n_, r_, q_, σ_, T_] :=JRAmericanOptionApp[-
S,n,σ,T,r,q,Max[K-#,0]&];
JRAmericanCallApp[S_, K_, n_, r_, q_, σ_, T_] :=JRAmericanOptionApp[-
S,n,σ,T,r,q,Max[#-K,0]&];
```

Evaluation and Verification of JR-Style Models

What we shall do is to compute sample values for a Call and a Put in both forms of the JR model:

```
JRdataC = Table[
  {k, FastJREuropeanCall[9, 10, k, 0.1, 0, 0.2, 1]}, {k, 15, 300}];
JRdataP = Table[{k, FastJREuropeanPut[9, 10, k, 0.1, 0, 0.2, 1]},
  {k, 15, 300}];
JRdataCApp = Table[{k, FastJREuropeanCallApp[
    9, 10, k, 0.1, 0, 0.2, 1]}, {k, 15, 300}];
JRdataPApp = Table[{k, FastJREuropeanPutApp[9,
    10, k, 0.1, 0, 0.2, 1]}, {k, 15, 300}];

JRPlotC =
 ListPlot[JRdataC, PlotJoined -> True, DisplayFunction -> Identity];
JRPlotP = ListPlot[JRdataP, PlotJoined -> True,
  DisplayFunction -> Identity];
JRPlotCApp = ListPlot[JRdataCApp, PlotJoined -> True,
  DisplayFunction -> Identity];
JRPlotPApp = ListPlot[JRdataPApp, PlotJoined -> True,
  DisplayFunction -> Identity];

Show[explot, JRPlotC, PlotRange -> {0.69, 0.70},
  DisplayFunction -> $DisplayFunction];
```

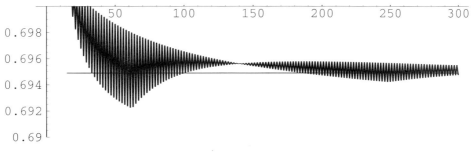

```
Show[explot, JRPlotCApp, PlotRange -> {0.69, 0.70},
  DisplayFunction -> $DisplayFunction];
```

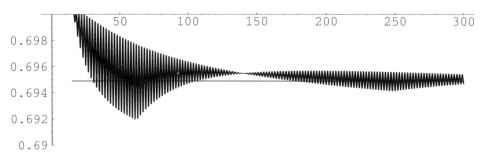

```
Show[explotP, JRPlotP,  PlotRange -> {0.74, 0.75},
  DisplayFunction -> $DisplayFunction];
```

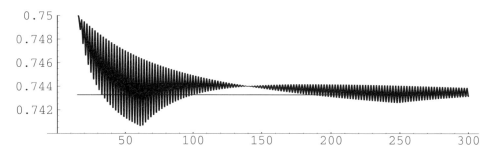

```
Show[explotP, JRPlotPApp,  PlotRange -> {0.74, 0.75},
  DisplayFunction -> $DisplayFunction];
```

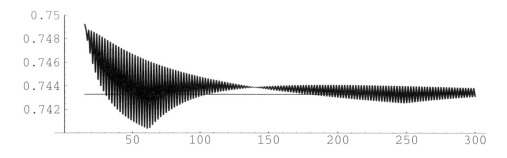

However, the convergence of both the exact and the approximate scheme remains problematic, being highly oscillatory, as before.

Magic Tree Sizes 2 - Approximate JR Models

A simple formula for the strike-straddling can be given in the approximate form of this model. This is

$$n = \frac{\left(m + \frac{1}{2}\right)^2 \sigma^2 T}{\left(\log(\frac{K}{S}) - \left(r - q - \frac{\sigma^2}{2}\right)\right)^2} \tag{4}$$

A *Mathematica* function for this case is

```
straddleJR[m_, σ_, T_, S_, K_, r_, q_] :=
  ((m + 1 / 2)^2*σ^2*T) / (Log[K / S] - (r - q - σ^2 / 2))^2
```

Let's evaluate these values (in real form - you would take the nearest integer in practice) for the parameters used above:

```
Table[straddleJR[m, 0.2, 1, 9, 10, 0.1, 0], {m, 1, 10}]

    {139.935, 388.708, 761.868, 1259.41, 1881.35,
      2627.67, 3498.37, 4493.47, 5612.95, 6856.81}
```

This formula explains the locations of the points in the JR plots compared to those in the CRR forms - only the first is visible in this particular example. The denominator, for this particular case, is smaller, resulting in the oscillation envelope having a longer frequency as a function of *n*. Note that the magic tree formula can go wrong when the expected value of the stock price at expiry coincides with the strike. The Jarrow-Rudd tree then always has a node right on the strike.

19.4 *Mathematica* Tree Models 3 - Supersymmetric Forms

We proceed to define the various functions to explore the supersymmetric form of a tree. We need the following functions also used in the CRR forms:

```
afunc[n_, r_, q_, T_] := N[Exp[(r - q) T / n]];
Pfunc[up_, down_, a_, disc_] := N[disc * (a - down) / (up - down)];
Qfunc[up_, down_, a_, disc_] := N[disc - Pfunc[up, down, a, disc]];
```

The up and down multipliers, in the exact and approximate forms of the model, are

```
SSupBin[n_, r_, q_, σ_, T_] := Module[{f = (1 + Exp[σ^2*T/n])},
Exp[(r-q)*T/n]*(f + Sqrt[f^2-4])/2];
SSdownBin[n_, r_, q_, σ_, T_] := Module[{f = (1 + Exp[σ^2*T/n])},
Exp[(r-q)*T/n]*2/(f + Sqrt[f^2-4])];

SSupBinApp[n_, r_, q_, σ_, T_] :=Exp[(r-q)*T/n+σ Sqrt[T/n]];
SSdownBinApp[n_, r_, q_, σ_, T_] := Exp[(r-q)*T/n-σ Sqrt[T/n]];
```

European options are dealt with by

```
FastSSEuropeanOption[S_, n_, σ_, T_, r_, q_, exercise_Function] :=
Module[{u = SSupBin[n,r,q,σ,T], d = SSdownBin[n,r,q,σ,T],
a = afunc[n,r,q,T],disc = Discount[n,r,T],P,Q},
P = Pfunc[u,d,a,disc];
Q = Qfunc[u,d,a,disc];
Sum[exercise[S u^j d^(n-j)] Binomial[n,j] P^j Q^(n-j), {j,0,n}]];
```

```
FastSSEuropeanOptionApp[S_, n_, σ_, T_, r_, q_, exercise_Function] :=
Module[{u = SSupBinApp[n,r,q,σ,T], d = SSdownBinApp[n,r,q,σ,T],
a = afunc[n,r,q,T],disc = Discount[n,r,T],P,Q},
P = Pfunc[u,d,a,disc];
Q = Qfunc[u,d,a,disc];
Sum[exercise[S u^j d^(n-j)] Binomial[n,j] P^j Q^(n-j), {j,0,n}]];


FastSSEuropeanPut[S_, K_, n_, r_, q_, σ_, T_] :=
FastSSEuropeanOption[S,n,σ,T,r,q,Max[K-#,0]&];
FastSSEuropeanCall[S_, K_, n_, r_, q_, σ_, T_] :=
FastSSEuropeanOption[S, n, σ, T, r, q, Max[# - K, 0]&];
FastSSEuropeanPutApp[S_, K_, n_, r_, q_, σ_, T_] :=
FastSSEuropeanOptionApp[S,n,σ,T,r,q,Max[K-#,0]&];
FastSSEuropeanCallApp[S_, K_, n_, r_, q_, σ_, T_] :=
FastSSEuropeanOptionApp[S, n, σ, T, r, q, Max[# - K, 0]&];
```

American options, as before, are given by the following. The same estimates of the Greeks are used as for the JR case.

```
SSAmericanOption[S_, n_, σ_, T_, r_, q_, exercise_Function] :=
Module[{u = SSupBin[n,r,q,σ,T], d = SSdownBin[n,r,q,σ,T],
a = afunc[n,r,q,T],
disc = Discount[n,r,T],
P,Q,TreeOp,value,delta,gamma,theta},
P = Pfunc[u,d,a,disc];
Q = Qfunc[u,d,a,disc];
TreeOp[node_,level_] := TreeOp[node,level] =
If[level==n,exercise[S d^node u^(level-node)],
Max[{P,Q}.{TreeOp[node,level+1],TreeOp[node+1,level+1]},
exercise[S d^node u^(level-node)]]];
value=TreeOp[0,0];
delta = (TreeOp[0,1] - TreeOp[1,1])/((u-d)*S);
gamma = ((TreeOp[0,2] - TreeOp[1,2])/(u^2-u*d) -
(TreeOp[1,2]-TreeOp[2,2])/(u*d-d^2))*2/(S^2 (u^2-d^2));
theta = (TreeOp[1,2]-(u*d-1)*S*delta - TreeOp[0,0])/(2*T/n);
Clear[TreeOp];
{value, delta, gamma, theta}];


SSAmericanOptionApp[S_, n_, σ_, T_, r_, q_, exercise_Function] :=
Module[{u = SSupBinApp[n,r,q,σ,T], d = SSdownBinApp[n,r,q,σ,T],
a = afunc[n,r,q,T],
disc = Discount[n,r,T],
P,Q,TreeOp,value,delta,gamma,theta},
P = Pfunc[u,d,a,disc];
Q = Qfunc[u,d,a,disc];
TreeOp[node_,level_] := TreeOp[node,level] =
If[level==n,exercise[S d^node u^(level-node)],
Max[{P,Q}.{TreeOp[node,level+1],TreeOp[node+1,level+1]},
exercise[S d^node u^(level-node)]]];
value=TreeOp[0,0];
delta = (TreeOp[0,1] - TreeOp[1,1])/((u-d)*S);
```

```
gamma = ((TreeOp[0,2] - TreeOp[1,2])/(u^2-u*d) -
(TreeOp[1,2]-TreeOp[2,2])/(u*d-d^2))*2/(S^2 (u^2-d^2));
theta = (TreeOp[1,2]-(u*d-1)*S*delta - TreeOp[0,0])/(2*T/n);
Clear[TreeOp];
{value, delta, gamma, theta}];


SSAmericanPut[S_, K_, n_, r_, q_, σ_, T_] :=
SSAmericanOption[S,n,σ,T,r,q,Max[K-#,0]&];
SSAmericanCall[S_, K_, n_, r_, q_, σ_, T_] :=
SSAmericanOption[S,n,σ,T,r,q,Max[#-K,0]&];
SSAmericanPutApp[S_, K_, n_, r_, q_, σ_, T_] :=
SSAmericanOptionApp[S,n,σ,T,r,q,Max[K-#,0]&];
SSAmericanCallApp[S_, K_, n_, r_, q_, σ_, T_] :=
SSAmericanOptionApp[S,n,σ,T,r,q,Max[#-K,0]&];
```

Evaluation and Verification of Supersymmetric Models

What we shall do is to compute sample values for a Call and a Put in the supersymmetric model:

```
SSdataC = Table[
  {k, FastSSEuropeanCall[9, 10, k, 0.1, 0, 0.2, 1]}, {k, 15, 300}];
SSdataP = Table[{k, FastSSEuropeanPut[9, 10, k, 0.1, 0, 0.2, 1]},
  {k, 15, 300}];


SSdataCApp =
 Table[{k, FastSSSEuropeanCallApp[9, 10, k, 0.1, 0, 0.2, 1]},
  {k, 15, 300}];
SSdataPApp = Table[
  {k, FastSSEuropeanPutApp[9, 10, k, 0.1, 0, 0.2, 1]}, {k, 15, 300}];

SSPlotC =
 ListPlot[SSdataC, PlotJoined -> True, DisplayFunction -> Identity];
SSPlotP = ListPlot[SSdataP, PlotJoined -> True,
  DisplayFunction -> Identity];


SSPlotCApp = ListPlot[SSdataCApp,
  PlotJoined -> True, DisplayFunction -> Identity];
SSPlotPApp = ListPlot[SSdataPApp, PlotJoined -> True,
  DisplayFunction -> Identity];


Show[explot, OTSPlotC, PlotRange -> {0.69, 0.70},
  DisplayFunction -> $DisplayFunction];
```

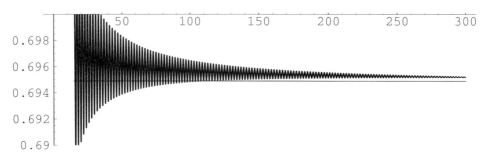

```
Show[explot, OTSPlotCApp, PlotRange -> {0.69, 0.70},
  DisplayFunction -> $DisplayFunction];
```

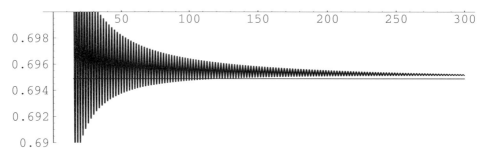

```
Show[explotP, OTSPlotP, PlotRange -> {0.74, 0.75},
  DisplayFunction -> $DisplayFunction];
```

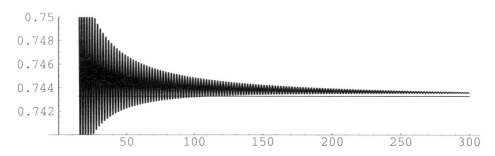

```
Show[explotP, OTSPlotPApp, PlotRange -> {0.74, 0.75},
  DisplayFunction -> $DisplayFunction];
```

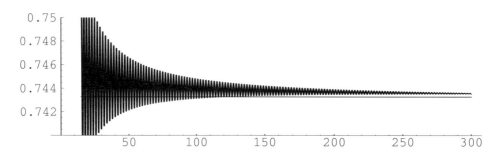

Magic Tree Sizes 3 - Approximate Supersymmetric Models

A simple formula for the strike-straddling can be given in the approximate form of this model. It is

$$n = \frac{(m + \frac{1}{2})^2 \sigma^2 T}{(\log(\frac{K}{S}) - (r - q))^2} \tag{5}$$

A *Mathematica* function for this case is:

```
straddleSS[m_, σ_, T_, S_, K_, r_, q_] :=
  ((m + 1 / 2)^2 * σ^2 * T) / (Log[K / S] - (r - q))^2
```

Let's evaluate these values (in real form - you would take the nearest integer in practice) for the parameters used above:

```
Table[straddleSS[m, 0.2, 1, 9, 10, 0.1, 0], {m, 1, 10}]

    {3132.05, 8700.15, 17052.3, 28188.5,
     42108.7, 58813., 78301.4, 100574., 125630., 153471.}
```

The denominator here is very small as the forward price is very close to the strike.

It should be appreciated that the character of the oscillations for these three (approximate) models depends very much on the closeness of:

(a) (CRR) The price to the strike: the denominator is $(\log(\frac{K}{S}))^2$.

(b) (JR) The expected price at maturity to the strike: the denominator is $\left(\log(\frac{K}{S}) - \left(r - q - \frac{\sigma^2}{2}\right)\right)^2$.

(c) (SS) The forward price to the strike: the denominator is $(\log(\frac{K}{S}) - (r - q))^2$.

The large-scale oscillation frequency is determined by the inverse squares of these quantities. It happens that in the particular numerical example given, the SS frequency is larger than the JR frequency, which is in turn larger than the CRR frequency. In other cases this situation can be permuted around. We can see that there is no one model that can be relied on in all situations. This is a fundamental problem with binomial trees. Both the supersymmetric model and the approximate JR models avoid the problem of negative weights or abscissas, but depending on the problem parameters, may or may not offer improved convergence characteristics. *One can of course pick the model with the largest denominator in the magic tree size formula, and work along the magic tree sizes that result until convergence is obtained - this is not a bad practical rule.* The averaging of odd and even n results - a technique in common use - smooths out the high-frequency oscillations, but cannot in general be relied on to give more accurate results.

19.5 Brief Investigation of the American Variants

To investigate the recursive formulae for American options, we need a good benchmark value. In Sections 16.7 and 16.8 we gave some data for American Put options with one year to expiry where we had close agreement between the integral equation approximation and the implicit finite-difference method. We shall use these as a benchmark. The strike was 10 units, the volatility 20%, the risk-free rate 5%, with zero yield and one year to expiry. The results from either method are summarized, for three values (in, at, out, the money):

$S = 9$, value 1.149, Delta −0.683, Gamma 0.312, Theta −0.141;
$S = 10$, value 0.609, Delta −0.411, Gamma 0.230, Theta −0.223;
$S = 11$, value 0.299, Delta, −0.224, Gamma 0.147, Theta −0.217.

In the following we output the results for value, Delta, Gamma, Theta, from each of our six binomial models.

CRR-exact results, $n = 100$:

```
MatrixForm[
  Table[CRRAmericanPut[S, 10, 100, 0.05, 0, 0.20, 1] , {S, 9, 11, 1}]]
```

$$\begin{pmatrix} 1.15024 & -0.683384 & 0.313254 & -0.142712 \\ 0.608719 & -0.411635 & 0.231217 & -0.226478 \\ 0.300102 & -0.223995 & 0.147282 & -0.218472 \end{pmatrix}$$

CRR-approximate results, $n = 100$:

```
MatrixForm[
  Table[CRRAmericanPutApp[S, 10, 100, 0.05, 0, 0.20, 1] , {S, 9, 11, 1}]]
```

$$\begin{pmatrix} 1.14986 & -0.683636 & 0.313591 & -0.142507 \\ 0.608235 & -0.411636 & 0.231395 & -0.22626 \\ 0.299688 & -0.223883 & 0.147337 & -0.218227 \end{pmatrix}$$

JR-exact results, $n = 100$:

```
MatrixForm[
  Table[JRAmericanPut[S, 10, 100, 0.05, 0, 0.20, 1] , {S, 9, 11, 1}]]
```

$$\begin{pmatrix} 1.14922 & -0.683449 & 0.313034 & -0.142833 \\ 0.610177 & -0.410759 & 0.229928 & -0.224678 \\ 0.298426 & -0.223286 & 0.147182 & -0.219025 \end{pmatrix}$$

JR-approximate results, $n = 100$:

```
MatrixForm[
  Table[JRAmericanPutApp[S, 10, 100, 0.05, 0, 0.20, 1], {S, 9, 11, 1}]]
```

$$\begin{pmatrix} 1.14909 & -0.68354 & 0.313151 & -0.142759 \\ 0.610003 & -0.41076 & 0.229992 & -0.224604 \\ 0.298284 & -0.223247 & 0.147201 & -0.218936 \end{pmatrix}$$

SS-exact results, $n = 100$:

```
MatrixForm[
  Table[SSAmericanPut[S, 10, 100, 0.05, 0, 0.20, 1], {S, 9, 11, 1}]]
```

$$\begin{pmatrix} 1.15006 & -0.681894 & 0.311989 & -0.142099 \\ 0.609478 & -0.410379 & 0.229748 & -0.22482 \\ 0.299629 & -0.22323 & 0.146609 & -0.217818 \end{pmatrix}$$

SS-approximate results, $n = 100$:

```
MatrixForm[
  Table[SSAmericanPutApp[S, 10, 100, 0.05, 0, 0.20, 1], {S, 9, 11, 1}]]
```

$$\begin{pmatrix} 1.15002 & -0.681927 & 0.312031 & -0.142071 \\ 0.609416 & -0.410378 & 0.229771 & -0.224791 \\ 0.299575 & -0.223216 & 0.146616 & -0.217787 \end{pmatrix}$$

These results can all be improved by taking more time-steps, bearing in mind that all the comments that related to the European case also apply here. Note also that the results for Delta, Gamma and Theta are all reasonable - indeed the results from the supersymmetric tree are excellent. This confirms our earlier assertion that it is easy to get the Delta, Gamma, Theta Greeks from any form of the tree model - there is nothing special about the CRR form, in contradiction to a common misconception. Any of these trees can also be started at earlier time-steps, to refine the estimates of the basic Greeks.

19.6 More Difficult Tree Pathology

We have so far investigated the numerical issues raised by the use of tree models in "vanilla" situations. A wonderful article by Boyle and Lau (1994) exposed some of the problems that may arise when exotics are considered. Similar issues were considered by Kat and Verdonk (1995). This section is meant to be a brief warning as to what can go wrong. Difficulties can arise when there are barriers, or caps to the option value. In such cases, the oscillations in the valuation, as a function of the number of nodes, can become massive compared to those we have already observed for the vanilla case.

We shall discuss the Barrier issue, as it raises the most severe numerical difficulties, and we shall follow Boyle and Lau in investigating a type of down-and-out Call. We already have functions to deal with the continuously sampled case, introduced in Chapter 8. Kat and Verdonk discuss the issues which also arise in a more realistic context with discrete barrier sampling. Go and load the functions if you wish to check the following:

```
? DownAndOutCall
```

```
Global`DownAndOutCall
```

```
DownAndOutCall[reb_, p_, k_, h_, sd_, r_, q_, t_] :=
  If[k >= h, intone[p, k, sd, r, q, t, 1] - intthree[p, k, h,
     sd, r, q, t, 1, 1] + intsix[reb, p, h, sd, r, q, t, 1],
  inttwo[p, k, h, sd, r, q, t, 1] - intfour[p, k, h, sd,
     r, q, t, 1, 1] + intsix[reb, p, h, sd, r, q, t, 1]]
```

Boyle's example is the case

```
DownAndOutCall[0, 95, 100, 90, 0.25, 0.1, 0, 1]
```

```
      5.99684
```

Another reason for discussing this case is that it illustrates just how easy it is to adapt a tree to cover new payoffs or boundary conditions. One might say that this is a case of "modify in haste, repent indefinitely".

```
CRRDownAndOutOptionApp[S_, n_, σ_, T_, r_, q_, H_, exercise_Function]
:=
Module[{u = CRRupBinApp[n,r,q,σ,T], d = CRRdownBinApp[n,r,q,σ,T],
a = afunc[n,r,q,T],
disc = Discount[n,r,T],
P,Q,TreeOp,value,delta,gamma,theta},
P = Pfunc[u,d,a,disc];
Q = Qfunc[u,d,a,disc];
TreeOp[node_,level_] := TreeOp[node,level] =
If[level==n,If[S d^node u^(level-node)> H, exercise[S d^node u^(level-
node)], 0],
If[S d^node u^(level-node)> H,
Max[{P,Q}.{TreeOp[node,level+1],TreeOp[node+1,level+1]},
exercise[S d^node u^(level-node)]],0]];
value=TreeOp[0,0];
delta = (TreeOp[0,1] - TreeOp[1,1])/((u-d)*S);
gamma = ((TreeOp[0,2] - TreeOp[1,2])/(u^2-1) -
(TreeOp[1,2]-TreeOp[2,2])/(1-d^2))*2/(S^2 (u^2-d^2));
theta = (TreeOp[1,2] - TreeOp[0,0])/(2*T/n);
Clear[TreeOp];
{value, delta, gamma, theta}];
```

```
CRRDownAndOutCallApp[S_, K_, H_, n_, r_, q_, σ_, T_] :=
  CRRDownAndOutOptionApp[S, n, σ, T, r, q, H, Max[# - K, 0] &];
```

Let's take a look at this function for tree sizes in multiples of 25 up to 200 - we just look at the value:

```
Table[
  {k, CRRDownAndOutCallApp[95, 100, 90, 25*k, 0.1, 0, 0.25, 1][[1]]},
  {k, 1, 8}]
```

```
          {{1, 8.84863}, {2, 7.24053}, {3, 6.30006}, {4, 7.50454},
           {5, 6.98519}, {6, 6.56115}, {7, 6.20959}, {8, 7.23071}}
```

These agree with the values given by Boyle and Lau, and illustrate the abominable convergence. Note that if we take $n = 21$, we get

```
CRRDownAndOutCallApp[95, 100, 90, 21, 0.1, 0, 0.25, 1][[1]]
```

```
    6.18582
```

which is not bad, and neither is the value for $n = 85$:

```
CRRDownAndOutCallApp[95, 100, 90, 85, 0.1, 0, 0.25, 1][[1]]
```

```
    6.01963
```

But look what happens if we increase the number of time intervals by one:

```
CRRDownAndOutCallApp[95, 100, 90, 22, 0.1, 0, 0.25, 1][[1]]
```

```
    9.21181
```

```
CRRDownAndOutCallApp[95, 100, 90, 86, 0.1, 0, 0.25, 1][[1]]
```

```
    7.8628
```

The following takes a while to evaluate, but gives the first part of the general pattern:

```
sawtooth =
 Table[
   {k, CRRDownAndOutCallApp[95, 100, 90, k, 0.1, 0, 0.25, 1][[1]]},
   {k, 2, 100}];
```

We plot it, bearing in mind that the exact solution is almost exactly 6, which is given by the lower bound of the plot:

```
ListPlot[sawtooth, PlotJoined -> True];
```

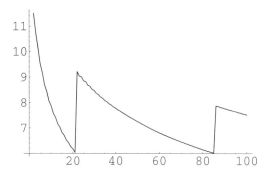

This reveals the massive nature of the error, and the particularly nasty jumps from 21 to 22 and from 85 to 86. This, like the rather smaller errors we encountered in the vanilla case, can be dealt with by careful control of the time-step - another case of magic tree sizes - and the calculation is a lot like the one we have already done. Boyle and Lau argue that to arrange things so that the barrier is close to, but just above, a layer of horizontal nodes in the tree, n must be the largest integer that is smaller than

$$\frac{m^2 \, \sigma^2 \, T}{\log^2(\frac{S}{H})} \tag{6}$$

This formula is obtained, as before, by counting the number of CRR price-steps needed to get to the barrier. We can work it out for the case under consideration.

```
BoyleLau[m_, σ_, T_, S_, H_] := Floor[T * (m * σ / Log[S / H]) ^ 2]

Table[BoyleLau[m, 0.25, 1, 95, 90], {m, 1, 5}]

    {21, 85, 192, 342, 534}
```

One point not considered by Boyle and Lau is the fact that the other types of tree cannot be remedied by a cunning choice of n. Let's take a very quick look at the JR model - the same problem arises with the supersymmetric form.

```
JRDownAndOutOptionApp[S_, n_, σ_, T_, r_, q_, H_, exercise_Function]
:=
Module[{u,d,disc = Discount[n,r,T],P,Q,TreeOp,value,delta,gamma,theta},
u = JRupBinApp[n,r,q,σ,T];
d = JRdownBinApp[n,r,q,σ,T];
P = JRUpWeight[disc];
Q = P;
TreeOp[node_,level_] := TreeOp[node,level] =
If[level==n,If[S d^node u^(level-node)> H, exercise[S d^node u^(level-
node)], 0],
If[S d^node u^(level-node)> H,
Max[{P,Q}.{TreeOp[node,level+1],TreeOp[node+1,level+1]},
exercise[S d^node u^(level-node)]],0]];
```

```
value=TreeOp[0,0];
delta = (TreeOp[0,1] - TreeOp[1,1])/((u-d)*S);
gamma = ((TreeOp[0,2] -TreeOp[1,2])/(u^2-u*d) -
(TreeOp[1,2]-TreeOp[2,2])/(u*d-d^2))*2/(S^2(u^2-d^2));
theta = (TreeOp[1,2]-(u*d-1)*S*delta - TreeOp[0,0])/(2*T/n);
Clear[TreeOp];
{value, delta, gamma, theta}];


JRDownAndOutCallApp[S_, K_, H_, n_, r_, q_, σ_, T_] :=
  JRDownAndOutOptionApp[S, n, σ, T, r, q, H, Max[# - K, 0] &];


jrtooth = Table[
  {k, JRDownAndOutCallApp[95, 100, 90, k, 0.1, 0, 0.25, 1][[1]]},
  {k, 2, 100}];
```

We plot this also - the behaviour of this plot is a reflection of the fact that with drift built into the tree, it is just not possible to align the constant barrier with the tree nodes - the result does not get anywhere near 6.

```
ListPlot[jrtooth, PlotJoined -> True];
```

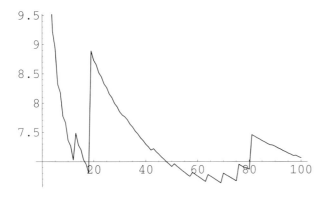

This illustrates one important advantage of the CRR model - the number of time slices in the tree can be tuned to accommodate precisely one constant Barrier. With more complex geometries there is a good case for abandoning simple binomial trees and moving to trinomial systems with a general price-step, or, better still, to finite-difference models with grids adapted to the boundaries. The reader may wish to adapt the finite-difference models described in Chapters 15 and 16 to see how well these models behave when the boundary of the grid is set to the location of barriers. There are of course additional complications in the presence of non-flat interest-rate term structures, but that is beyond the scope of this book.

The main point is to appreciate that with "standard" trees, there may be a severe mismatch between where the nodes go and any special locations induced by the option contract. If this is ignored, the values are not trustworthy. In simple vanilla cases, the best course of action is to pick a scheme where the nodes can be arranged to straddle the strike, using the methods we have described. With Barriers, the Boyle-Lau method can be used in some cases.

Chapter 19 Bibliography

Benninga, S., Steinmetz, R., and Stroughair, J., 1993, Implementing numerical option pricing models, *Mathematica Journal*, 3 issue 4, p. 66, 1993.

Boyle, P.P. and Lau, S.H., 1994, Bumping up against the barrier with the binomial method, *Journal of Derivatives*, Summer.

Cox, J., Ross, S. and Rubenstein, M., 1979, Option pricing: a simplified approach. *Journal of Financial Economics*, 7, p. 229.

Hull, J., 1996, *Options, Futures and Other Derivatives*, 3rd edition, Prentice-Hall.

Kat, H. and Verdonk, L., 1995, Tree surgery, *RISK Magazine*, February.

Jarrow, R.A. and Rudd, A., 1983, *Option Pricing*, Irwin.

Chapter 20.
Turbo-charged Trees with the *Mathematica* Compiler

Decorating the Tree with MapThread

20.1 Introduction

The release of version 3.0 of *Mathematica* eliminates one of the primary objections to its use as a serious derivatives modelling tool, in that the version 3.0 compiler permits the compilation of routines involving arrays of numerical data of any rank. Unlike some competing systems, you are not limited to vector or matrix arrays (ranks 1 and 2), allowing the set-up, in principle, of a finite-difference or tree approach to multi-factor modelling. We have already seen, in Chapters 13-17, the compiler applied to processing vector and matrix arrays for the solution of PDEs by finite-difference methods. Here we look at the application to tree modelling.

In what follows we shall explore in detail the compilation of routines using the "approximate" Cox-Ross-Rubenstein form of a binomial tree. There is clearly no difficulty in swapping in the structures appropriate to the Jarrow-Rudd or supersymmetric trees discussed in Chapters 18 and 19. We shall also look at a sample trinomial system based on the supersymmetric model.

Two styles of (compiled) computation will be explored. The first is procedural in form and uses a simple FOR loop. The second is more functional in style and exploits the list-processing functionality of *Mathematica*. The latter is about 50% more efficient on American options, and gives a factor of seven speed improvement over the uncompiled recursive programs introduced in Chapter 19.

20.2 Compiled *Mathematica* Tree Models 1 - Binomial CRR

Clearly there are several one-off computations that it is largely pointless to compile, and we re-use the following definitions without modification:

```
CRRupBinApp[n_, r_, q_, σ_, T_] := Exp[σ Sqrt[T/n]];
CRRdownBinApp[n_, r_, q_, σ_, T_] := Exp[-σ Sqrt[T/n]];
Discount[n_, r_, T_] := N[Exp[-r T/n]];
afunc[n_, r_, q_, T_] := N[Exp[(r-q) T/n]];
Pfunc[up_,down_, a_, disc_] := N[disc*(a-down)/(up-down)];
Qfunc[up_, down_, a_, disc_] := N[disc-Pfunc[up,down,a,disc]];
```

We shall want to perform verification and convergence checks on our models, so we wish to build both European and American versions based on the compiler. First we look at European style, for checking.

Evolution - European Style

First we build a compiled function to handle just one time-step:

```
OneTimeStepEuro =
 Compile[{{initial, _Real, 1}, {p, _Real, 0}, {q, _Real, 0}},
  Module[{wold = initial,
    wnew = Rest[initial], len = Length[initial] - 1, m},
   For[m = 1, m <= len, m++,
    wnew[[m]] = p * wold[[m]] + q * wold[[m + 1]]];
   wnew]];
```

When checking the operation of such functions, it is useful to give them symbolic inputs and numerical inputs:

```
OneTimeStepEuro[{a, b, c}, 1 / 3, 2 / 3]
```

```
CompiledFunction::cfts :
  Cannot use compiled code; Argument {a, b, c} at position
    1 should be a tensor of type Integer, Real, or Complex.
```

$$\{\frac{a}{3} + \frac{2\,b}{3}, \; \frac{b}{3} + \frac{2\,c}{3}\}$$

```
OneTimeStepEuro[{1 / 10, 1 / 100, 1 / 1000}, 1 / 3, 2 / 3]
```

```
{0.04, 0.004}
```

This is clearly behaving the way it should. Now we compile a routine that iterates this many times, back to the valuation point. The last time-step is handled differently in order to retain the information necessary to calculate Gamma and Theta, and is structured so that a proper *Mathematica* tensor is output.

```
ManyTimeStepEuro =
Compile[{{initial, _Real, 1},
    {p, _Real, 0}, {q, _Real, 0}, {n, _Integer, 0}},
Module[{new = initial, k, old, vold},
For[k = 1, k <= n - 1, k++,
(old = new; new = OneTimeStepEuro[old, p, q])];
vold = old; old = new;
new = OneTimeStepEuro[old, p, q];
{Join[new, old], vold}]];
```

We now just wrap this in the necessary beginning and end operations, and give the forms specific to Calls and Puts.

```
CRRTestEuroOption[S_, n_, σ_, T_, r_, q_, exercise_Function] :=
Module[{u = CRRupBinApp[n,r,q,σ,T], d = CRRdownBinApp[n,r,q,σ,T],
a = afunc[n,r,q,T],
disc = Discount[n,r,T],
P,Q,TreeOp,result,payoff},
P = Pfunc[u,d,a,disc];
Q = Qfunc[u,d,a,disc];
payoff = Table[exercise[S d^node u^(n-node)], {node, 0, n, 1}];
result = ManyTimeStepEuro[payoff,P,Q,n];
value = result[[1,1]];
delta = (result[[1,2]] - result[[1,3]])/((u-d)*S);
gamma = ((result[[2,1]] - result[[2,2]])/(u^2-1) - (result[[2,2]] -
result[[2,3]])/(1-d^2))*2/(S^2 (u^2-d^2));
theta = (result[[2,2]] - result[[1,1]])/(2*T/n);
{value, delta, gamma, theta}]


CRRTestEuroPut[S_, K_, n_, r_, q_, σ_, T_] :=
  CRRTestEuroOption[S, n, σ, T, r, q, Max[K - #, 0] &];
CRRTestEuroCall[S_, K_, n_, r_, q_, σ_, T_] :=
  CRRTestEuroOption[S, n, σ, T, r, q, Max[# - K, 0] &];
```

Let's test this for a Call option, with the same test parameters as discussed in Chapter 19, and 200 time-steps:

```
    Needs["Derivatives`BlackScholes`"]


{BlackScholesCall[9, 10, 0.2, 0.1, 0, 1],
 BlackScholesCallDelta[9, 10, 0.2, 0.1, 0, 1],
 BlackScholesCallGamma[9, 10, 0.2, 0.1, 0, 1],
 BlackScholesCallTheta[9, 10, 0.2, 0.1, 0, 1]}

    {0.694898, 0.529175, 0.221042, -0.764856}


data = CRRTestEuroCall[9, 10, 200, 0.1, 0, 0.2, 1]

    {0.694964, 0.52849, 0.221642, -0.765001}
```

We check again for a Put:

```
{BlackScholesPut[9, 10, 0.2, 0.1, 0, 1],
 BlackScholesPutDelta[9, 10, 0.2, 0.1, 0, 1],
 BlackScholesPutGamma[9, 10, 0.2, 0.1, 0, 1],
 BlackScholesPutTheta[9, 10, 0.2, 0.1, 0, 1]}
```

> {0.743272, -0.470825, 0.221042, 0.139982}

```
data = CRRTestEuroPut[9, 10, 200, 0.1, 0, 0.2, 1]
```

> {0.743339, -0.47151, 0.221642, 0.140289}

Evolution - American Style

First we build a compiled function to handle just one time-step, assuming that a payoff constraint has been supplied. We do it in two ways - a nasty FORTRAN-like form, and a cute functional way. (The author is actually rather fond of FORTRAN, especially F90, but has to say things like this to ingratiate himself with serious *Mathematica* hackers.) In the first case arrays are created using the inputs as templates; then they are updated. In the second **MapThread** is used to apply the payoff constraint and evolve the tree.

```
OneTimeStepAmer = Compile[{{initial, _Real, 1},
    {payoff, _Real, 1}, {p, _Real, 0}, {q, _Real, 0}},
  Module[{wold = initial, wnew = Rest[initial],
    len = Length[initial] - 1, m},
   For[m = 1, m <= len, m++,
    wnew[[m]] = Max[payoff[[m]], p * wold[[m]] + q * wold[[m + 1]]]];
   wnew]];
```

```
OneTimeStepAmerFun = Compile[{{initial, _Real, 1},
    {payoff, _Real, 1}, {p, _Real, 0}, {q, _Real, 0}},
  MapThread[Max,
    {payoff, p * Drop[initial, -1] + q * Drop[RotateLeft[initial], -1]}]];
```

To take the maximum, element by element, of two lists, we have used **MapThread**:

? **MapThread**

```
MapThread[f, {{a1, a2, ... }, {b1, b2, ... }, ... }] gives {f[a1,
    b1, ... ], f[a2, b2, ... ], ... }. MapThread[f, {expr1,
    expr2, ... }, n] applies f to the parts of the expri at level n.
```

We had better check that they are both working and give the same answers:

```
OneTimeStepAmer[{1, 1, 1}, {10, 0}, 1/6, 1/100]

   {10., 0.176667}


OneTimeStepAmerFun[{1, 1, 1}, {10, 0}, 1/6, 1/100]

   {10., 0.176667}
```

As **Compile** does not (yet) take functional arguments, we make functions specific to the exercise function to iterate the single-time-step routine. We give procedural and functional variants:

```
ManyTimeStepAmerCall =
Compile[{{initial, _Real, 1}, {p, _Real, 0},
   {q, _Real, 0}, {S, _Real, 0}, {u, _Real, 0},
   {d, _Real, 0}, {n, _Integer, 0}, {K, _Real, 0}},
Module[{new = initial, k, r, old, vold,
    payoff, finpay = {Max[S - K, 0]}},
For[k = 1, k <= n - 1, k++,
(payoff = Table[Max[S d^r u^(n - k - r) - K, 0], {r, 0, n - k}];
old = new; new = OneTimeStepAmer[old, payoff, p, q])];
vold = old; old = new;
new = OneTimeStepAmer[old, finpay, p, q];
{Join[new, old], vold}]];


ManyTimeStepAmerCallFun =
Compile[{{initial, _Real, 1}, {p, _Real, 0},
   {q, _Real, 0}, {S, _Real, 0}, {u, _Real, 0},
   {d, _Real, 0}, {n, _Integer, 0}, {K, _Real, 0}},
Module[{new = initial, k, r, old, vold,
    payoff, finpay = {Max[S - K, 0]}},
For[k = 1, k <= n - 1, k++,
(payoff = Table[Max[S d^r u^(n - k - r) - K, 0], {r, 0, n - k}];
 old = new;
 new = OneTimeStepAmerFun[old, payoff, p, q])];
vold = old; old = new;
new = OneTimeStepAmerFun[old, finpay, p, q];
{Join[new, old], vold}]];
```

```
ManyTimeStepAmerPut =
Compile[{{initial, _Real, 1}, {p, _Real, 0},
    {q, _Real, 0}, {S, _Real, 0}, {u, _Real, 0},
    {d, _Real, 0}, {n, _Integer, 0}, {K, _Real, 0}},
Module[{new = initial, k, r, old, vold,
    payoff, finpay = {Max[K - S, 0]}},
For[k = 1, k <= n - 1, k++,
(payoff = Table[Max[K - S d^r u^(n - k - r), 0], {r, 0, n - k}];
old = new; new = OneTimeStepAmer[old, payoff, p, q])];
vold = old; old = new;
new = OneTimeStepAmer[old, finpay, p, q];
{Join[new, old], vold}]];


ManyTimeStepAmerPutFun =
Compile[{{initial, _Real, 1}, {p, _Real, 0},
    {q, _Real, 0}, {S, _Real, 0}, {u, _Real, 0},
    {d, _Real, 0}, {n, _Integer, 0}, {K, _Real, 0}},
Module[{new = initial, k, r, old, vold,
    payoff, finpay = {Max[K - S, 0]}},
For[k = 1, k <= n - 1, k++,
(payoff = Table[Max[K - S d^r u^(n - k - r), 0], {r, 0, n - k}];
old = new; new = OneTimeStepAmerFun[old, payoff, p, q])];
vold = old; old = new;
new = OneTimeStepAmerFun[old, finpay, p, q];
{Join[new, old], vold}]];
```

Finally we wrap these four functions in the necessary end-matter:

```
CRRFastAmerCall[S_, K_, n_, r_, q_, σ_, T_] :=
Module[{u = CRRupBinApp[n,r,q,σ,T], d = CRRdownBinApp[n,r,q,σ,T],
a = afunc[n,r,q,T],
disc = Discount[n,r,T],
P,Q,TreeOp,result,payoffin},
P = Pfunc[u,d,a,disc];
Q = Qfunc[u,d,a,disc];
payoffin = Table[Max[S d^node u^(n-node)-K,0], {node, 0, n, 1}];
result = ManyTimeStepAmerCall[payoffin,P,Q,S,u,d,n,K];
value = result[[1,1]];
delta = (result[[1,2]] - result[[1,3]])/((u-d)*S);
gamma = ((result[[2,1]] - result[[2,2]])/(u^2-1) - (result[[2,2]] -
result[[2,3]])/(1-d^2))*2/(S^2 (u^2-d^2));
theta = (result[[2,2]] - result[[1,1]])/(2*T/n);
{value, delta, gamma, theta}]
```

```
CRRFastAmerPut[S_, K_, n_, r_, q_, σ_, T_] :=
Module[{u = CRRupBinApp[n,r,q,σ,T], d = CRRdownBinApp[n,r,q,σ,T],
a = afunc[n,r,q,T],
disc = Discount[n,r,T],
P,Q,TreeOp,result,payoffin},
P = Pfunc[u,d,a,disc];
Q = Qfunc[u,d,a,disc];
payoffin = Table[Max[K-S d^node u^(n-node),0], {node, 0, n, 1}];
result = ManyTimeStepAmerPut[payoffin,P,Q,S,u,d,n,K];
value = result[[1,1]];
delta = (result[[1,2]] - result[[1,3]])/((u-d)*S);
gamma = ((result[[2,1]] - result[[2,2]])/(u^2-1) - (result[[2,2]] -
result[[2,3]])/(1-d^2))*2/(S^2 (u^2-d^2));
theta = (result[[2,2]] - result[[1,1]])/(2*T/n);
{value, delta, gamma, theta}]


CRRFastAmerCallFun[S_, K_, n_, r_, q_, σ_, T_] :=
Module[{u = CRRupBinApp[n,r,q,σ,T], d = CRRdownBinApp[n,r,q,σ,T],
a = afunc[n,r,q,T],
disc = Discount[n,r,T],
P,Q,TreeOp,result,payoffin},
P = Pfunc[u,d,a,disc];
Q = Qfunc[u,d,a,disc];
payoffin = Table[Max[S d^node u^(n-node)-K,0], {node, 0, n, 1}];
result = ManyTimeStepAmerCallFun[payoffin,P,Q,S,u,d,n,K];
value = result[[1,1]];
delta = (result[[1,2]] - result[[1,3]])/((u-d)*S);
gamma = ((result[[2,1]] - result[[2,2]])/(u^2-1) - (result[[2,2]] -
result[[2,3]])/(1-d^2))*2/(S^2 (u^2-d^2));
theta = (result[[2,2]] - result[[1,1]])/(2*T/n);
{value, delta, gamma, theta}]


CRRFastAmerPutFun[S_, K_, n_, r_, q_, σ_, T_] :=
Module[{u = CRRupBinApp[n,r,q,σ,T], d = CRRdownBinApp[n,r,q,σ,T],
a = afunc[n,r,q,T],
disc = Discount[n,r,T],
P,Q,TreeOp,result,payoffin},
P = Pfunc[u,d,a,disc];
Q = Qfunc[u,d,a,disc];
payoffin = Table[Max[K-S d^node u^(n-node),0], {node, 0, n, 1}];
result = ManyTimeStepAmerPutFun[payoffin,P,Q,S,u,d,n,K];
value = result[[1,1]];
delta = (result[[1,2]] - result[[1,3]])/((u-d)*S);
gamma = ((result[[2,1]] - result[[2,2]])/(u^2-1) - (result[[2,2]] -
result[[2,3]])/(1-d^2))*2/(S^2 (u^2-d^2));
theta = (result[[2,2]] - result[[1,1]])/(2*T/n);
{value, delta, gamma, theta}]
```

Let's do some tests:

```
CRRFastAmerCall[9, 10, 200, 0.1, 0, 0.2, 1]
```

    {0.694964, 0.52849, 0.221642, -0.765001}

```
CRRFastAmerCallFun[9, 10, 200, 0.1, 0, 0.2, 1]
```

    {0.694964, 0.52849, 0.221642, -0.765001}

When the yield is zero, this is identical to the European result:

```
CRRTestEuroCall[9, 10, 200, 0.1, 0, 0.2, 1]
```

    {0.694964, 0.52849, 0.221642, -0.765001}

When the yield is non-zero, they are different:

```
CRRFastAmerCall[9, 10, 200, 0.1, 0.1, 0.2, 1]
```

    {0.33102, 0.310801, 0.191928, -0.277842}

```
CRRFastAmerCallFun[9, 10, 200, 0.1, 0.1, 0.2, 1]
```

    {0.33102, 0.310801, 0.191928, -0.277842}

```
CRRTestEuroCall[9, 10, 200, 0.1, 0.1, 0.2, 1]
```

    {0.325139, 0.303031, 0.183529, -0.264822}

To investigate the compiled formulae for American Put options, we use the same good benchmark value as before. The strike is 10 units, the volatility 20%, the risk-free rate 5%, with zero yield and one year to expiry. The results were

$S = 9$, value 1.149, Delta -0.683, Gamma 0.312, Theta -0.141;
$S = 10$, value 0.609, Delta -0.411, Gamma 0.230, Theta -0.223;
$S = 11$, value 0.299, Delta, -0.224, Gamma 0.147, Theta -0.217.

Let's work this out using the two forms of the faster code:

```
MatrixForm[
 Table[CRRFastAmerPut[S, 10, 100, 0.05, 0, 0.2, 1], {S, 9, 11, 1}]]
```

$$\begin{pmatrix} 1.14986 & -0.683636 & 0.313591 & -0.142507 \\ 0.608235 & -0.411636 & 0.231395 & -0.22626 \\ 0.299688 & -0.223883 & 0.147337 & -0.218227 \end{pmatrix}$$

```
MatrixForm[
 Table[CRRFastAmerPutFun[S, 10, 100, 0.05, 0, 0.2, 1], {S, 9, 11, 1}]]
```

$$\begin{pmatrix} 1.14986 & -0.683636 & 0.313591 & -0.142507 \\ 0.608235 & -0.411636 & 0.231395 & -0.22626 \\ 0.299688 & -0.223883 & 0.147337 & -0.218227 \end{pmatrix}$$

These are identical results to those obtained with the uncompiled recursive version developed in Chapter 19. However, the compiled forms go a lot faster!

```
Timing[CRRFastAmerPut[9, 10, 150, 0.05, 0, 0.2, 1]]
```

```
{0.883333 Second, {1.14956, -0.683474, 0.312893, -0.141592}}
```

```
Timing[CRRFastAmerPutFun[9, 10, 150, 0.05, 0, 0.20, 1]]
```

```
{0.716667 Second, {1.14956, -0.683474, 0.312893, -0.141592}}
```

```
$RecursionLimit=5000;
```

```
Timing[CRRAmericanPutApp[9, 10, 150, 0.05, 0, 0.20, 1]]
```

```
{5.08333 Second, {1.14956, -0.683474, 0.312893, -0.141592}}
```

The compiled functional form is about seven times as quick as the original recursive and uncompiled form developed by Benninga *et al* (1993), at least in the form given in Chapter 19. There may of course be other implementations, including compiled forms that follow a recursive method. As always, the author will be delighted to be told of improvements. In any case, it is now practical to ask for rather larger trees:

```
MatrixForm[
 Table[CRRFastAmerPutFun[S, 10, 500, 0.05, 0, 0.2, 1], {S, 9, 11, 1}]]
```

$$\begin{pmatrix} 1.14922 & -0.683433 & 0.312962 & -0.141916 \\ 0.608881 & -0.41117 & 0.230175 & -0.224262 \\ 0.298841 & -0.223662 & 0.146937 & -0.21759 \end{pmatrix}$$

Convergence of Trees - American Style

Now we have the power to investigate binomial tree convergence for American-style options. This still takes a while, but we can get results from almost 300 trees to play with reasonably quickly:

```
amdata = Table[{k, CRRFastAmerPutFun[9, 10, k, 0.05, 0, 0.2, 1][[1]]},
   {k, 15, 300}];
```

```
ListPlot[amdata, PlotJoined -> True];
```

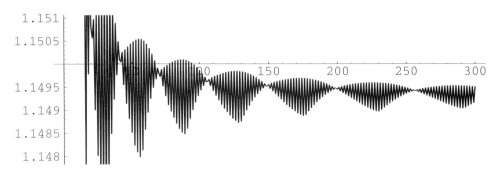

Recall that the strike-straddling formula is given by

```
straddleCRR[m_, σ_, T_, S_, K_] := (m + 1 / 2) ^ 2 σ ^ 2 T / Log[K / S] ^ 2
```

```
Table[Floor[straddleCRR[m, 0.2, 1, 9, 10]], {m, 1, 10}]
```

```
{8, 22, 44, 72, 109, 152, 202, 260, 325, 397}
```

This can be seen to be roughly the location of the envelope crossing-points, though the match is not quite as good as the European case. While I have not investigated the matter in any great detail, one has to appreciate that the convergence in the American case will also be influenced by the extent to which the moving early-exercise boundary matches to the locations of the nodes in the tree. Essentially it does not match, so the convergence is not so straightforward as in the European case.

20.3 Compiled Trinomial Trees

It is now no mystery surrounding how we apply the compiler to deal with trinomial trees. As an example, we put together a compiled form of the supersymmetric trinomial tree developed in Chapter 18. Most of the following will now, hopefully, be self-evident in its motivation.

```
Discount[n_, r_, T_] := N[Exp[-r T/n]];
afunc[n_, r_, q_, T_] := N[Exp[(r-q) T/n]];
bfunc[n_, σ_, T_] := N[Exp[σ^2 T/n]];


U[b_] := 1/2 (b*(b+1) + Sqrt[b^4 + 2 b^3 + b^2 - 4]);
up[a_, b_] := a U[b]
down[a_, b_] := a / U[b];
stay[a_, b_] := a;
pfunc[b_] := (b-1) / (1-U[b]) / (1/U[b] - U[b]);
qfunc[b_] := (b-1) / (1-1/U[b]) / (U[b] - 1/U[b]);
rfunc[b_] := 1 - pfunc[b] - qfunc[b];
```

We shall investigate the European case in some detail. First we need the function

```
OneTimeStepEuroTri = Compile[
  {{initial, _Real, 1}, {p, _Real, 0}, {q, _Real, 0}, {r, _Real, 0}},
 p*Drop[initial, -2] + q*Drop[RotateLeft[initial, 2], -2] +
   r*Drop[RotateLeft[initial], -2]];
```

The iteration for many time-steps is actually now simpler, as the computation of Greeks can be done with just two time-steps. In the following the output is the value of the option at the first time step, and the three values at the next time step - this is enough to build all the Greeks that can be computed from one call to the tree.

```
ManyTimeStepEuroTri =
Compile[{{initial, _Real, 1}, {p, _Real, 0},
    {q, _Real, 0}, {r, _Real, 0}, {n, _Integer, 0}},
Module[{new = initial, k, old},
For[k = 1, k <= n, k++,
(old = new; new = OneTimeStepEuroTri[old, p, q, r])];
Join[new, old]]];


ManyTimeStepEuroTri[{1, 2, 3, 4, 5}, 0.2, 0.3, 0.5, 2]


    {3.2, 2.1, 3.1, 4.1}
```

We write a function that just yields the valuation and all the Greeks from just one call to the tree - this is a good illustration of the use of the Rho and Vega identities applied in conjunction with the Greek extraction method used before in the Jarrow-Rudd or supersymmetric binomial schemes. As a reminder, the identities for European-style options are

$$\Lambda = S^2 \, \sigma \Gamma (T - t)$$
$$\rho = -(T - t)(V - S\Delta) \tag{1}$$

```
TrinomialCall[S_, K_, n_, r_, q_, σ_, T_] :=
Module[{a = afunc[n,r,q,T],b = bfunc[n, σ, T],
disc = Discount[n,r,T],u, d, s, P, Q, R, result,payoffin, value,
delta, gamma, theta, rho, vega},
u = up[a,b];
d = down[a,b];
s = stay[a,b];
P = disc*pfunc[b];
Q = disc*qfunc[b];
R = disc*rfunc[b];
payoffin =Join[
        Table[Max[S s^node u^(n-node)-K,0], {node, 0,n, 1}],
        Table[Max[S d^node s^(n-node)-K,0], {node, 1,n, 1}]];
result = ManyTimeStepEuroTri[payoffin, P, Q, R, n];
value = result[[1]];
delta = (result[[2]] - result[[4]])/((u-d)*S);
gamma = ((result[[2]] - result[[3]])/(u - s) -
(result[[3]] - result[[4]])/(s - d))*2/(S^2 (u-d));
theta = (result[[3]]-(s-1)*S*delta- result[[1]])*n/T;
rho = -T*(value - S*delta);
vega = T*S*S*gamma*σ;
{value, delta, gamma, theta, rho, vega}]
```

Before running it, we make sure that all the reorganization and compilation have preserved the six symmetry identities we used to build the system.

```
TrinomialCallSymmetryCheck[S_, K_, n_, r_, q_, σ_, T_] :=
Module[{a = afunc[n,r,q,T],b = bfunc[n, σ, T],u, d, s, P, Q, R},
u = up[a,b];
d = down[a,b];
s = stay[a,b];
P = pfunc[b];
Q = qfunc[b];
R = rfunc[b];
{P+Q+R-1, P*u+Q*d+R*s-a, P*u^2+Q*d^2+R*s^2-a^2*b,
P*u^3+Q*d^3+R*s^3-a^3*b^3, P/u+Q/d+R/s-b/a, P/u^2+Q/d^2+R/s^2-
b^3/a^2}];
```

```
TrinomialCallSymmetryCheck[9,10, 10, 0.1, 0, 0.2, 1]
```

$$\{0., 0., 0., 0., 0., -1.11022 \times 10^{-16}\}$$

So our implementation has preserved the supersymmetry. Here is a sample estimate of all the Greeks from the supersymmetric trinomial model with 300 slices:

```
TrinomialCall[9, 10, 300, 0.1, 0, 0.2, 1]
```

$$\{0.695003, 0.529322, 0.221107, -0.765413, 4.06889, 3.58193\}$$

Here are the exact answers:

```
Needs["Derivatives`BlackScholes`"];

{BlackScholesCall[9, 10, 0.2, 0.1, 0, 1],
 BlackScholesCallDelta[9, 10, 0.2, 0.1, 0, 1],
 BlackScholesCallGamma[9, 10, 0.2, 0.1, 0, 1],
 BlackScholesCallTheta[9, 10, 0.2, 0.1, 0, 1],
  BlackScholesCallRho[9, 10, 0.2, 0.1, 0, 1],
  BlackScholesCallVega[9, 10, 0.2, 0.1, 0, 1]}

   {0.694898, 0.529175, 0.221042, -0.764856, 4.06768, 3.58087}
```

Note that for this European-style option, for a non-CRR-style tree, we have accurately extracted all the Greeks with one call to the tree. Now we explore the convergence.

```
tridata = Table[
  {k, TrinomialCall[9, 10, k, 0.1, 0, 0.2, 1][[1]]}, {k, 15, 300}];

ListPlot[tridata, PlotJoined -> True];
```

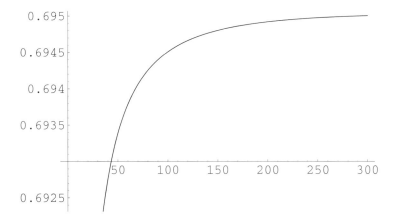

We have at least eliminated the rapid oscillations. There is a low-frequency oscillation in this particular example, but the tree does converge eventually. The American case is similar, except that re-computation is needed for accurate Rho and Vega.

Chapter 20 Bibliography

Benninga, S., Steinmetz, R., and Stroughair, J., 1993, Implementing numerical option pricing models, *Mathematica Journal*, 3, issue 4, p. 66.

Chapter 21.
Monte Carlo and Wozniakowski Sampling

Random and Pseudo-random Simulation of Paths and Events

21.1 Introduction

This chapter presents a sketch of the foundations of Monte Carlo sampling within *Mathematica*. We take it as a given that the values of options are given by the discounted (present) value of the expectation of the payoff under the risk-neutral distribution, as discussed in Chapter 3, and in detail by Baxter and Rennie (1996). One of our goals here is to explore the computations of such expectations.

Monte Carlo methods appear to be most useful for the computation of option values when the following conditions are met:

(a) there is no analytical characterization;
(b) the option is European in style, in the sense that early exercise is prohibited;
(c) the option involves path-dependent features, and/or many assets.

Path-dependent options may also be valued by the PDE approach, by suitably augmenting the variables to include a counter for the path-dependent features (see e.g. Wilmott *et al* 1993), but the additional complexity that this involves appears, to this author at least, not to be worthwhile unless there are American features, and is certainly computationally infeasible when there are several assets involved. Recent work also aims to extend the Monte Carlo approach to American-style options (see e.g. Broadie and Glasserman, 1996) but again at the price of a considerable increase in complexity. The view taken here is that Monte Carlo (abbreviated here on as MC) is technique of choice when conditions (a), (b), (c) above apply.

We shall consider the computation of hedge parameters by simulation, and we shall also give a brief introduction to the use of *non-random* samples for performing probabilistic integrals in option pricing.

21.2 Sampling the Normal Distribution

At the heart of the simulation method is the simulation of random numbers that are normally distributed. Computer programming systems usually come with a built-in random number generator that does some distribution or other, and we are required to use this to get the required normally distributed samples. *Mathematica* is no exception, but it turns out that the most obvious default approaches are not very useful.

To make the point, we shall play with random numbers for a little while, and shall begin by looking at the **Random** function.

? Random

> Random[ ] gives a uniformly distributed pseudorandom Real in the
> range 0 to 1. Random[type, range] gives a pseudorandom number
> of the specified type, lying in the specified range. Possible
> types are: Integer, Real and Complex. The default range
> is 0 to 1. You can give the range {min, max} explicitly;
> a range specification of max is equivalent to {0, max}.

Great! So we can get the uniform distribution really easily. How do we turn this into samples from the normal distribution? This is where matters become rather more interesting. A *little* knowledge can be rather counter-productive here. The elementary theory of probability tells us that if a (general) distribution has density function $f(x)$ and distribution function $F(x)$, with $f(x) = F'(x)$, then the variable

$$y = F(x) \tag{1}$$

is uniformly distributed on [0, 1]. So given a uniform distribution, to generate samples from such a distribution, we can simply invert this relationship, and set, for x uniformly distributed on [0, 1],

$$z = F^{-1}(x) \tag{2}$$

For the normal distribution, the cumulative distribution is related to the error function, and the required inverse mapping is

```
Convert[x_] := N[Sqrt[2] * InverseErf[2 * x - 1]]

{Convert[0.025], Convert[0.975]}

    {-1.95996, 1.95996}
```

In order to analyse this and other distributions, it is useful to have the statistics and graphics Packages to hand, so we load them, and make some samples:

```
Needs["Statistics`Master`"]; Needs["Graphics`Graphics`"];

ndataone = Map[(Convert[#] &), Table[Random[], {1000}]];
```

Our point will be that this takes too long, as you will have worried if you execute this in a Notebook. For the record, on a 266 MHz (G3) Power Mac:

```
Timing[Map[(Convert[#] &), Table[Random[], {1000}]];]

    {3.18333 Second, Null}
```

Let's take a look at the distribution as a histogram:

```
$DefaultFont = {"Courier", 7};

BarChart[BinCounts[ndataone, {-3.5, 3.5, 0.5}], PlotRange -> All,
BarLabels -> Map[ ToString, Table[-3.25 + 0.5 * k, {k, 0, 13}]]] ;
```

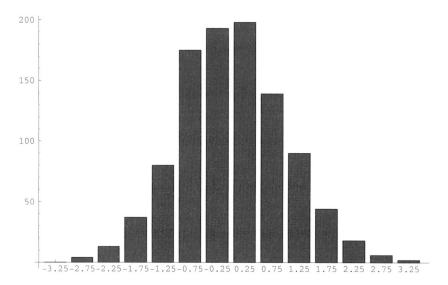

So far so good. We can use the statistics functions to output other information:

```
Mean[ndataone]
```

    0.0221019


```
Variance[ndataone]
```

    0.961322


```
StandardDeviation[ndataone]
```

    0.98047


```
StandardErrorOfSampleMean[ndataone]
```

    0.0310052

But the time taken to generate this sample is worrying. However, when we loaded the statistics Packages we secretly loaded another more interesting set of functions that relate specifically to the normal distribution:

```
? Statistics`NormalDistribution`*
```

ChiSquareDistribution NormalDistribution   PercentagePoint   Stu-
dentTDistribution
FRatioDistribution

This Package comes with a random number generator tailored to the normal distribution. This function has been improved in the transition from *Mathematica* 2.2 to 3.0. In versions prior to 3.0 it took the form (do not load this!)

```
NormalDistribution /:
  Random[NormalDistribution[mu_: 0, sigma_: 1]] :=
  N[mu + sigma Sqrt[-2 Log[Random[]]] Cos[2 Pi Random[]]]
```

This is one expression of the Box-Muller method for generating Gaussian deviates. In version 3.0 the Package contains a more efficient compiled form, with additional functionality (do not load these either - the Package has already done it!):

```
normal =
  Compile[{{mu, _Real}, {sigma, _Real}, {q1, _Real}, {q2, _Real}},
    mu + sigma Sqrt[-2 Log[q1]] Cos[2 Pi q2] ]
```

```
NormalDistribution /: Random[NormalDistribution[mu_: 0, sigma_: 1]] :=
  normal[mu, sigma, Random[], Random[]]
```

```
normalpair =
  Compile[{{mu, _Real}, {sigma, _Real}, {q1, _Real}, {q2, _Real}},
    mu + sigma Sqrt[-2 Log[q1]] {Cos[2 Pi q2], Sin[2 Pi q2]} ]
```

The last of these expresses the fact that the conversion from random sampling on a square to the normal distribution actually generates a pair of normal random numbers. Let's look at the efficiency of using the standard Package function (note that we are now using *ten* thousand calls):

```
nora[mu_, sigma_] := Random[NormalDistribution[mu, sigma]]
```

```
Timing[Table[nora[0, 1], {10000}];]
```

```
    {1.05 Second, Null}
```

In fact, re-naming it to save us typing later slowed us down slightly:

```
Timing[Table[Random[NormalDistribution[0, 1]], {10000}];]
```

```
{0.833333 Second, Null}
```

The most significant observation, at least with current versions (3.0) of *Mathematica*, is that matters can still be improved significantly, by direct compilation of the Package algorithm as a global function:

```
norb = Compile[{mu, sigma},
  mu + sigma * Sqrt[-2 Log[Random[]]] Cos[2 Pi Random[]]];
```

```
Timing[Table[norb[0, 1], {10000}];]
```

```
{0.25 Second, Null}
```

Now Press *et al* (1992) have argued that this type of approach is still not optimal, on the basis that it should be even more efficient to avoid the trigonometric computations altogether with the following routine, based on sampling with a square on {{−1, 1}, {−1, 1}}, and using only those points within the unit circle:

```
norm = Compile[{mu, sigma},
  Module[{va, vb, rad = 2.0, den},
    While[rad >= 1.00,
      (va = 2.0 * Random[] - 1.0;
       vb = 2.0 * Random[] - 1.0;
       rad = va * va + vb * vb)];
    den = Sqrt[-2.0 * Log[rad] / rad];
    mu + sigma * va * den]];
```

The benefits of this, in this particular form, *almost* outweigh the penalty of discarding the points outside the unit circle:

```
Timing[Table[norm[0, 1], {10000}];]
```

```
{0.333333 Second, Null}
```

When this routine is used in its most efficient guise, outputting pairs, it finally wins the speed war:

```
normpair = Compile[{mu, sigma}, Module[{va, vb, rad = 2.0, den},
    While[rad >= 1.00,
      (va = 2.0 * Random[] - 1.0; vb = 2.0 * Random[] - 1.0;
       rad = va * va + vb * vb)];
    den = Sqrt[-2.0 * Log[rad] / rad];
    {mu + sigma * va * den, mu + sigma * vb * den}]];
```

```
ndatatwo = Flatten[Table[normpair[0, 1], {5000}]];
```

```
Timing[Flatten[Table[normpair[0, 1], {5000}]];]
```

```
{0.233333 Second, Null}
```

If we make a similar algorithm out of the Package function, the two trigonometric calls appear to slow things down:

```
normalpair =
  Compile[{{mu, _Real}, {sigma, _Real}, {q1, _Real}, {q2, _Real}},
    mu + sigma Sqrt[-2 Log[q1]] {Cos[2 Pi q2], Sin[2 Pi q2]} ];
```

```
Timing[Flatten[Table[normalpair[0, 1, Random[], Random[]], {5000}]];]
```

```
{0.283333 Second, Null}
```

There are other tricks one can try, based on interpolations or rational function approximations to the inverse error function, but my own investigations suggest that these do not offer any speed advantages, and you pay a price in terms of approximation - as we shall see in the next chapter, the accuracy of simulation methods is a somewhat delicate matter, and we do not wish matters to be clouded further by approximations at the basic sampling level. In what follows we shall standardize on the algorithm recommended by Press *et al* (1992).

We can analyse this new data set **ndatatwo** in the same way as before:

```
BarChart[BinCounts[ndatatwo, {-3.5, 3.5, 0.5}], PlotRange -> All,
BarLabels -> Map[ ToString, Table[-3.25 + 0.5 * k, {k, 0, 13}]]];
```

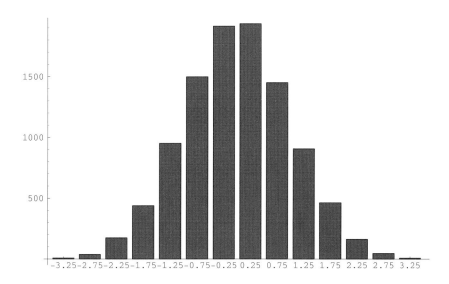

As before, we can use the statistics functions to output other information:

```
Mean[ndatatwo]
```

    -0.00393838

```
StandardDeviation[ndatatwo]
```

    0.99631

The functions **norm** and **normpair**, and some obvious antithetic extensions, will form the basis of our sampling routines henceforth. The author would be delighted to hear of still more efficient schemes.

Antithetic Variations

Here we generate pairs of samples, using the standardized normal variate of opposite sign:

```
normat = Compile[{mu, sigma},
  Module[{va, vb, rad = 2.0, den},
    While[rad >= 1.00,
      (va = 2.0 * Random[] - 1.0;
       vb = 2.0 * Random[] - 1.0;
       rad = va * va + vb * vb)];
    den = Sqrt[-2.0 * Log[rad] / rad];
    {mu + sigma * va * den, mu - sigma * va * den}]];

normpairat = Compile[{mu, sigma},
  Module[{va, vb, rad = 2.0, den},
    While[rad >= 1.00,
      (va = 2.0 * Random[] - 1.0;
       vb = 2.0 * Random[] - 1.0;
       rad = va * va + vb * vb)];
    den = Sqrt[-2.0 * Log[rad] / rad];
    {mu + sigma * va * den, mu - sigma * va * den,
     mu + sigma * vb * den, mu - sigma * vb * den}]];

{normat[0, 1], normat[1, 1]}
```

    {{0.130283, -0.130283}, {-0.206391, 2.20639}}

```
normpairat[0, 1]

    {-0.536588, 0.536588, -1.80106, 1.80106}
```

The output of four random numbers from one call to the routine is the most efficient method the author is aware of.

21.3 Simulation of Paths in Fine Detail

There are various types of simulation that we need to consider when dealing with a single underlying asset. In one case we wish to simulate the entire stock-price path in as much detail as possible, or just at a high frequency, as efficiently as possible. In the second case we wish to simulate asset values at a few particular points in time, and these may be regularly spaced or irregularly spaced. So we have three basic cases:

(a) fine detail, clockwork sampling;
(b) coarse but clockwork sampling;
(c) irregular sampling at any level of detail.

These three cases are best treated slightly differently, in order to optimize the efficiency of the process. We shall refer to the first case as "fine path simulation", the second as "coarse path simulation", and the third as "event simulation". The second method can be applied to the first case, and the third method can be applied to the first two.

Fine Path Simulation

In the following the parameters are:

s - starting asset price;
r - risk free rate;
q - dividend yield;
σ - volatility;
t - time to maturity;
n - number of sample points on the path;
navg - number of paths to be sampled.

We build the path by using the fundamental random walk characterization

$$\frac{dS}{S} = (r - q)\,\Delta T + Z\,\sigma\,\sqrt{\Delta T} \tag{3}$$

The simplest such path generator uses the simple **norm** function:

```
path[s_, r_, q_, σ_, t_, n_, navg_] :=
Module[
{mpath = 1 + (r-q)*t/(navg - 1), spath = σ Sqrt[t/(navg - 1)]},
Table[
NestList[# norm[mpath, spath]&, s, navg - 1], {i, n}] ];
```

We can instead use the independent pairs generated by **normpair** and reduce the number of paths tabulated separately by a factor of two. The function **pathpair** generates 2 *n* paths:

```
pathpair[s_, r_, q_, σ_, t_, n_, navg_] :=
Module[
{mpath = N[1 + (r-q)*t/(navg - 1)],
spath = N[σ Sqrt[t/(navg - 1)]]},
Flatten[
 Table[
  Transpose[
   NestList[# normpair[mpath, spath]&, {s, s}, navg - 1]],
  {i, n}],1]];
```

```
path[100, 0.1, 0, 0.2, 1, 4, 4]
```

    {{100, 114.776, 128.524, 133.198}, {100, 77.4593, 88.597, 86.3048},
     {100, 115.593, 98.266, 102.493}, {100, 112.346, 106.666, 129.608}}

```
pathpair[100, 0.1, 0, 0.2, 1, 2, 4]
```

    {{100, 75.4608, 82.3264, 87.4882}, {100, 129.312, 120.268, 135.048},
     {100, 107.242, 96.6208, 101.291}, {100, 111.79, 95.7271, 90.2203}}

Similarly, we can generate 4 *n* paths using two sets of antithetic pairs:

```
pathpairat[s_, r_, q_, σ_, t_, n_, navg_] :=
Module[
{mpath = N[1 + (r-q)*t/(navg - 1)],
spath = N[σ Sqrt[t/(navg - 1)]]},
Flatten[
 Table[
  Transpose[
   NestList[# normpairat[mpath, spath]&, {s, s, s, s}, navg - 1]],
  {i, n}],1]];
```

```
pathpairat[100, 0.1, 0, 0.2, 1, 1, 4]
```

    {{100, 116.675, 121.283, 110.841}, {100, 89.9912, 92.4371, 106.558},
     {100, 95.7166, 113.447, 132.907}, {100, 110.95, 97.7948, 87.539}}

Let's take a look at the relative timings, to obtain 100 paths of 100 sample points each:

```
Timing[u = path[100, 0.1, 0, 0.2, 1, 100, 100];]

    {0.516667 Second, Null}


Timing[v = pathpair[100, 0.1, 0, 0.2, 1, 50, 100];]

    {0.416667 Second, Null}


Timing[w = pathpairat[100, 0.1, 0, 0.2, 1, 25, 100];]

    {0.266667 Second, Null}


Map[Length, {u, v, w}]

    {100, 100, 100}
```

What Do Paths Look like?

Let's take a look at 20 paths of 40 steps each:

```
example = path[100, 0.1, 0, 0.2, 1, 20, 40];
plots = Table[ListPlot[example[[i]], PlotJoined -> True,
DisplayFunction -> Identity], {i, 1, Length[example]}];
Show[plots, DisplayFunction -> $DisplayFunction];
```

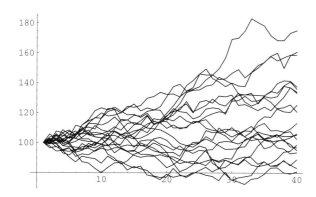

In this and the two subsequent investigations, note that any graphics you generate yourself will look different - these paths are random!

Looking at Antithetic Paths

If we take a look at two antithetic pairs, we can see the symmetry that is generated:

```
example = pathpairat[100, 0.1, 0, 0.2, 1, 1, 40];
```

```
plots = Table[ListPlot[example[[i]], PlotJoined -> True,
DisplayFunction -> Identity], {i, 1, Length[example]}];
```

```
Show[plots, DisplayFunction -> $DisplayFunction];
```

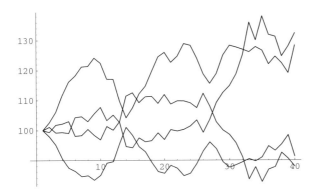

```
example = pathpairat[100, 0.1, 0, 0.2, 1, 5, 40];
plots = Table[ListPlot[example[[i]], PlotJoined -> True,
DisplayFunction -> Identity], {i, 1, Length[example]}];
Show[plots, DisplayFunction -> $DisplayFunction];
```

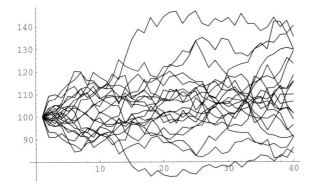

21.4 Coarse Paths and Event Simulation

Sometimes we shall wish to focus instead on a few regularly or irregularly spaced discrete events. In such circumstances it is more accurate to use the integrated form of the log-normal distribution to jump straight to each such event. We build three functions to generate the stock price at such discrete events, which parallel those we have used for the full path. When the times between events can vary, this is a good opportunity to use **FoldList**, the generalization of **NestList** that allows a separate parameter to be fed in at each stage. The stock price changes in scale between events by a multiplicative factor based on the log-normal distribution. The mean parameter is given by

$$\mu = \left(-\frac{\sigma^2}{2} - q + r\right) \Delta T \tag{4}$$

The standard deviation is given by $\sigma \sqrt{\Delta T}$. It is left as an exercise for the reader to show that the resulting path has the same *mean* drift as the one we have considered previously. Let's look first at the case of regular clockwork sampling. It is a simple matter to adjust the **path** function to use the integrated form of the log-normal distribution:

```
coarsepath[s_, r_, q_, σ_, t_, n_, navg_] :=
Module[
{mpath = N[(r-q-σ^2/2)*t/(navg - 1)], spath = N[σ Sqrt[t/(navg - 1)]]},
Table[
NestList[# Exp[norm[mpath, spath]]&, s, navg - 1], {i, n}] ];
```

Similarly, without bothering to write down the intermediate cases, we can generate $4n$ paths using two sets of antithetic pairs:

```
coarsepathpairat[s_, r_, q_, σ_, t_, n_, navg_] :=
Module[
{mpath = N[(r-q-σ^2/2)*t/(navg - 1)],
spath = N[σ Sqrt[t/(navg - 1)]]},
Flatten[
 Table[
  Transpose[
   NestList[# Exp[normpairat[mpath, spath]]&, {s, s, s, s}, navg -
1]],
  {i, n}],
 1]
];
```

Note that this is slower (0.27 seconds for the non-integrated form on the same computer):

```
Timing[w = coarsepathpairat[100, 0.1, 0, 0.2, 1, 25, 100];]

    {0.466667 Second, Null}
```

When the events are irregularly spaced, first of all we need the time-intervals between events. So, for example, suppose we have a list of times:

```
times = {ta, tb, tc};
```

We need the time intervals starting at zero - these are given by

```
Prepend[Drop[RotateLeft[times] - times, -1], First[times]]
```

    {ta, -ta + tb, -tb + tc}

The following function uses **FoldList** to incorporate the various time intervals into the sampling:

```
eventseq[s_, r_, q_, σ_, EventTimes_List, n_] :=
Module[{intervals = Prepend[Drop[RotateLeft[EventTimes]-EventTimes,
-1],
First[EventTimes]],muscale = (r - q - σ^2/2)},
Table[
FoldList[(#1*Exp[norm[muscale*#2, σ*Sqrt[#2]]])&, s, intervals], {i,
n}] ];
```

```
eventseq[100, 0.1, 0, 0.2, {0.5, 1}, 4]
```

    {{100, 107.185, 94.803}, {100, 121.38, 112.391},
     {100, 111.382, 134.198}, {100, 82.8332, 59.8648}}

Similarly, we can generate 4 *n* sets of events using our random pairs with sets of antithetic pairs:

```
eventseqpairat[s_, r_, q_, σ_, EventTimes_List, n_] :=
Module[
{intervals = Prepend[Drop[RotateLeft[EventTimes]-EventTimes, -1],
First[EventTimes]],
muscale = (r - q - σ^2/2)},
Flatten[Table[Transpose[FoldList[(#1*Exp[normpairat[muscale*#2,
σ*Sqrt[#2]]])&, {s,s,s,s}, intervals]], {i, n}],1]];
```

```
eventseqpairat[100, 0.1, 0, 0.2, {0.5, 1}, 1]
```

    {{100, 116.226, 158.816}, {100, 93.2052, 73.891},
     {100, 84.4241, 91.8794}, {100, 128.315, 127.723}}

21.5 Vanilla Options Re-visited

To check that all is working, and to gain some insight, we now look again at the pricing of simple vanilla Calls and Puts within the Monte Carlo framework. We make some comparisons with the standard analytical models, and begin by loading the required Package.

```
Needs["Derivatives`BlackScholes`"]

MonteCarloEuroCallA[S_, K_, σ_, r_, q_, t_, n_] :=
Map[(Max[0, Last[#] - K] &), eventseq[S, r, q, σ, {t}, n]] //
 Exp[-r * t] * {Mean[#], StandardErrorOfSampleMean[#]} &

MonteCarloEuroCallB[S_, K_, σ_, r_, q_, t_, n_] :=
Map[(Max[0, Last[#] - K] &),
   eventseqpairat[S, r, q, σ, {t}, Ceiling[n / 4]]] //
Exp[-r * t] * {Mean[#], StandardErrorOfSampleMean[#]} &

BlackScholesCall[110, 100, 0.2, 0.1, 0, 1]

    21.2488


Timing[MonteCarloEuroCallA[110, 100, 0.2, 0.1, 0, 1, 10000]]

    {2.03333 Second, {21.5809, 0.20097}}


Timing[MonteCarloEuroCallB[110, 100, 0.2, 0.1, 0, 1, 10000]]

    {1.46667 Second, {21.2068, 0.198906}}


Timing[MonteCarloEuroCallA[110, 100, 0.2, 0.1, 0, 1, 40000]]

    {8.18333 Second, {21.3162, 0.100076}}


Timing[MonteCarloEuroCallB[110, 100, 0.2, 0.1, 0, 1, 40000]]

    {6.05 Second, {21.3881, 0.101128}}


MonteCarloEuroPutA[S_, K_, σ_, r_, q_, t_, n_] :=
Map[(Max[0, K - Last[#]] &), eventseq[S, r, q, σ, {t}, n]] //
Exp[-r * t] * {Mean[#], StandardErrorOfSampleMean[#]} &

MonteCarloEuroPutB[S_, K_, σ_, r_, q_, t_, n_] :=
Map[(Max[0, K - Last[#]] &),
   eventseqpairat[S, r, q, σ, {t}, Ceiling[n / 4]]] //
Exp[-r * t] * {Mean[#], StandardErrorOfSampleMean[#]} &
```

```
BlackScholesPut[110, 100, 0.2, 0.1, 0, 1]
```

```
    1.73251
```

```
Timing[MonteCarloEuroPutA[110, 100, 0.2, 0.1, 0, 1, 10000]]
```

```
    {2.08333 Second, {1.6549, 0.0457261}}
```

```
Timing[MonteCarloEuroPutB[110, 100, 0.2, 0.1, 0, 1, 10000]]
```

```
    {1.51667 Second, {1.73589, 0.0477251}}
```

```
Timing[MonteCarloEuroPutA[110, 100, 0.2, 0.1, 0, 1, 40000]]
```

```
    {8.26667 Second, {1.7204, 0.0236014}}
```

```
Timing[MonteCarloEuroPutB[110, 100, 0.2, 0.1, 0, 1, 40000]]
```

```
    {6.16667 Second, {1.72123, 0.0235464}}
```

Note that these results suggest that the benefit of antithetic pair use is limited to the reduction in variance of the answer obtained from the use of twice the number of sample points, but we get a modest speed gain from the fact that we do not need to make so many calls to the random number generator.

Hedge Parameters

The key to getting good Monte Carlo estimates of hedge parameters is to understand that you have to isolate the effect of small changes in parameters from sample noise. The Golden Rule is: store the paths/event sequences, and use the same ones to work with the adjusted parameters. This is most easily illustrated using δ and Γ, but the same principles apply for ρ, Λ. To keep the *Mathematica* description simple, we work with the simplest simulation defined by **eventseq**, but with just one time, and extend this function to treat δ and Γ.

```
hedgeeventseq[s_, r_, q_, σ_, t_, n_] :=
Module[{muscale = (r - q - σ^2/2)},
Table[Exp[norm[muscale*t, σ*Sqrt[t]]]*s*{1,1.001,0.999}, {i, n}] ];
```

```
hedgeeventseq[100,0.1,0,0.2,1,5]
```

```
  {{109.734, 109.843, 109.624},
   {106.465, 106.572, 106.359}, {101.76, 101.861, 101.658},
   {88.3343, 88.4226, 88.2459}, {114.32, 114.434, 114.206}}
```

```
MonteCarloHedgeEuroCall[S_, K_, σ_, r_, q_, t_, n_] :=
Map[Exp[-r * t] * {Mean[#], StandardErrorOfSampleMean[#]} &,
  Transpose[Map[({#[[1]],
    (#[[2]] - #[[3]]) / (0.002 * S),
    (#[[2]] + #[[3]] - 2 * #[[1]]) / (0.001 * S) ^ 2} &),
   Map[(Max[0, # - K] &), hedgeeventseq[S, r, q, σ, t, n], {2}]]]]
```

Let's apply this function to our previous example. It outputs the value, δ and Γ together with their standard errors:

```
TableForm[MonteCarloHedgeEuroCall[110, 100, 0.2, 0.1, 0, 1, 40000]]
```

```
    21.158          0.0993959
    0.856831        0.00222584
    0.0121205       0.00125486
```

Here are the analytical results for comparison:

```
TableForm[{BlackScholesCall[110, 100, 0.2, 0.1, 0, 1],
  BlackScholesCallDelta[110, 100, 0.2, 0.1, 0, 1],
  BlackScholesCallGamma[110, 100, 0.2, 0.1, 0, 1]}]
```

```
    21.2488
    0.85916
    0.0101583
```

It should be noted that the computation of Γ by this method is particularly awkward, and this is reflected in the large standard error compared to the value. Most of the time the payoff is a linear function of the stock price at expiry, and most of the sampled values of Γ are zero! Only when the differencing is applied to an expiry value in the neighbourhood of the strike do we get a contribution, which is then very large. This is a critical and persistent difficulty.

Obviously the Puts work the same way:

```
MonteCarloHedgeEuroPut[S_, K_, σ_, r_, q_, t_, n_] :=
Map[Exp[-r * t] * {Mean[#], StandardErrorOfSampleMean[#]} &,
  Transpose[Map[({#[[1]], (#[[2]] - #[[3]]) / (0.002 * S),
    (#[[2]] + #[[3]] - 2 * #[[1]]) / (0.001 * S) ^ 2} &),
   Map[(Max[0, K - #] &), hedgeeventseq[S, r, q, σ, t, n], {2}]]]]
```

```
TableForm[MonteCarloHedgeEuroPut[110, 100, 0.2, 0.1, 0, 1, 40000]]
```

```
    1.78356          0.0241046
    -0.14318         0.00146466
    0.0104107        0.00113101
```

```
{BlackScholesPut[110, 100, 0.2, 0.1, 0, 1],
 BlackScholesPutDelta[110, 100, 0.2, 0.1, 0, 1],
 BlackScholesPutGamma[110, 100, 0.2, 0.1, 0, 1]}
```

```
    {1.73251, -0.14084, 0.0101583}
```

21.6 Wozniakowski Sampling

One of our methods for generating Gaussian deviates used pairs of random numbers within the unit square. We can reorganize our algorithm into two stages:

(a) first make points randomly distributed in the unit square;
(b) make samples of a normal distribution.

The first of these can be accomplished by a routine such as

```
randompoints[n_] := Table[{Random[], Random[]}, {k,1,n}];
```

The second part can be dealt with by taking the output of this and applying the function

```
hammernorm[mu_, sigma_, p_List] := mu+sigma*Sqrt[-2 Log[p[[1]]]] Cos[2
Pi p[[2]]]
```

One difficulty is that the points generated do not nicely cover the unit square - let's look at 400 of them:

```
Show[Graphics[{PointSize[0.015], Map[ Point[#]&,
randompoints[400]]}], AspectRatio -> 1];
```

There are both clumps and gaps! So it is of some interest to remedy this. Of course one can use a grid-like system to make the points more evenly distributed, but this requires a very large number of points in high dimensions. Wozniakowski (1991) has given an algorithm for numerical integration on a hypercube based on making a distribution of points that is as even as possible. The locations of the points he uses have been known for some time, but Wozniakowski was the first to demonstrate the power "on average" of the use of such points in integration. Although this illustration is only based on a square, a particularly important feature of his approach is that it remains viable in integration problems with very high dimension.

Central to the construction is the idea of representing an integer in the base of the *j*th prime number. *Mathematica* contains all the ingredients we need. The function **IntegerDigits** computes the digits of a number in a given base, while **Prime[j]** returns the *j*th prime. In two dimensions we only need the first prime, 2. Let's take a look at the **IntegerDigits** function for this case:

```
a[k_Integer] := IntegerDigits[k,2]
```

```
a[38]
```

```
    {1, 0, 0, 1, 1, 0}
```

This corresponds to the binary expansion:

```
1*32 + 0*16 + 0*8 + 1*4 + 1*2 + 0*1
```

```
    38
```

The function we need is the "radical inverse function", where we take the expansion and replace positive powers by negative powers.

```
phi[k_] := Module[{b=a[k], len},
len = Length[b];
Sum[b[[len-i]]/2^(i+1), {i,0,len-1}]]
```

```
phi[38]
```

$$\frac{25}{64}$$

This arises as

```
1/2^6 + 0/2^5 + 0/2^4 + 1/2^3 + 1/2^2 + 0/2
```

$$\frac{25}{64}$$

Using this function, we can build our nicely spread-out points thus:

```
X[n_] := Table[{1 - k/(n+1), phi[k]}, {k, 1, n}]
```

Let's take a look.

```
S = X[400];
```

```
Show[Graphics[{PointSize[0.015], Map[ Point[#]&,
S]}], AspectRatio -> 1];
```

We can easily use such distributions of points to define an option valuation technique for European options. First we define the distribution of final stock values:

```
WozSample[s_, r_, q_, σ_, t_, n_] :=
Module[{data = N[X[n]], mean = (r - q - σ^2/2)*t, sig = σ*Sqrt[t]},
Map[(s*Exp[hammernorm[mean,sig,#]]&),data]];
```

Now we define the corresponding option values:

```
WozEuroCall[S_, K_, σ_, r_, q_, t_, n_] :=
Map[(Max[0, # - K] &), WozSample[S, r, q, σ, t, n]] //
 Exp[-r*t] * {Mean[#], StandardErrorOfSampleMean[#]} &
```

```
WozEuroPut[S_, K_, σ_, r_, q_, t_, n_] :=
Map[(Max[0, K - #] &), WozSample[S, r, q, σ, t, n]] //
 Exp[-r*t] * {Mean[#], StandardErrorOfSampleMean[#]} &
```

```
BlackScholesCall[110, 100, 0.2, 0.1, 0, 1]
```

```
    21.2488
```

```
Timing[WozEuroCall[110, 100, 0.2, 0.1, 0, 1, 5000]]
```

```
    {10.3167 Second, {21.2637, 0.282562}}
```

```
BlackScholesPut[110, 100, 0.2, 0.1, 0, 1]

    1.73251


Timing[WozEuroPut[110, 100, 0.2, 0.1, 0, 1 , 5000]]

    {10.2833 Second, {1.72913, 0.0667961}}
```

Superficially, the method is slower than Monte Carlo, but we note that the resulting means are close to the true value, given the number of sample points, and it is also noteworthy that having decided on a sample size, the points only need to be computed once. In principle we could just store say 100,000 precomputed sample points on the square, and re-use them repeatedly.

This method can be used for a variety of applications, including Basket and path-dependent options, where the power of such a method when sampling in high dimensions is particularly pertinent. However, the details of this type of application are beyond the scope of this book.

Chapter 21 Bibliography

Baxter, M. and Rennie, A., 1996, *Financial Calculus - an Introduction to Derivative Pricing,* Cambridge University Press.

Broadie, M. and Glasserman, P., 1996, Pricing American-style securities by simulation, submitted to *Journal of Economic Dynamics and Control.*

Press, W.H., Teukolsky, S.A., Vetterling, W.T. and Flannery, B.P., 1992, *Numerical Recipes in C - the Art of Scientific Computing*, 2nd edition, Cambridge University Press.

Wilmott, P., Dewynne, J. and Howison, S., 1993, *Option Pricing - Mathematical Models and Computation,* Oxford Financial Press.

Wozniakowski, H., 1991, Average case complexity of multivariate integration, *Bulletin of the American Mathematical Society,* 24, p. 185.

Chapter 22.
Basic Applications of Monte Carlo

The Dangers in the Simulation of Simple Path-Dependent Options

22.1 Introduction

This chapter presents some applications of the Monte Carlo method to the simplest path-dependent options. It builds on the development given in the previous chapter to explore the simulation treatment of some common path-dependent options. All the options considered here actually have closed-form solutions, but are readily adapted to more realistic problems without closed-form descriptions. For those who may ask what the point is - we need to develop an understanding of the problems of simulation, particularly with regard to standard error control, and to demonstrate the effectiveness of control variates. It turns out that although the computation of path-dependent option values is conceptually trivial, practical issues make it difficult to secure precise and accurate answers, especially when the payoff involves extreme values along the path. Even in the case of options involving the average there are interesting subtleties in the choice of control variate. So our first applications of simulation will constitute something of a warning about potential problems.

We do, however, wish also to demonstrate the ease with which path-dependent options may be valued by using *Mathematica*. We merely generate paths and apply a payoff function. We begin by summarizing the required material from Chapter 21.

22.2 Summary of Required Technology

First we load the Package for computing means and standard errors:

```
Needs["Statistics`DescriptiveStatistics`"];
```

Next we define the function for sampling the normal distribution. We use the version with antithetic pairs, and that makes the most efficient use of the random number generator.

```
normpairat = Compile[{mu, sigma}, Module[{va, vb, rad = 2.0, den},
   While[rad >= 1.00, (va = 2.0 * Random[] - 1.0;
     vb = 2.0 * Random[] - 1.0; rad = va * va + vb * vb)];
   den = Sqrt[-2.0 * Log[rad] / rad];
   {mu + sigma * va * den, mu - sigma * va * den,
    mu + sigma * vb * den, mu - sigma * vb * den}]];
```

Now we recall our path-generation function suitable for coarse or fine paths with clockwork sampling:

```
coarsepathpairat[s_, r_, q_, σ_, t_, n_, navg_] :=
Module[{mpath = N[(r-q-σ^2/2)*t/(navg - 1)], spath = N[σ Sqrt[t/(navg
- 1)]]},
Flatten[Table[Transpose[
NestList[# Exp[normpairat[mpath, spath]]&, {s, s, s, s}, navg - 1]],
{i, n}],1]];
```

We shall use the latter for our experiments, though the reader may wish to consider alternative simulations using fine paths and/or non-antithetic pairs.

22.3 Monte Carlo Options on Extrema

In this section we define, as functions of the paths, valuation formulae for the Call on maximum and Put on minimum. In each case the most efficient variant using antithetic variables and both samples is the one actually used. Note that in each case explicit provision is made for the existence of a maximum or minimum between initiation and the valuation time. In each t is the time between valuation and expiry. The previous history serves only to supply the max or min of the path between initiation and the time of a re-valuation.

The Call on Maximum

In this case the payoff is given by Max[0, Max[path] − strike]:

```
MCCallOnMax[s_, smax_, k_, r_, q_, σ_, t_, n_, navg_] :=
(Exp[-t*r]*
Map[Max[0, Max[Prepend[#, smax]] - k]&,
coarsepathpairat[s, r, q, σ, t, n, navg]]) //
{Mean[#], StandardErrorOfSampleMean[#]}&
```

The Put on Minimum

In this case the payoff is given by Max[0, strike − Min[path]]:

```
MCPutOnMin[s_, smin_, k_, r_, q_, σ_, t_, n_, navg_] :=
(Exp[-t*r]*
Map[Max[0, k-Min[Prepend[#, smin]]]&,
coarsepathpairat[s, r, q, σ, t, n, navg]]) //
{Mean[#], StandardErrorOfSampleMean[#]}&
```

22.4 Monte Carlo Lookbacks

Here we give the corresponding formulae for the standard Lookbacks, but based on Monte Carlo.

Standard Lookback Call

In this case the payoff is given by FinalValue[path] − Min[path].

```
MCLookBackCall[s_, smin_, r_, q_, σ_, t_, n_, navg_] :=
(Exp[-t*r]*
Map[Last[#] - Min[Prepend[#, smin]]&,
coarsepathpairat[s, r, q, σ, t, n, navg]]) //
{Mean[#], StandardErrorOfSampleMean[#]}&
```

Standard Lookback Put

In this case the payoff is given by Max[path] − FinalValue[Path].

```
MCLookBackPut[s_, smax_, r_, q_, σ_, t_, n_, navg_] :=
(Exp[-t*r]*
Map[Max[Prepend[#, smax]]-Last[#]&,
coarsepathpairat[s, r, q, σ, t, n, navg]]) //
{Mean[#], StandardErrorOfSampleMean[#]}&
```

22.5 Monte Carlo Asian Options

To value an option based on paths, we simply introduce a payoff function of the path. In the example of an Asian Call, we take the mean stock price less the strike. The following formula does the case of the arithmetic average, using the basic antithetic path technique. It is readily generalized to treat the case of including an average to date, or to handle average strike options.

```
MCArithmeticAsianCall[s_, k_, r_, q_, σ_, t_, n_, navg_] :=
(Exp[-t*r]*
Map[Max[0, Mean[#] - k]&, coarsepathpairat[s, r, q, σ, t, n, navg]]) //
{Mean[#], StandardErrorOfSampleMean[#]}&
```

We can do the geometric average by a similar route:

```
MCGeometricAsianCall[s_, k_, r_, q_, σ_, t_, n_, navg_] :=
(Exp[-t*r]*
Map[Max[0, Exp[Mean[Log[#]]] - k]&,
coarsepathpairat[s, r, q, σ, t, n, navg]]) //
{Mean[#], StandardErrorOfSampleMean[#]}&
```

Similarly for the Puts:

```
MCArithmeticAsianPut[s_, k_, r_, q_, σ_, t_, n_, navg_] :=
(Exp[-t*r]*
Map[Max[0, k-Mean[#]]&,
coarsepathpairat[s, r, q, σ, t, n, navg]]) //
{Mean[#], StandardErrorOfSampleMean[#]}&
```

```
MCGeometricAsianPut[s_, k_, r_, q_, σ_, t_, n_, navg_] :=
(Exp[-t*r]*
Map[Max[0, k-Exp[Mean[Log[#]]]]&,
coarsepathpairat[s, r, q, σ, t, n, navg]]) //
{Mean[#], StandardErrorOfSampleMean[#]}&
```

Control Variate Model of the Arithmetic Asian

As an introduction to the benefits of control variate methods, we can also implement a function which gives the difference between the arithmetic and geometric cases, evaluated on the same set of paths.

```
MCAriGeoCallCorrection[s_, k_, r_, q_, σ_, t_, n_, navg_] :=
(Exp[-t*r]*
Map[(Max[0, Mean[#] - k]- Max[0, Exp[Mean[Log[#]]] - k])&,
coarsepathpairat[s, r, q, σ, t, n, navg]]) //
{Mean[#], StandardErrorOfSampleMean[#]}&

MCAriGeoPutCorrection[s_, k_, r_, q_, σ_, t_, n_, navg_] :=
(Exp[-t*r]*
Map[(Max[0, k-Mean[#]]- Max[0, k-Exp[Mean[Log[#]]]])&,
coarsepathpairat[s, r, q, σ, t, n, navg]]) //
{Mean[#], StandardErrorOfSampleMean[#]}&
```

There will be interesting results arising from the choice of precise implementation of such a control variate.

22.6 Verification Tests for Options Involving Extrema

We are now in a position to make some comparisons between the Monte Carlo method and the exact solution technique. This follows a similar approach to the one taken in Chapter 21, for the vanilla European options. However, with path-dependent options there are additional complications. In a real-world application, we may have the sampling frequency specified by the option contract to be at certain discrete intervals. Clearly we can set the parameter **navg** to represent the desired sampling frequency as long as it is regular. However, when we make comparisons with the continuous sampling limit, we face two convergence issues simultaneously:

(1) convergence rate controlled by the parameter **n** - the number of paths;
(2) convergence rate controlled by the parameter **navg** - the number of sample points on a path.

In the case of options involving extrema, it is obvious that the sampling rate within a path plays a critical role and can introduce a systematic bias. For example, the maximum sampled daily is necessarily no greater than the maximum sampled continuously.

In this section we shall make some basic comparisons between the analytical models and the Monte Carlo models to illustrate some of the convergence issues. We begin by re-examining the examples given in Chapter 9, Sections 9.4-9.7, for various types of Lookback option.

Standard Lookback Calls

We consider the example given in Section 9.4, and re-evaluate by Monte Carlo methods. Recall that with the given parameters, the exact solution for a Call, with minimum-to-date the same as the current price, was 8.6084. First we do an evaluation check with 10 paths and 10 time-steps per path.

```
MCLookBackCall[50, 50, 0.05, 0, 0.2, 1, 10, 10]
```

```
{8.0404, 1.36975}
```

This has a very high standard error. Next we use 1000 paths, but keep the number of time-steps per path at the very low figure of 10:

```
MCLookBackCall[50, 50, 0.05, 0, 0.2, 1, 1000, 10]
```

```
{7.19497, 0.11555}
```

Note that the answer has increased somewhat. Given the high standard error of the first figure, this is a fluke. Note that if we continue to keep the time-steps per path low, and increase the number of paths by a factor of 2, we lower the standard error, by about $1 / \sqrt{2}$, but the answer is no better!

```
MCLookBackCall[50, 50, 0.05, 0, 0.2, 1, 2000, 10]
```

```
{7.1541, 0.0810066}
```

Now we keep the number of paths at 2000, and raise the number of time-steps to 100:

```
MCLookBackCall[50, 50, 0.05, 0, 0.2, 1, 2000, 100]
```

```
{8.05555, 0.0792859}
```

Note that the standard error is about the same as the previous case, as one should expect, and the answer is higher, but still some way below the continuously sampled case. The minimum sampled 100 times per year is rather higher than the minimum sampled continuously, so that the payoff sampled at this discrete rate is rather lower.

This very simple example demonstrates some important points:

(a) a low standard error is no guarantee of an accurate result;
(b) the Monte Carlo valuation of options involving extrema requires a very large random sample to be used.

So we see that although the method is conceptually trivial, the process of getting a good answer may require of the order of millions of samples to be taken! The only exceptions to this are when the result is in fact dominated by the effect of a low minimum-to-date, or if the contract itself specifies low-frequency observation of the price. Recall that if the minimum-to-date was lowered to 40, the continuously sampled answer was 12.6849. If we repeat our analysis using 1000 paths of length 10, we obtain

```
MCLookBackCall[50, 40, 0.05, 0, 0.2, 1, 1000, 10]
```

```
{12.4065, 0.146316}
```

Which is a much better answer, but only because the answer is dominated by having a strike which for many paths will be 40.

Standard Lookback Puts

We consider the example given in Section 9.5, and re-evaluate by Monte Carlo methods. Recall that with the given parameters, the exact solution for a Put, with minimum-to-date the same as the current price, was 7.14528. First we do an evaluation check with 10 paths and 10 time-steps per path.

```
MCLookBackPut[50, 50, 0.05, 0, 0.2, 1, 10, 10]
```

```
{4.86046, 0.719}
```

This has a very high standard error. Next we use 1000 paths, but keep the number of time-steps per path at the very low figure of 10:

```
MCLookBackPut[50, 50, 0.05, 0, 0.2, 1, 1000, 10]
```

```
{5.27054, 0.078357}
```

```
MCLookBackPut[50, 50, 0.05, 0, 0.2, 1, 2000, 10]
```

```
{5.19474, 0.0545072}
```

As before, we keep the number of paths at 2000, and raise the number of time-steps to 100:

```
MCLookBackPut[50, 50, 0.05, 0, 0.2, 1, 2000, 100]
```

```
{6.52941, 0.0552326}
```

We are led to the same conclusions as for the Lookback Call.

Call on Maximum

We consider the example given in Section 9.6, and re-evaluate by Monte Carlo methods. Recall that with the given parameters, the exact solution for an at-the-money Call with maximum-to-date the same as the current price was 9.58381. First we do an evaluation check with 10 paths and 10 time-steps per path.

```
MCCallOnMax[50, 50, 50, 0.05, 0, 0.2, 1, 10, 10]

    {8.36809, 1.33235}
```

This has a very high standard error. Next we use 1000 paths, but keep the number of time-steps per path at 10:

```
MCCallOnMax[50, 50, 50, 0.05, 0, 0.2, 1, 1000, 10]

    {7.61404, 0.118183}
```

```
MCCallOnMax[50, 50, 50, 0.05, 0, 0.2, 1, 2000, 10]

    {7.59608, 0.0826023}
```

As before, we keep the number of paths at 2000, and raise the number of time-steps to 100:

```
MCCallOnMax[50, 50, 50, 0.05, 0, 0.2, 1, 2000, 100]

    {8.94624, 0.0856109}
```

We are led to the same conclusions as for the Lookback Call and Put.

Put on Minimum

We consider the example given in Section 9.7, and re-evaluate by Monte Carlo methods. Recall that with the given parameters, the exact solution for an at-the-money Put with minimum-to-date the same as the current price was 6.16987. First we do an evaluation check with 10 paths and 10 time-steps per path.

```
MCPutOnMin[50, 50, 50, 0.05, 0, 0.2, 1, 10, 10]

    {4.5928, 0.743312}
```

Next we use 1000 paths, but keep the number of time-steps per path at 10:

```
MCPutOnMin[50, 50, 50, 0.05, 0, 0.2, 1, 1000, 10]
```

```
{4.72631, 0.0717801}
```

```
MCPutOnMin[50, 50, 50, 0.05, 0, 0.2, 1, 2000, 10]
```

```
{4.71935, 0.0503521}
```

As before, we keep the number of paths at 2000, and raise the number of time-steps to 100:

```
MCPutOnMin[50, 50, 50, 0.05, 0, 0.2, 1, 2000, 100]
```

```
{5.69914, 0.0496797}
```

We are led to the same conclusions as for the previous three cases - when the option payoffs involve extrema the convergence to the continuously sampled limit is slow, and is controlled by the parameter **navg** giving the number of time-steps per path. The size of the standard error is a separate matter and is controlled by the number of paths, given here by **n**. The poor convergence to the continuously sampled limit is less of an issue when the contract specifies discrete sampling of the extrema, but, for example, with daily sampling, does require of the order of millions of total samples to obtain an accurate answer with a low standard error.

22.7 Verification Tests for Options Involving Averages

We can repeat our testing of the Monte Carlo method for the computation of values for arithmetic and geometric Asian options. The testing of such options offers a richer set of possibilities, as we can investigate:

(1) decrease in standard error as a function of **n**;
(2) accuracy as a function of **navg**;
(3) verification against the exact solution for the discretely sampled geometric case;
(4) influence of control variate on the standard error.

The reader is also referred to the work by Fu *et al* (1997), which describes extensive results, also, as here, using the Geman-Yor formula, as discussed in Chapter 10. Let's define a standard working example, with the following fixed parameters:

Call; strike, $K = 2$;
volatility, $\sigma = 0.5$;
risk-free rate, $r = 0.5$;
yield, $q = 0$;
time to expiry, $T = 1$;
option valued at initiation, no average to date.

We consider three price cases, $S = 1.9, 2.0, 2.1$ (out, at, in, the money), and the following situations, for which exact formulae are available:

(A) continuously sampled geometric averaging;
(B) discretely sampled geometric averaging;
(C) continuously sampled arithmetic averaging.

Case (A)

First we do the analytical valuations for the three price cases - we begin by loading the functions defined in Chapter 10 - here just the result of a query check is shown:

```
? GeoAsianPriceCall
```

```
    Global`GeoAsianPriceCall

    GeoAsianPriceCall[p_, k_, σ_, r_, q_, t_] :=
      BlackScholesCall[p, k, VolEff[σ], r, QEff[r, q, σ], t]
```

```
TableForm[Table[{S, GeoAsianPriceCall[S, 2.0, 0.5, 0.05, 0, 1]},
  {S, 1.9, 2.1, 0.1}]]
```

```
    1.9      0.17234
    2.       0.222788
    2.1      0.279743
```

10 Paths with 10 Time-Steps

```
MCGeometricAsianCall[1.9, 2, 0.05, 0.0, 0.5, 1, 10, 10]
```

```
    {0.182524, 0.0688138}
```

```
MCGeometricAsianCall[2.0, 2, 0.05, 0.0, 0.5, 1, 10, 10]
```

```
    {0.207094, 0.0571149}
```

```
MCGeometricAsianCall[2.1, 2, 0.05, 0.0, 0.5, 1, 10, 10]
```

```
    {0.286382, 0.0640291}
```

1000 Paths with 10 Time-Steps

```
MCGeometricAsianCall[1.9, 2, 0.05, 0.0, 0.5, 1, 1000, 10]
```

    {0.160651, 0.00516918}

```
MCGeometricAsianCall[2.0, 2, 0.05, 0.0, 0.5, 1, 1000, 10]
```

    {0.216129, 0.00579546}

```
MCGeometricAsianCall[2.1, 2, 0.05, 0.0, 0.5, 1, 1000, 10]
```

    {0.265238, 0.00669173}

1000 Paths with 100 Time-Steps

```
MCGeometricAsianCall[1.9, 2, 0.05, 0.0, 0.5, 1, 1000, 100]
```

    {0.174927, 0.00531081}

```
MCGeometricAsianCall[2.0, 2, 0.05, 0.0, 0.5, 1, 1000, 100]
```

    {0.218006, 0.00592822}

```
MCGeometricAsianCall[2.1, 2, 0.05, 0.0, 0.5, 1, 1000, 100]
```

    {0.280044, 0.00683811}

Case (B)

First we do the analytical valuations for the three price cases - we begin by loading the functions defined in Section 10.6 - here just the result of a query check is shown:

`? DiscreteGeoAsianPriceCall`

> Global`DiscreteGeoAsianPriceCall
>
> DiscreteGeoAsianPriceCall[p_, k_, σ_, r_, q_, ts_, T_, N_] :=
>   BlackScholesCall[p, k, VolEffD[σ, ts, T, N],
>   r, QEffD[r, q, σ, ts, T, N], T]

Let's explore the result of doing the averaging at 13 equally spaced points starting at initiation - so this is monthly sampling with equally spaced months.

```
TableForm[Table[
  {S, DiscreteGeoAsianPriceCall[S, 2.0, 0.5, 0.05, 0, 0, 1, 13]},
  {S, 1.9, 2.1, 0.1}]]
```

1.9	0.166861
2.	0.216852
2.1	0.273476

10 Paths with 12 Time-Steps

`MCGeometricAsianCall[1.9, 2, 0.05, 0.0, 0.5, 1, 10, 13]`

> {0.140096, 0.0421259}

`MCGeometricAsianCall[2.0, 2, 0.05, 0.0, 0.5, 1, 10, 13]`

> {0.212814, 0.0549732}

`MCGeometricAsianCall[2.1, 2, 0.05, 0.0, 0.5, 1, 10, 13]`

> {0.343834, 0.0909439}

1000 Paths with 12 Time-Steps

`MCGeometricAsianCall[1.9, 2, 0.05, 0.0, 0.5, 1, 1000, 13]`

> {0.157608, 0.00501741}

```
MCGeometricAsianCall[2.0, 2, 0.05, 0.0, 0.5, 1, 1000, 13]
```

```
{0.212613, 0.00586082}
```

```
MCGeometricAsianCall[2.1, 2, 0.05, 0.0, 0.5, 1, 1000, 13]
```

```
{0.27702, 0.0069206}
```

10000 Paths with 12 Time-Steps

```
MCGeometricAsianCall[1.9, 2, 0.05, 0.0, 0.5, 1, 10000, 13]
```

```
{0.167102, 0.00162602}
```

```
MCGeometricAsianCall[2.0, 2, 0.05, 0.0, 0.5, 1, 10000, 13]
```

```
{0.213316, 0.00185766}
```

```
MCGeometricAsianCall[2.1, 2, 0.05, 0.0, 0.5, 1, 10000, 13]
```

```
{0.272539, 0.00211113}
```

Case (C)

For the case of arithmetic averaging and continuous sampling the results are as given in Section 10.5, where we obtained the values of 0.193174, 0.246417, 0.306223 for the three cases. First we review the results for the arithmetic Asian with no control variate, for the two cruder of the three sample schemes:

10 Paths with 10 Time-Steps

```
MCArithmeticAsianCall[1.9, 2, 0.05, 0.0, 0.5, 1, 10, 10]
```

```
{0.216183, 0.097529}
```

```
MCArithmeticAsianCall[2.0, 2, 0.05, 0.0, 0.5, 1, 10, 10]
```

```
{0.278736, 0.0642441}
```

```
MCArithmeticAsianCall[2.1, 2, 0.05, 0.0, 0.5, 1, 10, 10]
```

```
{0.294881, 0.0642693}
```

1000 Paths with 10 Time-Steps

```
MCArithmeticAsianCall[1.9, 2, 0.05, 0.0, 0.5, 1, 1000, 10]
```

```
{0.185845, 0.00593817}
```

```
MCArithmeticAsianCall[2.0, 2, 0.05, 0.0, 0.5, 1, 1000, 10]
```

```
{0.241877, 0.00653948}
```

```
MCArithmeticAsianCall[2.1, 2, 0.05, 0.0, 0.5, 1, 1000, 10]
```

```
{0.300382, 0.00739763}
```

Addition of Control Variate

The addition of a control variate ought to make matters better. However, there are interesting complications that arise from whether one makes the choice of an unbiased control variate (these things are supposed to be zero mean!) using the discretely sampled geometric Asian, or the biased control variate arising from subtraction of the continuously sampled limit. Such a comparison has been made by Fu *et al* (1997), and the simple simulations below give credence to their view that the use of the unbiased control variate is allowing the accurate computation of the discretely sampled arithmetic Asian (which we cannot confirm in the absence of an exact solution), whereas the use of the biased control variate, somewhat obscurely, gives a better approximation to the value of the continuously sampled Asian.

10 Paths with 10 Time-Steps

```
MCAriGeoCallCorrection[1.9, 2, 0.05, 0.0, 0.5, 1, 10, 10]
```

```
{0.0242514, 0.00769201}
```

```
%[[1]] + {DiscreteGeoAsianPriceCall[1.9, 2.0, 0.5, 0.05, 0, 0, 1, 10],
DiscreteGeoAsianPriceCall[1.9, 2.0, 0.5, 0.05, 0, 0, 1, Infinity]}
```

```
{0.189458, 0.196591}
```

```
MCAriGeoCallCorrection[2.0, 2, 0.05, 0.0, 0.5, 1, 10, 10]
```

   `{0.0292901, 0.00899001}`

```
%[[1]] + {DiscreteGeoAsianPriceCall[2.0, 2.0, 0.5, 0.05, 0, 0, 1, 10],
DiscreteGeoAsianPriceCall[2.0, 2.0, 0.5, 0.05, 0, 0, 1, Infinity]}
```

   `{0.244348, 0.252078}`

```
MCAriGeoCallCorrection[2.1, 2, 0.05, 0.0, 0.5, 1, 10, 10]
```

   `{0.0293426, 0.00622787}`

```
%[[1]] + {DiscreteGeoAsianPriceCall[2.1, 2.0, 0.5, 0.05, 0, 0, 1, 10],
DiscreteGeoAsianPriceCall[2.1, 2.0, 0.5, 0.05, 0, 0, 1, Infinity]}
```

   `{0.300923, 0.309085}`

1000 Paths with 10 Time-Steps

```
MCAriGeoCallCorrection[1.9, 2, 0.05, 0.0, 0.5, 1, 1000, 10]
```

   `{0.0222768, 0.000776598}`

```
%[[1]] + {DiscreteGeoAsianPriceCall[1.9, 2.0, 0.5, 0.05, 0, 0, 1, 10],
DiscreteGeoAsianPriceCall[1.9, 2.0, 0.5, 0.05, 0, 0, 1, Infinity]}
```

   `{0.187483, 0.194617}`

```
MCAriGeoCallCorrection[2.0, 2, 0.05, 0.0, 0.5, 1, 1000, 10]
```

   `{0.0254163, 0.000793244}`

```
%[[1]] + {DiscreteGeoAsianPriceCall[2.0, 2.0, 0.5, 0.05, 0, 0, 1, 10],
DiscreteGeoAsianPriceCall[2.0, 2.0, 0.5, 0.05, 0, 0, 1, Infinity]}
```

   {0.240474, 0.248204}

```
MCAriGeoCallCorrection[2.1, 2, 0.05, 0.0, 0.5, 1, 1000, 10]
```

   {0.0297659, 0.000886579}

```
%[[1]] + {DiscreteGeoAsianPriceCall[2.1, 2.0, 0.5, 0.05, 0, 0, 1, 10],
DiscreteGeoAsianPriceCall[2.1, 2.0, 0.5, 0.05, 0, 0, 1, Infinity]}
```

   {0.301347, 0.309509}

We remind the reader that the continuously sampled values to 4 significant figures are 0.1932, 0.2464, 0.3062 for the three cases. With 1000 paths the unbiased control variate gives results of 0.188, 0.240, 0.301 with standard error of about 0.001. The biased control variate gives results of 0.195, 0.248, 0.319 with the same standard error. So the biased control variate is doing a better job of recovering the continuously sampled exact solution!

Finally let's take a look at what happens when we (a) increase the number of time steps keeping the number of paths fixed, and (b) increase the number of paths keeping the time-steps fixed.

1000 Paths with 100 Time-Steps

```
MCAriGeoCallCorrection[1.9, 2, 0.05, 0.0, 0.5, 1, 1000, 100]
```

   {0.0211872, 0.00071415}

```
%[[1]] +
  {DiscreteGeoAsianPriceCall[1.9, 2.0, 0.5, 0.05, 0, 0, 1, 100],
DiscreteGeoAsianPriceCall[1.9, 2.0, 0.5, 0.05, 0, 0, 1, Infinity]}
```

   {0.192818, 0.193527}

```
MCAriGeoCallCorrection[2.0, 2, 0.05, 0.0, 0.5, 1, 1000, 100]
```

   {0.0239402, 0.00079186}

```
%[[1]] +
 {DiscreteGeoAsianPriceCall[2.0, 2.0, 0.5, 0.05, 0, 0, 1, 100],
DiscreteGeoAsianPriceCall[2.0, 2.0, 0.5, 0.05, 0, 0, 1, Infinity]}
```

```
    {0.24596, 0.246728}
```

```
MCAriGeoCallCorrection[2.1, 2, 0.05, 0.0, 0.5, 1, 1000, 100]
```

```
    {0.0276019, 0.000809629}
```

```
%[[1]] +
 {DiscreteGeoAsianPriceCall[2.1, 2.0, 0.5, 0.05, 0, 0, 1, 100],
DiscreteGeoAsianPriceCall[2.1, 2.0, 0.5, 0.05, 0, 0, 1, Infinity]}
```

```
    {0.306534, 0.307345}
```

10000 Paths with 10 Time-Steps

```
MCAriGeoCallCorrection[1.9, 2, 0.05, 0.0, 0.5, 1, 10000, 10]
```

```
    {0.0226184, 0.000255898}
```

```
%[[1]] + {DiscreteGeoAsianPriceCall[1.9, 2.0, 0.5, 0.05, 0, 0, 1, 10],
DiscreteGeoAsianPriceCall[1.9, 2.0, 0.5, 0.05, 0, 0, 1, Infinity]}
```

```
    {0.187825, 0.194958}
```

```
MCAriGeoCallCorrection[2.0, 2, 0.05, 0.0, 0.5, 1, 10000, 10]
```

```
    {0.0259545, 0.00027229}
```

```
%[[1]] + {DiscreteGeoAsianPriceCall[2.0, 2.0, 0.5, 0.05, 0, 0, 1, 10],
DiscreteGeoAsianPriceCall[2.0, 2.0, 0.5, 0.05, 0, 0, 1, Infinity]}
```

```
    {0.241012, 0.248742}
```

```
MCAriGeoCallCorrection[2.1, 2, 0.05, 0.0, 0.5, 1, 10000, 10]
```

```
{0.0293193, 0.000291626}
```

```
%[[1]] + {DiscreteGeoAsianPriceCall[2.1, 2.0, 0.5, 0.05, 0, 0, 1, 10],
DiscreteGeoAsianPriceCall[2.1, 2.0, 0.5, 0.05, 0, 0, 1, Infinity]}
```

```
{0.3009, 0.309062}
```

These results suggest that not only can control variates assist with variance reduction, but if chosen to be suitably biased, they may also assist in remedying discretization errors. For further details see Fu *et al* (1997).

In summary, we see that it is very easy to value a wide range of path-dependent options using *Mathematica*, since payoffs that are functions of paths are easily written down. We also see, however, given *Mathematica*'s ability to deal with exact analytical solutions with equal ease, that it is easy to expose the basic difficulties that arise with Monte Carlo sampling. We have already noted that a low standard error may be no guarantee of accuracy, due to the nature of the discretization error. Time-discretization introduces biases, especially into options involving extrema when making comparisons with continuous sampling. However, when valuing contracts where the sampling of extrema or averages is required to be discrete, Monte Carlo sampling is a good method to use. The standard error may be reduced substantially by the introduction of a suitable control variate, but as our example with arithmetic Asians demonstrates, care is required in the choice of control variates.

Chapter 22 Bibliography

Fu, M.C., Madan, D.B. and Wang, T. *Pricing Asian Options: a Comparison of Analytical and Monte Carlo Methods*, University of Maryland College of Business and Management preprint, March 1997.

Chapter 23.
Monte Carlo Simulation of Basket Options

An Ideal Application of Simulation Technology

23.1 Introduction

This chapter looks at the valuation of options where the payoff depends on a combination of several assets. Sometimes called Rainbow options, a typical application is an option on a Basket. Where several correlated stochastic factors are present, the partial differential equation approach is particularly unwieldy and inappropriate, but it is a relatively straightforward matter to simulate the progress of the components of a Basket, even when the assets are correlated. Such methods are readily generalized to include Quanto effects, where the progress of associated currencies can also be simulated.

Sometimes Baskets are modelled in a crude way pretending that the total Basket may be treated as a single asset. We shall look at this approximation, and compare it with the Monte Carlo method. Such an approach fails to capture easily the appropriate hedge position for the Basket. It should be noted that to do the simulation at all in a risk-neutral environment requires that we construct an instantaneous hedge portfolio with weights that are the Δ_i for each asset i. Such weights can be captured by the Monte Carlo method quite easily. The log-normal approximation can, however, play a useful role in checking the gross levels of simulation results for errors. Also, the simulation of Baskets with many assets is a prime candidate for the use of pseudo-random numbers, though we shall not pursue this here.

A special case is afforded by the case of two assets, which can, in the case of the European instrument, easily be valued by other means, such as direct integration. If we allow one "weight" in the Basket to be negative, we obtain a spread option, and we can test the Monte Carlo method out by comparison with direct integration of the associated bivariate log-normal distribution. Of course, we could do standard Baskets by similar direct integration, and there is, for European instruments, no particular difficulty in doing so in the cases of just two or three assets. However, in many dimensions the Monte Carlo and pseudo-random approaches come into their own, where, like the PDE approach, direct integration becomes unwieldy.

This chapter will also introduce some useful techniques for memory management in *Mathematica*. The storage of many random sequences, particularly with many assets and/or entire path histories, becomes highly demanding on memory, particularly given the high-precision storage used by *Mathematica*. Here we shall present some devices for removing unwanted information from memory as the calculation proceeds. This allows more paths/assets/time-steps to be considered for a given amount of real or virtual memory.

Throughout this chapter our focus will be on options on a Basket B given by

$$B = \sum_{i=1}^{m} w_i \, S_i \qquad\qquad\qquad\qquad (1)$$

We shall do our experiments based on just two assets, but the results are applicable to general values of m.

23.2 Correlated Random Sequences

The first thing we do is to make it possible to stop *Mathematica* storing its "In" and "Out" references (to save memory), load a useful statistics Package, and stop it whining about spelling:

```
Unprotect[In, Out];
Needs["Statistics`DescriptiveStatistics`"];
Off[General::spell];
Off[General::spell1];
```

Next we define a function that builds a covariance matrix from a list of volatilities and correlations:

```
covariance[volvec_List, correl_List] :=
  Module[{len = Length[volvec]},
    Table[volvec[[i]] correl[[i, j]] volvec[[j]],
      {i, 1, len}, {j, 1, len}]];
```

As an example, suppose we have a pair of assets with

```
volassetA = 0.4;
volassetB = 0.15;
rhoAB = 0.1;
```

We set

```
volvec = {volassetA, volassetB};
correl = {{1, rhoAB}, {rhoAB, 1}};
```

```
MatrixForm[correl]
```

$$\begin{pmatrix} 1 & 0.1 \\ 0.1 & 1 \end{pmatrix}$$

```
cov = covariance[volvec, correl];
```

```
MatrixForm[cov]
```

$$\begin{pmatrix} 0.16 & 0.006 \\ 0.006 & 0.0225 \end{pmatrix}$$

Next we diagonalize the covariance matrix, as follows:

```
{evals, evecs} = Eigensystem[cov]
```

```
{{0.160261, 0.0222387},
 {{0.999053, 0.0435123}, {-0.0435123, 0.999053}}}
```

Now we form a diagonal matrix of pseudo-variances in a diagonal basis:

```
diag = Table[If[i == j, evals[[j]], 0], {i, 2}, {j, 2}];
```

```
MatrixForm[diag]
```

$$\begin{pmatrix} 0.160261 & 0 \\ 0 & 0.0222387 \end{pmatrix}$$

The associated volatilities of the independent processes are given by the square root:

```
MatrixForm[Sqrt[diag]]
```

$$\begin{pmatrix} 0.400327 & 0 \\ 0 & 0.149126 \end{pmatrix}$$

Next we define the function for sampling the normal distribution. We use the version that makes the most efficient use of the random number generator, bearing in mind that we are using just two assets here. This is easily generalized to handle n assets, throwing away one number if n is odd.

```
setgen := Module[{va, vb, rad = 2.0, den},
  While[rad >= 1.0, (va = 2.0 * Random[] - 1.0; vb = 2.0 * Random[] - 1.0;
    rad = va * va + vb * vb)]; den = Sqrt[-2.0 * Log[rad] / rad];
  {va * den, vb * den}];
```

Now we set a parameter to govern the size of the sample:

```
size = 50000;
vals = Table[setgen, {i, size}];
```

We make the antithetic sets, and then free the memory used by both **vals** and the In/Out variables:

```
avals = Join[vals, -vals];
vals =.;
Clear[In, Out];
```

At this point we have some independent pairs. Now we mix them up using the eigenvectors and eigenvalues, to generate suitably correlated sets. (See Appendix 23.1 for the mathematics of this - it is just elementary linear algebra.)

```
corrset = Transpose[Map[Transpose[evecs] . Sqrt[diag] . # &, avals]];
```

We clear out the memory again:

```
avals =.;
Clear[In, Out]
```

Now we are in a position to sample the correlated expiry values. Now we make up some parameters of interest, and compute the scale changes in the two correlated assets:

```
r = Log[1.06];
T = 225 / 365;
rootT = N[Sqrt[T]];

endassetA = Exp[(r - volassetA ^ 2 / 2) * T] * Exp[corrset[[1]] * Sqrt[T]];
endassetB = Exp[(r - volassetB ^ 2 / 2) * T] * Exp[corrset[[2]] * Sqrt[T]];

corrset =.;
```

We can now simulate the pricing of several options of interest. Let's consider first a Call on a Basket consisting of the assets equally weighted with initial prices both of 100 and unit weights. We make it an at-the-money Call and work out the possible payoffs:

```
callpayoffs = Map[Max[(# - 200), 0] &, 100 * (endassetA + endassetB)];
```

The value of the option is the discounted mean, and we also give its standard error:

```
callval = Exp[-r * T] * Mean[callpayoffs]
```

```
    17.3487
```

```
callse = Exp[-r * T] * StandardErrorOfSampleMean[callpayoffs]
```

```
    0.0827211
```

A spread is done just as easily. Consider the same two assets with a spread with strike 2:

```
sppayoffs = Map[Max[(# - 2), 0] &, 100 * (endassetA - endassetB)];
```

```
spval = Exp[-r * T] * Mean[sppayoffs]
```

12.1685

```
spse = Exp[-r * T] * StandardErrorOfSampleMean[sppayoffs]
```

0.0694514

Geobasket Control Variate Model

First we do the geometric basket in isolation:

```
geocallpayoffs =
  Map[Max[(# - 200), 0] &, 200 * Sqrt[(endassetA * endassetB)]];
```

```
geocallval = Exp[-r * T] * Mean[geocallpayoffs]
```

15.8974

```
geocallse = Exp[-r * T] * StandardErrorOfSampleMean[geocallpayoffs]
```

0.0748901

Now we work out the differences between the arithmetic and geometric Baskets, together with the standard error of the difference.

```
diffs = callpayoffs-geocallpayoffs;
```

```
diffsval =  Exp[-r*T]*Mean[diffs]
```

1.45133

```
diffse =  Exp[-r*T]*StandardErrorOfSampleMean[diffs]
```

0.0113503

Note that the standard error of the differences is well down on the original error - this is equivalent to using 55 times as many samples.

Geobasket Analytic Result

The use of the geometric average as a control variate requires the use of the Black-Scholes model with the drift and volatility set suitably. These are worked out in Appendix 23.3.

```
geovol[geoweights_, covariance_] :=
Sqrt[Sum[geoweights[[i]]*geoweights[[j]]*covariance[[i,j]],
{i,1,Length[geoweights]},{j,1,Length[geoweights]}]]
```

```
gv = geovol[{1/2, 1/2}, cov]
```

```
    0.220511
```

```
geodrift[rates_, yields_, geoweights_, covariance_] :=
Sum[geoweights[[i]]*(rates[[i]]-yields[[i]] -
(1/2)*covariance[[i,i]]), {i,1,Length[geoweights]}] +
1/2*geovol[geoweights, covariance]^2
```

```
gd = geodrift[{r, r}, {0,0}, {1/2, 1/2}, cov]
```

```
    0.0369564
```

```
Needs["Derivatives`BlackScholes`"]
```

```
geoan = BlackScholesCall[200, 200, gv, r, r-gd, T]
```

```
    15.8023
```

This compares well with the geometric average computed by simulation. We now have a control variate adjusted estimate of the arithmetic Basket, given by

```
geoan + diffsval
```

```
    17.2536
```

with a standard error of

```
diffse
```

```
    0.0113503
```

23.3 The Log-Normal Treatment of the Arithmetic Basket

In Appendix 23.2, formulae for the drift and volatility of a single asset with the same mean and variance as the expiry distribution of the basket are given. We now develop the corresponding *Mathematica* forms.

Drift Rates

First we treat the risk-neutral drift - this is given by

```
BasketDrift[rates_List, yields_List, weights_List, prices_List, T_]:=
Module[{len = Length[weights], top, bot},top =
Sum[weights[[i]]*prices[[i]]*
Exp[(rates[[i]]-yields[[i]])*T],{i,1,len}];
bot = Sum[weights[[i]]*prices[[i]],{i,1,len}];
PowerExpand[Log[top/bot]/T]]
```

First we confirm it is right for a single asset:

```
Clear[r,q];
BasketDrift[{r}, {q}, {1}, {S}, t]
```

```
    -q + r
```

For two assets, we have instead (converted to Traditional Form and numbered)

$$\frac{\log(e^{(ra-qa)t}\, Sa\, wa + e^{(rb-qb)t}\, Sb\, wb) - \log(Sa\, wa + Sb\, wb)}{t} \qquad (2)$$

For small times, this is given by

```
TraditionalForm[Simplify[Limit[%, t -> 0]]]
```

$$\frac{(ra - qa)\, Sa\, wa + (rb - qb)\, Sb\, wb}{Sa\, wa + Sb\, wb} \qquad (3)$$

which is just the suitably weighted average of the risk-neutral drifts.

Volatility

The formula for the effective volatility is

```
BasketVolSq[rates_List, yields_List, weights_List, prices_List,
cov_List, T_]:=
Module[{len = Length[weights], top, bot, drift},
drift = BasketDrift[rates, yields, weights, prices, T];
top = Sum[weights[[i]]*prices[[i]]*weights[[j]]*prices[[j]]*
Exp[(rates[[i]]-yields[[i]]+rates[[j]]-yields[[j]])*T]*
(Exp[cov[[i,j]]*T]-1),{i,1,len}, {j, 1, len}];
bot = Sum[weights[[i]]*prices[[i]],{i,1,len}];
(1/T)*Log[1+Exp[-2*drift*T]*top/(bot*bot)]
]
```

We check it for a single asset:

```
BasketVolSq[{r}, {q}, {1}, {S}, {{σ^2}}, t]
```

$$\frac{\text{Log}\left[1 + E^{-2\,(-q+r)\,t+(-2\,q+2\,r)\,t}\,(-1 + E^{t\,\sigma^2})\right]}{t}$$

```
PowerExpand[Simplify[%]]
```

$$\sigma^2$$

For two assets with weights summing to unity, and equal initial prices:

```
two = BasketVolSq[{ra,rb}, {qa,qb}, {w,1-w}, {1,1},
{{sa^2, ρ*sa*sb},
{ρ*sa*sb, sb^2}}, t];
```

```
TraditionalForm[PowerExpand[Simplify[two]]]
```

$$\frac{\log\left(\frac{e^{2\,(rb-qb)\,t}\,\left(-1+e^{sb^2\,t}\right)\,(w-1)^2-2\,e^{(-qa-qb+ra+rb)\,t}\,\left(-1+e^{sa\,sb\,t\,\rho}\right)\,w\,(w-1)+e^{2\,(ra-qa)\,t}\,\left(-1+e^{sa^2\,t}\right)\,w^2}{\left(e^{(rb-qb)\,t}\,(w-1)-e^{(ra-qa)\,t}\,w\right)^2}+1\right)}{t}$$

The small-time limit is perhaps more familiar:

```
Limit[%, t -> 0]
```

$$sb^2\,(-1+w)^2 + sa^2\,w^2 - 2\,sa\,sb\,(-1+w)\,w\,\rho$$

Application to Our Example

```
r = Log[1.06]; T = 225/365;
adrift = BasketDrift[{r,r}, {0,0}, {1/2,1/2}, {100,100}, T]

    0.0582689


avol = Sqrt[BasketVolSq[{r,r}, {0,0}, {1/2,1/2}, {100,100}, cov, T]]

    0.223456


BlackScholesCall[200, 200, avol, r, r-adrift, T]

    17.548
```

In this case the log-normal approximation has given quite a good estimate. One difficulty with such single-asset approximations is that they do not easily capture the hedge position. Risk-neutrality requires that we be delta-neutral with respect to each asset separately. These can be computed by varying the starting value for each asset separately, and using the same set of paths:

```
upAcallpayoffs =
  Map[Max[(# - 200), 0] &, 100 * (endassetA * 1.01 + endassetB)];

upAcallval = Exp[-r * T] * Mean[upAcallpayoffs]

    18.0014


deltaA = upAcallval - callval

    0.652683


upBcallpayoffs =
  Map[Max[(# - 200), 0] &, 100 * (endassetA + endassetB * 1.01)];

upBcallval = Exp[-r * T] * Mean[upBcallpayoffs]

    17.9081


deltaB = upBcallval - callval
```

```
0.559352
```

The single-asset approximation, if used naively, gives just one value of Δ:

```
BlackScholesCallDelta[200, 200, avol, r, r-adrift, T]
```

```
0.615031
```

Use of such an approximation in isolation would lead to an improperly hedged, and hence non-risk-neutral, position. The real point of this is that the hedge position is not strictly a multiple of the basket. The single-asset model, with some very careful use of the chain rule, can be used to derive the full risk-neutral hedge, but it is rather messy compared to direct calculation by simulation. A further point which the reader is invited to explore, is the fact that the single-asset log-normal model becomes less accurate as the overall basket volatility increases - it has this in common with the log-normal model approximation for arithmetically averaged Asian options.

23.4 Analytical Treatment of the Two-Asset Options

For a spread option the payoff at expiry is given by

$$\text{Max}[\text{So}[t] - \text{St}[t] - K,\ 0] \tag{4}$$

where the asset values at expiry are $\text{So}[t]$ and $\text{St}[t]$, the strike being K. In terms of samples x, y from a correlated log-normal distribution, this can be written as

$$\text{Max}\left[\text{So}\,e^{x\sqrt{t}-\frac{\sigma^2 t}{2}}\,e^{rt} - \text{St}\,e^{y\sqrt{t}-\frac{s^2 t}{2}}\,e^{rt} - K,\ 0\right] \tag{5}$$

When we integrate over the risk-neutral expiry distribution, the x-integration can be restricted to the range $(L,\ \infty)$, where L is given by setting the payoff to zero, i.e.,

$$L = \text{LowLimit}(\text{Fone}, \text{Ftwo}, K, \text{xtwo}, \sigma, s, t) := \frac{\sqrt{t}\,\sigma^2}{2} + \frac{\log\left(\frac{K}{\text{Fone}} + \frac{\text{Ftwo}\,e^{\text{xtwo}\sqrt{t}-\frac{s^2 t}{2}}}{\text{Fone}}\right)}{\sqrt{t}} \tag{6}$$

where Fo are the futures: Fo $=$ So $\text{Exp}[r*t]$ etc.

```
LowLimit[Fone_, Ftwo_, K_, xtwo_, sone_, stwo_, t_] :=
sone^2*Sqrt[t]/2 +1/Sqrt[t]*Log[K/Fone + Ftwo/Fone*Exp[xtwo*Sqrt[t] -
stwo^2*t/2]]
```

For a Call on a Basket rather than a spread, the corresponding integration limit is given by flipping a sign, i.e.

```
CallLowLimit[Fone_, Ftwo_, K_, xtwo_, sone_, stwo_, t_] :=
sone^2*Sqrt[t]/2 +1/Sqrt[t]*Log[Max[
K/Fone - Ftwo/Fone*Exp[xtwo*Sqrt[t] - stwo^2*t/2], 0.00001]]
```

Method (i): Direct Two-Dimensional Numerical Integration

The idea here is that we literally integrate the payoff numerically over the two-dimensional risk-neutral expiry distribution. In *Mathematica*, for the spread and the Basket Call respectively, the integrations are

```
Clear[So, St, K, σ, s, ρ, r, t, range, x, y]


SpreadIntegral[So_, St_, K_, σ_, s_, ρ_,r_,t_, range_] :=
Exp[-r*t]/(Sqrt[1-ρ^2]*2*Pi*σ*s)*NIntegrate[
Exp[-(x^2/σ^2+y^2/s^2 - 2*ρ*x*y/(σ*s))/(2*(1-ρ^2))]*
(So*Exp[r*t]*Exp[x*Sqrt[t] - σ^2*t/2] -
St*Exp[r*t]*Exp[y*Sqrt[t] - s^2*t/2]-K), {y, -range, range},
{x, LowLimit[So*Exp[r*t], St*Exp[r*t], K, y, σ, s, t], range}]


BasketCallIntegral[So_, St_, K_, σ_, s_, ρ_,r_,t_, range_] :=
Exp[-r*t]/(Sqrt[1-ρ^2]*2*Pi*σ*s)*NIntegrate[
Exp[-(x^2/σ^2+y^2/s^2 - 2*ρ*x*y/(σ*s))/(2*(1-ρ^2))]*(So*Exp[r*t]*
Exp[x*Sqrt[t] - σ^2*t/2] +
St*Exp[r*t]*Exp[y*Sqrt[t] - s^2*t/2]-K), {y, -range, range},
{x, CallLowLimit[So*Exp[r*t], St*Exp[r*t], K, y, σ, s, t], range}]
```

In a more explicit mathematical notation, the double integration can be written in the form

```
SpreadIntegral[So, St, K, σ, s, ρ,r,t,range]
```

$$\frac{1}{2\,\pi\,s\,\sqrt{1-\rho^2}\,\sigma}$$

$$\left(E^{-r\,t}\; \text{NIntegrate}\left[\text{Exp}\left[-\frac{\frac{x^2}{\sigma^2}+\frac{y^2}{s^2}-\frac{2\,\rho\,x\,y}{\sigma\,s}}{2\,(1-\rho^2)} \right] \left(\text{So Exp}[r\,t]\; \text{Exp}\left[x\,\sqrt{t}-\frac{\sigma^2\,t}{2} \right] - \right.\right.$$

$$\left. \text{St Exp}[r\,t]\; \text{Exp}\left[y\,\sqrt{t}-\frac{s^2\,t}{2} \right] - K \right), \{y, -\text{range, range}\},$$

$$\left. \{x, \text{LowLimit}[\text{So Exp}[r\,t], \text{St Exp}[r\,t], K, y, \sigma, s, t], \text{range}\} \right]$$

```
Timing[SpreadIntegral[100,100,2, 0.4, 0.15,0.1,Log[1.06],225/365,3]]


   {7.26667 Second, 12.0357}
```

```
BasketCallIntegral[100,100,200, 0.4, 0.15,0.1,Log[1.06],225/365,4]
```

```
      17.2325
```

Recall that the raw Monte Carlo estimate for the former was 12.17 with a standard error of 0.07, and the variance-reduced estimate for the latter was 17.25 with a standard error of 0.011. This confirms the quality of the Monte Carlo simulation, particularly when used with a control variate based on the geometric Basket.

Method (ii): One Integration Done Analytically

This is a bit of a pain for the Basket as the lower limit may be finite or infinite depending on parameters. I have therefore confined detailed attention to the spread case. The slightly nasty part is doing one integration analytically, leaving the second to be done numerically. The integral over x in the payoff of a spread option is given by linear combinations of the following function:

```
PartialG[L_, α_, β_] := Sqrt[2 Pi] Exp[β^2/(4*α)]/Sqrt[2α] (1 -
Norm[Sqrt[2α]*L - β/Sqrt[2α]])
```

The mathematical formula is

```
Simplify[PartialG[L,a,b]]
```

$$ - \frac{E^{\frac{b^2}{4a}} \sqrt{\pi} \left(-1 + \text{Erf}\left[\frac{-b+2\,a\,L}{2\sqrt{a}} \right] \right)}{2\sqrt{a}} $$

The significance of this is that its derivative is

```
ExpandAll[Simplify[-D[PartialG[L,a,b], L]]] /. L -> x
```

$$ E^{b\,x-a\,x^2} $$

and this function, with various values of a and b, is what appears in the integrand involving the correlated normal distribution and payoff. I build the result step by step. Some pen-and-paper work leads to the following expression for the once integrated payoff, with a general lower limit:

```
PartialH[y_, σ_, s_, ρ_, Fo_, Ft_, K_, t_, L_] :=
Fo*Exp[-t*σ^2/2]*PartialG[L,1/(2*σ^2*(1 - ρ^2)),Sqrt[t] +
2*y*ρ/(2*s*σ*(1-ρ^2))] -
(K+Ft*Exp[Sqrt[t]*y-t*s^2/2])*PartialG[L,1/(2*σ^2*(1 -
ρ^2)),2*y*ρ/(2*s*σ*(1-ρ^2))]
```

```
TraditionalForm[Simplify[PartialH[y,σ,s,ρ,Fo,Ft,K,t,L]]]
```

$$\frac{1}{2\sqrt{\frac{1}{2\pi\sigma^2-2\pi\rho^2\sigma^2}}}\left(e^{\frac{y^2\rho^2}{1-\rho^2}-s^4t}\left(e^{\sqrt{t}\,y}\,\text{Ft}+e^{\frac{s^2t}{2}}\,K\right)\left(\text{erf}\left(\frac{y\,\rho\,\sigma-Ls}{\sqrt{2}\,s\,(\rho^2-1)\,\sigma^2\,\sqrt{\frac{1}{\sigma^2-\rho^2\sigma^2}}}\right)-1\right)+\right.$$

(7)

$$\left.e^{\frac{1}{2}\sigma^2\left((1-\rho^2)\left(\frac{y\rho}{s\sigma-s\rho^2\sigma}+\sqrt{t}\right)^2-t\right)}\,\text{Fo}\left(\text{erf}\left(\frac{Ls+\sigma\left(s\sqrt{t}\,(\rho^2-1)\,\sigma-y\rho\right)}{\sqrt{2}\,s\,(\rho^2-1)\,\sigma^2\,\sqrt{\frac{1}{\sigma^2-\rho^2\sigma^2}}}\right)+1\right)\right)$$

Now we put in the lower limit, then put in the remaining p.d.f. for the distribution of y:

```
PartialJ[y_, σ_, s_, ρ_, Fo_, Ft_, K_, t_]:=
PartialH[y, σ, s, ρ, Fo, Ft, K, t, LowLimit[Fo, Ft, K, y, σ, s, t]];
PartialK[y_, σ_, s_, ρ_, Fo_, Ft_, K_, t_]= Simplify[Exp[-
y^2/(2*s^2*(1-ρ^2))]*PartialJ[y, σ, s, ρ, Fo, Ft, K, t]/
(2*Pi*σ*s*Sqrt[1-ρ^2])];
```

```
TraditionalForm[PartialK[y, σ, s, ρ, Fo, Ft, K, t]]
```

$$\left(e^{\frac{y^2}{2s^2(\rho^2-1)}}\left(\frac{1}{2}e^{\frac{1}{2}\sigma^2\left((1-\rho^2)\left(\frac{y\rho}{s\sigma-s\rho^2\sigma}+\sqrt{t}\right)^2-t\right)}\,\text{Fo}\left(\text{erf}\left(\frac{st(2\rho^2-1)\sigma^2-2\sqrt{t}\,y\rho\sigma+2s\log\left(\frac{e^{\sqrt{t}\,y-\frac{s^2t}{2}}\,\text{Ft}+K}{\text{Fo}}\right)}{2\sqrt{2}\,s\sqrt{t}\,(\rho^2-1)\sigma^2\,\sqrt{\frac{1}{\sigma^2-\rho^2\sigma^2}}}\right)+1\right)-\right.\right.$$

$$\left.\left.\frac{1}{2}e^{\frac{y^2\rho^2}{1-\rho^2}-s^4t}\left(e^{\sqrt{t}\,y}\,\text{Ft}+e^{\frac{s^2t}{2}}\,K\right)\left(\text{erf}\left(\frac{st\sigma^2-2\sqrt{t}\,y\rho\sigma+2s\log\left(\frac{e^{\sqrt{t}\,y-\frac{s^2t}{2}}\,\text{Ft}+K}{\text{Fo}}\right)}{2\sqrt{2}\,s\sqrt{t}\,(\rho^2-1)\sigma^2\,\sqrt{\frac{1}{\sigma^2-\rho^2\sigma^2}}}\right)+1\right)\right)\right)\right)/$$

(8)

$$\left(2\pi s\sqrt{1-\rho^2}\,\sigma\,\sqrt{\frac{1}{2\pi\sigma^2-2\pi\rho^2\sigma^2}}\right)$$

We can check that it looks sensible:

```
Plot[PartialK[y, 0.2, 0.1, 0.0,100,100,2,1], {y, -1, 1}, PlotRange ->
All];
```

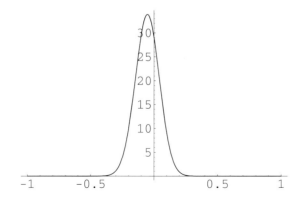

```
Dispay[Fo_, Ft_, K_, σ_, s_, ρ_, t_, r_] := Exp[-r*t]*NIntegrate[Par-
tialK[y, σ, s, ρ, Fo, Ft, K, t],
{y, -3.0, 3.0}]

Value[So_, St_, K_, σ_, s_, ρ_, t_, r_] :=Dispay[So*Exp[r*t], St*-
Exp[r*t], K, σ, s, ρ, t, r]
```

Re-calculation of our previous example gives consistent results, but much more quickly:

```
Timing[Value[100,100,2,0.4,  0.15,  0.1,225/365,Log[1.06]]]

     {0.383333 Second, 12.0357}
```

Appendix 23.1 Correlated Multivariate Samples

Suppose that we have m assets with prices or levels S_i and it is desired to construct random walks of the form

$$\frac{\Delta S_i}{S_i} = \sqrt{\Delta t}\ V_i + \Delta t\ \mu_i \tag{9}$$

where V_i, $i = 1, ..., m$, is a family of normally distributed random variables, each with zero mean, and covariance matrix

$$C_{ij} = \mathrm{Cov}(V_i\ V_j) = E[V_i\ V_j] \tag{10}$$

We use a matrix diagonalization routine based on the eigenvalues $\lambda_1{}^2, ..., \lambda_m{}^2$ (note that we write them as a square) and associated eigenvectors $e_{1i}, ..., e_{mi}$ of C. So we can write

$$C = X\Lambda X^T \tag{11}$$

where

$$\Lambda = \mathrm{Diag}(\lambda_1^2, ..., \lambda_m^2) \tag{12}$$

and X is orthogonal and built from the eigenvectors:

$$XX^T = I \tag{13}$$

Now let p_j be m independent $N(0, 1)$ random variables, and set

$$V_i = \sum_{j=1}^{m} X_{ij}\ \lambda_j\ p_j \tag{14}$$

It follows that

$$\mathrm{Cov}(V_i\ V_j) = C_{ij} \tag{15}$$

as required. In *Mathematica* the relevant diagonalization is trivial as a result of the existence of the built-in function `Eigensystem`. If you are coding this up from scratch in another language you will probably find the Cholesky decomposition most useful. The mathematics of this, and an implementation in C, are given by Press *et al* (1992), Section 2.9.

Appendix 23.2 Mean and Variance of a General Basket

The calculation of such quantities is useful in developing approximations based on single-asset log-normal approximations. We shall present an approach based on moment-generating functions that in principle allows all the moments of a basket to be computed, although here we shall just use the first two.

The Moment-Generating Function for Many Zero-Mean Normal Variables

Suppose that y_i, $i = 1, \ldots, m$, are zero-mean and normally distributed. Their covariance matrix is given by

$$W_{ij} = E[y_i\, y_j] = \Sigma_i\, \Sigma_j\, \rho_{ij} \tag{16}$$

where Σ_i is the standard deviation of y_i and ρ_{ij} is the correlation matrix. For assets it may help to remember that $\Sigma_i = \sigma_i \sqrt{T}$, where σ_i is the volatility and T the time-interval under consideration. A useful result from the theory of multivariate distributions gives us an explicit form for the moment-generating function (MGF) for this family of variables. Let

$$\omega = (\omega_1, \ldots, \omega_m) \tag{17}$$

be a vector of variables, so that the MGF is

$$M(\omega) = E[e^{\sum_{i=1}^{m} \omega_i\, y_i}] \tag{18}$$

Then the useful result we need is that

$$M(\omega) = e^{\frac{1}{2} \omega^T\, W\, \omega} \tag{19}$$

Computing Means and Variances with the MGF

Now suppose that $\omega_p = 0$ unless $p = i$, when $\omega_i = 1$. Applying this result gives us

$$E[e^{y_i}] = e^{\frac{\Sigma_i^2}{2}} = e^{\frac{\sigma_i^2\, T}{2}} \tag{20}$$

Now suppose that we set $\omega_p = 0$ unless $p = i$ or j, when $\omega_p = 1$. Applying this result and doing a little algebra gives us

$$E[e^{y_i + y_j}] = e^{\frac{\Sigma_i^2}{2} + \frac{\Sigma_j^2}{2} + \Sigma_i\, \Sigma_j\, \rho_{ij}} = e^{\frac{1}{2} T(\sigma_i^2 + 2\rho_{ij}\, \sigma_j\, \sigma_i + \sigma_j^2)} \tag{21}$$

Now we put back the means. Set $x_i = E[x_i] + y_i$, and define the asset prices or levels by

$$S_i = e^{x_i} \tag{22}$$

For a time-horizon T, if the initial value of S_i is $S_i(0)$, and the relevant risk-neutral drift rate is θ_i, typically given for equities by

$$\theta_i = r - q_i \tag{23}$$

then

$$E[x_i] = \left(\theta_i - \frac{\sigma_i^2}{2}\right) T + \log(S_i(0)) \tag{24}$$

Analysing the Basket

Let the Basket B be given by

$$B = \sum_{i=1}^{m} w_i S_i \tag{25}$$

Then the results given previously imply various results about the moments of B. First of all, at time T,

$$E[B] = \sum_{i=1}^{m} w_i S_i(0) e^{\theta_i T} \tag{26}$$

$$E[B^2] = \sum_{i,j=1}^{m} w_i w_j S_i(0) S_j(0) e^{(\theta_i + \theta_j + \sigma_i \sigma_j \rho_{ij}) T} \tag{27}$$

The variance is therefore given by

$$\text{Var}[B] = \sum_{i,j=1}^{m} w_i w_j S_i(0) S_j(0) e^{(\theta_i + \theta_j) T} (e^{\sigma_i \sigma_j \rho_{ij} T} - 1) \tag{28}$$

To develop the log-normal approximation to the basket, we assume that the mean and variance may be equated with the corresponding single asset forms, given by

$$E[B] = B(0) e^{\theta_B T} \tag{29}$$

$$\text{Var}[B] = B^2(0) e^{2\theta_B T} \left(e^{\sigma_B^2 T} - 1\right) \tag{30}$$

Equating the two means and two variances results in the formulae for the log-normal approximation given in the main body of this chapter.

Appendix 23.3 The Geometric Basket

The MGF may also be used to derive results about a related object of interest. For constant c and a vector of generalized weights γ_i we define a geometric Basket according to the prescription

$$G = c \prod_{i=1}^{m} S_i^{\gamma_i} \tag{31}$$

With the same notation as before

$$G = c \, e^{\sum_{i=1}^{m} \gamma_i \, x_i + \sum_{i=1}^{m} \gamma_i \, y_i} \tag{32}$$

Applying the MGF to the expectations of G and G^2 leads to

$$E[G] = G(0) \, e^{\frac{\gamma^T \, w_\gamma}{2} + T \sum_{i=1}^{m} \left(\theta_i - \frac{\sigma_i^2}{2}\right) \gamma_i} \tag{33}$$

$$E[G^2] = G^2(0) \, e^{2 \gamma^T \, w_\gamma + 2T \sum_{i=1}^{m} \left(\theta_i - \frac{\sigma_i^2}{2}\right) \gamma_i} \tag{34}$$

We identify these two moments with the corresponding single-asset variables (G is log-normal) by the relations

$$E[G] = G(0) \, e^{\theta_G \, T} \tag{35}$$

$$E[G^2] = G^2(0) \, e^{2 \theta_G \, T} \, e^{\sigma_G^2 \, T} \tag{36}$$

It follows that the volatility (squared) of G is given by

$$\sigma_G^2 = \sum_{i=1}^{m} \sum_{j=1}^{m} \gamma_i \, \gamma_j \, \sigma_i \, \sigma_j \, \rho_{ij} \tag{37}$$

and the risk-neutral drift of the geometric basket is

$$\theta_G = \sum_{i=1}^{m} \left(\theta_i - \frac{\sigma_i^2}{2}\right) \gamma_i + \frac{1}{2} \sum_{i=1}^{m} \sum_{j=1}^{m} \gamma_i \, \gamma_j \, \sigma_i \, \sigma_j \, \rho_{ij} \tag{38}$$

Various prescriptions for the generalized weights may be considered. Frequently the underlying arithmetic basket will represent the average return of m assets. In this case the simple geometric mean, with $\gamma_i = 1/m$, is appropriate.

Chapter 23 Bibliography

Press, W.H., Teukolsky, S.A., Vetterling, W.T. and Flannery, B.P., 1992, *Numerical Recipes in C, the Art of Scientific Computing*, 2nd edition, Cambridge University Press.

Chapter 24.
Getting Jumpy over Dividends

The Cash Dividend Conundrum - What Is the Underlying?

24.1 Introduction

We are now, finally, in a position to discuss a certain type of discontinuity in the solution of option-pricing problems. This relates to the payment of cash dividends on an underlying asset, at discrete points in time, and involves certain types of junction condition, in the partial differential equation framework. It is, under certain circumstances, readily modelled by Monte Carlo simulation, so we are now in a position to discuss this - it is a good application of the form of simulation where we look at the underlying at a few discrete points in time.

Throughout this chapter the term "dividend" is used slightly sloppily, but in accordance with common practice, to refer to the jump downwards in the stock price at the dividend date. There are additional complications when the dividend stream, as in real life, is uncertain, but here we shall be concerned with the case of a dividend stream that is known. There are three types of known dividend in common consideration:

(1) continuous-yield model, as captured by the q parameter used throughout this text - this is a good model for FX options, and a useful approximation for equity options;
(2) discrete-yield model, where a certain yield is paid at discrete points in time, with value QS with S the stock price at the dividend date, and Q the value of the discrete yield;
(3) discrete-cash model, where a certain cash value D is paid at the dividend date.

Type (1) has been dealt with explicitly in several models, and we shall deal quickly with type (2) presently. It is (3) which causes headaches, as the author found when trying to verify a finite-difference model against the analytical models for cash dividends.

The reader may have wondered why we have not given a proper discussion of the dividend issue before this point. In fact, the payment of cash dividends, type (3), is a surprisingly subtle matter. Some standard texts give results based on a prescription where the underlying is no longer the asset price, but instead is the variable

$$S^* = S - \sum_{i=1}^{n} D_i \, e^{-r t_i} \tag{1}$$

where the stock price is depressed by the present values of dividends. This is a well-established view (see for example, Hull, 1996, Section 11.2). One assumes that S^* follows a standard log-normal random walk, and it becomes the underlying for the option-pricing problem. One has to come to a view as to the volatility of S^* and how it is related to that of S, with which S^* happens to coincide between the time of the last dividend payment before expiry, and expiry.

Another perfectly acceptable formulation is given by Wilmott *et al* (1993, Section 8.3), in terms of jump conditions, who argue that the option price should be a continuous function of the dividend-jump-adjusted stock price. We shall spell this out mathematically presently. This view is of course consistent with using S^* as the underlying, and demanding that the option price V should be a continuous function of S^* - no problem there. But such a jump condition can be imposed in other representations of the process, for example, with the real stock price S as the underlying. Then the jump condition plays a more active role in relating the solution of the Black-Scholes equation just after the dividend payment, to the one just before the payment.

In case (3), that of known cash payments, the difficulty is that the view where S^* is the underlying, and following a log-normal random walk, is fundamentally inconsistent with the view that the real stock price S is the underlying, also following a log-normal random walk. This point should not be confused with some practical details, such as:

(a) how to relate the volatility of S to that of S^*;
(b) how to stop the stock price going negative in a strict cash payment model.

One can invent various schemes to try to cope with (a), and (b) is easily handled by capping the cash payment by a yield limit. We shall see examples of both of these.

Perhaps the best way of illustrating these points is by being a little careless. We shall assume the same volatility for S and S^*, fix a cash dividend, and observe the chaos. The inconsistencies are best appreciated by making the dividend large, but they are there for any size dividend. Our purpose here is first to expose the problem, and then to try to reconcile the jump-condition picture, with the real stock price as underlying, with the S^* result given by (1), and also with Monte Carlo simulation. We begin the next section with the latter.

Our last point before proceeding to the detail is to note that the management of cash dividends introduces an element of "model risk". The price you get depends on what model you assume, and in particular which underlying is regarded as log-normal. This being so, the introduction of cash dividend modelling has much in common with interest-rate derivatives, where there are several models to choose from. This is one reason for discussing dividends at this point - it forms a convenient bridge to the next chapter where we completely open up the range of distributions for the underlying.

24.2 A Monte Carlo Approach

The first thing we shall do is to build a simple and clear picture through Monte Carlo simulation. We shall add to the stock-price path a discrete jump at a series of dividend dates. We shall keep it simple and work with our standard samples **norm**:

```
norm = Compile[{mu, sigma},
Module[{va, vb, rad = 2.0, den}, While[rad >= 1.00,
    (va = 2.0 * Random[] - 1.0; vb = 2.0 * Random[] - 1.0;
    rad = va * va + vb * vb)]; den = Sqrt[-2.0 * Log[rad] / rad];
    mu + sigma * va * den]];
```

We write a function that generates a sequence of discrete events, where the stock price takes a discrete downward jump. The last event is the valuation point, where no jump is applied.

```
diveventseq[s_, r_, q_, σ_, EventTimes_List, divs_List, n_] :=
Module[{intervals = Prepend[Drop[RotateLeft[EventTimes]-EventTimes,
-1],
First[EventTimes]],muscale = (r - q - σ^2/2),extdivs,genint},
extdivs = Append[divs, 0];
genint = Transpose[{intervals, extdivs}];
Table[
FoldList[(#1*Exp[norm[muscale*#2[[1]], σ*Sqrt[#2[[1]]]]]-#2[[2]])&, s,
genint], {i, n}] ];
```

Here is just one path with a very large payment half-way through the year to expiry:

```
diveventseq[100, 0.1, 0, 0.2, {0.5, 1}, {50}, 1]
```

```
{{100, 29.7899, 31.4569}}
```

Here are four such paths:

```
diveventseq[100, 0.1, 0, 0.2, {0.5, 1}, {50}, 4]
```

```
{{100, 57.814, 57.3904}, {100, 76.8379, 92.6669},
 {100, 63.4352, 72.1161}, {100, 66.4361, 58.0004}}
```

We need some standard Package functions, which we now load:

```
Unprotect[In, Out];
Needs["Statistics`DescriptiveStatistics`"];
Off[General::spell];
Off[General::spell1];
```

Now suppose that the list **divinfo** supplied to the following function consists of pairs {time, payment}. Then it is a trivial matter to write a function to value a European Call option.

```
MonteCarloDivEuroCall[S_, K_, σ_, r_, q_, t_, divinfo_, n_] :=
 Module[{times, divs},
  times = First[Transpose[divinfo]];
  AppendTo[times, t];
  divs = Last[Transpose[divinfo]];
Map[(Max[0, Last[#] - K] &),
    diveventseq[S, r, q, σ, times, divs, n]] //
 Exp[-r * t] * {Mean[#], StandardErrorOfSampleMean[#]} &]
```

```
MonteCarloDivEuroCall[100, 100,
  0.2, Log[1.05], 0, 1, {{0.5, 20}}, 10000]
```

```
    {2.52413, 0.0725404}
```

Let's get the variance down without doing anything complicated - 250,000 samples should do the trick

```
MonteCarloDivEuroCall[100, 100,
  0.2, Log[1.05], 0, 1, {{0.5, 20}}, 250000]
```

```
    {2.58777, 0.014875}
```

So now we have one pretty unambiguous estimate for the value. Let's go for another, but before we do, let's get the result with no dividend, using an otherwise identical function call:

```
MonteCarloDivEuroCall[100, 100,
  0.2, Log[1.05], 0, 1, {{0.5, 0}}, 10000]
```

```
    {10.75, 0.150772}
```

24.3 The S^* Analytic Approach

An approach that was introduced in Chapter 11 without critical comment is the dividend-adjusted analytic model, as described by equation (1). This is based on the idea that one reduces the stock price by the present value of all the dividends, where the discounting is done at the risk-free rate. The variable S^* is regarded as the fundamental underlying asset. There is no difficulty in applying this method.

```
Needs["Derivatives`BlackScholes`"]
```

```
AnalyticalDivEuroCall[S_, K_, σ_, r_, q_, t_, divinfo_List] :=
  Module[{adjprice},
    adjprice = S - Sum[divinfo[[k, 2]] * Exp[-r * divinfo[[k, 1]]],
      {k, 1, Length[divinfo]}];
    BlackScholesCall[adjprice, K, σ, r, q, t]]
```

```
AnalyticalDivEuroCall[100, 100, 0.2, Log[1.05], 0, 1, {{0.5, 20}}]
```

```
    1.94853
```

With no dividend, but an otherwise identical computation, we get

```
AnalyticalDivEuroCall[100, 100, 0.2, Log[1.05], 0, 1, {{0.5, 0}}]
```

   10.3863

It should be clear that the dividend-adjusted result is completely contradictory with the Monte Carlo result, while the result with no dividend is consistent. We had better try it another way!

24.4 The Jump-Adjusted Finite-Difference Approach

In Chapter 4, Section 4.7, we implemented a transformation of the Black-Scholes equation valid for time-dependent yields, that could be applied to *discrete* dividends with a certain known yield. This results in the jump condition relating the values of an option just before and just after the dividend payment:

$$V(S, t_i -) = V(S\, e^{-q_i}, t_i +) \tag{2}$$

The option is a continuous function of the jump-adjusted price. Although we cannot infer the corresponding cash dividend rule directly from the particular change of variables used in Chapter 4, it makes sense to use the rule

$$V(S, t_i -) = V(S - D_i, t_i +) \tag{3}$$

to express continuity. A detailed argument as to why this holds in general is given by Wilmott *et al* (1993). However, there is the separate matter of whether the stock price S, or the adjusted price S^*, is to be regarded as the underlying. In the latter case there is nothing left to do, but in the former case some extra work is needed. We can use such jump conditions directly with any of our finite-difference models. However, it is more straightforward to use a standard two-time-level scheme, as the three-time-level scheme needs a careful re-start after the dividend payment. We shall therefore use the two-time-level Douglas scheme. This is chosen deliberately as we do not wish our conclusions to be undermined by any scepticism on the part of the reader regarding the author's judgement as to good (i.e. exotic) and bad FD schemes! The dividend jump condition will be applied making good use of *Mathematica*'s built-in interpolation routines. We shall do enough to find out which of the previous estimates is consistent with the FD approach with S as the underlying - readers may wish to clean up this scheme into something more palatable, or explore the effect of adjusting volatilities between the pre- and post-dividend-payment phases. We shall set it up in a form tailored to the particular problem at hand, and stay resolutely European, in order to keep it simple.

First a reminder of the solver we shall use:

```
CompTridiagSolve =      Compile[
  {{a, _Real, 1}, {b, _Real, 1},      {c, _Real, 1}, {r, _Real, 1}},
  Module[{len = Length[r], solution = r,
    aux = 1 / (b[[1]]),     aux1 = r, a1 = Prepend[a, 0.0], iter},
   solution[[1]] = aux * r[[1]];    Do[aux1[[iter]] = c[[iter - 1]] aux;
     aux = 1 / (b[[iter]] - a1[[iter]] * aux1[[iter]]);
     solution[[iter]] = (r[[iter]] - a1[[iter]] solution[[iter - 1]]) aux,
   {iter, 2, len}];
   Do[solution[[iter]] -= aux1[[iter + 1]] solution[[iter + 1]],
   {iter, len - 1, 1, -1}]; solution]];
```

Here are the standard functions expressing the change of variables:

$$\text{NonDimExpiry}[T\_, \sigma\_] := \frac{\sigma^2 \, T}{2};$$

$$\text{kone}[r\_, \sigma\_] := \frac{2 \, r}{\sigma^2};$$

$$\text{ktwo}[r\_, q\_, sd\_] := \frac{2 \, (r - q)}{sd^2};$$

ValuationMultiplier[strike_, r_, q_, x_, tau_, sd_] := strike

$$\text{Exp}\left[-\frac{1}{2} \, (\text{ktwo}[r, q, sd] - 1) \, x - \left(\frac{1}{4} \, (\text{ktwo}[r, q, sd] - 1)^2 + \text{kone}[r, sd]\right) \text{tau}\right]$$

Here is the initial condition:

CallExercise[x_, r_, q_, sd_] :=

$$\text{Max}\left[\text{Exp}\left[\frac{1}{2} \, (\text{ktwo}[r, q, sd] - 1) \, x\right] \, (\text{Exp}[x] - 1), \, 0\right];$$

Here are the boundary conditions:

f[x_, tau_, r_, q_, sd_] := 0;
g[x_, tau_, r_, q_, sd_] := Exp[

$$\frac{1}{2} \, (\text{ktwo}[r, q, sd] + 1) \, x + \frac{1}{4} \, ((\text{ktwo}[r, q, sd] - 1)^2 + 4 \, \text{kone}[r, sd]) \, \text{tau}\right] -$$

$$\text{Exp}\left[\frac{1}{2} \, (\text{ktwo}[r, q, sd] - 1) \, x + \frac{1}{4} \, ((\text{ktwo}[r, q, sd] - 1)^2) \, \text{tau}\right];$$

Here are the grid parameters:

```
M=40; nminus = 320; nplus = 320;
dx = 0.01; dtau = 0.0005; alpha = dtau/dx^2
```

    5.

We have matched the dimensionless time, the time-steps and the number of them:

```
{NonDimExpiry[1, 0.2], M*dtau}
```

    {0.02, 0.02}

Now we set up, evolve and post-process the result in the usual way:

```
initial = Table[CallExercise[(k - 1 - nminus) dx, Log[1.05], 0, 0.2],
   {k, nminus + nplus + 1}];
lower = Table[f[-nminus dx, (m - 1) dtau, Log[1.05], 0, 0.2],
   {m, 1, M + 1}];
upper = Table[g[+nplus dx,  (m - 1) dtau, Log[1.05], 0, 0.2],
   {m, 1, M + 1}];
wold = initial;
wvold = wold; wnew = wold;


DougCMatrix[alpha_, nminus_, nplus_] :=
Sequence[Table[1 - 6 * alpha, {nplus + nminus - 2}],
Table[10 + 12 * alpha, {nplus + nminus - 1}],
Table[1 - 6 * alpha, {nplus + nminus - 2}]];


DougDMatrix[alpha_, vec_List] := Module[{temp},
temp = (10 - 12 * alpha) * vec +
    (1 + 6 * alpha) * (RotateRight[vec] + RotateLeft[vec]);
temp[[1]] = Simplify[First[temp] - (1 + 6 * alpha) * Last[vec] / 2];
temp[[-1]] = Simplify[Last[temp] - (1 + 6 * alpha) * First[vec] / 2];
temp];


CMat = DougCMatrix[alpha, nminus, nplus];


For[m=2, m<=M+1, m++,
(wvold = wold; wold = wnew;
rhs = DougDMatrix[alpha, Take[wold, {2, -2}]]+
Table[Which[
k==1, alpha*(lower[[m-1]] + lower[[m]])/2,
k== nplus + nminus-1,  alpha*(upper[[m-1]] + upper[[m]])/2,
True, 0],
{k, 1, nplus + nminus-1}];
temp = CompTridiagSolve[CMat, rhs];
wnew = Join[{lower[[m]]}, temp, {upper[[m]]}])]
```

```
interpoldatab = Table[{(k - nminus - 1)*dx, wnew[[k]]}, {k, 1, nminus
+ nplus + 1}];
ufuncb = Interpolation[interpoldatab, InterpolationOrder -> 3];
Valuation[strike_, r_, q_, S_, T_, sd_] :=
  ValuationMultiplier[strike, r, q, Log[S/strike], (sd^2*T)/2, sd]*
    ufuncb[Log[S/strike]]
```

We plot the error, and tabulate the results and errors to ensure that things are under control:

```
Plot[Valuation[100, Log[1.05], 0, S, 1, 0.2] -
  BlackScholesCall[S, 100, 0.2, Log[1.05], 0, 1],
 {S, 80, 120}, PlotPoints → 50, PlotRange → All];
```

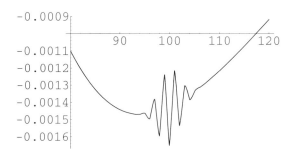

```
samples = TableForm[Join[{{"S", "Doug(2T)", "Exact", "Error"}},
 Table[(PaddedForm[N[#1], {6, 4}] &) /@
  {S, Valuation[100, Log[1.05], 0, S, 1, 0.2],
     BlackScholesCall[S, 100, 0.2, Log[1.05], 0, 1],
     Valuation[100, Log[1.05], 0, S, 1, 0.2] -
      BlackScholesCall[S, 100, 0.2, Log[1.05], 0, 1]},
 {S, 90, 110, 2}]]]
```

S	Doug(2T)	Exact	Error
90.0000	5.0492	5.0507	-0.0014
92.0000	5.9477	5.9491	-0.0015
94.0000	6.9325	6.9340	-0.0015
96.0000	8.0021	8.0036	-0.0015
98.0000	9.1538	9.1554	-0.0016
100.0000	10.3846	10.3863	-0.0017
102.0000	11.6908	11.6923	-0.0015
104.0000	13.0678	13.0692	-0.0014
106.0000	14.5111	14.5124	-0.0013
108.0000	16.0159	16.0171	-0.0013
110.0000	17.5773	17.5785	-0.0012

Jump Adjustment for Cash Dividend

We have got the FD model working well for the standard case with no dividend payment. Now let's look at the adjustments needed for dividends. We need to implement the jump condition at the dividend date, and make an adjustment to the upper boundary condition prior (in real time) to the payment. Let's pursue the detailed set-up.

Here are the grid parameters:

```
M=40;
nminus = 320;
nplus = 320;
dx = 0.01;
dtau = 0.0005;
alpha = dtau/dx^2;

divm = 21; (* discrete dividend paid half-way *)
cashdiv = 20; (* big to test it properly! *)
strike = 100;
ko = kone[Log[1.05],0.2];
kt = ktwo[Log[1.05],0,0.2];
```

Here is the dividend jump correction to the upper boundary condition:

```
DivCorrUBC[strike_, r_, q_, tau_, sd_, cashdiv_, divtime_] :=
  cashdiv * Exp[kone[r, sd] * (sd^2 * divtime / 2 - tau)] /
    ValuationMultiplier[strike, r, q, nplus * dx, tau, sd]
```

The following function implements part of the jump condition, ensuring that the resulting price stays within range, i.e. positive. This is done at the lower end of the grid by capping the yield at 70%. The reader can investigate the effect of varying this cap - it does not materially alter our conclusions, but makes very small shifts to the details of the valuation.

```
FixedInterp[k_, dx_, nminus_, kt_, cashdiv_, strike_, interfunc_] :=
Module[{q, new},
q = If[cashdiv*Exp[-(k-nminus-1)*dx]/strike < 0.7, -Log[1 - cashdiv*-
Exp[-(k-nminus-1)*dx]/strike], -Log[1-0.7]];
new = Which[
q > 99999,
0,
(k-nminus-1)*dx - q < -(nminus-1)*dx,
0,
True,
Exp[0.5*(kt-1)*q]*interfunc[(k-nminus-1)*dx -q]
];
new]
```

Now we set up, evolve and post-process the result in the usual way:

```
initial = Table[CallExercise[(k - 1 - nminus) dx, Log[1.05], 0, 0.2],
   {k, nminus + nplus + 1}];
lower = Table[f[-nminus dx, (m - 1) dtau, Log[1.05], 0, 0.2],
   {m, 1, M + 1}];
upper = Table[g[+nplus dx,  (m - 1) dtau, Log[1.05], 0, 0.2],
   {m, 1, M + 1}];
wold = initial;
wvold = wold;
wnew = wold;
points = Table[(k - nminus - 1) * dx,  {k, 2, nplus + nminus}];


For[m=2, m<=M+1, m++,
(wvold = wold;
wold = wnew;
rhs = DougDMatrix[alpha, Take[wold, {2, -2}]]+
Table[
Which[
k==1, alpha*(lower[[m-1]] + lower[[m]])/2,
k== nplus + nminus-1,  alpha*(upper[[m-1]] + upper[[m]])/2,
True, 0],
{k, 1, nplus + nminus-1}];
temp = CompTridiagSolve[CMat, rhs];

If[m==divm,
(Print["Doing Dividend Jump Adjustments"];
interpoldata = Transpose[{points,temp}];
divfunc = Interpolation[interpoldata, InterpolationOrder -> 3];
vnew = Table[
FixedInterp[k,dx,nminus,kt,cashdiv,strike,divfunc], {k, 2,
nplus+nminus}];
temp = vnew;
(* having corrected interior, do upper boundary condition *)
Do[upper[[m]] -= DivCorrUBC[100, Log[1.05], 0, (m-1)*dtau, 0.20
,cashdiv,0.5],
{m, divm, M+1}];
),];
wnew = Join[{lower[[m]]}, temp, {upper[[m]]}])]


    Doing Dividend Jump Adjustments

interpoldatab = Table[{(k - nminus - 1)*dx, wnew[[k]]}, {k, 1, nminus
+ nplus + 1}];
ufuncb = Interpolation[interpoldatab, InterpolationOrder -> 3];
Valuation[strike_, r_, q_, S_, T_, sd_] :=
  ValuationMultiplier[strike, r, q, Log[S/strike], (sd^2*T)/2, sd]*
  ufuncb[Log[S/strike]]
```

We plot and tabulate the results in the neighbourhood of the at-the-money value of interest:

```
samples = TableForm[Join[{{"S", "Doug(2T) with Cash Div"}},
  Table[(PaddedForm[N[#1], {6, 4}] &) /@
    {S, Valuation[100, Log[1.05], 0, S, 1, 0.2]},
  {S, 98, 102, 1}]]]
```

S	Doug(2T) with Cash Div
98.0000	2.0874
99.0000	2.3174
100.0000	2.5645
101.0000	2.8293
102.0000	3.1120

The value of interest is 2.5645, which compares well with the Monte Carlo simulated value of 2.5877 with a standard error of 0.015. The analytical model has been out-voted in this naive set of calculations.

The FD implementation of a jump condition can also be used when dealing with American-style options. It is simpler to code the Monte Carlo version for the European case. But we really need to find out what is causing the difference between the results.

24.5 Analytical Models - a Second Look

To reconcile the two numerical schemes and the simple analytical treatment we must take a closer look at the random processes involved. We will do so for just one dividend payment between initiation and expiry, though the principle is the same for many cash dividends. We shall also keep the volatility fixed at a constant, though one can also experiment with two different volatilities for the two stages. Suppose that the underlying price at initiation is S. Suppose that the dividend D is paid at time t_1. Just before time t_1, the underlying price, under standard assumptions, is given by

$$S\, e^{x\sigma\sqrt{t_1} + \left(r - \frac{\sigma^2}{2}\right)t_1} \tag{4}$$

where x is distributed $N(0, 1)$. If the dividend is defined to be the effective jump in the stock price, the price just after the dividend is paid is

$$S\, e^{x\sigma\sqrt{t_1} + \left(r - \frac{\sigma^2}{2}\right)t_1} - D \tag{5}$$

Assuming that this then follows a new random walk, the price at expiry will be

$$S_T = S\, e^{\left(r - \frac{\sigma^2}{2}\right)T + y\sigma\sqrt{T - t_1} + x\sigma\sqrt{t_1}} - D\, e^{y\sigma\sqrt{T - t_1} + \left(r - \frac{\sigma^2}{2}\right)(T - t_1)} \tag{6}$$

where y is also $N(0, 1)$. This is in a form that is very similar to a spread option on two assets. Indeed, the fact that this expiry variable is not log-normal, and hypothetically could become negative, arises for the same reasons as discussed in Chapter 23. In practice, D is much smaller than S, so we can discard most of the concerns that apply to a general spread option. But we can value it accurately by similar methods. To save space, let us define

$$\tilde{S} = S\, e^{\left(r - \frac{\sigma^2}{2}\right)T}$$

$$\tilde{D} = D\, e^{\left(r - \frac{\sigma^2}{2}\right)(T - t_1)} \tag{7}$$

so that

$$S_T = \tilde{S}\, e^{y\sigma\sqrt{T-t_1}\,+x\sigma\sqrt{t_1}} - \tilde{D}\, e^{y\sigma\sqrt{T-t_1}} \tag{8}$$

and the payoff for a call is given by

$$\mathrm{Max}\!\left[\tilde{S}\, e^{y\sigma\sqrt{T-t_1}\,+x\sigma\sqrt{t_1}} - \tilde{D}\, e^{y\sigma\sqrt{T-t_1}} - K,\, 0\right] \tag{9}$$

This means that we only need to do the integration from the lower limit $x_{\min}$ to infinity, where

$$x_{\min} = \frac{\log\!\left(K\, e^{-y\sigma\sqrt{T-t_1}} + \tilde{D}\right) - \log(\tilde{S})}{\sigma\sqrt{t_1}} \tag{10}$$

We can express this as a *Mathematica* function:

```
DivCallLimit[y_, S_, K_, D_, r_, σ_, td_, T_] :=
  (Log[K Exp[-y σ Sqrt[T - td]] + D Exp[(r - σ^2 / 2) (T - td)]] -
     Log[S Exp[(r - σ^2 / 2) T]]) / (σ Sqrt[td])
```

Now we do the valuation as for a spread:

```
DivCallIntegral[S_, K_, D_, r_, σ_, td_, T_, range_] :=
  Exp[-r * T] / (2 Pi) NIntegrate[Exp[- (x^2 + y^2) / 2] *
    (S * Exp[(r - σ^2 / 2) * T + y * σ * Sqrt[T - td] + x * σ * Sqrt[td]] -
      D * Exp[y * σ * Sqrt[T - td] + (r - σ^2 / 2) * (T - td)] - K), {y,
    -range, range}, {x, DivCallLimit[y, S, K, D, r, σ, td, T], range}]
```

Here is the value of our running example.

```
DivCallIntegral[100, 100, 20, Log[1.05], 0.2, 0.5, 1, 8]
```

```
     2.56562
```

This agrees well with the Monte Carlo result and the FD result, as it should. Another independent check on this model is that if the dividend is paid an instant before expiry, it is as though the strike were increased by the dividend - we can check that this is the case.

```
DivCallIntegral[100, 100, 20, Log[1.05], 0.2, 1, 1, 10]
```

```
     3.21677
```

```
BlackScholesCall[100, 100 + 20, 0.2, Log[1.05], 0, 1]
```

```
    3.21677
```

So it has to be appreciated that if we allow the real stock price to follow a log-normal random walk at all times to the dividend payment, and then thereafter, the resulting expiry distribution is not in fact log-normal at all, and is actually bivariate.

Readers should follow through the above calculation when the model is a type (2) model, and the dividend is of the form $Q\,S$. Then the log-normal character is preserved, and the option can be valued as though the initial price were $S(1 - Q)$. The results from this analytical model then agree with Monte Carlo or FD results, as the reader may also wish to check.

Moment Matching in the Log-Normal Approximation

So far, in the view that the real stock price is the underlying, we have found that the expiry distribution is akin to that for a spread. In general, spreads do not admit a good log-normal distribution, but as the dividend is typically small compared to the stock price, we can go for a log-normal model. The approach is the same as we have taken for Asian and Basket options, and we match the first and second moments of the expiry distribution in the usual way. This leads to an interesting result. Matching of the means says that the variable to be used is just S^*! But this is not enough on its own. The effective price is just the standard adjustment, but there is also a shift in the effective volatility. It is a simple matter to check that the effective volatility is given by

$$\sqrt{\sigma^2 + \frac{\log\left(1 + \frac{(1-e^{-\sigma^2\,\text{td}})(2\,S\,D\,e^{-r\,\text{td}} - D^2\,e^{-2\,r\,\text{td}})}{(S-D\,e^{-r\,\text{td}})^2}\right)}{T}} \tag{11}$$

where the time to the only dividend payment is td. It is a simple matter to implement this correction:

```
VolEff[σ_, T_, td_, r_, S_, D_] :=
  Sqrt[σ^2 + 1/T * Log[1 + (2 * S * D * Exp[-r * td] - D^2 Exp[-2 * r * td]) *
      (1 - Exp[-σ^2 td]) / (S - D * Exp[-r * td])^2]]
```

When the underlying has a volatility of 20%, the effective volatility to use in the log-normal approximation is given by

```
VolEff[0.2, 1, 0.5, Log[1.05], 100, 20]
```

```
    0.225191
```

or about 22.5%. We can now create a Black-Scholes analytical approximation to the case where the real stock price is the underlying:

```
AccAnalyticalDivEuroCall[S_, K_, σ_, r_, q_, t_, td_, D_] :=
 Module[{adjprice, adjvol},
  adjprice = S - D*Exp[-r*td];
  adjvol = VolEff[σ, t, td, r, S, D];
  BlackScholesCall[adjprice, K, adjvol, r, q, t]]
```

Here are the results of this simplified model with the dividend paid half-way, and just before expiry:

```
AccAnalyticalDivEuroCall[100, 100, 0.2, Log[1.05], 0, 1, 0.5, 20]
```

    2.58741

```
AccAnalyticalDivEuroCall[100, 100, 0.2, Log[1.05], 0, 1, 1, 20]
```

    3.2775

These should be compared with the pure effective price models:

```
AnalyticalDivEuroCall[100, 100, 0.2, Log[1.05], 0, 1, {{0.5, 20}}]
```

    1.94853

```
AnalyticalDivEuroCall[100, 100, 0.2, Log[1.05], 0, 1, {{1, 20}}]
```

    2.0584

These are considerably closer to the numerically computed values. For smaller dividends we have closer results - here are the exact, the vol-adjusted and the folklore estimates for a 5% dividend:

```
DivCallIntegral[100, 100, 5, Log[1.05], 0.2, 0.5, 1, 8]
```

    7.71977

```
AccAnalyticalDivEuroCall[100, 100, 0.2, Log[1.05], 0, 1, 0.5, 5]
```

    7.717

The first two agree, but the last is rather lower:

```
AnalyticalDivEuroCall[100, 100, 0.2, Log[1.05], 0, 1, {{0.5, 5}}]
```

```
    7.52292
```

The difference is amplified if the dividend is paid just before expiry.

```
DivCallIntegral[100, 100, 5, Log[1.05], 0.2, 1, 1, 8]
```

```
    7.96557
```

```
AccAnalyticalDivEuroCall[100, 100, 0.2, Log[1.05], 0, 1, 1, 5]
```

```
    7.9562
```

```
AnalyticalDivEuroCall[100, 100, 0.2, Log[1.05], 0, 1, {{1, 5}}]
```

```
    7.58625
```

But note that with a 5% yield there is a material (5%) difference in the option value computed by the two methods. The simple PV-dividend rule simply cannot be used with any confidence in the view where the stock price is the underlying, but if used with the volatility adjustment also in place, good results can be obtained, again, if you believe the fundamental log-normal assumptions apply to the asset price and not to S^*.

All of these considerations have American-style counterparts, only this time the problem is matching the numerical results for S the underlying with the RGW model of American Calls discussed previously, in Chapter 11. If one implements an FD model of Americans using a jump condition, using the straightforward modification of the FD example given here, the results will differ markedly from the RGW values applied with the same volatility.

The use of an effective volatility is also briefly discussed by Hull (1996). That author remarks that the volatility of the risky component is approximately equal to the volatility of the whole stock price multiplied by

$$\frac{S}{S - V} \tag{12}$$

where V is the present value of the dividends. If we apply this as a fixed constant to the 20% dividend, we would get a volatility of

```
20 * 100 / 80 // N
```

```
    25.
```

and a value of

```
AnalyticalDivEuroCall[100, 100, 0.25, Log[1.05], 0, 1, {{0.5, 20}}]
```

```
   3.25642
```

which is now too high. For the more reasonable case of 5 units of cash, we get a value of

```
AnalyticalDivEuroCall[100, 100, 0.25, Log[1.05], 0, 1, {{0.5, 5}}]
```

```
   9.40942
```

which again is too high. It may be possible to get better agreement, and indeed a consistent view, by building a volatility term structure based on (12). The only *simple* way to get good agreement between the analytical models and the numerical models, with the stock price as the underlying, is to regard the variable S^* as a component of a log-normal approximation, and to make the volatility adjustment we have given in equation (11).

24.6 Remarks

Although we have, hopefully, explained the different answers which result when S and S^* are treated as the underlying, we have not given a view as to the "right" choice. It is hard to come to a firm view. The personal view of the author is that treating the real-world stock price as the underlying gives a more realistic view of the dependence of the option value on the time at which the dividend is paid, as can be seen by considering two limiting cases. If the dividend is paid immediately, we would expect to use the Black-Scholes formula with S replaced by essentially $S - D$. At the other extreme, if the dividend is paid just before the expiry of a (Call) option, we would expect it to be valued by the Black-Scholes formula with the strike K replaced by $K + D$. These limits are captured perfectly in the picture in which S is the underlying, but not at all, so far as this author can see, by the S^* picture.

Fans of trees may wonder what happens if binomial trees are used. If S^* is the underlying, there are no surprises and everything ties up with the S^* analytical model if it exists. As described by Section 15.3 of Hull (1996), the fixed cash dividend model applied directly to a tree in the S-picture results in a non-recombining tree. This is precisely because of the deviation from log-normality. If you have the patience to build a non-re-combining tree, you will find the same result as obtained here in Monte Carlo and finite-difference models. In fact, you can use a standard recombining tree with a jump condition implemented by interpolation, in much the same way as in our finite-difference model, and again get the same result. The difference between the use of S and S^* is a fundamental matter of principle, and not affected by whether we model analytically, by finite differences, by Monte Carlo, or by trees - it must be appreciated.

Chapter 24 Bibliography

Hull, J.C., 1996, *Options, Futures, and Other Derivatives*, 3rd edition, Prentice-Hall.

Wilmott, P., Dewynne, J. and Howison, S., 1993, *Option Pricing - Mathematical Models and Computation*, Oxford Financial Press.

Chapter 25.
Simple Deterministic and Stochastic Interest-Rate Models

*Bonds and Yield Curves for Deterministic
and Simple Random Interest Rates*

25.1 Introduction

Our goal in this chapter is to understand the valuation of bonds within two interest-rate frameworks:

(a) deterministic time-dependent interest rates;
(b) simple stochastic models of interest rates.

In the latter case, we need to capture the basic properties of interest-rate random walks, including non-log-normal behaviour and mean-reversion, without sacrificing manifest positivity. We need a framework within which one can calibrate to and price:

bonds;
options on bonds;
swaps, caps and floors;
swaptions, captions and floortions;
etc., etc.

and other more exotic objects. This is a subject of ongoing research. An outstanding survey of interest-rate models is given in the text by Rebonato (1996) - the reader is referred to that work for a definitive mathematical and financial discussion of interest-rate option modelling. An excellent overview is also given in Section 5 of Baxter and Rennie (1996). In this chapter we shall develop a unified treatment of several common models of interest, based on the technology prior to HJM and variants. Notation for individual models follows Hull (1996), whereas families of models are treated within the notation of Wilmott *et al* (1993).

It is not possible to give, within the scope of this text, a definitive overview and set of implementations - at least, not to the extent we have considered equity options, or by changes of variables, FX options. What we shall do is to give a taste of what can be done with *Mathematica*, within a partially unified, if not state-of-the-art, framework. This chapter will consider basic definitions and the pricing of bonds within various frameworks. The deterministic case will be described only as a device for fixing notation and definitions. We shall then survey the pre-HJM models and the pricing of bonds. Simple methods for building yield curves and associated rates will be considered in Chapter 26. The pricing of very simple interest-rate options will be consider in Chapter 27.

25.2 Bonds with Deterministic Time-Dependent Interest Rates

This topic was studied briefly in Chapter 4, where we considered how to transform the Black-Scholes equation with time-dependent deterministic interest rates into the diffusion equation. Let's review the details as they apply to a straight bond. Our terminology is as follows:

V is the value of the bond;
$r(t)$ is the time-dependent spot rate;
K is the coupon payment on the bond, and will normally be a discrete function.

The deterministic bond equation is then

$$\frac{\partial V}{\partial t} - r(t)\,V + K[t] = 0 \tag{1}$$

We can solve this using an integrating factor to obtain

$$V_0 = e^{-\int_t^T r(s)\,ds}\left(Z + \int_t^T K[s]\,e^{\int_s^T r(p)\,dp}\,ds\right) \tag{2}$$

Now the coupon payments are typically discrete, so we write, accordingly,

$$K[t] = \sum_{t \le t_i \le T} K_i\,\delta(t - t_i) \tag{3}$$

representing the coupon payments as a general cash flow of payments K_i at times t_i .

$$V_0 = e^{-\int_t^T r(s)\,ds}\left(Z + \sum_{t \le t_i \le T} K_i\,e^{\int_{t_i}^T r(p)\,dp}\right) \tag{4}$$

This boils down to

$$V_0 = Ze^{-\int_t^T r(s)\,ds} + \sum_{t \le t_i \le T} K_i\,e^{-\int_t^{t_i} r(p)\,dp} \tag{5}$$

If the interest rate were constant at a value r_0, this would simplify further to

$$V_0 = Ze^{-r_0(T-t)} + \sum_{t \le t_i \le T} K_i\,e^{-r_0(t_i - t)} \tag{6}$$

but we do not assume this here. In general, we require knowledge of the integrals of the form

$$\int_{t_i}^T r(p)\,dp \tag{7}$$

and this must be based on yield-curve information.

25.3 The Yield Curve

This matter is easy mathematically, but is messed up by terminological hang-ups. Given the discount factor structure

$$P(t, T) \tag{8}$$

describing the price at time t of a zero-coupon (or "discount") bond maturing at time T, with $Z = 1$, the "yield" as defined by Baxter and Rennie (1996) or Wilmott *et al* (1996), but called by Rebonato (1996) the "continuously compounded discrete spot rate of maturity T", is a function

$$R(t, T) = -\frac{\log(P(t, T))}{T - t} \tag{9}$$

Clearly, inverting this relationship:

$$P(t, T) = e^{-(T-t)R(t,T)} \tag{10}$$

It follows that the yield for a fixed t and T is the interest rate that discounts the bond correctly to its present value as though it were constant.

Simply Compounded Analogues

The formulae above are implicitly (explicitly in the case of Rebonato's terminology) based on continuously compounded variables. One can invent numerous compounding conventions. The corresponding relations for interest rates defined by simple compounding are

$$R_S[t, T] = \frac{\frac{1}{P(t,T)} - 1}{T - t} \tag{11}$$

so that

$$P(t, T) = \frac{1}{1 + (T - t)R_S[t, T]} \tag{12}$$

The Yield with Deterministic Interest Rates

If interest rates are deterministic, we have the yield expressed as the time-average:

$$R(t, T) = \frac{\int_t^T r(p)\,dp}{T - t} \tag{13}$$

25.4 Some Stochastic Model History

An interesting category of interest-rate (IR) models is obtained by considering the family of (risk-neutral) random walks, on the spot rate, given by

$$dr = m\,dt + s\,dz \tag{14}$$

A large part of the recent history of single-factor stochastic IR models can be written down by making simple choices for m and s. (We shall consider the exceptional case of the Black model in Chapter 27.) Some of the better known ones include:

Rendleman-Bartter

This model, given by R. Rendleman and B. Bartter (1980), assumes a process that is the same for an asset:

$$m = \mu r \quad s = \sigma r \tag{15}$$

The main drawback of this model is a lack of mean-reversion, observed in real-world rates. The advantage is that we can use all of the existing equity technology.

Vasicek

This model, given by O.A. Vasicek (1977), assumes a process that has mean-reversion:

$$m = a(b - r) \quad s = \sigma \tag{16}$$

The main drawback of this model is the presence of a normal distribution, rather than a log-normal form, so that the interest rate can become negative.

Cox-Ingersoll-Ross

This model, given by Cox, Ingersoll and Ross (1985), assumes a process that also has mean-reversion, but which deals with the positivity issue:

$$m = a(b - r) \quad s = \sigma \sqrt{r} \tag{17}$$

Ho-Lee

This model, given by Ho and Lee (1986), assumes a process where one of the parameters may be time-dependent. In the continuous-time description, it may be regarded as

$$m = \theta(t) \quad s = \sigma \tag{18}$$

This again has no mean-reversion, but the presence of a time-dependent drift allows more flexibility in fitting the model to a given term structure.

Hull-White

This model, given by Hull and White (1990), assumes a process where one of the parameters may be time-dependent. In the continuous-time description, it may be regarded as

$$m = (\theta(t) - a\,r) \quad s = \sigma \tag{19}$$

This has mean-reversion, in addition to flexibility in fitting the model to a given term structure.

Black-Derman-Toy and Variants

This model, given by Black, Derman and Toy (1990), assumes a process which combines elements of log-normality with the ability to fold in time-dependent features:

$$d \log (r) = d\,t \left(\theta (t) + \frac{\log (r)\, \sigma' (t)}{\sigma (t)} \right) + d\,z\, \sigma (t) \tag{20}$$

Note that when σ and θ are time-independent this model, which we abbreviate as BDT, reduces to the log-normal model. The variant developed by Black and Karasinski (1991) is more general (we call it BK):

$$d \log(r) = d\,t\, (\theta(t) - \log(r)\, a(t)) + d\,z\, \sigma(t) \tag{21}$$

There are therefore two identifiable branches of this family of models. The log-normal and generalizations of it such as BDT, BK are one easily identifiable family. The second family is most easily characterized by the nature of the solution to the bond-pricing equation. This will be considered next.

25.5 The Log-Linear Bond Family of Models

There are other variants of these models, which we shall not pursue explicitly. What is interesting is that from the point of view of bond valuation, there is very little difference between several of these models. The observation, probably made first by N. Pearson and T.-S. Sun at MIT, but more accessibly discussed in the work by Wilmott *et al* (1993), is that all of these models, with the exception of the Rendleman-Bartter asset-like model, and the Black-Derman-Toy family, have a simple "similarity" solution for the bond price. If we assume a certain functional form for the price of a bond, we can show that the parameters of the interest-rate random walk must satisfy certain constraints, which happen to be satisfied by all of the models above except the asset-like models. Under these assumptions, we can derive differential equations for the parameters of the bond price, and construct explicit solutions in many cases of interest. We take the bond-pricing equation to be

$$\frac{1}{2}\, q\,(r,\, t)\, \frac{\partial^2 V\,(r,\, t)}{\partial r^2} + p\,(r,\, t)\, \frac{\partial V\,(r,\, t)}{\partial r} - r\, V\,(r,\, t) = 0 \tag{22}$$

where $q(r,\, t) = s^2(r,\, t)$, $p(r,\, t) = \mu(r,\, t) - \lambda(r,\, t)\, s(r,\, t)$ and λ is the market price of risk associated with the real-world drift μ. This equation may be derived by constructing a hedge position from two bonds of different maturities, in a similar manner to the original Black-Scholes equation. Let's write the bond-pricing equation in symbolic form:

```
Clear[p,q,r,t]


BondBSE[V_, r_, t_, p_, q_] := D[V[r,t],t] + 1/2 q[r,t] D[V[r,t], {r,
2}] + p[r,t] D[V[r,t], r] - r V[r,t]
```

The log-linear bond family is defined by assuming a solution family of the form (we suppress an implicit dependence on the maturity T)

```
U[r_, t_] = Z*Exp[A[t] - r*B[t]];


TraditionalForm[U[r,t,T]]
```

$$e^{A(t)-r\,B(t)}\,Z \tag{23}$$

We substitute this into the bond-pricing equation and simplify:

```
Simplify[2 BondBSE[U,r,t,p,q]]
```

$$-E^{A[t]-r\,B[t]}\,Z\,(2\,B[t]\,p[r,\,t]-B[t]^2\,q[r,\,t]+2\,(r-A'[t]+r\,B'[t]))$$

Next we divide through by the hypothesized bond price and differentiate twice w.r.t. r:

```
eqn = Simplify[%/U[r,t]]
```

$$-2\,B[t]\,p[r,\,t]+B[t]^2\,q[r,\,t]-2\,(r-A'[t]+r\,B'[t])$$

```
Simplify[D[%, {r,2}]/B[t]]
```

$$-2\,p^{(2,0)}[r,\,t]+B[t]\,q^{(2,0)}[r,\,t]$$

This combination must vanish. Since B alone depends on T, it follows that the second r-derivatives of $q(r, t) = s^2(r, t)$, and of $p(r, t) = m(r, t) - \lambda(r, t)\,s(r, t)$, are zero separately. So these must be linear functions of r. This includes the models of Vasicek, Ho and Lee, Hull and White, and Cox, Ingersoll and Ross.

25.6 Ordinary Differential Equations for the Bond Price

We write down a linear form for the coefficients of the bond-pricing equation:

```
Clear[α, β, η, γ];
q[r_, t_] := α[t] r - β[t];
p[r_, t_] := η[t] - r γ[t];
```

The A, B functions must satisfy the condition

```
eqn == 0
```

$$B[t]^2\,(r\,\alpha[t]-\beta[t])-2\,B[t]\,(-r\,\gamma[t]+\eta[t])-2\,(r-A'[t]+r\,B'[t])==0$$

We can obtain two ordinary differential equations for the bond price by equating the coefficients of powers of r.

```
aeqn = Coefficient[eqn/2, r, 0];
```

```
TraditionalForm[aeqn==0]
```

$$-\frac{1}{2}\beta(t)(B(t))^2 - \eta(t)B(t) + A'(t) = 0 \tag{24}$$

```
beqn = -Coefficient[eqn/2, r, 1];
```

$$-\frac{1}{2}\alpha(t)(B(t))^2 - \gamma(t)B(t) + B'(t) + 1 = 0 \tag{25}$$

This pair of coupled ODEs determine the bond price as a function of the functions α, β, η, γ.

25.7 Simple Symbolic Solutions of the Bond-Pricing Equation

The bond ODEs can now be solved subject to the initial conditions that the bond value is Z when $t = T$. This requires that

$$A(T) = 0 = B(T) \tag{26}$$

Let's explore the several cases of interest.

Vasicek Bond Formula

In this case the special nature of the interest-rate random walk allows us to set

```
α[t_] := 0;
β[t_] := -σ^2;
η[t_] := a b;
γ[t_] := a;
```

The differential equations governing the bond price are obtained by setting the following quantities to zero:

```
aeqn
```

$$-a\,b\,B[t] + \frac{1}{2}\,\sigma^2\,B[t]^2 + A'[t]$$

```
beqn
```

$$1 - a\,B[t] + B'[t]$$

Solving for A and B

This pair of ODEs is sufficiently simple that we can get *Mathematica* to solve the pair together.

```
soln = DSolve[{beqn==0, B[T]==0, aeqn==0, A[T]==0}, {B[t], A[t]},
t][[1]]
```

$$\left\{B[t] \to \frac{1}{a} - \frac{E^{a\,t-a\,T}}{a}, \; A[t] \to -\frac{E^{2\,a\,t-2\,a\,T}\,\sigma^2}{4\,a^3} - \frac{E^{a\,t-a\,T}\,(a^2\,b - \sigma^2)}{a^3} + \right.$$
$$\left. \frac{t\,(2\,a^2\,b - \sigma^2)}{2\,a^2} + \frac{1}{4}\left(\frac{4\,b}{a} - 4\,b\,T - \frac{3\,\sigma^2}{a^3} + \frac{2\,T\,\sigma^2}{a^2}\right)\right\}$$

```
BB[t_, T_] = B[t] /. soln;
```

```
AA[t_, T_] = A[t] /. soln;
```

```
TraditionalForm[BB[t, T]]
```

$$\frac{1}{a} - \frac{e^{a\,t-a\,T}}{a}$$

```
TraditionalForm[AA[t, T]]
```

$$-\frac{e^{2\,a\,t-2\,a\,T}\,\sigma^2}{4\,a^3} - \frac{e^{a\,t-a\,T}\,(a^2\,b - \sigma^2)}{a^3} +$$
$$\frac{t\,(2\,a^2\,b - \sigma^2)}{2\,a^2} + \frac{1}{4}\left(\frac{2\,T\,\sigma^2}{a^2} - \frac{3\,\sigma^2}{a^3} + \frac{4\,b}{a} - 4\,b\,T\right)$$

This gives the exact formula for the bond price in terms of the parameters a, b, σ.

Cox-Ingersoll-Ross Bond Formula

We set

```
α[t_] := σ^2;
β[t_] := 0;
η[t_] := a b;
γ[t_] := a;
```

```
aeqn
```

$$-a\,b\,B[t] + A'[t]$$

beqn

$$1 - a\, B[t] - \frac{1}{2}\, \sigma^2\, B[t]^2 + B'[t]$$

This pair of equations is a little more awkward, so we first solve for *B* and then solve for *A*.

Solving for *B*

bsoln = DSolve[{beqn == 0, B[T]==0}, B[t], t][[1]]

$$\left\{ B[t] \rightarrow \frac{1}{\sigma^2}\left(-a + \sqrt{-a^2 - 2\sigma^2}\right.\right.$$
$$\left.\left. \mathrm{Tan}\left[\frac{1}{2}\left(t\sqrt{-a^2 - 2\sigma^2} - T\sqrt{-a^2 - 2\sigma^2} + 2\,\mathrm{ArcTan}\left[\frac{a}{\sqrt{-a^2 - 2\sigma^2}}\right]\right)\right]\right)\right\}$$

We have an answer, but in a slightly awkward form. Let's make a standard substitution, and do some further simplification:

PowerExpand[bsoln /. -a^2 - 2 σ^2 -> - g^2]

$$\left\{ B[t] \rightarrow \frac{-a + I\,g\,\mathrm{Tan}\left[\frac{1}{2}\left(I\,g\,t - I\,g\,T - 2\,I\,\mathrm{ArcTanh}\left[\frac{a}{g}\right]\right)\right]}{\sigma^2}\right\}$$

bbsoln = Simplify[%];

BB[t_, T_] = B[t] /. bbsoln

$$-\frac{a + g\,\mathrm{Tanh}\left[\frac{1}{2}\,g\,(t - T) - \mathrm{ArcTanh}\left[\frac{a}{g}\right]\right]}{\sigma^2}$$

We check the boundary condition:

Simplify[BB[T, T]]

0

Solving for _A_

```
asoln = Simplify[DSolve[{(aeqn /. bbsoln)==0, A[T]==0}, A[t], t][[1]]];
```

Here is the full form of the answer:

```
AA[t_, T_] = A[t] /. asoln
```

$$
\frac{1}{g\,\sigma^2}\left(a\,b\left((a-g)\,\text{Log}\left[1-\frac{a}{g}\right]-(a+g)\,\text{Log}\left[-\frac{a+g}{g}\right]\right]+\right.
$$

$$
a\,\text{Log}\left[-1+\text{Tanh}\left[\frac{1}{2}\,g\,(t-T)-\text{ArcTanh}\left[\frac{a}{g}\right]\right]\right]+
$$

$$
g\,\text{Log}\left[-1+\text{Tanh}\left[\frac{1}{2}\,g\,(t-T)-\text{ArcTanh}\left[\frac{a}{g}\right]\right]\right]-
$$

$$
a\,\text{Log}\left[1+\text{Tanh}\left[\frac{1}{2}\,g\,(t-T)-\text{ArcTanh}\left[\frac{a}{g}\right]\right]\right]+
$$

$$
\left.g\,\text{Log}\left[1+\text{Tanh}\left[\frac{1}{2}\,g\,(t-T)-\text{ArcTanh}\left[\frac{a}{g}\right]\right]\right]\right)
$$

Again, we check the boundary condition:

```
Simplify[AA[T,T]]
```

0

Hull-White Model

We set

```
α[t_] := 0;
β[t_] := -σ^2;
η[t_] := θ[t];
γ[t_] := a;
```

```
aeqn
```

$$
\frac{1}{2}\,\sigma^2\,B[t]^2 - B[t]\,\theta[t] + A'[t]
$$

```
beqn
```

```
1 - a B[t] + B'[t]
```

Solving for *B*

We give the ODE to *Mathematica* as it is:

```
bsoln = DSolve[{beqn == 0, B[T]==0}, B[t], t][[1]]
```

$$\left\{ B[t] \to \frac{1 - E^{a\,t - a\,T}}{a} \right\}$$

Solving for *A*

We can proceed as far as the differential equation:

```
aaeqn = aeqn /. bbsoln
```

$$\frac{\left(a + g\,\text{Tanh}\left[\frac{1}{2}\,g\,(t - T) - \text{ArcTanh}\left[\frac{a}{g}\right]\right]\right)^2}{2\,\sigma^2} +$$

$$\frac{\left(a + g\,\text{Tanh}\left[\frac{1}{2}\,g\,(t - T) - \text{ArcTanh}\left[\frac{a}{g}\right]\right]\right)\,\theta[t]}{\sigma^2} + A'[t]$$

Although *B* is identical to its form in the Vasicek model, this equation for *A* cannot be solved without a knowledge of $\theta(t)$, which is usually based on matching to the initial term structure. The reader is referred to Chapter 17 of Hull (1996) for details of how to do this - similar issues arise in the model of Ho and Lee.

Remarks

In each case we are able to find formulae for the price of a zero-coupon bond in terms of the model parameters, which may be constant or time-dependent. Given a family of zero-coupon bonds in the market, each model can, at least hypothetically, be calibrated to the market data. Other instruments can then be priced (or used for further calibration) within the model. This plan raises a host of practical issues. Some very important ones are:

(a) there is not usually an ample supply of zero-coupon bonds;
(b) what bonds there are may not be government or Treasury bonds, and may contain spread effects arising from default probabilities, as well as coupons and embedded options;
(c) the results you will get for the prices of all other instruments will depend on the model chosen.

The issue in (c) is in some ways the most disconcerting - interest-rate options are subject to "model risk". The same really is true of equity options, but the log-normal picture has such a wide acceptance that the issue of equity derivative model risk is frequently swept under the carpet. The trouble arising with cash dividends exposes this to some extent, but changing the model is perhaps more interesting, and is discussed in the equity context in Chapter 28. In Chapter 27 we shall see how options on bonds can be priced in some different interest-rate models. Our next task is to figure out how to build yield curves.

It is a useful exercise to use these *Mathematica* representations of bond prices to explore the types of yield curve that can be modelled within each interest-rate model - the reader is encouraged to follow this up.

Chapter 25 Bibliography

Baxter, M. and Rennie, A., 1996, *Financial Calculus, an Introduction to Derivative Pricing*, Cambridge University Press.

Black, F., Derman, E. and Toy, W., 1990, A one-factor model of interest rates and its applications to Treasury bond options, *Financial Analysts Journal*, Jan-Feb., p. 33

Black, F. and Karasinski, P., 1991, Bond and option pricing when short rates are lognormal, *Financial Analysts Journal*, July-Aug., p. 52.

Cox, J.C., Ingersoll, J.E. and Ross, S.A., 1985, A theory of the term structure of interest rates, *Econometrica*, 53, p. 385.

Ho, T.S.Y. and Lee, S.-B., 1986, Term structure movements and pricing interest rate contingent claims, *Journal of Finance*, 41, p. 1011.

Hull, J.C., 1996, *Options, Futures, and Other Derivatives*, 3rd edition, Prentice-Hall.

Hull, J. and White, A., 1990, Pricing interest rate derivative securities, *Review of Financial Studies*, 3, p. 573.

Rebonato, R., 1996, *Interest-Rate Option Models*, Wiley.

Rendleman, R. and Bartter, B., 1980, The pricing of options on debt securities, *Journal of Financial and Quantitative Analysis*, 15, p. 11.

Vasicek, O.A., 1977, An equilibrium characterization of the term structure, *Journal of Financial Economics*, 5, p. 177.

Wilmott, P., Dewynne, J. and Howison, S., 1993, *Option Pricing - Mathematical Models and Computation*, Oxford Financial Press.

Chapter 26.
Building Yield Curves from Market Data

Useful Tools, and a Case Study Using LIBOR and Swap Data

26.1 Introduction

Our purpose in this chapter is to discuss some simple approaches to the construction of the yield curve from market data. The first question is - What market data are available? The idealized case consists of an ample family of zero-coupon Treasury (more carefully, zero-default-risk) bonds, with diverse maturities, to which some form of interpolation can be applied. These rarely exist outside the major markets and are usually incomplete. The available data usually consist of some cocktail of:

(a) zero-coupon bonds;
(b) short term borrowing data such as LIBOR;
(c) FRAs;
(d) futures;
(e) swaps;
(f) coupon bonds and par bonds in particular.

The process of reconciling such diverse sources of information, with all the real-world complications of calendar management and local market conventions for day counts, is rather beyond the scope of this text. Most of the tools you need are built into *Mathematica* or its Packages, and what we shall do here is to point you to some relevant functions that you may find useful, and then implement the simplest possible yield-curve construction technique where we fuse LIBOR data and swap data in a combination of bootstrapping, interpolation and extrapolation. Although limited in scope, this is a valuable prototype for more complicated schemes.

26.2 Functions You May Find Useful

Let's quickly review some very useful functions built in to *Mathematica*:

? **Fit**

> Fit[data, funs, vars] finds a least-squares fit to a list
>   of data as a linear combination of the functions funs
>   of variables vars. The data can have the form {{x1,
>   y1, ... , f1}, {x2, y2, ... , f2}, ... }, where the
>   number of coordinates x, y, ... is equal to the number
>   of variables in the list vars. The data can also be of
>   the form {f1, f2, ... }, with a single coordinate assumed
>   to take values 1, 2, ... The argument funs can be any
>   list of functions that depend only on the objects vars.

For example, let's make up some data and use this function. Note that you can supply a set of basis functions as a list, and **Fit** determines the best *linear* combination of these basis functions.

```
data = {{1, 2}, {2, 3}, {3, 4.5}, {4, 4.8}, {5, 6}, {7, 8}};
f[x_] = Fit[data, {1, x}, x]
```

> 1.11286 + 0.982857 x

```
Plot[f[x], {x, 0, 7}, PlotRange -> {0, 9},
  Epilog -> {PointSize[0.03], Map[Point, data]}];
```

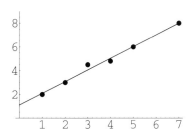

We can also interpolate with a built-in function:

? **Interpolation**

> Interpolation[data] constructs an InterpolatingFunction
>   object which represents an approximate function that
>   interpolates the data. The data can have the forms {{x1,
>   f1}, {x2, f2}, ... } or {f1, f2, ... }, where in the
>   second case, the xi are taken to have values 1, 2, ...

Let's do it:

```
g = Interpolation[data]
```

```
InterpolatingFunction [{{1., 7.}}, <>]
```

```
Plot [g[x], {x, 1, 7}, PlotRange -> {0, 9},
  Epilog -> {PointSize[0.03], Map[Point, data]}];
```

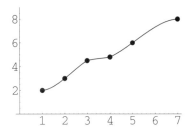

`Interpolation` has an interesting option:

```
Options [Interpolation]
```

```
{InterpolationOrder → 3}
```

Let's look at the effect of varying it. Linear interpolation is order 1:

```
g = Interpolation [data, InterpolationOrder -> 1];
Plot [g[x], {x, 1, 7}, PlotRange -> {0, 9},
  Epilog -> {PointSize[0.03], Map[Point, data]}];
```

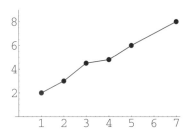

Piecewise constant is order zero:

```
g = Interpolation [data, InterpolationOrder -> 0];
Plot [g[x], {x, 1, 7}, PlotRange -> {0, 9},
  Epilog -> {PointSize[0.03], Map[Point, data]}];
```

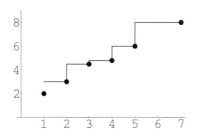

If you want it the other way round (and this may be the default in different versions of *Mathematica*), it is easily done:

```
datafix = Table[
  {Last[data][[1]] - data[[i, 1]], data[[i, 2]]}, {i, Length[data]}];

g = Interpolation[datafix, InterpolationOrder -> 0];
Plot[g[Last[data][[1]] - x], {x, 1, 7}, PlotRange -> {0, 9},
  Epilog -> {PointSize[0.03], Map[Point, data]}];
```

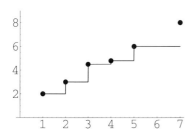

Quadratic interpolation, not surprisingly, is order 2:

```
g = Interpolation[data, InterpolationOrder -> 2];
Plot[g[x], {x, 1, 7}, PlotRange -> {0, 9},
  Epilog -> {PointSize[0.03], Map[Point, data]}];
```

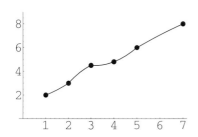

Other types of fitting are also possible. We saw how to implement a robust L^1 fit in Chapter 17. Frequently, the fitting process may involve a model non-linear in the unknown parameters. There is a Package to do this:

```
Needs["Statistics`NonlinearFit`"]
```

```
? NonlinearFit
```

> NonlinearFit[data, model, vars, params, (opts)] searches for a
>   least-squares fit to a list of data according to the model
>   containing the variables vars and the parameters params.
>   Parameters may be expressed as a list of symbols or a list
>   of lists, where each parameter is listed with starting value(
>   s) and bounds in one of several different ways: {symbol,
>   start} or {symbol, min, max} or {symbol, {start1, start2}}
>   or {symbol, {start1, start2}, min, max}. The data can have
>   the form {{x1, y1, ..., f1}, {x2, y2, ..., f2}, ...},
>   where the number of coordinates x, y, ... is equal to the
>   number of variables in the list vars. The data can also
>   be of the form {f1, f2, ...}, with a single coordinate
>   assumed to take values 1, 2, ... Method specifies the
>   LevenbergMarquardt or FindMinimum search methods. AccuracyGoal
>   and PrecisionGoal specify the number of digits of absolute
>   and relative error allowed in the residual sum of squares.

```
exptdata = Table[
  {x, 1 / ((x - 10) ^ 2 + 0.1) + 1 / ((x - 13) ^ 2 + 0.5) + 0.3 * Random[]},
  {x, 0, 20, 0.1}];
```

```
Options[NonlinearFit]
```

> {AccuracyGoal → Automatic, Gradient → Automatic,
> MaxIterations → 30, Method → LevenbergMarquardt,
> PrecisionGoal → Automatic, ShowProgress → False,
> Tolerance → Automatic, Weights → Equal, WorkingPrecision → 16}

> Clear[a,b,c,d,e,f,g]

```
NonlinearFit[exptdata, a + b / ((x - c) ^ 2 + d) + e / ((x - f) ^ 2 + g),
  x, {a, b, {c, 11}, d, e, {f, 14}, g}, MaxIterations -> 40]
```

> {a → 0.150673, b → 1.00533, c → 10.0004,
>   d → 0.100412, e → 1.00294, f → 12.9977, g → 0.493341}

Note that some early releases of *Mathematica* 3.0 contained a bug in this package - if it does not work essentially as shown you can download a working version from *MathSource,* at http://www.wolfram.com on the World Wide Web. This of course is a good place to visit anyway! For a more extensive discussion of *Mathematica's* data modelling functions, see Shaw and Tigg (1993), for collections of data import and analysis routines, as well as techniques for time series analysis, robust regression, and two and three-dimensional graphics. You should also look at the calendar Package, which can be loaded with

```
Needs["Miscellaneous`Calendar`"]
```

and contains several useful functions - **DaysBetween** is particularly useful as a simple converter from dates to time.

```
? Miscellaneous`Calendar`*
```

Calendar	EasterSundayGreekOrthodox	Monday
CalendarChange	Friday	Saturday
DayOfWeek	Gregorian	Sunday
DaysBetween	Islamic	Thursday
DaysPlus	JewishNewYear	Tuesday
EasterSunday	Julian	Wednesday

26.3 Building the Yield Curve from LIBOR and Swap Data

In this section we will take raw market data for LIBOR and swap data and extract various associated measures of the yield curve. In each step we have taken the simplest approach possible, in order to ensure that the overall logic is correct, in order to illustrate the general principles without getting involved in detail. No account of day count conventions is taken, for example.

It is perfectly possible to replace parts of this calculation by more sophisticated versions. For example, a more sophisticated interpolation of swap rates could be used. We could also use a combination of LIBOR, futures and swaps to model short-, medium- and longer-term information in the term structure. An excellent and explicit discussion of curve construction is given by Miron and Swanell (1991).

Note also that choices have to be made about at least two sets of issues:

(1) are the swap data based on swaps paid annually or semi-annually?
(2) are interest-rate data based on continuous compounding or an annual representation?

Plan of Attack

We develop a six-stage analysis program:

(1) interpolate raw swap data to annual or semi-annual values;
(2) build discount factors for short term using LIBOR data;
(3) create dummy swap values for cross-over domain based on LIBOR data;
(4) bootstrap with real and dummy swap data to get medium- and long-term discount factors;
(5) create single unified discount factor structure;
(6) build yields and forward rates for interest rates.

26.4 Swap Data Interpolation

The swap data are typically only available for certain lifetimes, so we interpolate to generate intermediate values, depending on whether the swap is annual or semi-annual. In our working example we invent data for 2, 3, 4, 5, 7, and 10 years.

```
rawswapdata =
  {{2, 10.4}, {3, 10.25}, {4, 10.15}, {5, 10.15}, {7, 9.8}, {10, 9.7}}
```

```
    {{2, 10.4}, {3, 10.25}, {4, 10.15}, {5, 10.15}, {7, 9.8}, {10, 9.7}}
```

```
rawplot = ListPlot[rawswapdata,
  PlotStyle -> PointSize[0.03], PlotRange -> {{0, 11}, {9, 11}}];
```

Now we build a function that linearly interpolates the data. In *Mathematica* this can all be done with one line, with the **InterpolationOrder** option set to unity characterizing the fact that we are using a linear scheme.

```
swapfunc = Interpolation[rawswapdata, InterpolationOrder -> 1]
```

```
    InterpolatingFunction[{{2., 10.}}, <>]
```

Filling in Data for Annual Swap

In this case the data are filled in for all the missing annual values:

```
AnnSwapOriginalRange = Table[{i, N[swapfunc[i], 5]}, {i, 2, 10, 1}]
```

```
    {{2, 10.4}, {3, 10.25}, {4, 10.15}, {5, 10.15},
     {6, 9.975}, {7, 9.8}, {8, 9.7667}, {9, 9.7333}, {10, 9.7}}
```

So we can plot the interpolated data along with the original points, to illustrate the nature of the interpolation:

```
annplot = ListPlot[AnnSwapOriginalRange,
   PlotStyle -> PointSize[0.01], PlotRange -> {{0, 11}, {9, 11}}];
```

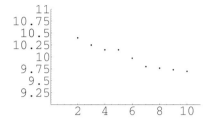

```
Show[rawplot, annplot];
```

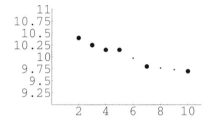

Filling in Data for Semi-annual Swap

In this case the data are filled in for value every six months. (Note that we are now re-interpreting the same test data set as being based on a swap with semi-annual payments.) We plot them as before:

```
SemiAnnSwapOriginalRange =
  Table[{i, N[swapfunc[i], 5]}, {i, 2, 10, 0.5}];
```

```
semiannplot = ListPlot[SemiAnnSwapOriginalRange,
   PlotStyle -> PointSize[0.01], PlotRange -> {{0, 11}, {9, 11}}];
```

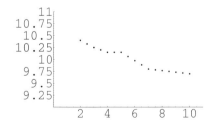

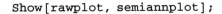

```
Show[rawplot, semiannplot];
```

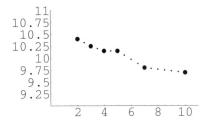

26.5 Short-Term LIBOR Discount Factors

At this stage we have real or interpolated swap data for annual or semi-annual values beginning at two years. The issue here is that swap rates are not available for shorter time scales, so one uses other instruments. In a more sophisticated scheme, LIBOR and futures data might be used for short and medium term rates, with futures being used to cover the join between LIBOR and swap data. In the following exercise we are interested in joining just the LIBOR and swap data.

We assume that we have short-term interest-rate data for the following time-intervals. These calenderization conventions are taken for simplicity - you will probably want to be more accurate:

overnight = 1/365 Y
1 week = 1/52 Y
1 month = 1/12 Y
3 month = 1/4 Y
6 month = 1/2 Y
12month = 1 Y

```
LiborTimes = {1 / 365, 1 / 52, 1 / 12, 1 / 4, 1 / 2, 1} // N

    {0.00273973, 0.0192308, 0.0833333, 0.25, 0.5, 1.}
```

```
LiborValues = {9, 8.875, 8.3125, 8.375, 8.4375, 8.4375}

    {9, 8.875, 8.3125, 8.375, 8.4375, 8.4375}
```

```
Libor = Transpose[{LiborTimes, LiborValues}]

    {{0.00273973, 9}, {0.0192308, 8.875}, {0.0833333, 8.3125},
     {0.25, 8.375}, {0.5, 8.4375}, {1., 8.4375}}
```

```
ListPlot[Libor, PlotStyle -> PointSize[0.03],
  PlotRange -> {8, 10}];
```

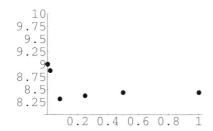

Calculation of Short-Term Discount Factors

The discount factors associated with the short-term LIBOR data are calculated using the formulae

(a) if input data are continuously compounded,

$$\{t, r\} \rightarrow \left\{t,\ e^{\frac{1}{100}(-rt)}\right\} \tag{1}$$

(b) if input data are annually compounded,

$$\{t, r\} \rightarrow \left\{t,\ \frac{1}{(1 + \frac{r}{100})^{t}}\right\} \tag{2}$$

A *Mathematica* implementation of this mapping on pairs of values is

```
LiborDisc[libordata_List, compounding_String] :=
  Which[
    compounding == "cc",
    Map[{#[[1]], Exp[-#[[1]] * #[[2]] / 100]} &, libordata],
     compounding == "an",
    Map[{#[[1]], 1 / (1 + #[[1]] / 100) ^ #[[2]]} &, libordata],
    True, "This compounding not supported"]

cclibordiscs = LiborDisc[Libor, "cc"]

    {{0.00273973, 0.999753},
     {0.0192308, 0.998295}, {0.0833333, 0.993097},
     {0.25, 0.97928}, {0.5, 0.95869}, {1., 0.919087}}


anlibordiscs = LiborDisc[Libor, "an"]
```

```
{{0.00273973, 0.999753}, {0.0192308, 0.998295}, {0.0833333, 0.9931},
 {0.25, 0.979306}, {0.5, 0.958791}, {1., 0.919472}}
```

```
ListPlot[anlibordiscs, PlotStyle -> PointSize[0.03]];
```

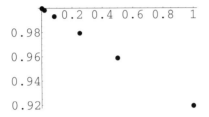

From here on the LIBOR data are taken to be in an annual representation.

26.6 Calculation of Dummy Swaps for Short Term and Crossover

Annually Paid Swaps

In this case we have just one dummy swap value, based on the discount factor for one year. The swap rate for one year is just given in terms of a single discount factor, P, being the discount factor appropriate for one year.

```
P = Last[anlibordiscs][[2]]
```

```
    0.919472
```

The dummy (made-up) swap rate for a one-year swap is then just

```
X = 100 * (1 - P) / P
```

```
    8.75809
```

Semi-annually Paid Swaps

In this case we have to find three dummy swap values based on the discount factors for 0.5 year and 1 year (to get the first two dummy values) with interpolation being used to get a third dummy value for 1.5 years. Let's extract the discount factors appropriate to six months (Pone) and one year (Ptwo):

```
Ponepair = anlibordiscs[[5]]
```

> {0.5, 0.958791}

```
Pone = Ponepair[[2]]
```

> 0.958791

The dummy (made-up) swap rate for a six-month semi-annual swap is then just

```
Xone = 100 * (1 - Pone) / (0.5 * Pone)
```

> 8.59608

```
Ptwopair = Last[anlibordiscs]
```

> {1., 0.919472}

```
Ptwo = Ptwopair[[2]]
```

> 0.919472

The dummy (made-up) swap rate for a 12-month semi-annual swap is then just

```
Xtwo = 100 * (1 - Ptwo) / (0.5 * (Pone + Ptwo))
```

> 8.57476

The dummy swap rate for 1.5 years is given by interpolating the one-year dummy value with the real two-year value (taking the data in this case to be for semi-annual payments):

```
Xthree = 0.5 * (Xtwo + First[SemiAnnSwapOriginalRange][[2]])
```

> 9.48738

Creating Extended Swap Series

Recall that we had the following two interpolated swap series, for semi-annual and annual swap data:

```
AnnSwapOriginalRange
```

```
    {{2, 10.4}, {3, 10.25}, {4, 10.15}, {5, 10.15},
     {6, 9.975}, {7, 9.8}, {8, 9.7667}, {9, 9.7333}, {10, 9.7}}
```

```
SemiAnnSwapOriginalRange
```

```
    {{2, 10.4}, {2.5, 10.325}, {3., 10.25}, {3.5, 10.2}, {4., 10.15},
     {4.5, 10.15}, {5., 10.15}, {5.5, 10.063}, {6., 9.975},
     {6.5, 9.8875}, {7., 9.8}, {7.5, 9.7833}, {8., 9.7667},
     {8.5, 9.75}, {9., 9.7333}, {9.5, 9.7167}, {10., 9.7}}
```

Now we join the dummy swap values to the real swap values. First the annual case:

```
AnnSwap = Join[ {{1.0, X}}, AnnSwapOriginalRange]
```

```
    {{1., 8.75809}, {2, 10.4}, {3, 10.25}, {4, 10.15}, {5, 10.15},
     {6, 9.975}, {7, 9.8}, {8, 9.7667}, {9, 9.7333}, {10, 9.7}}
```

Then the semi-annual version:

```
SemiAnnSwap = Join[{{0.5, Xone}, {1.0, Xtwo}, {1.5, Xthree}},
  SemiAnnSwapOriginalRange]
```

```
    {{0.5, 8.59608}, {1., 8.57476}, {1.5, 9.48738}, {2, 10.4},
     {2.5, 10.325}, {3., 10.25}, {3.5, 10.2}, {4., 10.15},
     {4.5, 10.15}, {5., 10.15}, {5.5, 10.063}, {6., 9.975},
     {6.5, 9.8875}, {7., 9.8}, {7.5, 9.7833}, {8., 9.7667},
     {8.5, 9.75}, {9., 9.7333}, {9.5, 9.7167}, {10., 9.7}}
```

We can plot the joined-up series:

```
ListPlot[SemiAnnSwap, PlotStyle -> PointSize[0.03],
  PlotRange -> {{0, 11}, {8, 11}}];
```

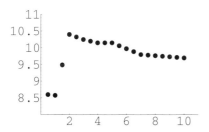

26.7 Bootstrapping the Swap-Discount Relations

For completeness we give the swap rates as a function of discount factors - this is a simple ratio with a summation in the denominator:

```
SwapTable[discfacs_List, frequencies_List] :=
    Table[100*(1 -discfacs[[k]])/Apply[Plus,
              Take[discfacs*frequencies, k]],
       {k, 1, Length[discfacs]}]


SimpleSwap[discfacs_List, freq_] :=
    Module[{dummy = Table[1,{ Length[discfacs]}]},
    SwapTable[discfacs, dummy/freq]]
```

Note that that the **frequencies** variable is defined reciprocally - it is 0.5 for a semi-annual swap.

Swap Rates to Discount Rates

The following is a *Mathematica* implementation of the bootstrapping algorithm to extract discount factors from swap rates:

```
Discounts[swapdata_List, frequency_] :=
    Module[{len = Length[swapdata], running = 0, newdisc, disclist =
{}, y, k=1},

        Do[(newdisc =
          (1 - swapdata[[k]]*running*frequency/100)/(
            1+swapdata[[k]]*frequency/100);
             running+=newdisc;
             k+=1;
             AppendTo[disclist, newdisc]), {k, 1, len}];
        disclist]
```

Applied to Annual Swap Data

Let's just extract the swap rates for 1 to 10 years:

```
{anntimedata, annswapdata} = Transpose[AnnSwap]

    {{1., 2, 3, 4, 5, 6, 7, 8, 9, 10}, {8.75809, 10.4,
       10.25, 10.15, 10.15, 9.975, 9.8, 9.7667, 9.7333, 9.7}}
```

Now we calculate the discount factor every year for 1 to 10 years:

```
andisc = Discounts[annswapdata, 1]

    {0.919472, 0.81918, 0.745386, 0.678956, 0.616392,
       0.566498, 0.522863, 0.477819, 0.43706, 0.400172}
```

or, in pairs, timestamped:

```
tandisc = Transpose[{anntimedata, Discounts[annswapdata, 1]}]

    {{1., 0.919472}, {2, 0.81918}, {3, 0.745386},
       {4, 0.678956}, {5, 0.616392}, {6, 0.566498},
       {7, 0.522863}, {8, 0.477819}, {9, 0.43706}, {10, 0.400172}}
```

```
ListPlot[tandisc, PlotStyle -> PointSize[0.03]];
```

Applied to Semi-annual Swap Data

Let's just extract the swap rates for 0.5 to 10 years:

```
{satimedata, saswapdata} = Transpose[SemiAnnSwap]

    {{0.5, 1., 1.5, 2, 2.5, 3., 3.5, 4., 4.5,
       5., 5.5, 6., 6.5, 7., 7.5, 8., 8.5, 9., 9.5, 10.},
     {8.59608, 8.57476, 9.48738, 10.4, 10.325, 10.25,
       10.2, 10.15, 10.15, 10.15, 10.063, 9.975, 9.8875,
       9.8, 9.7833, 9.7667, 9.75, 9.7333, 9.7167, 9.7}}
```

Now we calculate the discount factors every six months from 0.5 year to 10 years:

```
semiandisc = Discounts[saswapdata, 0.5]
```

   {0.958791, 0.919472, 0.869648, 0.814742, 0.776016,
    0.739732, 0.705044, 0.672368, 0.639893, 0.608987,
    0.583024, 0.558781, 0.536146, 0.515015, 0.491783,
    0.469711, 0.44874, 0.428814, 0.409878, 0.391884}

```
tsadisc = Transpose[{satimedata, Discounts[saswapdata, 0.5]}]
```

   {{0.5, 0.958791}, {1., 0.919472}, {1.5, 0.869648}, {2, 0.814742},
    {2.5, 0.776016}, {3., 0.739732}, {3.5, 0.705044}, {4., 0.672368},
    {4.5, 0.639893}, {5., 0.608987}, {5.5, 0.583024}, {6., 0.558781},
    {6.5, 0.536146}, {7., 0.515015}, {7.5, 0.491783}, {8., 0.469711},
    {8.5, 0.44874}, {9., 0.428814}, {9.5, 0.409878}, {10., 0.391884}}

```
ListPlot[tsadisc, PlotStyle -> PointSize[0.03]];
```

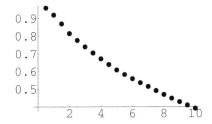

26.8 Joining the LIBOR Discount Factors to the Swap Discount Factors

Having used the swap data to create discount factors out to ten years, we now join it up with the short-term discount factors based on LIBOR data.

Joining Up with Annual Swap Data

Here are the long-term discount factors:

`tandisc`

```
{{1., 0.919472}, {2, 0.81918}, {3, 0.745386},
 {4, 0.678956}, {5, 0.616392}, {6, 0.566498},
 {7, 0.522863}, {8, 0.477819}, {9, 0.43706}, {10, 0.400172}}
```

Here are the short-term ones:

`anlibordiscs`

```
{{0.00273973, 0.999753}, {0.0192308, 0.998295}, {0.0833333, 0.9931},
 {0.25, 0.979306}, {0.5, 0.958791}, {1., 0.919472}}
```

They are consistent where they join up, as they should be, so we just join them up with the duplicate point deleted:

`andiscs = Join[anlibordiscs, Rest[tandisc]]`

```
{{0.00273973, 0.999753}, {0.0192308, 0.998295}, {0.0833333, 0.9931},
 {0.25, 0.979306}, {0.5, 0.958791}, {1., 0.919472}, {2, 0.81918},
 {3, 0.745386}, {4, 0.678956}, {5, 0.616392}, {6, 0.566498},
 {7, 0.522863}, {8, 0.477819}, {9, 0.43706}, {10, 0.400172}}
```

`ListPlot[andiscs, PlotStyle -> PointSize[0.02]];`

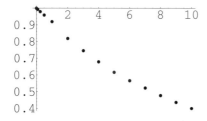

Joining Up with Semi-annual Swap Data

Here are the long-term discount factors:

```
tsadisc
```

```
{{0.5, 0.958791}, {1., 0.919472}, {1.5, 0.869648}, {2, 0.814742},
 {2.5, 0.776016}, {3., 0.739732}, {3.5, 0.705044}, {4., 0.672368},
 {4.5, 0.639893}, {5., 0.608987}, {5.5, 0.583024}, {6., 0.558781},
 {6.5, 0.536146}, {7., 0.515015}, {7.5, 0.491783}, {8., 0.469711},
 {8.5, 0.44874}, {9., 0.428814}, {9.5, 0.409878}, {10., 0.391884}}
```

Here are the shor- term ones (same as before)

```
anlibordiscs
```

```
{{0.00273973, 0.999753}, {0.0192308, 0.998295}, {0.0833333, 0.9931},
 {0.25, 0.979306}, {0.5, 0.958791}, {1., 0.919472}}
```

They are consistent where they join up, and overlap, as they should be, so we just join them up with the duplicate points deleted:

```
sadiscs = Join[Take[anlibordiscs, 4], tsadisc];
```

```
sadplot = ListPlot[sadiscs, PlotStyle -> PointSize[0.02]];
```

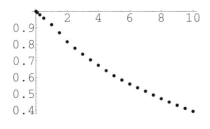

At this stage we have a representation of the yield curve as a time series from very small times out to 10 years, obtained by fusing LIBOR and swap data. From now on the working data set will be the time-stamped discount factors based on fusing LIBOR data with semi-annual payment swap data, as embodied in the data set, **sadiscs**, consisting of a list of pairs (time, discount_factor(time))

26.9 The Yield Curve and Forwards Calculation

Here we give a *Mathematica* implementation of the mappings for continuously compounded and annual variables:

Work with $t = 0$ (now!) so only maturities enter relationship.

```
YieldCC[{maturity_, discount_} ] :=
    {maturity, -100*Log[discount]/maturity}
```

```
YieldAnn[{maturity_, discount_} ] :=
    {maturity, 100*((1/discount)^(1/maturity) - 1)}
```

```
YieldAnn[sadiscs[[3]]]
```

```
    {0.0833333, 8.664}
```

```
yieldsa = Map[YieldAnn, sadiscs]
```

```
    {{0.00273973, 9.41729}, {0.0192308, 9.27981},
     {0.0833333, 8.664}, {0.25, 8.72434}, {0.5, 8.78081},
     {1., 8.75809}, {1.5, 9.7584}, {2, 10.7873}, {2.5, 10.6755},
     {3., 10.5712}, {3.5, 10.5011}, {4., 10.4329}, {4.5, 10.43},
     {5., 10.4278}, {5.5, 10.3068}, {6., 10.186}, {6.5, 10.0649},
     {7., 9.94326}, {7.5, 9.92509}, {8., 9.90593},
     {8.5, 9.88585}, {9., 9.86492}, {9.5, 9.84319}, {10., 9.8207}}
```

The data in the set **yieldsa** extends to ten years. We extrapolate to times larger than ten years by taking the yield to be constant thereafter. The simplest way of sorting this is to add a data point at, say, 30 years, and duplicate the 10-year value into it:

```
extrayield = {30, Last[yieldsa][[2]]}
```

```
    {30, 9.8207}
```

Similarly, in order to treat very short times, we take the yield at time zero to be the same as that at the ON rate:

```
firstyield = {0, First[yieldsa][[2]]}
```

```
    {0, 9.41729}
```

```
yields = Prepend[Append[yieldsa, extrayield], firstyield]
```

```
    {{0, 9.41729}, {0.00273973, 9.41729}, {0.0192308, 9.27981},
     {0.0833333, 8.664}, {0.25, 8.72434}, {0.5, 8.78081},
     {1., 8.75809}, {1.5, 9.7584}, {2, 10.7873}, {2.5, 10.6755},
     {3., 10.5712}, {3.5, 10.5011}, {4., 10.4329}, {4.5, 10.43},
     {5., 10.4278}, {5.5, 10.3068}, {6., 10.186}, {6.5, 10.0649},
     {7., 9.94326}, {7.5, 9.92509}, {8., 9.90593}, {8.5, 9.88585},
     {9., 9.86492}, {9.5, 9.84319}, {10., 9.8207}, {30, 9.8207}}
```

So, finally, we get the yield curve based on the fused, interpolated and extrapolated information:

```
ycurve = ListPlot[yields, PlotRange -> {{0, 20}, All},
    PlotJoined -> True];
```

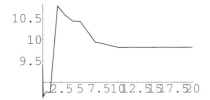

Yield Curve Interpolation

For our purposes it may be useful to define an interpolated function for the yield:

```
yieldfunc = Interpolation[yields, InterpolationOrder -> 1]

    InterpolatingFunction[{{0, 30.}}, <>]


yieldfunc[1.2]

    9.15822


Plot[yieldfunc[T], {T, 0, 15}];
```

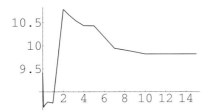

Discount Factors to Forward LIBOR Rates Calculation

Given the discount factor structure we have defined a function `yieldfunc[T]`, based on linear interpolation of the known yield values. This allows us to define an interpolated discount function

```
DiscFunc[T_] := 1 / (1 + yieldfunc[T] / 100) ^ T
```

```
Plot[DiscFunc[T], {T, 0, 30}];
```

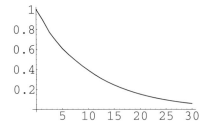

Note that this gives us a suitable interpolation of the discrete discount factor points already calculated, with a suitable exponential decay for times greater than 10 years, with the decay factor being that already computed at 10 years. We can see this by overlaying the plot of the interpolated discount function with the plot given previously:

```
Show[%, sadplot];
```

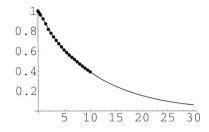

The Interest-Rate Forward

Given the function `DiscFunc[T]`, we can define the continuously compounded discrete forward rate function:

$$CCDFwd[T\_, dT\_] := -\frac{100 \; (Log[DiscFunc[dT + T]] - Log[DiscFunc[T]])}{dT}$$

The corresponding annually compounded discrete forward rate function is

$$ACDFwd[T\_, dT\_] := 100 \left(\left(\frac{DiscFunc[T]}{DiscFunc[dT + T]} \right)^{1/dT} - 1 \right)$$

```
ACDFwd[1, 0.25]
```

```
11.282
```

```
Plot[ACDFwd[T, 0.25], {T, 0, 15}, PlotRange -> All];
```

Note: for times T greater than 10 years, the forward rate is now a constant independent of $T > 10$ and dT:

```
{ACDFwd[12, 0.1], ACDFwd[14, 0.1], ACDFwd[14, 2]}
```

```
   {9.8207, 9.8207, 9.8207}
```

Remarks

This is the simplest possible modelling process, but is complete within its limitations. More complex schemes can be built using many instruments and the fitting techniques we have described. Different types of interpolation can be used and the effects of this on the details of the yield curve and forward rates are well worth investigating. A good place to start, including more real-world effects, is the text by Miron and Swannell (1991).

Chapter 26 Bibliography

Miron, P. and Swannell, P., 1991, *Pricing and Hedging Swaps*, Euromoney.

Rebonato, R., 1996, *Interest-Rate Option Models*, Wiley.

Shaw, W.T. and Tigg, J., 1993, *Applied Mathematica*, Addison-Wesley.

Chapter 27.
Simple Interest-Rate Options

Analytical Models of Swaptions and Bond Options

27.1 Introduction

Readers with a special interest in interest-rate options may be disappointed that the discussion of interest-rate options comes almost last and is confined to just one chapter. In fact, a text larger than this one could be written on the subject of interest-rate options alone, and a full treatment is well beyond the scope of this text. The project that led to this book was in fact a test of all *equity* derivative models, which is the reason for the presentational bias. It may well be the case that material for a text on *Mathematica* applied specifically to interest-rate options becomes available as a result of further work in this area. For now, I will make the following points.

(1) Much of the technology we have developed for the equity environment is immediately transferable to the interest-rate environment with very little change other than a re-interpretation of underlyings, to use something that is approximately log-normal. We shall discuss this very specifically with reference to the Black model. Furthermore, when the PDEs associated with the interest-rate world are analysed carefully, the single-factor models can all be reduced to the standard heat equation with advection and source terms added, so that much of the numerical technology developed here can be carried over.

(2) The number of interesting interest-rate options admitting simple closed-form analytics is relatively small, so there is less emphasis on the symbolic calculus aspects of *Mathematica*, and more on the numerical methods.

(3) The issues we have raised on the equity side, with regard to strong stability of FD schemes, the manner in which trees are constructed, and the convergence properties of Monte Carlo simulations, all apply to the interest-rate side. Whereas we have proposed some specific solutions relevant to the equity side, the nature of the problems on the interest-rate side, and the details of how they may be fixed, are not quite as straightforward. My hope is that some of the issues raised here for the vanilla Black-Scholes PDE (i.e. the diffusion equation with nasty boundary conditions) will stimulate investigation into the stability and robust differentiability of numerically obtained solutions on the interest-rate side.

(4) There are many more complications on the interest-rate side. We have already discussed the family of interest-rate diffusion models, which leads to a high component of model risk, and the need to build yield curves. Interest-rate derivatives force one also to confront the issue of the term structure of volatility, which is now of course a hot topic in equity derivative modelling.

The previous chapters have, hopefully, equipped you with the resources to build trees, FD models and Monte Carlo simulations within *Mathematica*. It is up to you to develop the implementations and applications to other derivative instruments. There are many interesting projects. One that is obvious is to adapt the tree technology we have developed to include the Hull-White and Black-Derman-Toy formalisms. Another is to build a two-factor finite-difference model that is adaptable to equities and interest rates (e.g. for convertibles), or to equities and volatility, for real-world equity option pricing, or to any other pair of factors.

Our main purpose in this chapter is first to point out the simple cases that can be modelled by use of the existing equity technology, and then to explore briefly the Vasicek and Cox-Ingersoll-Ross models as applied to the pricing of bond options. This makes particularly good use of *Mathematica*'s special-function capabilities, and in the case of Vasicek, exposes the way in which log-normal models re-emerge from within more complicated schemes.

The references for this chapter include those of Chapter 25. In particular, the reader is referred to Chapters 16 and 17 of Hull (1996), and to works by Jamshidian, on elegant analytical models of bond derivatives. One which is easily accessible is Jamshidian (1989/1996), which is an exact formula for options on bonds in the Vasicek world. The corresponding formula in the Cox-Ingersoll-Ross model is Cox, Ingersoll and Ross (1985/1996). These latter papers are part of a wonderful collection of articles put together by Hughston (1996), which is an excellent complement to the indispensable text by Rebonato (1996).

27.2 Interest-Rate Options in the Black World

For certain classes of European-style interest-rate options, it is commonplace, indeed, almost a market standard, to bypass the interest-rate diffusion models discussed in Chapter 25, and to use Black's (1976) model. The idea is to treat the forward value of a generic underlying as being log-normally distributed at the expiry of the associated (European) option. Typical applications include (rates/prices are always forwards, and the term "underlying" is used solely to indicate a choice of log-normal variable):

(a) swaptions - the swap rate at option maturity is the underlying;
(b) caplets - the interest rate at maturity is the underlying;
(c) bond options - the bond price at maturity is the underlying.

The Black formula was originally based on options of the futures price of a commodity. Being based on the futures price, it is implicit that we are dealing with an underlying whose risk-neutral drift is zero, so to make use of the equity technology developed previously, we need to set $q = r$ in the option-pricing formulae. It is used in general for any option where the underlying has zero risk-neutral drift, and can be assumed to have a log-normal standard deviation $\sigma \sqrt{T}$ at maturity T. The interest rate used for discounting, r, is the risk-free rate for time T, i.e., is taken off the yield curve.

It is important to realize that the Black model, as applied to any one of (a), (b), (c) above in isolation, is an extremely useful approximation. It happens to have deeper justifications in particular cases, but it is not logically consistent to assume that swaptions, caplets and bond options can all be priced on the assumption of a log-normal world. For a detailed discussion, see Chapter 1 of Rebonato (1996), however, since a swap is a linear combination of forward rates, it is not possible (precisely as in the case of the dividend model discussed in Chapter 24) for the forward interest rates and the forward swap rate to both be log-normal. Log-normal distributions do not add to log-normal. But in the equity case, it was extremely useful to make approximations for Baskets, Asian options, and options on underlyings with dividends, based on a log-normal view. This is also the case in the interest-rate world.

The analogy with equity log-normal approximation schemes is rather more useful still. The volatility one uses in the Black model is just a parameter σ with the property that $\sigma \sqrt{T}$ is the standard deviation of the log of the forward price. It does not (cf. log-normal approximations in the equity case) have to be related to a local or constant volatility of a random walk followed by an interpretable underlying.

When used for European options, then, it is just a matter of defining the underlying suitably as a forward, taking r off the yield curve, setting $q = r$, and choosing the right volatility parameter. We shall explore this in more detail presently for the particular case of a swaption. There is little point here in roaming through all the different applications of Black's model. The reader is encouraged to consult Chapter 16 of Hull (1996), and to plug in relevant parameters to the formulae we have given in Chapter 7 for the vanilla Black-Scholes model. For the case of bond options, we shall see explicitly how a Black model emerges from the Vasicek world through the work of Jamshidian (1989/1996). That author has also looked at deeper justifications for the assumptions of a Black environment (Jamshidian, 1993).

Stretching the Black Model

The following remarks will almost certainly have no support in the academic community, but are sometimes very useful in getting rough estimates for the values of more complex instruments, and checking more complex interest-rate models, by providing an approximate comparison. It is tempting to interpret the σ parameter as indeed being a local volatility, and to regard the option value as being the solution of the PDE, possibly with constraints for early exercise,

$$\frac{\partial V}{\partial t} + \frac{1}{2} \sigma^2 F^2 \frac{\partial^2 V}{\partial F^2} - r V = 0 \tag{1}$$

where F is the appropriate forward rate. This allows one to use the FD and tree models to value, for example, Bermudan and American swaptions, using all of the numerical technology developed for the equity case. One just sets $q = r$ and writes down the payoff constraints very carefully. If σ is allowed to be time-dependent, this approach can produce estimates that are not completely unreasonable, in spite of it having no serious theoretical justification.

European Swaptions in a Black World

The application of Black's model, at least if one has the input data, is little more than the application of some interesting decoration to the Black-Scholes model discussed in Chapter 7. It will probably be helpful to see how to do this once, and to then encourage the reader to play with other cases. We shall look at the case of a European swaption, and follow the route taken by Hull (1996), Section 16.8. We build the model from scratch, in case you have dived directly into the section of this book that deals with interest rates. The cumulative normal distribution is given by

```
Norm[z_ ? NumberQ] := N[0.5 Erf[ z/√2 ] + 0.5];
```

```
Norm[x_] := 1/2 (1 + Erf[ x/√2 ])
```

In terms of the forward swap rate, strike, effective volatility and time, the relevant auxiliary functions for the Black formula are

$$\text{swapdone}[f\_, c\_, \sigma\_, t\_] := \frac{\text{Log}[\frac{f}{c}]}{\sigma \sqrt{t}} + \frac{\sigma \sqrt{t}}{2};$$

$$\text{swapdtwo}[f\_, c\_, \sigma\_, t\_] := \frac{\text{Log}[\frac{f}{c}]}{\sigma \sqrt{t}} - \frac{\sigma \sqrt{t}}{2};$$

We build formulae for payer's swaptions ("Calls") and receiver's swaptions ("Puts") in two stages. The raw option component is given by the pair of functions:

```
RawOptPaySwap[f_, c_, σ_, t_, years_, r_] :=
  Exp[-r*t] (f Norm[swapdone[f, c, σ, t]] - c Norm[swapdtwo[f, c, σ, t]])
RawOptReceiveSwap[f_, c_, σ_, t_, years_, r_] :=
  Exp[-r*t] (c Norm[-swapdtwo[f, c, σ, t]] - f Norm[-swapdone[f, c, σ, t]])
```

If we build in all the relevant discounted cash flows, we arrive at the following *Mathematica* form of the Hull (1996) implementation of the Black swaption model:

```
HullPaySwap[f_, c_, σ_, t_, mfreq_, years_, principal_, rates_] :=
```
$$\frac{1}{\text{mfreq}} \left(\left(\sum_{i=1}^{\text{mfreq years}} \text{Exp}\left[-\text{rates}[\![i]\!] \left(t + \frac{i}{\text{mfreq}} \right) \right] \right) \text{principal} \right.$$

$$\left. (f \text{ Norm}[\text{swapdone}[f, c, \sigma, t]] - c \text{ Norm}[\text{swapdtwo}[f, c, \sigma, t]]) \right)$$

```
HullReceiveSwap[f_, c_, σ_, t_, mfreq_, years_, principal_, rates_] :=
```
$$\frac{1}{\text{mfreq}} \left(\left(\sum_{i=1}^{\text{mfreq years}} \text{Exp}\left[-\text{rates}[\![i]\!] \left(t + \frac{i}{\text{mfreq}} \right) \right] \right) \text{principal} \right.$$

$$\left. (c \text{ Norm}[-\text{swapdtwo}[f, c, \sigma, t]] - f \text{ Norm}[-\text{swapdone}[f, c, \sigma, t]]) \right)$$

We can do a quick symbolic parity check on this model:

```
TraditionalForm[Simplify[HullPaySwap[f, c, σ, t, mfreq, years, princi-
pal, rates] - HullReceiveSwap[f, c, σ, t, mfreq, years, principal,
rates]]]
```

$$- \frac{(c - f) \text{ principal} \sum_{i=1}^{\text{mfreq years}} e^{-\text{rates}[\![i]\!] \left(\frac{i}{\text{mfreq}} + t \right)}}{\text{mfreq}}$$

We can also check that the model agrees with published data. Consider the example given by Hull (1996). The yield curve is taken to be flat with 6% continuously compounded interest rate. The swaption expires in five years and has a strike of 6.2% for a three-year semi-annual swap starting at the expiry of the options - the swaption holder has the right to enter into a three-year swap, after five years, where they pay

6.2%. The effective volatility for the forward swap rate is 0.20, and for a three-year semi-annual swap with a *flat* yield curve, the swap forward rate, F, is given by

```
r = 0.06;
F = 2 * (Exp[r / 2] - 1)
```

> 0.0609091

For a principal of 100 currency units, the value of the payer's swaption is then

```
HullPaySwap[0.060901, 0.062, 0.2, 5, 2, 3, 100, 0.06 * {1, 1, 1, 1, 1, 1}]
```

> 2.07006

This agrees with Hull's result of 2.07.

27.3 Bond Options in a Vasicek World

The Black model for bond options can be justified on the basis of several of the interest-rate models. The Vasicek, Ho-Lee and Hull-White models introduced briefly in Chapter 25 all have the property that the interest-rate distribution is itself Gaussian. Given the mathematical exponential relationship between bond prices and interest rates, it should not come as too much of a surprise that the distribution of bond prices within these models, at the expiry of an option, can be regarded as log-normal, with results that map nicely into the Black model.

We shall explore this correspondence in detail for the Vasicek interest-rate world, where there are very simple closed-form analytics for discount bond options, which have been spelled out in detail by Jamshidian (1989/1996). That paper also contains a useful and generally applicable method for dealing with options on bonds with coupons.

Recall first that in Chapter 25 we got *Mathematica* to derive the discount bond-pricing formula for the Vasicek world. We have the functions A, B expressed by the following *Mathematica* expressions (go and re-load the calculation in Chapter 25 to avoid re-keying here) where we now make these functions explicitly dependent on the Vasicek model parameters.

$$\text{VasB}[t\_, \; T\_, \; a\_, \; b\_] \; := \; \frac{1}{a} - \frac{E^{a\,t - a\,T}}{a}$$

$$\text{VasA}[t\_, \; T\_, \; a\_, \; b\_, \; \sigma\_] \; := \; -\frac{E^{2\,a\,t - 2\,a\,T}\,\sigma^2}{4\,a^3} -$$

$$\frac{E^{a\,t - a\,T}\,(a^2\,b - \sigma^2)}{a^3} + \frac{t\,(2\,a^2\,b - \sigma^2)}{2\,a^2} + \frac{1}{4}\left(\frac{4\,b}{a} - 4\,b\,T - \frac{3\,\sigma^2}{a^3} + \frac{2\,T\,\sigma^2}{a^2}\right)$$

This gives the exact formula for the bond price in terms of the parameters a, b, σ. The bond price is then given explicitly as

```
VasP[r_, t_, T_, a_, b_, σ_] :=
 Exp[VasA[t, T, a, b, σ] - r*VasB[t, T, a, b]]
```

Readers may wish to explore this issue of calibrating this model against real yield-curve data, such as those obtained by the processing of the data in Chapter 26. We shall proceed without this step, to the option case. In what follows, the option is being valued at time t, and expires at time T, and is on a bond maturing at time s. The first variable is the log-normal standard deviation parameter, usually denoted by σ_P. A *Mathematica* representation is

```
sigp[t_, T_, s_, a_, b_, σ_] :=
 (σ / a) * (1 - Exp[-a * (s - T)]) Sqrt[(1 - Exp[-2 * a * (T - t)]) / (2 a)]
```

The analogue of the d-functions is the variable h, given, for principal L and option strike X, by

```
h[L_, X_, r_, t_, T_, s_, a_, b_, σ_] :=
 (1 / sigp[t, T, s, a, b, σ]) *
   Log[(L * VasP[r, t, s, a, b, σ]) / (X * VasP[r, t, T, a, b, σ])] +
 sigp[t, T, s, a, b, σ] / 2
```

The Call and Put options on a zero-coupon bond in the Vasicek world are then given, respectively, by

```
VasDiscCall[L_, X_, r_, t_, T_, s_, a_, b_, σ_] :=
 L * VasP[r, t, s, a, b, σ] * Norm[h[L, X, r, t, T, s, a, b, σ]] -
   X * VasP[r, t, T, a, b, σ] *
   Norm[h[L, X, r, t, T, s, a, b, σ] - sigp[t, T, s, a, b, σ] ]
```

```
VasDiscPut[L_, X_, r_, t_, T_, s_, a_, b_, σ_] :=
 L * VasP[r, t, T, a, b, σ] *
   Norm[-h[L, X, r, t, T, s, a, b, σ] + sigp[t, T, s, a, b, σ]] -
   X * VasP[r, t, s, a, b, σ] * Norm[-h[L, X, r, t, T, s, a, b, σ]]
```

27.4 Interest Rates and Bond Options in a Cox-Ingersoll-Ross World

Recall first that the CIR world is governed by an interest-rate diffusion of the form

$$d\,r = a(b - r)\,d\,t + \sigma \sqrt{r}\,d\,z \tag{2}$$

This describes a mean-reverting process without negative rates. The existence of a simple closed form for bond options is not confined to the Vasicek environment. Corresponding results exist also for the CIR world. Before we can develop these we need to do ground work on some new special functions related to the distribution of interest rates in a CIR world. We load the *Mathematica* Package describing a range of interesting distributions.

```
Needs["Statistics`ContinuousDistributions`"]
```

Here are the contents, which are useful for all kinds of applications:

?Statistics`ContinuousDistributions`*

BetaDistribution	LogNormalDistribution
CauchyDistribution	NoncentralChiSquareDistribution
ChiDistribution	NoncentralFRatioDistribution
ExponentialDistribution	NoncentralStudentTDistribution
ExtremeValueDistribution	ParetoDistribution
GammaDistribution	RayleighDistribution
HalfNormalDistribution	UniformDistribution
LaplaceDistribution	WeibullDistribution
LogisticDistribution	

Here is some information about the one we want:

? NoncentralChiSquareDistribution

> NoncentralChiSquareDistribution[n, lambda] represents
> the non-central chi-square distribution with n degrees
> of freedom and non-centrality parameter lambda.

We have a formula for the probability density function:

PDF[NoncentralChiSquareDistribution[n,λ], x]

$$2^{-n/2} \, E^{\frac{1}{2}(-x-\lambda)} \, x^{-1+\frac{n}{2}} \, \text{Hypergeometric0F1Regularized}\left[\frac{n}{2}, \frac{x\,\lambda}{4}\right]$$

The cumulative distribution function is in general an integral:

CDF[NoncentralChiSquareDistribution[n,λ], x]

$$2^{-n/2} \int_{0}^{x} E^{\frac{1}{2}(-x1-\lambda)} \, x1^{-1+\frac{n}{2}} \, \text{Hypergeometric0F1Regularized}\left[\frac{n}{2}, \frac{x1\,\lambda}{4}\right] \, dx1$$

The relationship of this distribution to the probability distribution of interest rates is obtained by introducing some auxiliary functions. Our discussion is based on CIR (1985/1996) but with the notation (for the random walk) of Hull (1996). We set

```
q[a_, b_, σ_] := 2 * a * b / σ ^ 2 - 1;
c[a_, σ_, s_, t_] := 2 * a / (σ ^ 2 * (1 - Exp[-a * (s - t)]));
u[rt_, c_, a_, s_, t_] := c * rt * Exp[-a * (s - t)];
v[rs_, c_] := c * rs;
```

The role of these parameters is as follows:

(1) the variable q, which depends on the random walk parameters, is related to the number of degrees of freedom $n = 2q + 2$;

(2) rt is the value of the interest rate at the current time t; rs is the probabilistic value at some later time s;

(3) the quantity $2u$ is the parameter of non-centrality;

(4) the quantity $2v = 2crs$ is the standardized random variable, in that the distribution function is

```
CDF[NoncentralChiSquareDistribution[2 * q + 2, 2 u], 2 v]
```

It follows that the density function, as a function of rs, is given by

```
DensA[c_, q_, u_, v_] =
  Simplify[2 * c * PDF[NoncentralChiSquareDistribution[2 * q + 2, 2 u], 2 v]]
```

$$c\,E^{-u-v}\,v^q\,\text{Hypergeometric0F1Regularized}[1 + q, u\,v]$$

Let's consider an alternative formulation of this function, which may be more familiar to you. We set

```
DensB[c_, q_, u_, v_] =
  c Exp[-u - v] (v / u) ^ (q / 2) BesselI[q, 2 Sqrt[u * v]];

TraditionalForm[DensB[c, q, u, v]]
```

$$c\,e^{-u-v}\left(\frac{v}{u}\right)^{q/2} I_q\!\left(2\sqrt{u\,v}\right) \tag{3}$$

These two representations are equivalent, as may be verified by consulting special-function tables. You can reassure yourself of this by making some simple checks - for example:

```
{DensA[1, 0.1, 2, 3], DensB[1, 0.1, 2, 3]}

    {0.170955, 0.170955}
```

The representation in terms of Bessel functions is the one originally given by CIR (1985). These distributions of interest rates are very revealing. We can use the symbolic power of *Mathematica* to investigate them. For example, it is easy to demonstrate the mean-reverting character. Let's first work out the mean of the distribution, using the powerful tools built in to the statistics Packages. We need to re-scale by $2c$ to undo the standardization of the random variables.

```
Simplify[Mean[NoncentralChiSquareDistribution[2 * q + 2, 2 u]] / (2 * c)]
```

$$\frac{1 + q + u}{c}$$

Now let's substitute for the actual variables, and simplify (you should run the following twice):

```
Expand[Simplify[
  % /. {u -> u[rt, c, a, s, t], c -> c[a, σ, s, t], q -> q[a, b, σ]}]]
```

$$b - b \, E^{a \, (-s+t)} + E^{a \, (-s+t)} \, rt$$

```
Collect[%, b]
```

$$b \, (1 - E^{a \, (-s+t)}) + E^{a \, (-s+t)} \, rt$$

So for s close to the current time t, we get rt back, whereas for large times s the mean tends to b.

Real World vs Risk-Neutral World

The literature on interest-rate models is made a little confusing by different defaults on whether to talk in terms of real-world variables or risk-neutral variables. The notation employed by Hull (1996) refers to risk-neutral variables. So in the random walk specification

$$d r = a(b - r) d t + \sigma \sqrt{r} \, d z \tag{4}$$

the variables a and b are based on a risk-neutral description. In contrast, the papers by CIR (1985/1996) refer to random walks of the form

$$d r = \kappa(\theta - r) d t + \sigma \sqrt{r} \, d z \tag{5}$$

and κ, θ are *real-world* variables. These different relations are linked by the variable representing the *market price of risk*, usually called λ, by the conditions

$$a = \kappa + \lambda$$
$$b = \frac{\kappa \theta}{\kappa + \lambda} \tag{6}$$

For example, the bond-pricing formula given by Hull is in terms of risk-neutral variables, and is equivalent to the one we gave in Chapter 25. CIR give the bond-pricing formula in terms of real-world variables. The two formulae are linked by equation (5). Clearly we can interpret our remarks on the distribution of interest rates as applying to real or risk-neutral coordinates, but when we price instruments we can hypothetically work with either risk-neutral variables, or real-world variables with explicit management of the market price of risk. However, there is an additional issue that must be faced. We cannot observe the values of κ, θ, λ directly from the market - we can only observe the values of the combinations $\kappa + \lambda$ and $\kappa \theta$. This is most easily seen by examining the bond-pricing formula which only involves a and b. In particular, when we calibrate our model against supplied yield curve data, we can only expect to obtain values for $\kappa + \lambda$ and $\kappa \theta$.

Pricing of Discount Bond Options in CIR

The question of the pricing of instruments can now be addressed. Note that in Chapter 25 we did bonds in a risk-neutral world by manipulations with the bond-pricing equations. Now we turn our attention to an option on a bond. The results for this neatly parallel the results for equity options, except that we have to work in terms of the cumulative non-central χ^2 distribution, rather than the ubiquitous N for the log-normal case. We consider the case of pure discount bonds - there are no coupons. First we re-define the A and B functions governing the price of a discount bond. We use the forms derived by *Mathematica* from the reduced form of the bond-pricing equation - readers may wish to satisfy themselves that these are the same functions as those derived by CIR, even though they look completely different. Note that the CIR A is the exponential of our original A, so in order to follow CIR, we exponentiate our previous definitions. We introduce variables kpl $= \kappa + \lambda$ and kth $= \kappa\theta$ so that we work only in terms of observable quantities:

```
B[t_, T_, kpl_, σ_] := Module[{g = Sqrt[kpl^2 + 2 σ^2]},
```

$$- \frac{\text{kpl} + g\,\text{Tanh}\left[\frac{1}{2}\,g\,(t-T) - \text{ArcTanh}\left[\frac{\text{kpl}}{g}\right]\right]}{\sigma^2}\Bigg]$$

```
A[t_, T_, kpl_, σ_, kth_] := Exp[Re[Module[{g = Sqrt[kpl^2 + 2 σ^2]},
```

$$\frac{1}{g\,\sigma^2}\left(\text{kth}\left((\text{kpl}-g)\,\text{Log}\left[1-\frac{\text{kpl}}{g}\right] - (\text{kpl}+g)\,\text{Log}\left[-\frac{\text{kpl}+g}{g}\right] + \right.\right.$$

$$\text{kpl Log}\left[-1 + \text{Tanh}\left[\frac{1}{2}\,g\,(t-T) - \text{ArcTanh}\left[\frac{\text{kpl}}{g}\right]\right]\right] +$$

$$g\,\text{Log}\left[-1 + \text{Tanh}\left[\frac{1}{2}\,g\,(t-T) - \text{ArcTanh}\left[\frac{\text{kpl}}{g}\right]\right]\right] -$$

$$\text{kpl Log}\left[1 + \text{Tanh}\left[\frac{1}{2}\,g\,(t-T) - \text{ArcTanh}\left[\frac{\text{kpl}}{g}\right]\right]\right] +$$

$$g\,\text{Log}\left[1 + \text{Tanh}\left[\frac{1}{2}\,g\,(t-T) - \text{ArcTanh}\left[\frac{\text{kpl}}{g}\right]\right]\right]\right)\Bigg]\Bigg]\Bigg]$$

Now we write the bond price explicitly as

```
Clear[A, B, P];
P[r_, t_, s_, kpl_, σ_, kth_] :=
  A[t, s, kpl, σ, kth] Exp[-r * B[t, s, kpl, σ]]
```

Now we can prescribe the CIR bond option formula. We give two forms. The first is a function of some intermediate parameters, and the second is fully explicit and makes use of the first.

```
PartialCall[r_, t_, T_, s_, K_, rstar_, φ_, ψ_, g_, kpl_, σ_, kth_] :=
  P[r, t, s, kpl, σ, kth] * CDF[NoncentralChiSquareDistribution[
      4 * kth / σ ^ 2, 2 φ ^ 2 r Exp[g * (T - t)] / (φ + ψ + B[T, s, kpl, σ])],
    2 * rstar * (φ + ψ + B[T, s, kpl, σ])] -
  K * P[r, t, T, kpl, σ, kth] * CDF[NoncentralChiSquareDistribution[
      4 * kth / σ ^ 2, 2 φ ^ 2 r Exp[g * (T - t)] / (φ + ψ)], 2 * rstar * (φ + ψ)]
```

The parameters in the final form are

r - current interest rate;
t - current time;
T - option expiry date;
s - bond maturity date;
K - strike price;

and the CIR model parameters make up the rest.

```
CIRBondCall[r_, t_, T_, s_, K_, kpl_, σ_, kth_] :=
  Module[{rstar, φ, ψ, g},
    rstar = Log[A[T, s, kpl, σ, kth] / K] / B[T, s, kpl, σ];
    g = Sqrt[kpl ^ 2 + 2 σ ^ 2];
    ψ = (kpl + g) / σ ^ 2;
    φ = 2 g / (σ ^ 2 (Exp[g * (T - t)] - 1));
    PartialCall[r, t, T, s, K, rstar, φ, ψ, g, kpl, σ, kth]]
```

Readers should note the structural similarity to the log-normal formula.

Chapter 27 Bibliography

See the Bibliography for Chapter 25, and also

Black, F., 1976, The pricing of commodity contracts, *Journal of Financial Economics*, 3, p. 167.

Cox, J.C., Ingersoll, J.E. and Ross, S.A., 1985/1996, A theory of the term structure of interest rates, *Econometrica*, 53, p. 385, reprinted in Hughston (1996).

Hughston, L.P. (editor), 1996, *Vasicek and Beyond - Approaches to Building and Applying Interest Rate Models*, RISK-EuroBrokers.

Jamshidian, F., 1989/1996, An exact bond option formula, *Journal of Finance*, 44, p. 205, reprinted in Hughston (1996).

Jamshidian, F., 1993, Options and futures evaluation with deterministic volatilities, *Mathematical Finance*, 3, p. 149.

Chapter 28.
Modelling Volatility by Elasticity

The CEV Family Explained and Explored

28.1 Introduction

Attempts to improve the treatment of volatility have been developed by many authors. This can be done in several contexts, and our focus here will be on improving the modelling of the volatility of equity options. Different methods can be used to treat volatility of FX and interest-rate options. Within the equity area, possible approaches include, but are not necessarily limited to:

(a) volatility as a deterministic function of the (stochastic) stock price ("level-dependent" models);
(b) stochastic volatility as a separate factor from the stock price;
(c) GARCH approaches.

This chapter addresses a subset of models of type (a), known as the "constant elasticity of variance" (CEV) models, which capture in a simple way the notion that the volatility increases as the asset price decreases. There are good reasons for considering this case:

(1) the model nicely captures the leverage effect associated with asset prices;
(2) there is research suggesting that the pricing of warrants in particular is improved by the use of this model;
(3) it is analytically tractable;
(4) it predicts skews and, with some additional adjustments for spreads, smiles;
(5) one member of the family is equivalent to a non-normal and non-log-normal interest rate model (CIR), allowing us to explore aspects of the interest-rate mathematics simultaneously with the equity mathematics.

This last point explains in part why this last chapter is placed directly after the discussion of elementary interest-rate models - the CEV theory applied to equity derivatives is closely related to the CIR model. Furthermore, placing the discussion here allows us to end this book at a point closely related to where we began, with a discussion of some of the effects that can be seen in implied volatility, in an application where it does make sense!

There are quite a few papers now published on elasticity models, some of them with minor glitches in them - part of the purpose of this chapter is to straighten out some of the technology and to give a comprehensive presentation. Also, some of the formulae employed for Call options or warrants turn out to be rather complicated. The general CEV Call option formula involves an infinite series of Gamma-functions, which is somewhat awkward to evaluate, even in *Mathematica*, let alone in C/C++ directly. There is an approximation for the cases of the SRCEV model (where the volatility varies as the inverse square root of the asset price - it is this model that is equivalent to the CIR interest-rate model without the mean-reversion) known as the Cox-Beckers approximation. Unfortunately there are at least three different and inconsistent representations of this formula in the literature.

One of the purposes of this chapter is to derive new simplified forms of the exact pricing formula. In the particular case of the SRCEV model, we give a simple formula involving an integration from 0 to π of the normal distribution, which is readily implementable in *Mathematica* or C/C++. While it may be possible to derive similar formulae for other elasticities, I have not done so, but the values of options can always be done directly in terms of integrals of Bessel functions. This can be coded up in C or C++ using standard routines, for example those given by Press *et al* (1992). Other methods for evaluation are given by Schroder (1989), which an interesting paper that among other things gives the clearest exposition, of the approximation of the non-central chi-square by normal distributions, that I have seen.

The running example considered in detail is that of a Call option - this is of particular interest in the development of improved warrant-pricing models.

The analysis presented suggests that at least part of the volatility structure observed in market prices may be captured by a simple level-dependent model, and one interesting issue relates to how to bolt on further stochasticity or GARCH behaviour to this type of approach. This approach, although simplistic, nicely captures the "leverage effect", so it should be considered as at least a part of the overall picture. The role of two-factor or GARCH models should perhaps be more properly defined as explaining those effects not captured by the type of model presented here.

28.2 CIR Interest Rates and Asset Elasticity Models

The type of stochastic process under consideration is probably more familiar in the CIR interest-rate model, as discussed in the previous chapter. The CIR world is governed by an interest-rate diffusion of the form

$$d\,r = a(b - r)\,d\,t + \sigma \sqrt{r}\,d\,z \tag{1}$$

This describes a mean-reverting process. It has nice analytical properties, such as the closed-form solution for bond options we have briefly outlined in the previous chapter. On the other hand, if we set $b = 0$, and $r \to S$, $a \to (q - r)$, we get a process of the form

$$d\,S = (r - q)\,S\,d\,t + \sigma \sqrt{S}\,d\,z \tag{2}$$

We can write (2) in the form

$$\frac{d\,S}{S} = (r - q)\,d\,t + \frac{\sigma}{\sqrt{S}}\,d\,z \tag{3}$$

which is an equity-like process with a volatility varying as the inverse square root of the stock price. This is commonly known as an SRCEV model - "Square root constant elasticity of variance". There are more general CEV models, where the process takes the form

$$\frac{d\,S}{S} = (r - q)\,d\,t + \frac{\sigma}{S^{1-\alpha}}\,d\,z \tag{4}$$

where α is between 0 (normal) and 1 (log-normal). So we see that a special form of the CIR interest-rate model is equivalent to a special case of stochastic volatility model for the equity world. The only difference is in the details and instruments valued - options on bonds (payoffs exponential functions of r) cf.

options on S (payoffs typically piecewise linear).

A brief review of the asset versions of these models was done recently by their principal inventor, J. Cox, (1996) in an article entitled: "The constant elasticity of variance option pricing model". It is well worth a detailed read, but note that the formula for the probability distribution is missing a term. Cox goes on to develop the infinite series of Gamma-functions as the price of a Call - it is this that we wish to try to simplify to make it more straightforward as a model. We shall also look at the associated Black-Scholes PDE in some detail, and explore its solution for a case slightly more general than that considered by Cox, where we have a time-dependent interest-rate and yield structure. Our approach will be based on the partial differential equation method, and parallels that given for the ordinary Black-Scholes equation, and it is here that the detailed development begins. Readers may wish to explore the development of an equivalent martingale formulation, and of the corresponding interest-rate analogues with mean reversion.

28.3 Standardized Form of the CEV Black-Scholes Equation

The Black-Scholes equation with time-dependent interest rates and dividend yield, and volatility $\sigma(S)$, is

$$\frac{\partial V}{\partial t} - r(t) V + \frac{1}{2} \sigma^2(S) S^2 \frac{\partial^2 V}{\partial S^2} + (r(t) - q(t)) S \frac{\partial V}{\partial S} = 0 \tag{5}$$

First we make the standard time-reversing transformation $t^* = T - t$, with the option expiring at time T. This gives

$$\frac{\sigma^2(S)}{2} S^2 \frac{\partial^2 V}{\partial S^2} + (r(T - t^*) - q(T - t^*)) S \frac{\partial V}{\partial S} = \frac{\partial V}{\partial t^*} + r(T - t^*) V \tag{6}$$

Next we make the standard discounting transformation (Section 4.6) $V = e^{-B(t^*)} V_1$ with B arranged so that

$$\frac{\partial B}{\partial t^*} = r(T - t^*) \tag{7}$$

which reduces the equation to

$$\frac{\sigma^2(S)}{2} S^2 \frac{\partial^2 V_1}{\partial S^2} + (r(T - t^*) - q(T - t^*)) S \frac{\partial V_1}{\partial S} = \frac{\partial V_1}{\partial t^*} \tag{8}$$

Note that the appropriate solution to the differential equation for B is indeed the discount factor:

$$e^{-B(t^*)} = e^{-\int_t^T r(t_1)\, dt_1} \tag{9}$$

The next task is to eliminate the drift term, and this again makes use of an approach similar to that discussed in Chapter 4 for CBs. We introduce new variables

$$S = S' \, e^{F(t')}$$
$$t^* = t' \tag{10}$$

and change variables in the PDE (just use the chain rule carefully if you want to check this), obtaining

$$\frac{\sigma^2(S)}{2} S'^2 \frac{\partial^2 V_1}{\partial S'^2} + \left(\frac{\partial F}{\partial t'} + r(T - t') - q(T - t') \right) S' \frac{\partial V_1}{\partial S'} = \frac{\partial V_1}{\partial t'} \tag{11}$$

Note that the volatility remains expressed in terms of the original underlying. So we can eliminate the drift term by choosing F to satisfy the ordinary differential equation

$$\frac{\partial F}{\partial t'} + r(T - t') - q(T - t') = 0 \tag{12}$$

We can write the solution formally in terms of an integral of the spot rate and yield. We shall write the answer in terms of a function Q that is the reciprocal of $e^{F(t')}$, i.e.,

$$Q(t') = e^{+\int_0^{t'} (r(T-s)-q(T-s))\,ds} \tag{13}$$

Then

$$S = \frac{S'}{Q(t')} \tag{14}$$

The PDE is now

$$\frac{\sigma^2 (S'/Q)}{2} S'^2 \frac{\partial^2 V_1}{\partial S'^2} = \frac{\partial V_1}{\partial t'} \tag{15}$$

Note that at maturity $S = S'$, whereas at initiation

$$S' = S\, e^{+\int_0^T (r(s)-q(s))\,ds} \tag{16}$$

It is difficult to proceed further without an assumption about the form of the volatility. It is here that we make the elasticity assumption:

$$\sigma(S) = \sigma_0\, S^{\alpha-1} \tag{17}$$

This allows us to collect the independent variables on both sides of the equation, obtaining

$$Q(t')^{2(\alpha-1)} \frac{\partial V_1}{\partial t'} = \frac{1}{2} \sigma_0^2\, S'^{2\alpha} \frac{\partial^2 V_1}{\partial S'^2} \tag{18}$$

We re-parametrize the elasticity variable by writing $2\alpha = 2 - 1/v$, so that the PDE is now

$$\frac{1}{Q(t')^{1/v}} \frac{\partial V_1}{\partial t'} = \frac{1}{2} \sigma_0^2\, S'^{2-1/v} \frac{\partial^2 V_1}{\partial S'^2} \tag{19}$$

We are almost at a recognizable PDE! It is just a matter of making one further transformation on each of the time and price variables. To simplify the right side, we introduce

$$X = S'^{1/v} \tag{20}$$

This gets us to

$$\frac{2v^2}{\sigma_0^2\, Q^{\frac{1}{v}}(t')} \frac{\partial V_1}{\partial t'} = X \frac{\partial^2 V_1}{\partial X^2} + (1 - v) \frac{\partial V_1}{\partial X} \tag{21}$$

The left side is now simplified by picking a new time coordinate τ with the property that

$$d\tau = \frac{1}{2\,v^2}\,\sigma_0^2\,Q^{\frac{1}{v}}(t')\,d\,t' \tag{22}$$

This differential equation can be solved formally as

$$\tau = \frac{\sigma_0^2}{2\,v^2}\,\int_0^{t'} e^{\frac{1}{v}\int_0^s (r(T-s')-q(T-s'))\,ds'}\,d\,s \tag{23}$$

Note that this new time is zero at expiry. If r and q are constant, this simplifies to

$$\tau = \frac{\sigma_0^2}{2\,v\,(r-q)}\,\left(e^{\frac{(r-q)(T-t)}{v}} - 1\right) \tag{24}$$

which is the reciprocal of the parameter called c or k by Cox and other authors. Whether or not the r, q parameters are constant, we arrive at the standardized CEV equation in the form

$$\frac{\partial V_1}{\partial \tau} = X\,\frac{\partial^2 V_1}{\partial X^2} + (1-v)\,\frac{\partial V_1}{\partial X} \tag{25}$$

The elasticity parameter v appears now in the first order term - this vanishes in this view of the model when $v = 1$, which is the special square root CEV model. This PDE is akin to that for purely radial diffusion in $-v$ dimensions! Students of probability may recognize it as that corresponding to a Bessel-squared process. We note that a futher transformation can reduce this to the ordinary diffusion equation with a non-zero drift - a fact that is of considerable importance, and that also applies to all of the interest-rate models, when source terms are also admitted.

In principle we could proceed to analyse this equation in much the same way as the ordinary one-dimensional diffusion equation. As the details of this are hard - rather beyond the scope of this text - we shall just quote the Green's function analogous to that given in Chapter 4. The Green's function corresponding to a delta-function source located at Y at time $\tau = 0$ is:

$$G(X, Y, \tau) = \frac{1}{\tau}\,e^{-\frac{(X+Y)}{\tau}}\,\left(\frac{X}{Y}\right)^{v/2}\,I_v\!\left(\frac{2\,\sqrt{X\,Y}}{\tau}\right) \tag{26}$$

If you wish to satisfy yourself that this is correct, then use *Mathematica* to check that it solves the differential equation, and use the fact that for large z,

$$I_v(z) \sim \frac{e^z}{\sqrt{2\,\pi\,z}} \tag{27}$$

to verify the behaviour for small τ. Normalization issues are discussed presently. Taking G as the Green's function allows us to write down the option valuation and interpret it in terms of risk-neutral valuation. Let the option payoff be $P(S_T)$. Because $S' = S$ at maturity, the value of X at maturity, denoted by X_T, corresponding to the stock price S_T is just

$$S_T = X_T^v \tag{28}$$

The non-discounted value at an earlier time τ is then

$$V'(X, \tau) = \int P(X_T^v)\,G(X, X_T, \tau)\,d\,X_T \tag{29}$$

Making a change of variable to S_T, we are led, finally, to

$$V'(X, \tau) = \int P(S_T) f(S_T) \, d S_T \tag{30}$$

where, if we set $X = x \tau$, $X_T = x_T \tau$,

$$f(S_T) = \frac{1}{\nu} \tau^{-\nu} \left(x \, x_T^{\frac{2}{\nu}-3} \right)^{\nu/2} e^{-x-x_T} I_\nu \left(2 \sqrt{x \, x_T} \right) \tag{31}$$

The variable X is then related to the price at initiation by

$$X = S^{1/\nu} \, e^{\frac{\int_0^T (r(s)-q(s)) \, ds}{\nu}} \tag{32}$$

The one catch in dealing with such models is that the expiry distribution does not integrate to unity. The density function for X_T is

$$U(X_T) = \frac{1}{\tau} \, e^{-\frac{(X+X_T)}{\tau}} \left(\frac{X}{X_T} \right)^{\nu/2} I_\nu \left(\frac{2 \sqrt{X \, X_T}}{\tau} \right) \tag{33}$$

This function has the property that

$$\int_0^\infty U(X_T) \, d X_T = 1 - \frac{\Gamma(\nu, \frac{X}{\tau})}{\Gamma(\nu)} \tag{34}$$

So the probability that $X_T = 0$ is just

$$\frac{\Gamma(\nu, \frac{X}{\tau})}{\Gamma(\nu)} \tag{35}$$

This has to be taken into account when valuing options whose payoff is non-zero at the origin - a Put being the most obvious case. This distribution is exceptional in having both a discrete portion and a continuous portion. There are many other interesting probabilistic aspects to CEV models, which we shall not consider here. A final reminder - the special cases are:

(a) $\nu = 1/2$ - the absolute Gaussian model;
(b) $\nu = 1$ - the SRCEV model;
(c) $\nu = \infty$ - the log-normal model.

28.4 CEV Call Option Formulae

By risk-neutral valuation, the value of a European Call with strike K is just

$$e^{-rt} \int_K^\infty (S_T - K) f(S_T) \, d S_T \tag{36}$$

There are no known "nice" formulae for this. Cox (1996) gives the result as an infinite series of Gamma-functions. Cox's model is the standard exact solution representation, and can be coded up in *Mathematica*, but requires careful treatment with respect to both evaluation of the Gamma-functions and also summation of the infinite series. A more down-to-earth representation is the direct integration approach involving the

Bessel functions (use the routines in Numerical Recipes if working in C) and the evaluation of (36). The integral can be simplified a little in terms of a standard integral. If we define:

$$\text{CEVfunc}(x, a, v, n) = \int_a^\infty e^{-\frac{z^2}{4x}} z^n I_v(z)\, dz \tag{37}$$

then the value of a Call, with strike K, $c = \tau^{-1}$, $x = X/\tau$, is given by

$$e^{-rt-x}\left(\frac{\text{CEVfunc}\left(x, 2\sqrt{c\,x\,K^{1/v}}, v, v+1\right)}{x\,c^v\,2^{v+1}} - K(2x)^{v-1}\,\text{CEVfunc}\left(x, 2\sqrt{c\,x\,K^{1/v}}, v, 1-v\right)\right) \tag{38}$$

It is left to the reader to explore an interesting set of special cases when the object CEVfunc can be evaluated in closed-form. This occurs when v is half an odd integer (cf Schroder, 1989), and provides a useful basis for approximating the answer by interpolating bewteen the values obtained for half-odd-integer v. Our interest will focus on the case of general v, and the particular case $v = 1$.

Mathematica Implementation

```
CEVfunc[x_, a_, v_, n_, upper_] :=
  NIntegrate[Exp[-(z^2/(4*x))]*z^n*BesselI[v, z], {z, a, upper}]
```

Call Formula

What we do is to directly integrate to a multiple of the strike, set by the variable "scale". Note that the σ parameter is set by assuming the volatility is that for the current spot price.

```
CEVCall[S_, K_, r_, q_, vol_, t_, v_, scale_] :=
Module[{σ = vol * S ^ (1 / (2 v)), x, c, a},
  c = (2 * v * (r - q)) / (σ ^ 2 * (Exp[((r - q) * t) / v] - 1));
  x = c * S ^ (1 / v) * Exp[((r - q) * t) / v];
  a = 2 Sqrt[c x K ^ (1 / v)];
   Exp[-r * t - x] *
     (CEVfunc[x, a, v, v + 1, scale * a] / (x * c ^ v * 2 ^ (v + 1)) -
   K * ((2 * x) ^ (v - 1)) * CEVfunc[x, a, v, 1 - v, scale * a])]
```

Here it is evaluated for a high value of $v = 10$, which should be close to the Black-Scholes value:

```
CEVCall[100, 100, 0.1, 0, 0.2, 1, 10, 10]

    13.269711019
```

Is it close?

```
Needs["Derivatives`BlackScholes`"]
```

```
BlackScholesCall[100, 100, 0.2, 0.1, 0, 1]
```

```
    13.2697
```

The answer is yes. As we lower the parameter the value increases for these at-the-money examples:

```
CEVCall[100, 100, 0.1, 0, 0.2, 1, 5, 10]
```

```
    13.269814334
```

Here is the SRCEV case:

```
CEVCall[100, 100, 0.1, 0, 0.2, 1, 1, 10]
```

```
    13.2731
```

28.5 A Simplified Exact Model for SRCEV

There are special cases where the integral can be simplified significantly. In addition to the case of v equal to half an odd integer, where closed-form answers exist, I believe a useful simplification can be obtained for integer values also, and the following explores the simplest case of $v = 1$, corresponding to the SRCEV model. The evaluation can be simplified dramatically by expressing the Bessel functions using an integral identity, and then doing the integration above. This still leaves an integral. However, the integrand now involves much simpler functions, and the remaining integral is typically over a simple finite interval such as $[0, \pi]$. The integration of periodic functions over such intervals is particularly simple, due to the special nature of the numerical error formula for periodic integrals. It can be evaluated to any desired degree of precision, but in practice only a few samples are needed. For example, in the case $v = 1$, we can write

$$I_1(z) = \frac{1}{\pi} \int_0^{\pi} \cos(x)\, e^{z\cos(x)}\, dx \tag{39}$$

Integral identities similar to this do exist for other values of v. The difficult integral from the strike to infinity can be done in closed form in terms of the standard cumulative Normal distribution. The result is as follows - let

$$\alpha = 2\,c\,K;\ \beta = 2\,x; \tag{40}$$

$$f(\alpha, \beta, \theta) = \left(\beta\cos^2(\theta) - \alpha + 1\right)\left(1 - N\left(\sqrt{\alpha} - \cos(\theta)\sqrt{\beta}\right)\right)\sqrt{\frac{\beta}{2\pi}} +$$

$$\frac{\sqrt{\beta}\ e^{-\frac{1}{2}\left(\sqrt{\alpha} - \cos(\theta)\sqrt{\beta}\right)^2}\left(\sqrt{\alpha} + \cos(\theta)\sqrt{\beta}\right)}{2\pi} \tag{41}$$

Then the value of a call option is just

$$
\frac{1}{c} e^{-rt - \frac{\beta}{2}} \int_0^\pi \cos(\theta)\, e^{\frac{1}{2}\beta \cos^2(\theta)}\, f(\alpha, \beta, \theta)\, d\theta \tag{42}
$$

Let's first build a *Mathematica* implementation:

```
interfunc[α_, β_, θ_] := Sqrt[β / (2 Pi)] * (1 - α + β Cos[θ] ^ 2)
    (1 - Norm[Sqrt[α] - Cos[θ] * Sqrt[β]]) + (Sqrt[β] / (2 * Pi)) *
    (Sqrt[α] + Cos[θ] * Sqrt[β]) * Exp[-1 / 2 (Sqrt[α] - Cos[θ] * Sqrt[β]) ^ 2]


SRCEVCall[S_, K_, r_, q_, σ_, t_] :=
 Module[{sig = σ * Sqrt[S], α, β, c, int},
  c = 2 * (r - q) / (sig ^ 2 * (Exp[(r - q) * t] - 1));
  α = 2 * c * K;
  β = 2 * c * S * Exp[(r - q) * t];
  int = NIntegrate[
    Cos[θ] Exp[β Cos[θ] ^ 2 / 2] interfunc[α, β, θ], {θ, 0, Pi}];
  Exp[-r * t - β / 2] * int / c]


SRCEVCall[100, 100, 0.1, 0, 0.2, 1]

    13.2731
```

This evaluates rather more quickly, and rather than involving Bessel functions and an infinite integration, involves only the normal distribution and a finite integration. Because of the nature of periodic functions, the integration can be done very efficiently using the trapezoidal rule and relatively few points. We halve the integration done from $[0, 2\pi]$ and proceeding by just summing the values of the integrand sampled at $(0, 2\pi/n, 4\pi/n, \ldots k*2\pi/n, \ldots 2\pi(1 - 1/n))$:

```
NumericalSRCEVCall[S_, K_, r_, q_, σ_, t_, samples_] :=
 Module[{sig = σ * Sqrt[S], α, β, c, sum},
  c = 2 * (r - q) / (sig ^ 2 * (Exp[(r - q) * t] - 1));
  α = 2 * c * K;
  β = 2 * c * S * Exp[(r - q) * t];
  sum =
   (2 Pi) / samples Sum[Cos[θ] Exp[β Cos[θ] ^ 2 / 2] interfunc[α, β, θ],
     {θ, 0, 2 * Pi (1 - 1 / samples), 2 Pi / samples}];
  Exp[-r * t - β / 2] * sum / (2 * c)]
```

60 terms are enough to get agreement to six decimal places, and it evaluates very quickly.

```
NumericalSRCEVCall[100, 100, 0.1, 0, 0.2, 1, 60]
```

```
13.2731
```

This is therefore a trivial addition to any derivatives modelling system where the cumulative normal distribution is already coded up, and is exact within the framework of the model - any desired degree of precision may be obtained by taking more terms in the summation.

28.6 The Cox-Beckers Approximation to SRCEV

This just involves a nest of substitutions and what looks like a normal approximation to the non-central chi-square distribution whose pdf we have given. It is just a bunch of formulae, but implementation is complicated by the fact that the literature has, what appears to this author at least, three different and inconsistent representations of this formula. If you wish to make your own evaluation of these comments, four papers all containing a version of this approximation are Beckers (1980), Jarrow and Rudd (1983), Lauterbach and Schultz (1990), Hauser and Lauterbach (1996). In the following, y is essentially c times the underlying. The implementation in here is based on Jarrow and Rudd's (1983) book, which in the humble opinion of this author is the correct version, but this is based on the subsequent comparison with the exact solution. I cannot claim to have gone into the approximation of the required distributions by normal distributions in any detail, but the following appears also to be consistent with the remarks by Schroder (1989) who also gives more general results that apply for non-zero yield. In any case it is probably more efficient to use the exact new model given above, and the following is largely included for completeness. Here are the formulae:

```
y[S_, r_, σ_, t_] := 4 * r * S / (σ^2 (1 - Exp[-r * t]));
z[K_, r_, σ_, t_] := 4 * r * K / (σ^2 (Exp[r * t] - 1));
p[v_, y_] := (v + 2 y) / (v + y) ^2;
h[v_, y_] := 1 - 2 * (v + y) * (v + 3 y) / 3 / (v + 2 y) ^2;
q[v_, h_, p_, z_, y_] :=
  (1 + h * (h - 1) * p - 1 / 2 h (h - 1) (2 - h) (1 - 3 h) p^2 - (z / (v + y)) ^h) /
  Sqrt[2 * h^2 p (1 - (1 - h) (1 - 3 h) p)];
```

The Call formula is then

$$S\,N(q(4, h(4, Y), p(4, Y), Z, Y)) - X\,e^{-rt}\,N(q(0, h(0, Y), p(0, Y), Z, Y)) \tag{43}$$

The *Mathematica* representation of this is

```
CBAppSRCEVCall[S_, X_, r_, σ_, t_] :=
  Module[{sigone, Y, Z, Pzero, Pfour, Hzero, Hfour, Qzero, Qfour},
    sigone = σ * Sqrt[S]; Y = y[S, r, sigone, t];
    Z = z[X, r, sigone, t]; Pzero = p[0, Y]; Pfour = p[4, Y];
    Hzero = h[0, Y]; Hfour = h[4, Y];
    Qzero = q[0, Hzero, Pzero, Z, Y];
    Qfour = q[4, Hfour, Pfour, Z, Y];
    S * Norm[Qfour] - X Exp[-r * t] * Norm[Qzero]]
```

Comparisons with Exact Solution

Let's look at the approximate solution vs the exact solution for ATM, ITM, OTM cases:

```
{CBAppSRCEVCall[100, 100, 0.1, 0.2, 1],
 SRCEVCall[100, 100, 0.1, 0, 0.2, 1]}
```

```
    {13.2763, 13.2731}
```

```
{CBAppSRCEVCall[110, 100, 0.1, 0.2, 1],
 SRCEVCall[110, 100, 0.1, 0, 0.2, 1]}
```

```
    {21.3727, 21.3699}
```

```
{CBAppSRCEVCall[90, 100, 0.1, 0.2, 1],
 SRCEVCall[90, 100, 0.1, 0, 0.2, 1]}
```

```
    {6.76843, 6.76697}
```

28.7 Cox's Gamma-Function Series

For completeness the exact solution given by Cox is presented. We can then compare this with the direct integration of the discounted risk-neutral expiry distribution. We need first to define some Gamma-functions.

```
?Gamma
```

> Gamma[z] is the Euler gamma function. Gamma[a, z] is the
>   incomplete gamma function. Gamma[a, z0, z1] is the generalized
>   incomplete gamma function Gamma[a, z0] - Gamma[a, z1].

```
g[n_, z_] := Exp[-z] z^(n-1)/Gamma[n]
```

```
G1[n_, w_] = Integrate[g[n,w], w]
```

$$-\ \frac{\text{Gamma}[n,\ w]}{\text{Gamma}[n]}$$

This G is the one defined by Cox - it is the complementary cumulative distribution function.

```
G[n_, w_] := Gamma[n, w]/Gamma[n]
```

Next we need a series as given by Cox:

```
CoxSeries[a_, b_, x_, y_, m_] := Sum[g[n+a,x]*G[n+b,y], {n, 1, m}]

PartialCoxSeries[a_, b_, x_, y_, kk_, m_] := Sum[g[n+a,x]*G[n+b,y],
{n, kk, m}]
```

Here is what the series looks like for a few terms:

```
CoxSeries[a,b,x,y,2]
```

$$\frac{E^{-x}\, x^a\, \text{Gamma}[1+b,\ y]}{\text{Gamma}[1+a]\ \text{Gamma}[1+b]} + \frac{E^{-x}\, x^{1+a}\, \text{Gamma}[2+b,\ y]}{\text{Gamma}[2+a]\ \text{Gamma}[2+b]}$$

In the following formula the elasticity parameter corresponds to α, and the yield is zero.

```
SeriesCEVOption[p_, k_, vol_, r_, t_, elas_, trunc_] :=
Module[{lambda = 0.5/(1-elas), sig = vol*p^(1-elas), x, y, rtl},
rtl = Exp[r(t/lambda)];
x = 2 lambda r p^(1/lambda) rtl/((rtl - 1)*sig^2);
y = 2 lambda r k^(1/lambda)/((rtl - 1)*sig^2);
p*CoxSeries[0,lambda, x, y, trunc] - k*Exp[-r*t]*Cox-
Series[lambda,0,x,y, trunc]]
```

```
SeriesCEVOption[S, K, σ, r, t, e, n]
```

$$-E^{-0.06\,t}\,K\sum_{n=1}^{n} g\left[n+\frac{0.5}{1-e},\ \frac{0.06\,E^{0.12\,(1-e)\,t}\,S^{-2+2.\,(1-e)+2e}}{(1-e)\,(-1+E^{0.12\,(1-e)\,t})\,\sigma^2}\right]$$

$$G\left[n+0,\ \frac{0.06\,K^{2.\,(1-e)}\,S^{-2+2e}}{(1-e)\,(-1+E^{0.12\,(1-e)\,t})\,\sigma^2}\right] +$$

$$S\sum_{n=1}^{n} g\left[n+0,\ \frac{0.06\,E^{0.12\,(1-e)\,t}\,S^{-2+2.\,(1-e)+2e}}{(1-e)\,(-1+E^{0.12\,(1-e)\,t})\,\sigma^2}\right]$$

$$G\left[n+\frac{0.5}{1-e},\ \frac{0.06\,K^{2.\,(1-e)}\,S^{-2+2e}}{(1-e)\,(-1+E^{0.12\,(1-e)\,t})\,\sigma^2}\right]$$

With enough terms in the series, we can compute a stable result. Here we give the SRCEV values.

```
SeriesCEVOption[100, 100, 0.20, 0.1, 1, 0.5, 140]
```

```
    13.2731
```

Just to check that we have enough terms:

```
SeriesCEVOption[100, 100, 0.20, 0.1, 1, 0.5, 150]
```

```
    13.2731
```

Now we can look closer to Black-Scholes, by increasing the elasticity to be closer to unity. This requires very many more terms. The formula is singular when the elasticity is unity, so we set it to 0.95, to compare with the Bessel function form. Counting in thousands, 6000 terms are both necessary and sufficient for 6 SF:

```
SeriesCEVOption[100, 100, 0.20, 0.1, 1, 0.95, 5000]
```

```
    1.69942
```

```
SeriesCEVOption[100, 100, 0.20, 0.1, 1, 0.95, 6000]
```

```
    13.2697
```

```
SeriesCEVOption[100, 100, 0.20, 0.1, 1, 0.95, 7000]
```

```
    13.2697
```

28.8 Skews and Smiles by "Real-World" Option Pricing

The idea in this section is to explore what happens when the price computed by an allegedly more realistic model (here a CEV model) is fed backwards through Black-Scholes, to work out the implied volatility. The point to be made is that there is a simple mechanism for the emergence of a skew, based on an elasticity approach. Note that this is not due to the volatility going up as the price at initiation goes down, for we are fixing the volatility as a fixed constant at time zero. The skew arises as a result of subsequent evolution of the volatility as the asset price changes.

```
Needs["Derivatives`BlackScholes`"]
```

```
? BlackScholesCallImpVol
```

```
    BlackScholesCallImpVol [price, strike,
      riskfree, divyield, expiry, optionprice]   returns
      the implied volatility of a vanilla European Call.
```

So we fix the volatility, price with the SRCEV model, then infer the implied volatility from the Black-Scholes model. As before, the σ parameter is set simply according to the scale of variation of the underlying.

```
ForwardBackward [S_, K_, r_, q_, σ_, t_] :=
  BlackScholesCallImpVol [S, K, r, q, t, SRCEVCall[S, K, r, q, σ, t]]
```

```
{ForwardBackward [100, 90, 0.1, 0, 0.2, 1],
 ForwardBackward [100, 100, 0.1, 0, 0.2, 1],
 ForwardBackward [100, 110, 0.1, 0, 0.2, 1]}
```

```
    {0.20538, 0.200104, 0.195409}
```

A Skew or Half a Smile

This model does capture a volatility skew typical of major indices under certain circumstances - the volatility at time zero is always 20%, but the volatility implied from the BS model varies from nearly 22% with a strike at 70, to under 19% with a strike at 130. Note that the out-of-the-money end is at the right of this plot.

```
Plot [ForwardBackward [100, K, 0.05, 0, 0.2, 1], {K, 70, 130}];
```

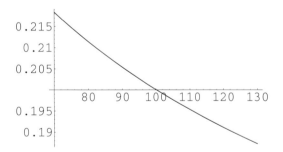

28.9 The Effect of Spread

Real-world trading can also involve a bid-offer spread. Suppose we add 2.5% to the elasticity result and again compute the implied volatility:

```
ForwardBackward[S_, K_, r_, q_, σ_, t_, spread_] :=
 BlackScholesCallImpVol[S, K, r, q, t,
  (1 + spread / 100) * SRCEVCall[S, K, r, q, σ, t]]

Plot[ForwardBackward[100, K, 0.05, 0, 0.2, 1, 2.5], {K, 70, 130},
 PlotLabel -> "Skew for Elasticity\n+ Spread Model of Call\n"];
```

```
        Skew for Elasticity
        + Spread Model of Call

 0.3
0.28
0.26
0.24
0.22

    80   90  100  110  120  130
```

28.10 Puts and Smiles

The discussion of Puts is complicated by the fact that there is a non-zero probability of the stock price reaching zero and staying there, for finite values of v. For example, in our test problem, it is very small:

```
ZeroProb[S_, r_, q_, vol_, t_, v_] :=
Module[{σ = vol * S ^ (1 / (2 v)), x, c},
 c = (2 * v * (r - q)) / (σ^2 * (Exp[((r - q) * t) / v] - 1));
 x = c * S ^ (1 / v) * Exp[((r - q) * t) / v];
Gamma[v, x] / Gamma[v]]

ZeroProb[100, 0.05, 0, 0.2, 1, 1]

5.4687 × 10^-23
```

As v increases (i.e. we move closer to standard log-normal) it becomes smaller:

```
ZeroProb[100, 0.05, 0, 0.2, 1, 10]

6.8533110773 × 10^-2150
```

It can be significant in the Gaussian case, for longer times and smaller drifts:

```
ZeroProb[100, 0.02, 0, 0.2, 5, 0.5]
```

```
    0.0188362
```

What we shall do is model a Put which includes the contribution for the final asset price being zero, so that it only remains to consider the continuous part of the distribution, which proceeds as before. (You should check that this ensures Put-Call parity applies.) The model then becomes the pair of functions

```
CEVPutfunc[x_, a_, v_, n_] :=
  NIntegrate[Exp[-(z^2/(4*x))]*z^n*BesselI[v, z], {z, 0, a}]
```

```
CEVPut[S_, K_, r_, q_, vol_, t_, v_] :=
Module[{σ = vol*S^(1/(2 v)), x, c, a},
  c = (2*v*(r - q))/(σ^2*(Exp[((r - q)*t)/v] - 1));
  x = c*S^(1/v)*Exp[((r - q)*t)/v];
  a = 2 Sqrt[c x K^(1/v)];
    Exp[-r*t - x] (K*((2*x)^(v - 1))*CEVPutfunc[x, a, v, 1 - v] -
        CEVPutfunc[x, a, v, v + 1]/(x*c^v*2^(v + 1))) +
    K*Exp[-r*t]*ZeroProb[S, r, q, vol, t, v]]
```

```
CEVPut[100, 100, 0.05, 0, 0.2, 1, 1]
```

```
    5.57683
```

As before, as v gets large we approach the BS values:

```
BlackScholesPut[100, 100, 0.2, 0.05, 0, 1]
```

```
    5.57353
```

```
CEVPut[100, 100, 0.05, 0, 0.2, 1, 100]
```

```
    5.5735264
```

We can do corresponding implied volatility computations, limiting attention, as before, to SRCEV:

```
PutForwardBackward[S_, K_, r_, q_, σ_, t_, spread_] :=
  BlackScholesPutImpVol[S, K, r, q, t,
    (1 + spread/100)*CEVPut[S, K, r, q, σ, t, 1/2]]
```

With no spread, the skew obtained is as follows - note that the out-of-the-money case is at the left of this plot:

```
Plot[PutForwardBackward[100, K, 0.05, 0, 0.2, 1, 0], {K, 70, 130}];
```

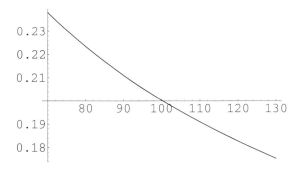

This time folding in the spread effect turns the skew into the beginning of a smile.

```
Plot[PutForwardBackward[100, K, 0.05, 0, 0.2, 1, 2.5], {K, 70, 130},
  PlotLabel -> "Smile for Elasticity + Spread Model of Put\n"];
```

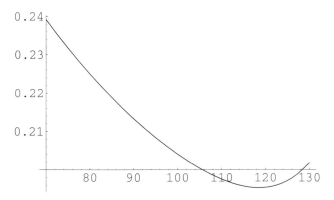

These are qualitatively similar to what one sees in equity markets (note that we regard the FX case as totally different - it is not a candidate for this type of elasticity treatment). We see that the effect of the elasticity in combination with the spread is to dramatically raise the OTM implied volatilities, and to flatten or curve the ITM region. It is clear that the SRCEV model does capture part of the skew effect seen in equity markets. The reader should note the discussion by R.J. Brenner, "Volatility is not constant", for further comments on CEV and related level-dependent volatility models in the book by Nelken (1990). The time-dependence of the elasticity and spread-adjusted implied volatility is that the smile becomes more of a skew, and the overall level drifts down - here is the implied volatility surface for 0.5 to 5 years:

```
Plot3D[PutForwardBackward[100, K, 0.05, 0, 0.2, t, 2.5],
  {K, 70, 130}, {t, 0.5, 5},
  PlotLabel -> "Vol Surface for Elasticity\n+ Spread Model of Put\n"];
```

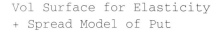

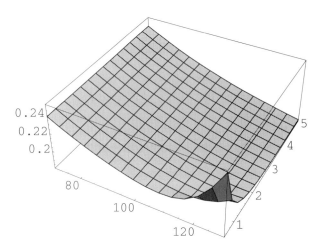

If, in addition, one adds in the effect of a time-dependent yield curve, but uses a constant interest rate in assessing BS implied volatility, one can generate a fictitious term structure for the implied volatility. The lesson that should be learned is that one should subtract out all the known time-dependent and asset-level-dependent effects before attempting to build two-factor stochastic or GARCH models of volatility.

28.11 Closing Remarks

A book has to end somewhere, and this is where matters are brought to a close. There is much more that could be said about CEV and its cousins in interest-rate modelling. An important point is that the sequence of transformations made to analyse the PDE can also be applied, with minor modifications, to the family of interest-rate models discussed in Chapters 25-27. All such single-factor models, and the CEV family, can be reduced to the standard diffusion equation with advection and source terms, allowing most of the numerical machinery developed in Chapters 13-20 to be brought to bear, in a single unified framework. Another point not discussed here is how to characterize the probability distributions that arise when v is not unity and there is also mean reversion. It is left as an exercise for the reader to follow through the sequence of steps of Section 28.3 with the mean reversion included. The resulting PDE becomes nice (in a sense that becomes obvious when you try it) only if $v = 1/2$, 1 or the mean-reversion is excluded. What the enveloping distribution is when mean-reversion is present and v is general is not known to this author.

Chapter 28 Bibliography

Beckers, S., 1980, The constant elasticity of variance model and its implications for option pricing *Journal of Finance*, 35, June, p. 661.

Cox, J., 1996, The constant elasticity of variance option pricing model, *Journal of Portfolio Management*, special issue.

Hauser, S., and Lauterbach, B., 1996, Tests of warrant pricing models: the trading profits perspective,

Journal of Derivatives, Winter, p. 71.

Jarrow, R.A. and Rudd, A., 1983, *Option Pricing* (Section 11.6), Irwin.

Lauterbach, B. and Schultz, P., 1990, Pricing warrants: an empirical study of the Black-Scholes model and its alternatives, *Journal of Finance,* 45 (Sept.), p. 1181;

Nelken, I. (editor), 1990, *The Handbook of Exotic Options*, Irwin.

Press, W.H., Teukolsky, S.A., Vetterling, W.T. and Flannery, B.P., 1992, *Numerical Recipes in C - the Art of Scientific Computing*, 2nd edition. Cambridge University Press.

Schroder, M., 1989, Computing the constant elasticity of variance option pricing formula, *Journal of Finance,* 44, March, p. 221.

Index

Note: this is not a comprehensive index of *Mathematica* commands built in to *Mathematica* – see *The Mathematica Book* also.